Space–Time Coding for Broadband Wireless Communications

The Wiley Bicentennial–Knowledge for Generations

Each generation has its unique needs and aspirations. When Charles Wiley first opened his small printing shop in lower Manhattan in 1807, it was a generation of boundless potential searching for an identity. And we were there, helping to define a new American literary tradition. Over half a century later, in the midst of the Second Industrial Revolution, it was a generation focused on building the future. Once again, we were there, supplying the critical scientific, technical, and engineering knowledge that helped frame the world. Throughout the 20th Century, and into the new millennium, nations began to reach out beyond their own borders and a new international community was born. Wiley was there, expanding its operations around the world to enable a global exchange of ideas, opinions, and know-how.

For 200 years, Wiley has been an integral part of each generation's journey, enabling the flow of information and understanding necessary to meet their needs and fulfill their aspirations. Today, bold new technologies are changing the way we live and learn. Wiley will be there, providing you the must-have knowledge you need to imagine new worlds, new possibilities, and new opportunities.

Generations come and go, but you can always count on Wiley to provide you the knowledge you need, when and where you need it!

William J. Pesce
President and Chief Executive Officer

Peter Booth Wiley
Chairman of the Board

Space–Time Coding for Broadband Wireless Communications

GEORGIOS B. GIANNAKIS
ZHIQIANG LIU
XIAOLI MA
SHENGLI ZHOU

BICENTENNIAL 1807 WILEY 2007 BICENTENNIAL

WILEY-INTERSCIENCE

A JOHN WILEY & SONS, INC., PUBLICATION

Published by John Wiley & Sons, Inc., Hoboken, New Jersey.
Published simultaneously in Canada.

Library of Congress Cataloging-in-Publication Data:

Space–time coding for broadband wireless communications / Georgios B. Giannakis . . . [et al.].
p. cm.
Includes index.
ISBN-13: 978-0-471-21479-3 (cloth : alk. paper)
ISBN-10: 0-471-21479-5 (cloth : alk. paper)
1. Coding theory. 2. Wireless communication systems. 3. Mobile communication systems. I. Giannakis, Georgios B., 1958–.
TK5102.92.S72 2006
621.384—dc22 2006014214

10 9 8 7 6 5 4 3 2 1

To my parents, Sofia and Vassili
G. B. G.

To Guangmei and Gwyneth
Z. L.

To Xiangqian and my parents
X. M.

To Juanjuan and Daniel
S. Z.

Contents

	Preface	*xv*
	Acronyms	*xix*
1	*Motivation and Context*	*1*
	1.1 Evolution of Wireless Communication Systems	2
	1.2 Wireless Propagation Effects	3
	1.3 Parameters and Classification of Wireless Channels	5
	1.3.1 Delay Spread and Coherence Bandwidth	6
	1.3.2 Doppler Spread and Coherence Time	7
	1.4 Providing, Enabling, and Collecting Diversity	11
	1.4.1 Diversity Provided by Frequency-Selective Channels	11
	1.4.2 Diversity Provided by Time-Selective Channels	13
	1.4.3 Diversity Provided by Multi-Antenna Channels	15
	1.5 Chapter-by-Chapter Organization	18
2	*Fundamentals of ST Wireless Communications*	*23*
	2.1 Generic ST System Model	23
	2.2 ST Coding viz Channel Coding	27
	2.3 Capacity of ST Channels	29
	2.3.1 Outage Capacity	30

2.3.2 Ergodic Capacity 34
2.4 Error Performance of ST Coding 39
2.5 Design Criteria for ST Codes 43
2.6 Diversity and Rate: Finite SNR viz Asymptotics 44
2.7 Classification of ST Codes 48
2.8 Closing Comments 50

3 Coherent ST Codes for Flat Fading Channels 51
3.1 Delay Diversity ST Codes 51
3.2 ST Trellis Codes 53
3.2.1 Trellis Representation 53
3.2.2 TSC ST Trellis Codes 55
3.2.3 BBH ST Trellis Codes 56
3.2.4 GFK ST Trellis Codes 58
3.2.5 Viterbi Decoding of ST Trellis Codes 60
3.3 Orthogonal ST Block Codes 61
3.3.1 Encoding of OSTBCs 61
3.3.2 Linear ML Decoding of OSTBCs 63
3.3.3 BER Performance with OSTBCs 65
3.3.4 Channel Capacity with OSTBCs 66
3.4 Quasi-Orthogonal ST Block Codes 68
3.5 ST Linear Complex Field Codes 70
3.5.1 Antenna Switching and Linear Precoding 71
3.5.2 Designing Linear Precoding Matrices 72
3.5.3 Upper Bound on Coding Gain 72
3.5.4 Construction Based on Parameterization 73
3.5.5 Construction Based on Algebraic Tools 74
3.5.6 Decoding ST Linear Complex Field Codes 76
3.5.7 Modulus-Preserving STLCFC 79
3.6 Linking OSTBC, QO-STBC, and STLCFC Designs 82
3.6.1 Embedding MP-STLCFCs into the Alamouti Code 82
3.6.2 Embedding 2×2 MP-STLCFCs into an OSTBC 83
3.6.3 Decoding QO-MP-STLCFC 84
3.7 Closing Comments 85

4 Layered ST Codes 87
4.1 BLAST Designs 88
4.1.1 D-BLAST 88

8.5.3 Numerical Examples 205
8.6 Closing Comments 206

9 ST Codes for Time-Varying Channels 209
9.1 Time-Varying Channels 210
9.1.1 Channel Models 211
9.1.2 Time-Frequency Duality 214
9.1.3 Doppler Diversity 215
9.2 Space-Time-Doppler Block Codes 216
9.2.1 Duality-Based STDO Codes 219
9.2.2 Phase Sweeping Design 222
9.3 Space-Time-Doppler FDFR Codes 227
9.4 Space-Time-Doppler Trellis Codes 227
9.4.1 Design Criterion 228
9.4.2 Smart-Greedy Codes 229
9.5 Numerical Examples 229
9.6 Space-Time-Doppler Differential Codes 231
9.6.1 Inner Codec 233
9.6.2 Outer Differential Codec 234
9.7 ST Codes for Doubly Selective Channels 235
9.7.1 Numerical Examples 237
9.8 Closing Comments 239

10 Joint Galois- and Linear Complex-Field ST Codes 241
10.1 GF-LCF ST Codes 242
10.1.1 Separate Versus Joint GF-LCF ST Coding 243
10.1.2 Performance Analysis 245
10.1.3 Turbo Decoding 248
10.2 GF-LCF Layered ST Codes 251
10.2.1 GF-LCF ST FDFR Codes: QPSK Signaling 251
10.2.2 GF-LCF ST FDFR Codes: QAM Signaling 253
10.2.3 Performance Analysis 256
10.2.4 GF-LCF FDFR Versus GF-Coded V-BLAST 259
10.2.5 Numerical Examples 260
10.3 GF-LCF Coded MIMO OFDM 263
10.3.1 Joint GF-LCF Coding and Decoding 263
10.3.2 Numerical Examples 265
10.4 Closing Comments 265

11 MIMO Channel Estimation and Synchronization 269
11.1 Preamble-Based Channel Estimation 270
11.2 Optimal Training-Based Channel Estimation 271
11.2.1 ZP-Based Block Transmissions 274
11.2.2 CP-Based Block Transmissions 283
11.2.3 Special Cases 288
11.2.4 Numerical Examples 290
11.3 (Semi-)Blind Channel Estimation 293
11.4 Joint Symbol Detection and Channel Estimation 294
11.4.1 Decision-Directed Methods 294
11.4.2 Kalman Filtering-Based Methods 295
11.5 Carrier Synchronization 299
11.5.1 Hopping Pilot-Based CFO Estimation 300
11.5.2 Blind CFO Estimation 305
11.5.3 Numerical Examples 307
11.6 Closing Comments 310

12 ST Codes with Partial Channel Knowledge: Statistical CSI 313
12.1 Partial CSI Models 315
12.1.1 Statistical CSI 315
12.2 ST Spreading 319
12.2.1 Average Error Performance 321
12.2.2 Optimization Based on Average SER Bound 323
12.2.3 Mean Feedback 324
12.2.4 Covariance Feedback 328
12.2.5 Beamforming Interpretation 330
12.3 Combining OSTBC with Beamforming 331
12.3.1 Two-Dimensional Coder-Beamformer 333
12.4 Numerical Examples 335
12.4.1 Performance with Mean Feedback 335
12.4.2 Performance with Covariance Feedback 339
12.5 Adaptive Modulation for Rate Improvement 344
12.5.1 Numerical Examples 347
12.6 Optimizing Average Capacity 350
12.7 Closing Comments 351

13 ST Codes with Partial Channel Knowledge: Finite-Rate CSI 353
13.1 General Problem Formulation 354

13.2 Finite-Rate Beamforming 356
13.2.1 Beamformer Selection 357
13.2.2 Beamformer Codebook Design 357
13.2.3 Quantifying the Power Loss 362
13.2.4 Numerical Examples 364
13.3 Finite-Rate Precoded Spatial Multiplexing 366
13.3.1 Precoder Selection Criteria 367
13.3.2 Codebook Construction: Infinite Rate 369
13.3.3 Codebook Construction: Finite Rate 371
13.3.4 Numerical Examples 374
13.4 Finite-Rate Precoded OSTBC 380
13.4.1 Precoder Selection Criterion 381
13.4.2 Codebook Construction: Infinite Rate 381
13.4.3 Codebook Construction: Finite Rate 382
13.4.4 Numerical Examples 382
13.5 Capacity Optimization with Finite-Rate Feedback 383
13.5.1 Selection Criterion 383
13.5.2 Codebook Design 384
13.6 Combining Adaptive Modulation with Beamforming 385
13.6.1 Mode Selection 386
13.6.2 Codebook Design 386
13.7 Finite-Rate Feedback in MIMO OFDM 387
13.8 Closing Comments 388

14 ST Codes in the Presence of Interference 391
14.1 ST Spreading 392
14.1.1 Maximizing the Average SINR 393
14.1.2 Minimizing the Average Error Bound 394
14.2 Combining STS with OSTBC 396
14.2.1 Low-Complexity Receivers 399
14.3 Optimal Training with Interference 399
14.3.1 LS Channel Estimation 400
14.3.2 LMMSE Channel Estimation 401
14.4 Numerical Examples 401
14.5 Closing Comments 408

15 ST Codes for Orthogonal Multiple Access 409
15.1 System Model 410

15.1.1 Synchronous Downlink 410
15.1.2 Quasi-synchronous Uplink 411
15.2 Single-Carrier Systems: OSTBC-CIBS-CDMA 413
15.2.1 CIBS-CDMA for User Separation 413
15.2.2 OSTBC Encoding and Decoding 417
15.2.3 Attractive Features of OSTBC-CIBS-CDMA 418
15.2.4 Numerical Examples 421
15.3 Multi-Carrier Systems: STF-OFDMA 425
15.3.1 OFDMA for User Separation 425
15.3.2 STF Block Codes 426
15.3.3 Attractive Features of STF-OFDMA 426
15.3.4 Numerical Examples 428
15.4 Closing Comments 431

References 433

Index 461

Preface

The next generation of broadband wireless communication systems is envisioned to offer broadband multimedia services, including high-speed wireless Internet access, wireless television, and mobile computing, to a large population of mobile users. The rapidly growing demand for these services is driving the communication technology toward higher data rates, higher mobility, and in certain systems even higher carrier frequencies, thus raising the requirements for reliable communications over the shared air interface. To support reliable broadband wireless communications, tremendous research efforts are currently undertaken to develop advanced coding, modulation, signal processing, and multiple-access schemes for improving the quality and spectral efficiency of wireless links. Among them, particularly promising is the development of space-time (ST) coding and processing algorithms for multi-antenna systems operating over multi-input multi-output (MIMO) channels.

Exploiting the extra degrees of freedom provided by MIMO channels, ST systems can potentially offer significant gains in error performance and spectral efficiency. To this end, ST coding and processing techniques are expected to play a key role in the success of future-generation broadband wireless mobile systems. However, to realize the potential of ST coded multi-antenna systems, designers must deal with various performance-limiting challenges that include but are not limited to time- and frequency-selective fading, multiuser interference as well as size, power, and cost constraints. Research and development efforts to address these challenges in academia, industry, and government laboratories have resulted in a wealth of exciting results over the past eight years. As standardization efforts proceed swiftly, research in this area is expected to remain active in the years to come.

The objective of this research monograph is to present the state-of-the-art in ST coded multi-antenna systems operating over broadband wireless mobile channels. Although benefits of multi-antenna systems permeate to higher layers in the protocol stack, this book is devoted to algorithms and performance analysis of ST coded systems at the physical layer. Chapter-by-chapter contents are outlined in Section 1.5. Background on wireless communication basics, diversity techniques, and capacity of MIMO channels is provided in Chapters 1 and 2 to allow for self-containment. For deeper exposition to these subjects the interested reader may consult standard citations [214,218,236,242] and recent books in these areas [97,263]. The present book is intended for researchers in academia and practicing engineers with an elementary background in linear algebra, undergraduate-level knowledge in wireless communications and statistical signal processing, and masters-level expertise in digital communications.

A number of research monographs [8,16,119,126,130,208,281] and edited books [85,268] have been published recently on the subject of ST coded multi-antenna systems. The present book complements these contributions and is unique in several aspects. It offers a unified and comprehensive exposition of the subject. Besides ST codes for flat fading MIMO channels it presents in sufficient depth ST codes for the gamut of frequency-, time-, and doubly selective MIMO channels which are encountered with broadband wireless mobile links. It further covers recent advances in complex field coded transmissions, sphere decoding algorithms, closed-loop ST coded and multi-user multi-antenna systems. While attempting to do justice in overviewing the state-of-the-art in ST coded systems, page limitations and the authors' subjective view must have led to inevitable omissions or insufficient coverage of certain contributions. Besides apologizing, we must confess at the outset to one more bias. The contents of this book were motivated by the research performed in the Signal Processing in Networking and Communications (SPiNCOM) group directed by the lead author from 1999 to 2005, a period when the three co-authors were carrying out their graduate studies at the University of Minnesota. This fact explains why the contents naturally lean toward subjects explored by SPiNCOM researchers.

Writing this book has been a lengthy but rewarding experience. The authors as well as the contents of this book have benefited in various forms from a number of people who contributed their generous help and support directly or indirectly. On the technical side, it is our pleasure to acknowledge contributions and thank former and current members of the SPiNCOM group: Prof. Zhengdao Wang for influencing various ideas in this book through our joint research and brainstorming sessions; Drs. Wanlun Zhao and Renqiu Wang for their contributions to an early version of Chapter 5; and Alfonso Cano, Alejandro Ribeiro, Yingqun Yu, Profs. Xiaodong Cai, Geert Leus, Shuichi Ohno, Yan Xin, Liuqing Yang, Yinwei Yao, and Drs. Qingwen Liu and Pengfei Xia for many helpful discussions. The exposition of the book improved markedly from constructive comments we received from friends and colleagues who reviewed various chapters on a short notice. Our sincere thanks go to P. Banelli, H. Bölcskei, X. Cai, T. Davidson, F. Digham, M. Kisaliou, J. Kleider, G. Leus, S. Ohno, L. Rugini, B. Sadler, E. Serpedin, S. Shahbazpanahi, N. Sidiropoulos, A. Swami, C. Tepedelenlioglu, Z. Tian, Z. Wang, Y. Xin, Y. Yao, and W. Zhao.

The research in the book was funded primarily by the Army Research Laboratory (ARL) under the Collaborative Technology Alliance on Communications and Networking (CTA-CN grant DAAD19-01-2-0011); and in part by the Army Research Office (ARO grants DAAG55-98-1-0336, W911NF-04-1-0338, W911NF-05-01-0283, and W911NF-06-1-0090); by the National Science Foundation (NSF Wireless Initiative 9979443 and CCF-0515032); by ETRI, Korea; and by I2R, Singapore. The authors are also grateful to their institutions for facilitating their efforts in writing this research monograph.

On the personal side, we are all grateful to our families for their understanding, patience, and unwavering love during the course of writing this book.

Georgios B. Giannakis, *University of Minnesota*
Zhiqiang Liu, *University of Iowa*
Xiaoli Ma, *Georgia Institute of Technology*
Shengli Zhou, *University of Connecticut*

Acronyms

1G	First Generation
2G	Second Generation
3G	Third Generation
a.k.a.	also known as
AMPS	Advanced Mobile Phone System
AP	Affine Precoding
AWGN	Additive White Gaussian Noise
BCDD	Block Circular Delay Diversity
BEM	Basis Expansion Model
BER	Bit Error Rate
BPSK	Binary Phase Shift Keying
BS	Base Station
CC	Convolutional Coding
CDF	Cumulative Distribution Function
CDMA	Code Division Multiple Access
cf.	confer (Latin: Compare)
CFO	Carrier Frequency Offset

CIBS	Chip-Interleaved Block Spread
CM	Constant Modulus
CP	Cyclic Prefix
CSI	Channel State Information
D-BLAST	Diagonal Bell Laboratories Layered Space-Time
DFE	Decision-Feedback Equalizer
DFT	Discrete Fourier Transform
DPS	Digital Phase Sweeping
DS	Direct Sequence
EDGE	Enhanced Date Rates for GSM Evolution
EVD	Eigenvalue Decomposition
FDD	Frequency Division Duplex
FDFR	Full Diversity Full Rate
FDMA	Frequency Division Multiple Access
FFT	Fast Fourier Transform
GDD	Generalized Delay Diversity
GMSK	Gaussian Minimum Shift Keying
GSTF	Grouped Space Time Frequency
IBI	Interblock Interference
IFFT	Inverse Fast Fourier Transform
i.i.d.	independent identically distributed
ISI	Inter-Symbol Interference
LCFC	Linear Complex-Field Code
LOS	Line-Of-Sight
LS	Least-Squares
MC	Multi-Carrier
MGF	Moment Generating Function
MIMO	Multi-Input Multi-Output
MISO	Multi-Input Single-Output
ML	Maximum Likelihood
MLSE	Maximum Likelihood Sequence Estimation
MMSE	Minimum Mean-Square Error
MP	Modulus-Preserving
MRC	Maximum Ratio Combining

MUI	Multiuser Interference
NC	Nulling-Cancelling
OFDM	Orthogonal Frequency-Division Multiplexing
OFDMA	Orthogonal Frequency-Division Multiple Access
PAM	Pulse Amplitude Modulation
PAR	Peak-to-Average Power Ratio
pcu	per channel use
PDA	Probabilistic Data Association
pdf	probability density function
PEP	Pairwise Error Probability
OSTBC	Orthogonal Space-Time Block Code
QAM	Quadrature Amplitude Modulation
QO-STBC	Quasi-Orthogonal Space-Time Block Code
QPSK	Quadrature Phase Shift Keying
SDA	Sphere Decoding Algorithm
SER	Symbol Error Rate
SF	Space Frequency
SISO	Single-Input Single-Output
siso	soft-input soft-output
SNR	Signal-to-Noise Ratio
ST	Space-Time
STBC	Space-Time Block Code
STF	Space-Time-Frequency
STS	Space-Time Spreading
STTC	Space-Time Trellis Code
SU	Subscriber Unit
SVD	Singular Value Decomposition
TCM	Trellis Coded Modulation
TDD	Time Division Duplex
TDMA	Time Division Multiple Access
TU	Typical Urban
TV	Time-Varying
UP	Unitary Precoding
V-BLAST	Vertical Bell Laboratories Layered Space-Time

ZF	Zero-Forcing
ZP	Zero-Padding

1

Motivation and Context

The goal of this book is to present recent advances in space-time (ST) coded multi-antenna systems operating over broadband wireless mobile channels. To appreciate these advances, we start this chapter by tracing the evolution of wireless communications through past, present, and future-generation systems, which motivates the deployment of multiple antennas to meet today's and tomorrow's needs for high-performance multimedia services on the move. Since high-performance multi-antenna transceivers must account for the wireless interface, we proceed to overview briefly the characteristics and detrimental effects of fading that have to be mitigated at the design phase. Subsequently, we classify wireless fading channels according to the coherence and selectivity they exhibit in diversifying waveforms transmitted in the time, frequency, and space dimensions. This diversification can be exploited to combat fading effects; and depending on the channel type, diversity appears in the multipath, Doppler, and/or space domains.

Transmission and reception by multiple antennas model, respectively, the excitation and response of multi-input multi-output (MIMO) channels, which provide spatial diversity in the form of transmit-diversity or receive-diversity when multiple antennas are deployed, respectively, at the transmitter or receiver. Both forms of spatial diversity are particularly attractive because at the cost of deploying multiple antennas, they do not necessarily incur loss in spectral efficiency and power efficiency or a considerable increase in complexity. In this chapter we outline the benefits of receive-diversity and allude to the challenges emerging with transmit-diversity systems which motivate the need for judicious transmitted waveform designs through the use of ST coding. Based on intuitive arguments, we further describe two major innovations that ST coded multi-antenna systems demonstrate over their single-antenna counterparts. We close this chapter with a road map of the book's contents.

1.1 EVOLUTION OF WIRELESS COMMUNICATION SYSTEMS

The history of wireless communications is relatively short. In fact, the debut of wireless communications could be dated back to 1901, when the first telegraph was sent across the Atlantic Ocean from Cornwall to St. John's Newfoundland [192]. However, cellular wireless communications have experienced unprecedented growth over the past thirty years and have already revolutionized the way that we communicate and live. Thus far, cellular wireless communications have evolved through three generations.

- **1G systems**

  First-generation (1G) cellular systems were introduced between the late 1970s and the early 1980s. All 1G systems were analog and were designed for narrowband voice services. The multiple-access technology used in 1G systems is frequency division multiple access (FDMA). Examples of 1G cellular systems include the Advanced Mobile Phone System (AMPS) in North America, the European Total Access Communications System (ETACS), and the Nippon Telephone and Telegraph (NTT) system in Japan.

- **2G systems**

  Second-generation (2G) cellular systems were deployed in the early 1990s, and are still in service in most countries. 2G systems are all-digital and employ either time division multiple access (TDMA) or code division multiple access (CDMA). Relative to the 1G systems, 2G systems offer better spectral efficiency, enhanced system capacity, and improved quality-of-service. Examples of 2G cellular systems include the Global System for Mobile Communication (GSM) in Europe, the Personal Communication Service (PCS) IS-95 system in North America, and the Personal Digital Cellular (PDC) system in Japan.

- **3G systems**

  Although data services are offered by 2G systems, they are predominantly narrowband. Driven by the growing demand for broadband wireless services such as high-speed wireless Internet access, wireless television and mobile computing, cellular wireless communication systems are now evolving into their third generation (3G) under the names International Mobile Telecommunications 2000 (IMT-2000) and CDMA-2000. IMT-2000 employs wideband direct-sequence CDMA (DS-CDMA), and CDMA-2000 is based on multi-carrier CDMA (MC-CDMA). Compared to 2G systems, 3G systems are capable of supporting much higher transmission rates and user mobility. 3G systems are currently under development and are designed to support transmission rates up to several megabits per second at end-user speeds as high as a few hundred kilometers per hour. Compared to 1G and 2G systems, the emphasis in 3G is on high-quality broadband multimedia services. Fourth-generation (4G) systems are on the horizon and are envisioned to offer broadband services with even higher data rates and user mobility.

The ultimate goal of wireless communications is to provide "anywhere, anytime, anymedia" wireless access at a reasonably low price. How to achieve this ambitious goal with limited bandwidth and power resources at affordable complexity while adhering to various implementation constraints brings about tremendous challenges to researchers and developers. The contents of this book provide a step toward addressing these challenges through ST coding techniques for use by broadband wireless mobile systems with multiple antennas deployed at the transmitter and/or receiver side.

Before we move on to argue as to the potential benefits that multi-antenna systems have over their single-antenna counterparts, it is instructive to overview the main source of these challenges, which can be traced back to the only element the designer does not have full control over: the wireless propagation channel.

1.2 WIRELESS PROPAGATION EFFECTS

When designing high-performance wireless systems, it is important to understand the challenges posed by wireless propagation. To this end, we summarize in this section the basic factors influencing propagation over wireless point-to-point links and their effects on the transmitted waveforms.

- **Path loss**

  It is well known that as electromagnetic waves propagate in free space with the speed of light, their magnitude decays with distance. Specifically, a waveform transmitted with a certain power from one point is received at a distance d with power proportional to $1/d^n$, where the exponent n depends on terrain contours and the environment (urban versus rural and outdoor versus indoor). For typical wireless links, n ranges between 2 and 4, whereas values of 4 to 6 are encountered within stadiums, buildings and other indoor environments. The reason for this loss is simple. Even if propagation encounters no obstacle, the radius of the transmitted spherical wave grows with distance. And since the energy is fixed, the received power at any point on the sphere is reduced along with the square of the distance. Besides distance, the power decay also depends on the wavelength, the transmit-antenna gain, the propagation medium itself, and the receive-antenna gain. As far as path loss is concerned, the harshness of the propagation medium constitutes the major difference between wireline and wireless communications and explains why the exponent n exceeds 2 in wireless links [242]. The obvious practical implication is that designers of wireless communication systems have to pay more attention to power efficiency.

- **Shadowing, reflecting, diffracting, and scattering**

  Even though waveforms travel along straight lines in free space, wireless communications do not take place in free space. If, on the other hand, the straight line between transmitter and receiver is "obstacle-free," line-of-sight (LOS) wireless communication is possible. When obstacles are present, whether wireless transmissions can penetrate them depends on the operating carrier frequency; low-frequency transmissions are easier to penetrate. The higher the frequency, the closer the transmitted waveforms resemble light in their propagation characteristics. Thus, even small obstacles such as a wall or a tree may block wireless transmissions, as they block light. This severe form of attenuation is called the shadowing effect. When the size of obstacles is much larger than the transmitted wavelength, reflections occur. Since the obstacles absorb part of the incident power, the reflected signal has reduced power. If the size of an obstacle is on the order of the transmitted wavelength or less, then scattering takes place (i.e., the incoming signal is scattered into a bunch of weaker waveforms). At obstacle edges in particular, diffraction happens pretty much as with light. Reflected, scattered, and diffracted renditions of the transmitted waveform can cause severe fading of the wireless link; but when constructively combined, they can facilitate wireless communications, especially when LOS is not available.

- **Multipath propagation**

  Whether LOS is absent or present to allow for the direct transmitter-receiver path, a waveform propagating through the wireless interface may be reflected, scattered, or diffracted before reaching the receiver through various indirect paths. Traveling over the direct and all indirect paths constitutes what is known as multipath propagation. As the transmitted waveform travels through all these paths, multiple versions of it reach the receiver at different times, since the speed of light is finite and multiple paths have variable length. This is a direct manifestation of multipath propagation and causes what is known as time dispersion or delay spread of the transmitted waveform. The delay spread deforms a narrow pulse into a broader one. As a result, one symbol waveform may spill over adjacent symbol waveforms, inducing what is called inter-symbol interference (ISI). But even when waveforms propagating through different paths arrive at the receiver almost simultaneously, they may differ in amplitude, phase, or carrier frequency offset. These differences may add constructively or destructively to amplify or attenuate the received power.

- **Fading effects**

  While multipath induces delay spread, the situation becomes even worse if the propagation medium changes or if transmitter and receiver are in relative motion, which is common in wireless mobile communications. In these cases, the power of the received waveform changes considerably over time because the transmitted waveform experiences time-varying paths. Experimental measurements at the output of real-world wireless channels confirm that instantiations of the receive-power over time resemble realizations of a random process. As power fluctuates randomly, it may approach vanishing levels, in which case we say that the wireless channel is fading. The random nature of the wireless fading channel constitutes its major difference with the wireline channel, which is modeled as deterministic. In general, fading effects caused by wireless propagation can be roughly categorized in three scales: (i) large-scale path loss effects modeled through an envelope that decays with distance; (ii) large- to medium-scale slowly varying shadowing effects modeled by a random channel amplitude following log-normal distribution; and (iii) small-scale fast-varying effects modeled as a random channel amplitude adhering to a Rice distribution if LOS is present, to a Rayleigh distribution if LOS is absent, or more generally, to a Nakagami distribution which can approximate a Rayleigh or Rice distribution and also capture additional fading effects [236, 242].

The brief account of all these propagation characteristics testifies to the harshness of wireless point-to-point links and illuminates the challenges that wireless systems face relative to their wireline counterparts. Detailed treatment of fading effects encountered with single-antenna links can be found in standard monographs devoted to wireless communication systems and their performance analysis [214, 218, 236, 242].

The multi-antenna wireless systems considered in this book comprise a collection of point-to-point wireless links, each corresponding to one pair of transmit-receive antennas. The fading effects summarized in this section are thus also present in the multi-antenna channel [97, 263]. In addition, depending on how the distance between neighboring antenna elements compares with the wavelength, the multiple channels may exhibit variable degrees of correlation. The latter will also affect the performance of multi-antenna wireless systems [87, 136, 235].

But before we proceed to discuss the challenges and potential benefits brought by multi-antenna channels, it is instructive to highlight the key parameters that define even single-antenna wireless channels. Based on these parameters and the time-bandwidth extent of the transmission, we will be able to classify wireless channels as we summarize next.

1.3 PARAMETERS AND CLASSIFICATION OF WIRELESS CHANNELS

In this section, we introduce key parameters characterizing the wireless fading channel. These parameters affect two basic features of random channels: coherence and selectivity. On a per realization basis, coherence pertains to the extent over which the channel's impulse or frequency response remains approximately constant, whereas selectivity refers to variation in the channel's impulse or frequency response. As an example, Figure 1.1 depicts coherence segments and selectivity features in a realization of the wireless channel gain varying across time, frequency, or space.

Besides controlling coherence and selectivity, the four major parameters we specify next lead to a natural classification of wireless fading channels that we present subsequently.

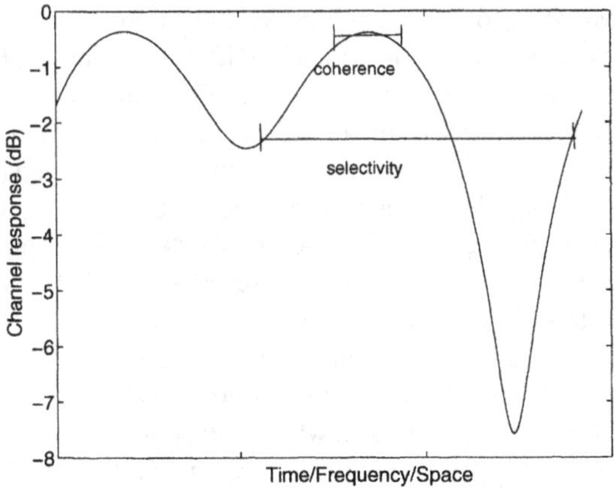

Figure 1.1 Coherence vs. selectivity

1.3.1 Delay Spread and Coherence Bandwidth

Let us consider a linear time-invariant random channel with impulse response $h(t;\tau) = h(\tau)$, $\forall t$. With $\star$ denoting convolution, the channel output $y(t)$ is related to the channel input $x(t)$ via $y(t) = (h \star x)(t)$. If through this wireless channel we transmit a pulse of infinitesimal width (delta-like input) and what we receive at the channel output spreads over τ_d seconds, we say that the channel is time-dispersive with delay spread τ_d. In terms of the impulse response realization $h(\tau)$, the delay spread is defined as the lag τ_d for which $h(\tau) = 0$, $\forall \tau > \tau_d$. In the ensemble sense, we can let $\phi_h(\tau) := E[|h(\tau)|^2]$ denote the variance of the nonstationary channel process with respect to (w.r.t.) τ, which we assume to be zero-mean, and define the delay spread as the lag τ_d for which $\phi_h(\tau) = 0$, $\forall |\tau| > \tau_d$. For average error performance analysis, the ensemble definition is useful. However, the sample definition of τ_d is also useful since it is necessary to cope with the delay spread in each realization.

The finite nonzero support of $h(\tau)$ and $\phi_h(\tau)$ is only approximately true, in general, and is satisfied by the popular tap-delay line model with a finite number of taps [214]. We can also interpret τ_d through the clusters of rays arriving coherently to form each tap (path) in the multipath propagation model discussed in Section 1.2. In this case, τ_d represents the maximum or the root-mean-square (rms) of the cluster arrival times, which correspond to the relative delays among paths [218].

Clearly, if the pulse used to transmit every information-bearing symbol has duration T_p and the period with which symbols are transmitted is T_s, the duration of each received symbol waveform through the channel $h(\tau)$ with delay spread τ_d will be $T_p + \tau_d$, and overlap among received pulses will be avoided if $T_s \geq T_p + \tau_d$. If on the other hand, $T_s < T_p + \tau_d$, then received symbol pulses will overlap, giving rise to ISI.

It is also useful to consider the effect of ISI channels in the frequency domain. The Fourier transform of $\phi_h(\tau)$ defines the channel's power spectral density (PSD)

$\Phi_h(f)$, whose nonzero support (bandwidth B_h) is infinite since $\phi_h(\tau)$ has finite support. Of interest is the coherence bandwidth $B_{\text{coh}} < B_h$, which is defined as the frequency spread over which the channel PSD can be considered approximately flat. The uncertainty principle asserts that frequency spread and delay spread must be inversely proportional:

$$B_{\text{coh}} \approx \frac{1}{\tau_d}.$$

To appreciate where the term coherence comes from, it is useful to recall that two waveforms, $x(t)$ and $y(t)$, are called coherent if they are equal up to a (generally complex) scale; i.e., if $y(t) = \alpha x(t)$. Notice that if these two coherent waveforms are realizations of two stationary random processes, their cross-correlation has constant magnitude and their correlation coefficient (a.k.a. coherence) has magnitude one. Let us also recall that according to its Fourier expansion, a waveform consists of complex exponentials. Consider one such exponential with frequency f_0 as the channel input. If f_c denotes the carrier frequency and $f_0 \in [f_c - B_{\text{coh}}/2, f_c + B_{\text{coh}}/2]$, the channel output $y_0(t)$ is coherent with the channel input $x_0(t) = A_0 \exp(j2\pi f_0 t)$ because complex exponentials are eigenfunctions of linear time-invariant channels and thus $y_0(t) = H(f_0)x_0(t)$, where $H(f_0)$ is the channel's frequency response at f_0. If the transmitted waveform consists of complex exponentials with frequencies within the coherence bandwidth of the channel, all these exponentials will be affected by basically the same $H(f_0)$ since $H(f)$ is approximately flat over B_{coh}; and thus the channel output will be coherent with the channel input. This explains why B_{coh} is called the coherence bandwidth.

On the other hand, if the input contains frequencies outside B_{coh}, then the scale $H(f_0)$ will generally be different for different f_0's and the transmitted waveform will not be coherent with the received one. In this case, where the transmission bandwidth exceeds the channel's coherence bandwidth, we say that the channel exhibits frequency-selectivity since it affects selectively the various "frequency components" of the channel input.

In a nutshell, how time-dispersive a linear time-invariant wireless channel is can be assessed in the time(lag) domain by the delay spread τ_d of its impulse response; and depending on how the delay spread compares with the symbol period and the duration of the symbol pulse, decides the severity of the ISI. In the frequency domain, the amount of time dispersion and ISI correspond to the degree of frequency selectivity, which is quantified by the channel's coherence bandwidth B_{coh}, and depending on how the latter compares with the transmission bandwidth decides how pronounced the frequency selectivity is.

1.3.2 Doppler Spread and Coherence Time

Suppose now that the wireless channel is linear but time-varying and non-dispersive in time with impulse response $h(t;\tau) = h(t), \forall\tau$. With $x(t)$ and $y(t)$ denoting as before the channel input and output, such a channel obeys a multiplicative input-output relationship: $y(t) = h(t)x(t)$. Using as input a finite-duration sinusoidal waveform with infinitesimal bandwidth (delta-like in the frequency domain), the time-varying channel $h(t)$ will disperse the bandwidth of this input waveform by an amount B_d, pretty much as frequencies shift due to the Doppler effect when a source emitting harmonic signals is moving. Such a channel clearly causes frequency dispersion of its input waveforms by an amount B_d that we naturally call Doppler spread.

In a dual fashion to the time-dispersive channel, if $H(\nu)$ denotes the Fourier transform of a frequency-dispersive channel $h(t)$ over an infinite time horizon, the Doppler spread is defined as the frequency for which $H(\nu) = 0, \forall |\nu| > B_d$. As before, for a realization-independent definition, we can let $\phi_h(t) := E[h(t_1)h^*(t_1 + t)]$ denote the autocorrelation of the channel process (* denotes conjugation), which is assumed zero-mean stationary w.r.t. t_1, and define the Doppler spread as the frequency B_d for which the PSD $\Phi_h(\nu) = 0, \forall |\nu| > B_d$. In general, the finite nonzero support of $H(\nu)$ and $\Phi_h(\nu)$ is only approximately true and is satisfied by the finite-parameter basis expansion model introduced in Chapter 9.

If information symbols are transmitted on successive frequency bands each with bandwidth B_p and successive bands are B_s Hertz (Hz) far apart from each other, each received symbol waveform will have bandwidth $B_p + B_d$ and the overlap in the frequency domain will be avoided if $B_s \geq B_p + B_d$. But if $B_s < B_p + B_d$, the received symbol bands will overlap in the frequency domain giving rise to what one could term Doppler intersymbol interference (D-ISI).

As with ISI, it is useful to consider how D-ISI manifests itself in the time domain. Clearly, since $\Phi_h(\nu)$ is bandwidth-limited, $\phi_h(t)$ has infinite support in the time domain. However, it is of interest to define the finite time horizon T_{coh} over which $h(t)$ and thus $\phi_h(t)$ can be considered as approximately constant functions of time. This time interval is called coherence time and in accordance with the uncertainty principle, it is inversely proportional to the Doppler spread:

$$T_{\text{coh}} \approx \frac{1}{B_d}.$$

Clearly, an input pulse $x_0(t)$ of infinitesimal duration centered at t_0 will result in a coherent channel output $y_0(t) = h(t_0)x_0(t)$. And if the symbol waveform consists of such pulses with $t_0 \in [0, T_{\text{coh}}]$ the entire received symbol waveform will remain coherent with the transmitted waveform since $h(t)$ is flat over T_{coh} seconds. This explains why T_{coh} is called coherence time. Except for cases of high mobility, T_{coh} typically spans a number of symbol periods, allowing the channel to be considered as time-invariant over a time horizon of multiple symbols. Certainly, if the channel input $x(t)$ is considered over a duration exceeding T_{coh}, the scale $h(t_0)$ will generally be different for different t_0's, and the transmitted waveform will not be coherent with the received one. In this case we say that the channel exhibits time selectivity since it affects selectively the various "time components" of the channel input.

To summarize: How frequency-dispersive a linear time-varying wireless channel is can be assessed in the frequency domain by the Doppler spread B_d of its frequency response; and depending on how the Doppler spread compares with the symbol rate and the bandwidth of the transmitted waveform, decides the severity of D-ISI. In the time domain, the amount of frequency dispersion and of D-ISI correspond to the degree of time selectivity, which is quantified by the channel's coherence time

T_{coh}, and depending on how the latter compares with the duration of the transmitted waveform decides how pronounced the time selectivity is.

A wireless channel can generally be flat, individually time- or frequency-selective, but also jointly time- and frequency-selective. However, with the B_{coh} and T_{coh} of a channel given, which form of selectivity is present and to what degree depends on the transmission bandwidth and the time window over which the channel input (and thus the channel output) is considered. Based on the size of this time-bandwidth extent, the wireless channels considered in this book fall into four categories, depending on the amount of time selectivity and frequency selectivity they exhibit.

- **Quasi-static flat and block fading channels**

  The random channels in this class are linear time-invariant (i.e., not time-selective) and also not frequency-selective; i.e., both ISI and D-ISI are absent over the time interval and bandwidth considered. This occurs when (i) the transmission bandwidth is smaller than the coherence bandwidth, or equivalently, the symbol period is larger than the delay spread plus the duration of the transmitted symbol waveform; and (ii) the time horizon considered is smaller than the coherence time of the channel. Their impulse response is flat (a random constant complex-valued coefficient) in both the time and the lag variable; i.e., $h(t;\tau) = h$. Because a random constant is a non-ergodic[1] process, these channels are considered as time-invariant only over a finite time horizon. Thus, they are quasi-static in the sense that their impulse response is a random constant assuming a certain value over a coherence time interval of duration T_{coh}, but may change from one coherence interval to the next. Quasi-static flat fading channels considered over multiple blocks of size exceeding T_{coh} are called block fading channels.

- **Time-invariant frequency-selective fading channels**

  In this class, the channels are linear time-invariant (i.e., not time-selective) over the time horizon considered, but they exhibit frequency-selectivity; i.e., ISI is present but D-ISI is absent. This happens when (i) the transmission bandwidth exceeds the coherence bandwidth, or equivalently, the symbol period is smaller than the delay spread plus the transmit pulse duration; and (ii) the time window considered is within the channel coherence time. The class of frequency-selective channels arises when the mobility is relatively low but the data rate is high, as in broadband fixed wireless applications. As we mentioned earlier, these channels are modeled using a tapped-delay line with a finite number of complex tap coefficients [214]. The values of these coefficients remain constant over intervals of size T_{coh}, but they may change from one coherence interval

[1]Non-ergodic channels yield non-ergodic received processes. Since detection and estimation tasks at the receiver must be performed on a per realization basis, sample statistics can be used instead of ensemble statistics only if the received process is ergodic; for example, to obtain the autocorrelation sequence or the average capacity from a single realization, the received process must be ergodic.

to the next. These channels are special cases of their multivariate counterparts dealt with in Chapters 7, 8, and 11.

- **Time-varying time-selective fading channels**

  This class of channels is dual to the preceding one. They are time-selective over the time horizon under study but not frequency-selective over the bandwidth considered. As a result, they induce D-ISI but not ISI. This class arises when (i) the time interval under consideration exceeds the channel coherence time; and (ii) the transmission bandwidth is within the channel's coherence bandwidth, or equivalently, the symbol period (minus the pulse duration) is larger than the channel delay spread. The class of time-selective but frequency-flat channels appears when mobility is high but data rates are not very high. They are modeled using the basis expansion model, which is discussed in Chapter 9.

- **Doubly selective fading channels**

  In the most general and challenging case, wireless channels can be both time- and frequency-dispersive. Since these channels can exhibit both frequency- and time-selectivity, they are often referred to as doubly selective channels. Depending on the delay spread and Doppler spread, a doubly selective channel may exhibit variable degrees of ISI and D-ISI. Clearly, the preceding two classes are subsumed as special cases of the doubly selective class. Both forms of selectivity emerge when: (i) the time horizon under consideration exceeds the channel coherence time; and (ii) the transmission bandwidth exceeds the channel coherence bandwidth. Channels in this class appear in many applications since they encompass cases where both data rates and mobility are high. They can be modeled using a tapped-delay line with time-varying taps whose variations adhere to the basis expansion model detailed in Chapter 9.

- **Multi-input multi-output fading channels**

  Multi-input multi-output (MIMO) fading channels are multivariate generalizations of the single-input single-output (SISO) random channels considered so far. In fact, a MIMO channel is a collection of SISO channels, each of which can fall into any of the four classes we discussed earlier, depending on the selectivity it exhibits in the time and/or the lag variable of its impulse response. MIMO channels arise in various scenarios: with single-user single-antenna systems entailing block transmissions, with multi-user single-antenna systems, and with multi-antenna systems in point-to-point or multi-access communications. Since multi-antenna communication systems provide the context of this book, MIMO fading channels will be encountered throughout. Multi-antenna transmitter and receiver designs will be sought for operation over quasi-static flat fading as well as frequency- and time-selective MIMO channels. Besides time and frequency selectivity, the multiple antennas deployed at the transmitter and/or receiver ends can provide an additional form of selectivity: space selectivity. Furthermore, transmitting over multiple antennas induces extra interference at the receiver side.

The most general wireless channel is a doubly selective MIMO channel, where ISI, D-ISI, and interference in space can be present simultaneously. For these sources of interference to be removed prior to demodulation, the doubly selective MIMO fading channel must be estimated at the receiver. Channel estimation and demodulation can be challenging due to the "curse of dimensionality." But the presence of interference in lag, Doppler, and/or spatial dimensions can be exploited to the designer's advantage. This is possible because over space and over a time-bandwidth window, these sources of interference provide multiple renditions of the information symbol, which can be used to mitigate fading effects by combining these "diversified" transmissions coherently.

As we will see soon, this diversification can improve the performance of MIMO communications through techniques that enable (at the transmitter) and collect (at the receiver) the diversity that these channels are capable of providing. The different flavors of diversity and diversity techniques that can be used to turn the "curse" into a "blessing" are overviewed in the next section.

1.4 PROVIDING, ENABLING, AND COLLECTING DIVERSITY

We define and quantify diversity analytically in Chapter 2. Our goal in this section is to provide an intuitive explanation of diversity and outline techniques for enabling and collecting the diversity to improve rate and error performance of wireless communications over fading channels. Diversity is usually defined in the literature as the number of independent (or at least uncorrelated) copies of the information-bearing signal available at the receiver and is often attributed to operations such as channel coding, interleaving, or frequency hopping performed at the transmitter. This is at times confusing since diversity indeed amounts to the signal copies provided by the channel, but these copies do not have to be independent or uncorrelated. Furthermore, diversity is provided inherently by the channel. The transmission scheme can only enable the diversity provided by the channel (or part of it), while the receiver processing can collect the diversity (or part of it). We can enable only part of the diversity with a suboptimum transmitter and also collect part of it with a suboptimum receiver but can never achieve more diversity than what the channel can provide us with.

Keeping these remarks in mind, we consider next the flavors of diversity provided by the various classes of channels described in Section 1.3. For each class we outline transceiver paradigms to enable and collect the channel diversity.

1.4.1 Diversity Provided by Frequency-Selective Channels

When ISI is present, received symbol waveforms overlap in the time domain, which implies that each symbol is replicated multiple times (each scaled with a different channel tap) in any time window exceeding the channel delay spread in size. These symbol replicas can be combined at the receiver to improve the instantaneous signal-

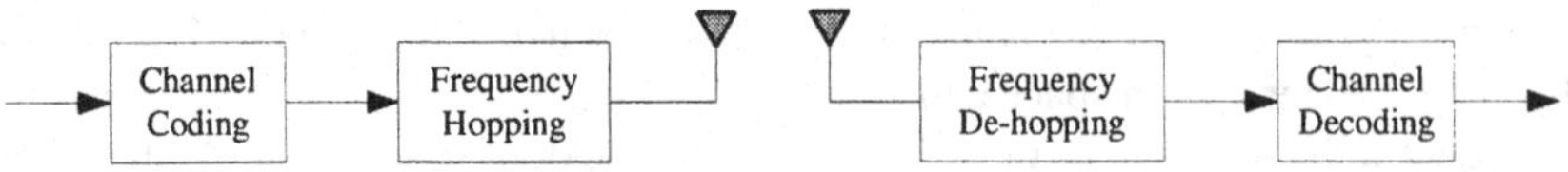

Figure 1.2 Multipath (or frequency) diversity enabled via coding and frequency hopping

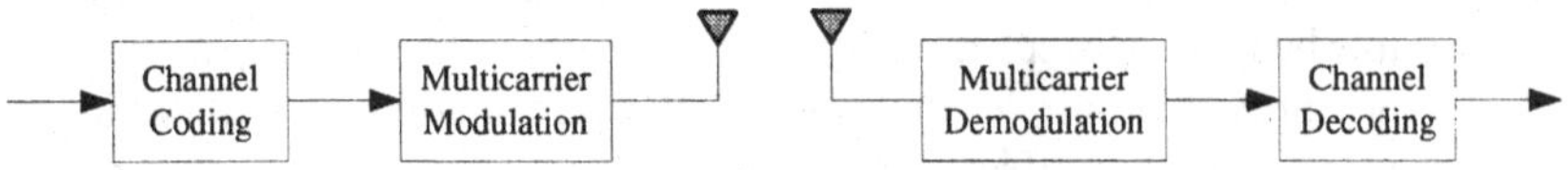

Figure 1.3 Multipath (or frequency) diversity enabled via coding and multi-carrier modulation

to-noise-ratio (SNR) and in turn, the average error performance. Intuitively speaking, this combination adds degrees of freedom to the probability density function (pdf), which renders the pdf of the combined SNR "less fading."[2] In the case of frequency-selective channels, these degrees of freedom are related directly to what we will henceforth call multipath diversity.

A synonymous term often used in place of multipath diversity is frequency diversity. We prefer the term multipath diversity because it captures better the cause behind the creation of symbol replicas. The term frequency diversity refers to the enabling transmission that can take advantage of the frequency selectivity and can be explained if we consider the frequency response $H(f)$ of an ISI channel $h(\tau)$. Suppose that we transmit the same symbol on multiple carriers that are pairwise separated in the frequency domain by at least as much as the coherence bandwidth of the channel. Each symbol replica on a different carrier will then experience a different fading coefficient and combining the replicas at the receiver, using, e.g., a maximum ratio combiner (MRC), will improve the SNR and thus the average error performance.

Transmitting the same symbol on multiple carriers can be implemented either by periodic interleaving or by frequency hopping with a hop interval at least as large as the coherence bandwidth. Furthermore, instead of repetition we can use more powerful channel codes prior to frequency hopping. A system combining channel coding with frequency hopping to enable the multipath (or frequency) diversity of an ISI channel is depicted in Figure 1.2. Channel coding can also be combined with multicarrier modulation as illustrated in Figure 1.3. Clearly, even if fading nulls a number of bits in each channel codeword, the remaining bits, which are likely to experience less severe fading, can lead to reliable recovery of the information-bearing symbol. Maximum likelihood (ML) decoding at the receiver can certainly collect the maximum available diversity. However, the available diversity (or part of it) can be collected even with suboptimum detectors.

[2]When replicas are received through, e.g., uncorrelated complex Gaussian fading channels, the pdf of the combined SNR is chi-square with as many degrees of freedom as the number of replicas. We will see in Section 2.6 that this combined pdf is "less fading" than that of individual pdfs because it is smoother around the origin.

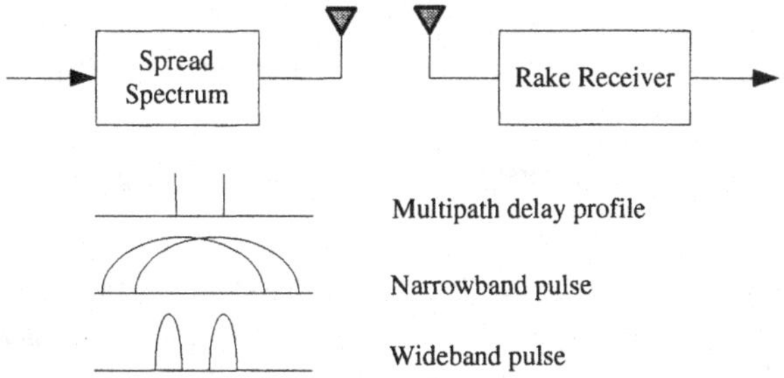

Figure 1.4 Multipath (or frequency) diversity enabled via SS transmission and collected using Rake reception

The amount of multipath diversity provided by an ISI channel depends on the degree of frequency selectivity it exhibits. The more pronounced that frequency-selectivity is, the higher multipath diversity it provides and this diversity can be enabled by shorter codewords, which in turn imply smaller decoding delay and lower complexity at the receiver. On the other hand, frequency selectivity depends on the transmission bandwidth. This means that even flat fading channels can be rendered frequency-selective if we increase the transmission bandwidth. Spread-spectrum (SS) transmissions aim at precisely this objective. Relying on spreading codes with good autocorrelation properties, a Rake receiver can effectively combine the multipath components and collect the available multipath diversity, as in the system depicted in Figure 1.4.

A recent approach to enabling multipath diversity is through the use of linear complex field coding (a.k.a. linear precoding). As we will see in Chapters 7, 8, and 10, this approach does not introduce redundancy and nicely complements the use of redundant channel coding in combating frequency-selective fading effects.

1.4.2 Diversity Provided by Time-Selective Channels

When D-ISI is present, symbol waveforms in the Doppler domain overlap, which means that over a range of Doppler frequencies a symbol can appear multiple times. For this reason, although this form of diversity is often referred to as time diversity, a more appropriate term which we use in this book, is Doppler diversity since signal copies appear in the Doppler domain. The potential benefits of time selectivity can also be appreciated from the input-output relationship in the time domain. If $h(t)$ varies over a symbol duration, it is likely that it will not stay in a deep fade over the duration of the entire symbol waveform, thus allowing for symbol recovery.

Recall that even if a channel is flat over a time window (e.g., of duration equal to one symbol period), it can become time-selective if we expand the time window under consideration. One simple approach to enabling the diversity provided in this case is by repeating the same symbol with period at least equal to the channel coherence time. Knowing the channel at the receiver end, we can employ an MRC

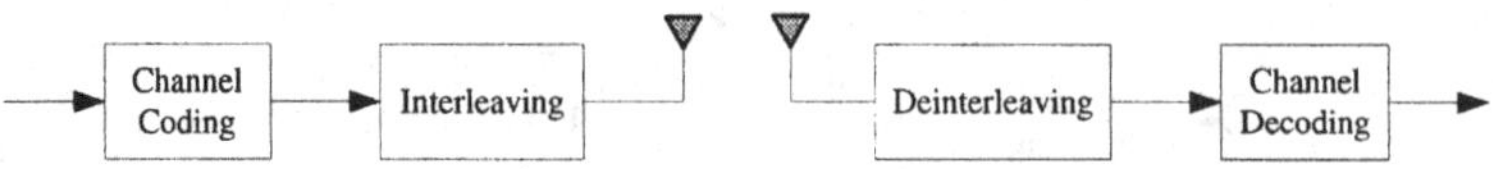

Figure 1.5 Doppler (or time) diversity enabled via channel coding and interleaving

receiver to combine the symbol replicas, enhance the SNR, and thus reduce the average probability of error due to fading. Clearly, periodic insertion of the same symbol can be implemented through an interleaving operation with depth equal to the period of insertion. Furthermore, since channel codes more powerful than repetition are available, it is possible to enable the diversity provided by a time-selective channel using channel coding and interleaving, as depicted in Figure 1.5. At the receiver end, the Doppler diversity can be collected if after deinterleaving, an ML detector is used for decoding; e.g., via Viterbi's algorithm if convolutional coding is employed. The intuitive reason why this approach can improve error performance is that even if fading nulls a number of bits in each codeword, the remaining bits can allow for reliable recovery of the information-bearing symbol.

Notice that coding followed by interleaving can improve error performance even when the channel is quasi-static (block fading), provided that the interleaver depth is larger than the channel coherence time. This is because a flat channel considered over an expanded time horizon can be rendered time-selective and thus provide Doppler diversity. Of course, the faster a channel varies, the higher Doppler diversity it provides, and the transmitter can enable it with an interleaver of smaller depth. This is important because as the interleaver depth increases, the decoding delay and complexity at the receiver side may become intolerable.

Decoding delay and complexity also depend on the amount of redundancy introduced by the channel code. It is possible, however, to eliminate this redundancy if instead of encoding over the Galois field, we encode symbols over the complex field. We will see in Chapter 9 that linear complex field coding offers an alternative means of enabling the Doppler diversity provided by time-selective channels. Galois field codes, on the other hand, offer higher coding gains than do complex field codes. These considerations suggest their joint use over fading channels as detailed in Chapter 10.

As the reader might have already guessed, a general doubly selective channel can provide both multipath and Doppler diversity. In fact, we will see in Chapter 9 that since the degrees of freedom in these general channels equal the product of those provided by time and frequency selectivity, the resulting multipath-Doppler diversity effects are multiplicative. To enable this large amount of diversity, one has to rely on channel coding and/or precoding applied both across frequency (or multipath components) and across time (or Doppler components). Furthermore, to make a quasi-static flat fading channel exhibit frequency and time selectivity, we should clearly increase the transmission bandwidth beyond the channel's coherence bandwidth and also expand the time window at the receiver to exceed the channel's coherence time. By expanding both time and bandwidth, complexity at the receiver will increase along with decoding delay while spectral efficiency will drop considerably, especially if Galois field coding is adopted.

One form of selectivity not requiring time or bandwidth expansion is the one provided by multi-antenna channels, which we consider next.

1.4.3 Diversity Provided by Multi-Antenna Channels

With multiple antennas deployed at the transmitter and/or at the receiver end, the resulting MIMO fading channel possesses degrees of freedom providing what we term space diversity. At the cost of deploying extra antennas even when the MIMO channel is flat, space diversity can improve error performance without increasing the transmission bandwidth to induce frequency selectivity or expanding the observation window to effect time selectivity. Certainly, if these forms of selectivity are already available by the MIMO channel, coded or precoded multi-antenna systems can provide the combined (multiplicative) form of multipath, Doppler, and space diversity.

Space diversity also comes in two flavors: transmit- and receive-antenna diversity. Receive-antenna diversity has been well studied for decades; transmit-antenna diversity has received increasing attention since the late 1990s.

1.4.3.1 Receive-Diversity SIMO Systems Figure 1.6 depicts a receive-diversity system with N_r receive-antennas and a single transmit-antenna giving rise to a single-input multiple-output (SIMO) channel. At each receive-antenna we obtain a copy of the transmitted symbol, s, denoted as $y_i = h_i s + v_i$, where h_i is the flat fading coefficient corresponding to the channel linking the transmit-antenna with the ith receive-antenna; and v_i denotes additive white Gaussian noise (AWGN). Even intuitively it is evident that with the same transmit power, the N_r receive-antennas collect N_r times more power than does a single receive-antenna. This clearly increases the instantaneous receive SNR at the output of a combiner that can be invoked to demodulate the symbol s using a weighted superposition of the received symbols (with corresponding weights w_i) as

$$\hat{s} = \sum_{i=1}^{N_r} w_i y_i = \sum_{i=1}^{N_r} w_i h_i s + \sum_{i=1}^{N_r} w_i v_i.$$

Commonly used combiners include the MRC, where $w_i = h_i^*$, the equal-gain combiner (EGC), where $w_i = h_i^*/|h_i|$, or the selective combiner (SC), where $w_j = 1, j = \arg\max_i |h_i|$, and $w_i = 0, \forall i \neq j$. These three combiners present different trade-offs among required channel knowledge, complexity, and average error probability achieved; but they all collect the maximum receive-antenna diversity even when the channels h_i are correlated and have different pdfs [289]. Again, the underlying reason is that the pdf of the combined SNR at each combiner's output possesses the sum of the degrees of freedom of each individual branch's pdf, which renders it smoother around the origin and thus more resilient to fading effects [289].

A receive-diversity system with multiple receive-antennas and a single transmit-antenna is well suited for the uplink since the access point (base station) can afford the deployment of multiple antennas. However, it may not be a good fit for the downlink since packing multiple antennas at mobile stations may not be cost-effective, and

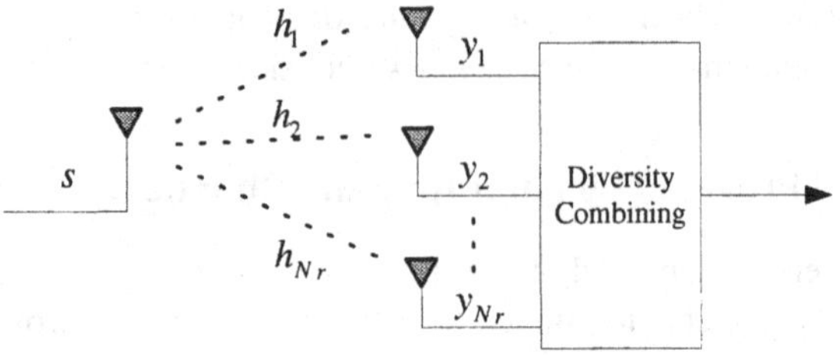

Figure 1.6 Space diversity provided by multiple receive antennas

certainly it is less feasible as the size of the handsets decreases. A more attractive solution for the downlink is to deploy multiple transmit-antennas to enable transmit-antenna space diversity.

1.4.3.2 Transmit-Diversity MISO Systems In a transmit-diversity system, the signal copies providing the diversity come from channels linking multiple (say, N_t) transmit-antennas to a single receive-antenna. However, different from receive-diversity systems, where the power available for demodulation can increase by a factor equal to the number of receive-antennas (N_r), the transmit-power in a transmit-diversity system has to be divided by the N_t transmit-antennas, while the receive-power is identical to that of a SISO system. Nonetheless, it will turn out that even a multi-input single-output (MISO) channel can provide degrees of freedom that can be used at the receiver to collect the available space diversity if the transmitted signals are also designed properly. Collecting the diversity provided by MISO channels is not as easy as in the SIMO case since the signals originating from the multiple transmit-antennas are superimposed at the receive-antenna. But even more challenging is the design of transmissions enabling the available space diversity.

To illustrate the importance of signal design in enabling the available transmit-diversity of MISO channels, let us consider the example system depicted in Figure 1.7, with two transmit-antennas and a single receive-antenna. Assuming flat fading channels where the two channel coefficients h_1 and h_2 are modeled as two zero-mean uncorrelated complex Gaussian random variables with unit variance, the received symbol can be expressed as

$$y = h_1 \frac{1}{\sqrt{2}} s_1 + h_2 \frac{1}{\sqrt{2}} s_2 + v,$$

where the $1/\sqrt{2}$ scale is present since the transmit-power is divided equally between the two antennas.

Consider now repeating the same symbol in both antennas (i.e., $s_1 = s_2$), which yields the input-output relationship

$$y = \left(\frac{h_1 + h_2}{\sqrt{2}} \right) s_1 + v.$$

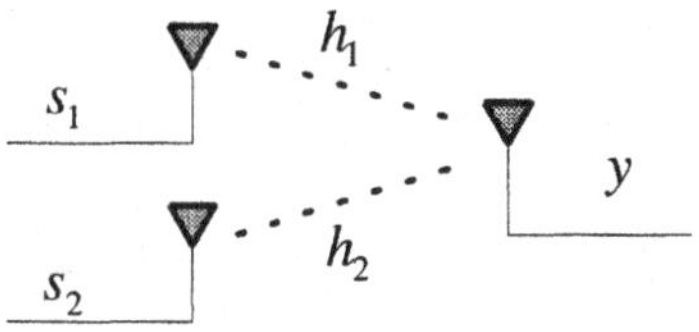

Figure 1.7 Space diversity provided by two transmit-antennas

Since h_1 and h_2 are uncorrelated Gaussian distributed, $(h_1 + h_2)/\sqrt{2}$ is distributed according to the same pdf as h_1. As a result, this multi-antenna system behaves like a SISO system, and no advantage can be enabled from the use of multiple transmit-antennas.

This simple example illustrates the need for smart signal designs in order to benefit from the transmit-diversity provided by a MISO fading channel. The main emphasis in this book is on signal designs that rely on what we will call ST coding techniques. ST codes presume the deployment of multiple transmit-antennas and will entail channel coding over the Galois field and possibly over the complex field in the form of spatial multiplexing across transmit-antennas.

1.4.3.3 Transmit/Receive-Diversity MIMO Systems For a more general system with multiple transmit-antennas and multiple receive-antennas, recent research has shown that ST coded transmissions over the resulting MIMO channel can effect diversity as high as the number of transmit-antennas times the number of receive-antennas ($N_t N_r$); see also Figure 1.8. As we will see in Chapter 3, this can provably enhance the average error performance of multi-antenna systems at sufficiently high SNR [6, 253, 289].

But furthermore, we show in Chapter 3 that the degrees of freedom that become available with a MIMO channel can boost the capacity (and thus data rates) of multi-antenna systems well beyond those available with single-antenna links over SISO channels [69, 256, 326]. This can be roughly guessed if we notice that with, for example, two transmit-antennas we can transmit two symbols simultaneously for every channel use. Critical to this enhanced rate performance is the use of ST coded transmissions over multiple transmit-antennas, which intuitively speaking, provide multiple "information pipes" for symbols to flow through the MIMO channel.

Besides error and rate performance, the design of ST coded multi-antenna transmissions must take into account a number of practical issues. One pertains to complexity, which becomes critical if ST coding is employed in the downlink, where the receiver can be a handset. Since a handset typically has limited power and size, one has to design ST coding schemes properly so that the corresponding receiver complexity is affordable. Another issue relates to data rate and mobility. As ST coded systems are envisioned to support high data rates possibly at high mobility, the MIMO fading channel is likely to be frequency- and/or time-selective. Hence, ST codes must be designed to account for these challenging propagation channels. Furthermore, ST

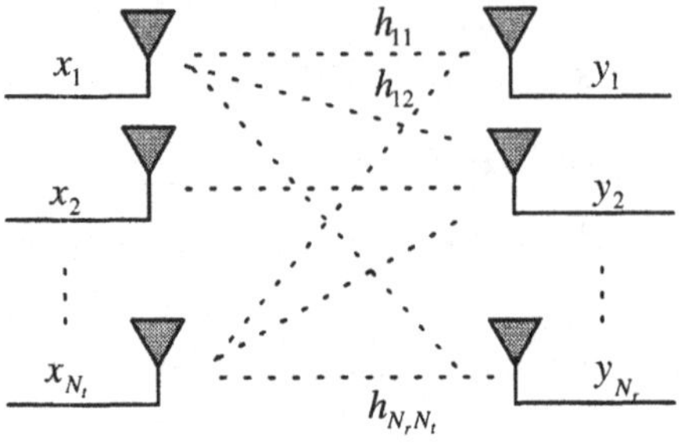

Figure 1.8 Space diversity provided by multiple transmit-receive antennas

coded systems should be robust, to handle various sources of interference in addition to fading and AWGN. These could include co-channel or intentional interference as well as multiuser interference in a multi-access setup. Finally, a major issue that must be taken into account when designing ST codes is how much knowledge of the channel state information (CSI) is available at which end of the link (receiver and possibly transmitter via, e.g., a feedback channel).

1.5 CHAPTER-BY-CHAPTER ORGANIZATION

Design of ST coded multi-antenna systems for operation over general broadband mobile wireless MIMO channels is the subject of this book. The ST designs we cover in the ensuing 14 chapters can be classified based on a variety of criteria, as we will see in Section 2.7. A common objective is to design systems flexible enough to reach desirable trade-offs among error performance, complexity, and spectral efficiency. As is often the case in practice, there is no universally "best ST code" and the designer's choice certainly depends on application-specific constraints. A road map of the contents in Chapters 2–15 is outlined next.

- *Fundamentals of ST Wireless Communications (Chapter 2)* In this chapter we describe fundamental capacity and error performance metrics pertaining to ST coded multi-antenna transmissions over quasi-static flat fading MIMO channels. Through these metrics it is established quantitatively that multi-antenna MIMO systems can improve considerably the capacity and error probability of SISO systems. After laying out a unifying system model for multi-antenna ST coded links, capacity and error performance bounds are presented to serve also as criteria for designing and analyzing ST coded systems — one based on the diversity gain and the other on the coding gain. Differences and similarities between channel coding and ST coding are presented, trade-offs between diversity and spatial multiplexing are delineated, and the definition of spectral efficiency used for comparing ST coded systems is specified.

- *Coherent ST Codes for Flat Fading Channels (Chapter 3)* In this chapter we present several classes of ST codes designed for quasi-static flat fading MIMO channels. The classes include ST trellis codes, orthogonal and quasi-orthogonal ST block codes, ST linear complex field codes, and unifying designs. They are coherent in the sense that they all require knowledge of the MIMO channel at their decoding stage. The designs aim to maximize the space diversity provided by the MIMO channel, but the resulting ST codes end up exhibiting different merits in rate, coding gain, and (de)coding complexity. Common to all these classes is that they improve error performance relative to their single-antenna counterparts, but in spectral efficiency they do not exceed rates achievable by SISO systems.

- *Layered ST Codes (Chapter 4)* The classes of coherent ST codes in this chapter aim to attain rates higher than those achievable by SISO systems, and they do so by ST coding and multiplexing groups of symbols which are called layers. These ST codes are capable of approaching the capacity gains that MIMO channels can provide. In this chapter, hybrid ST codes are further introduced to trade off diversity for spectral efficiency, thus bridging the rate-oriented layered ST codes with the classes of error performance-oriented ST codes of Chapter 3. Finally, ST codes are presented to achieve an interesting combination of full diversity at full spectral efficiency while being flexible enough to achieve desirable trade-offs in complexity, error, and rate performance.

- *Sphere Decoding and (Near-)Optimal MIMO Demodulation (Chapter 5)* This chapter deals with decoding of ST codes based on the so-called sphere decoding algorithm (SDA). SDA is useful for coherent demodulation of most ST codes designed in this book for flat, frequency-selective, time-selective, or doubly selective MIMO channels. Besides multi-antenna systems, SDA also finds applications in demodulation of single-antenna single-user block (coded or precoded) transmissions as well as in detection of multiuser transmissions. Zero-forcing, linear minimum mean-square error, decision feedback, and recent quasi-ML demodulators are also mentioned briefly in this chapter. But the emphasis on SDA is well justified because it outperforms these lower-complexity alternatives while being able to reach near-ML or exact-ML optimality for a number of SNR values and problem dimensions of practical interest. Surprisingly, SDA can achieve exact- or near-ML optimality at polynomial average complexity, which is approximately cubic in the problem dimension. (Recall that exact ML by enumeration incurs exponential complexity in the problem dimension.) In addition to hard decoding, this chapter shows how SDA can facilitate soft decoding of ST coded multi-antenna transmissions, which is necessary to approach the performance dictated by the enhanced capacity of MIMO channels.

- *Noncoherent and Differential ST Codes for Flat Fading Channels (Chapter 6)* In this chapter we present noncoherent and differential ST codes for quasi-static flat fading MIMO channels. As these two classes of ST codes do not require channel knowledge at their decoding stage, they are suitable for applications where channel estimation is impossible or cannot be afforded. They are also well motivated when the MIMO channel undergoes slow time variations. Non-coherent and differential ST codes have their own pros and cons, and one class could be preferred over the other, depending on application-specific trade-offs among error performance, complexity, and spectral efficiency. The theme of this chapter is to present the basic ideas behind these schemes, analyze their error performance, and compare them with their coherent counterparts.

- *ST Codes for Frequency-Selective Fading Channels: Single-Carrier Systems (Chapter 7)* The coherent ST codes designed in this chapter for single-carrier systems as well as those developed in Chapter 8 for multi-carrier systems are tailored for broadband wireless applications where frequency selectivity must be accounted for. Upon recognizing that the input-output relationship of a frequency-selective MIMO channel can be brought into an equivalent flat fading MIMO form, this chapter shows how to generalize the ST codes of Chapter 3 to enable the joint space-multipath diversity provided by frequency-selective MIMO channels. Besides diversity, these ST codes are compared on the basis of their outage capacity, and they all attain rates not exceeding those of a SISO channel.

- *ST Codes for Frequency-Selective Fading Channels: Multi-Carrier Systems (Chapter 8)* The ST codes in this chapter are also intended for frequency-selective MIMO channels but different from the ST codes in Chapter 7, they are designed for multi-carrier systems employing orthogonal frequency division multiplexing (OFDM) modulation. The classes considered include trellis codes, orthogonal block codes, and linear complex field codes based on digital phase sweeping (DPS) operations. All these classes of ST codes enable the maximum space-multipath diversity provided by the frequency-selective MIMO channel and relative error performance is assessed using their outage capacity. ST coded MIMO OFDM systems offer improved error performance relative to single-antenna OFDM systems at the same spectral efficiency. In addition, linear complex-field coded MIMO OFDM transmissions are designed in this chapter to enable maximum diversity at spectral efficiencies higher than those that SISO OFDM systems can attain. This feature, along with the lower-complexity that MIMO OFDM receivers can afford, favors multi-carrier multi-antenna systems over their single-carrier counterparts of Chapter 7, especially for MIMO channels with long delay spreads. The brief comparison between multi-carrier and single-carrier systems included at the end of this chapter gives an edge to single-carrier systems only when it comes to deciding on the basis of a low peak-to-average-power-ratio and insensitivity to carrier frequency offsets.

- *ST Codes for Time-Varying Channels (Chapter 9)* Mobility effects, drifts, and mismatch between transmit-receive oscillators all give rise to time-varying (TV) MIMO channels. To design ST codes for these channels, we rely on a parsimonious basis expansion model, based on which it becomes possible to establish a duality between time- and frequency-selective MIMO channels. Several classes of ST codes designed for frequency-selective channels are then mapped based on this duality, to obtain ST codes capable of enabling the joint space-Doppler diversity provided by time-selective MIMO channels. Furthermore, ST codes are designed in this chapter for doubly selective channels and are shown to enable the joint space-multipath-Doppler diversity. Besides coherent designs, differential ST codes are also developed.

- *Joint Galois-Field and Linear Complex-Field ST Codes (Chapter 10)* The reliability of MIMO fading links improves considerably with space-diversity but can at best reach the performance of uncoded transmissions over single-antenna AWGN channels, even if one can afford the complexity and cost of deploying multiple antennas. In this chapter we deal with concatenated Galois-field (GF) and LCF codes, which allow ST coded systems to approach the error performance dictated by the capacity of MIMO fading channels. In enabling the diversity, LCF codes turn out to be less complex and more spectrally efficient than GF codes; while GF codes bring larger coding gains than LCF codes, which is critical as the channel comes closer to an AWGN one. It is demonstrated that the combination of GF-LCF ST codes is powerful in dealing with MIMO fading channels when a turbo decoder with relatively low complexity is employed at the receiver. Besides flat fading MIMO channels, it is further shown that the GF-LCF ST system is applicable to time- and frequency-selective MIMO channels.

- *MIMO Channel Estimation and Synchronization (Chapter 11)* In this chapter we introduce carrier synchronization and MIMO channel estimation algorithms with universal applicability in decoding coherent ST coded transmissions. Along with preamble-based schemes, optimal training patterns are designed for channel estimation to optimize estimation performance jointly with transmitter resources (power and bandwidth). Estimators are derived for both single-carrier and multi-carrier transmissions over frequency-selective MIMO channels, but they specialize to flat MIMO channels, and through the duality established in Chapter 9, they also apply to time-selective and doubly selective MIMO channels. Decision-directed, Kalman-filtering based, and (semi-)blind alternatives are also outlined for joint channel estimation, tracking, and demodulation. Training patterns are further designed in this chapter for carrier frequency offset (CFO) estimation, ensuring full acquisition range. For acquisition ranges not exceeding half-subcarrier spacing, a low-complexity blind CFO estimator is also developed for MIMO OFDM systems.

- *ST Codes with Partial Channel Knowledge: Statistical CSI (Chapter 12)* Different from other chapters, in this chapter and Chapter 13 we consider ST systems where the multi-antenna transmitter has partial channel state information (CSI). Partial CSI here is modeled statistically as a multivariate complex Gaussian vector representing the rendition of the true channel as perceived by the transmitter. This form of statistical CSI can be either made a priori available to the transmitter through sounding experiments, or, it can reach the transmitter through feedback from the receiver. Either way, it is used in this chapter to design ST spread-spectrum systems, transmit-beamformers, and ST coder-beamformer hybrids operating with fixed or adaptive modulations. The designs are optimized to minimize appropriate error bounds and in certain cases to maximize the capacity of multi-antenna systems.

- *ST Codes with Partial Channel Knowledge: Finite-Rate CSI (Chapter 13)* In this chapter we present closed-loop ST designs where partial CSI becomes available to the transmitter only through feedback from the receiver in the form of a finite number of bits. Capitalizing on this finite-rate CSI, multi-antenna transmitters are designed based on beamforming with and without adaptive modulation, on unitary precoded ST multiplexing to further increase transmission rates, and on ST coder-precoder combinations. All designs are cast in a vector quantization framework which entails off-line construction of a codebook to specify a finite number of optimal transmission modes (beamformer or precoder codewords). One codeword for each feedback cycle is selected from this codebook in online operation to "best" adapt the ST transmitter to the quantized MIMO channel codeword conveyed by the feedback bits. Average error probability and in some cases capacity are adopted as criteria of optimality in both the codebook design and in the codeword selection rule.

- *ST Codes in the Presence of Interference (Chapter 14)* In this chapter we introduce spread-spectrum and transmit-beamforming-based schemes to enhance the robustness of ST coded multi-antenna systems in the presence of intentional or unintentional interference. The ST transmitter designs exploit the second-order spatial statistics of the MIMO channel and temporal statistics of the interference. Because ML decoding can be prohibitively complex in the presence of interference, low-complexity quasi-ML demodulators are devised. Since the latter require reliable channel estimators, optimal training sequences are also designed for channel estimation in the presence of interference by exploiting knowledge of the interference covariance matrix at the transmitter.

- *ST Codes for Orthogonal Multiple Access (Chapter 15)* Aiming at orthogonal ST multiple access, in this chapter we show how ST coding can be combined with signature spreading sequences to permeate single-user benefits to multiuser communications stemming from the use of multiple antennas. Spreading sequences are judiciously designed for quasi-synchronous system operation with either single-carrier or multi-carrier transmissions over frequency-selective MIMO channels. The single-carrier system relies on chip-interleaving and block-spreading operations to suppress multiuser interference deterministically. The multi-carrier system is based on MIMO OFDM, where non-overlapping subcarriers are assigned to individual users. With deterministic multiuser separation, ST (de)coding can be performed on a per user basis.

2

Fundamentals of ST Wireless Communications

In this chapter we outline basic notions and fundamental limits of ST coding. After laying out a unifying system model for multi-antenna ST coded links, we introduce basic figures of merit involved in ST coding, including capacity and error performance bounds. These figures of merit lend themselves naturally to criteria for designing ST codes. For the purpose of illustrating these basic notions, we restrict our discussion to the simplest, yet most important class of wireless channel models, which exhibit flat Rayleigh fading effects. More challenging channel models are considered in later chapters.

2.1 GENERIC ST SYSTEM MODEL

This section introduces a generic system model that is applicable to most existing ST coding schemes, e.g., [68, 70, 114, 120, 249, 251, 253]. Such a model is instrumental in understanding the basic issues involved when designing ST codes and plays an important role in linking ST transmissions with block transmissions.[1] Although this model assumes flat Rayleigh fading channels, most of the principles and analytical results derived under this model provide important guidelines for the design of ST codes in the presence of frequency-selective and/or time-selective fading channels, treated in Chapters 7–9.

[1] In this book, serial transmissions are treated as special cases of block transmissions with blocks having infinite length.

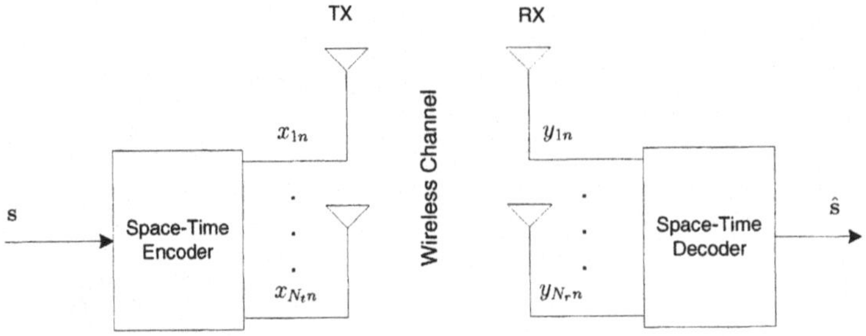

Figure 2.1 Generic space-time system model

Figure 2.1 depicts a wireless system equipped with N_t transmit-antennas and N_r receive-antennas. As a convention throughout the book, a wireless communication system is referred to as an ST system when more than one transmit-antenna is deployed; otherwise, it is called a non-ST system. Note that multiple receive-antennas are optional in both ST and non-ST systems. Different from a non-ST system, where information data are transmitted in a block-by-block fashion, data transmission in an ST system is carried out in two dimensions, the space dimension and the time dimension, as the acronym ST suggests. The space dimension is spanned by multiple transmit-antennas while the time dimension is spanned by multiple time intervals over which multiple blocks are transmitted. Although a block transmission from a single transmit-antenna can be thought of as a special form of an ST transmission, in the sequel we distinguish between these two transmission modalities.

As depicted in Figure 2.1, the key module of an ST system is the ST encoder, whose function is to convert one-dimensional block transmissions into two-dimensional ST transmissions. Let $\mathbb{S}$ be the signal constellation to which information symbols belong. The ST encoder takes a block $\boldsymbol{s} \in \mathbb{S}^{N_s \times 1}$ of N_s information symbols as input and uniquely outputs an ST code matrix:

$$\boldsymbol{X} = \begin{bmatrix} x_{11} & x_{12} & \cdots & x_{1N_x} \\ x_{21} & x_{22} & \cdots & x_{2N_x} \\ \vdots & \vdots & \ddots & \vdots \\ x_{N_t1} & x_{N_t2} & \cdots & x_{N_tN_x} \end{bmatrix} \in \mathbb{C}^{N_t \times N_x}.$$

Matrix $\boldsymbol{X}$ is generally complex valued ($\boldsymbol{X} \in \mathbb{C}^{N_t \times N_x}$) and is often normalized to satisfy a transmit-power constraint:

$$\frac{1}{N_x N_t} E\left[\operatorname{tr}(\boldsymbol{X}\boldsymbol{X}^{\mathcal{H}})\right] = 1, \tag{2.1}$$

where $E[\cdot]$ denotes statistical expectation and $\operatorname{tr}(\cdot)$ stands for the trace of a matrix. From a conceptual viewpoint, ST coding is nothing but a one-to-one mapping from $\boldsymbol{s}$ to $\boldsymbol{X}$:

$$\mathcal{M}_{\text{ST}}: \quad \boldsymbol{s} \longleftrightarrow \boldsymbol{X}. \tag{2.2}$$

Let $\mathbb{X}$ denote the set of all possible ST code matrices $\boldsymbol{X}$, which henceforth we refer to as ST constellation. Since the mapping (2.2) is one-to-one, the ST constellation $\mathbb{X}$ has cardinality

$$|\mathbb{X}| = |\mathbb{S}|^{N_s}. \tag{2.3}$$

Note that in writing (2.3), we have implicitly assumed an ST system without error control coding (or precoding) where information symbols are statistically independent.

The N_x columns of $\boldsymbol{X}$ are generated in N_x successive time intervals each of duration T_s, while each of the N_t entries in a given column is forwarded to one of the N_t transmit-antennas. At the μth transmit-antenna, $x_{\mu n}$ is first pulse-shaped and then transmitted during the nth time interval. The transmitted waveforms from N_t transmit-antennas are sent simultaneously. Because N_s information symbols in $\boldsymbol{s}$ are transmitted in N_x time slots, it is reasonable to define the ST code rate as

$$\eta_{\text{ST}} = \frac{N_s}{N_x} \text{ symbols per channel use.} \tag{2.4}$$

Notice that η_{ST} is different from the rate of a channel code that is defined over the Galois field. Since each information symbol carries $\log_2 |\mathbb{S}|$ information bits, the transmission rate is given by

$$R_{\text{ST}} = \frac{N_s}{N_x} \log_2 |\mathbb{S}| = \eta_{\text{ST}} \log_2 |\mathbb{S}| \text{ bits per channel use.} \tag{2.5}$$

For a non-ST system, i.e., when ST coding is not used, (2.4) becomes

$$\eta_{\text{non-ST}} = 1 \text{ symbol per channel use.} \tag{2.6}$$

At the receiving end, each of the N_r antennas collects the superposition of N_t transmitted signals corrupted by fading propagation effects as well as additive noise. The received waveform is first passed through a filter which is matched to the pulse-shaping filter at the transmitter, and the resulting output is sampled at rate $1/T_s$. We will assume that:

A2.1) The fading channel between the μth transmit-antenna and the νth receive-antenna is quasi-static, flat over N_x symbol intervals. As a result, this channel can be represented as a complex zero-mean Gaussian random variable $h_{\nu\mu} \in \mathbb{C}$ which assumes a certain value over the block duration of one ST code matrix but may change from block to block. This assumption is satisfied if the channel delay spread is smaller than T_s but the channel coherence time T_{coh} is larger than $N_x T_s$.

A2.2) The channel coefficients $h_{\nu\mu}$ are i.i.d., complex Gaussian having statistically independent real and imaginary parts each with zero mean and variance 0.5. This assumption corresponds to the case of flat Rayleigh fading and is satisfied when the transmit- and receive-antennas are sufficiently well separated so that correlations emerging due to coupling effects can be ignored.

Under assumption A2.1), the nth received sample $y_{\nu n}$ at the νth receive-antenna can be expressed as

$$y_{\nu n} = \sum_{\mu=1}^{N_t} \sqrt{\frac{\bar{\gamma}}{N_t}} h_{\nu\mu} x_{\mu n} + w_{\nu n}, \quad \forall \, \nu \in [1, N_r], \quad n \in [1, N_x], \tag{2.7}$$

where $\bar{\gamma}$ is introduced to control the transmission power and $w_{\nu n}$ denotes the zero-mean complex additive white Gaussian noise (AWGN) . Upon defining the channel matrix

$$\boldsymbol{H} = \begin{bmatrix} h_{11} & h_{12} & \cdots & h_{1N_t} \\ h_{21} & h_{22} & \cdots & h_{2N_t} \\ \vdots & \vdots & \ddots & \vdots \\ h_{N_r 1} & h_{N_r 2} & \cdots & h_{N_r N_t} \end{bmatrix} \in \mathbb{C}^{N_r \times N_t}, \tag{2.8}$$

equation (2.7) can be expressed in matrix form as

$$\boldsymbol{Y} = \sqrt{\frac{\bar{\gamma}}{N_t}} \boldsymbol{H}\boldsymbol{X} + \boldsymbol{W}, \tag{2.9}$$

where $\boldsymbol{Y}$ with entries $[\boldsymbol{Y}]_{\nu n} := y_{\nu n}$ forms the $N_r \times N_x$ matrix of received noise-free samples and $\boldsymbol{W}$ with entries $[\boldsymbol{W}]_{\nu n} = w_{\nu n}$ is the corresponding noise matrix. The average signal-to-noise ratio (SNR) per receive-antenna is [cf. (2.9)]

$$\overline{\text{SNR}} = \frac{\bar{\gamma}}{N_t} \frac{E[\text{tr}(\boldsymbol{H}\boldsymbol{X}\boldsymbol{X}^{\mathcal{H}}\boldsymbol{H}^{\mathcal{H}})]}{E[\text{tr}(\boldsymbol{W}\boldsymbol{W}^{\mathcal{H}})]}. \tag{2.10}$$

Unless specified otherwise, $x_{\mu n}$ and $w_{\nu n}$ are normalized throughout to have unit variance. Based on this normalization and under assumption A2.2), it can easily be verified from (2.10) that

$$\overline{\text{SNR}} = \frac{E[\text{tr}(\boldsymbol{X}\boldsymbol{X}^{\mathcal{H}})]}{N_t N_x} \bar{\gamma}. \tag{2.11}$$

It is clear that if $E[\text{tr}(\boldsymbol{X}\boldsymbol{X}^{\mathcal{H}})] = N_x N_t$, then $\overline{\text{SNR}} = \bar{\gamma}$. In other words, $\bar{\gamma}$ represents the average SNR per receive-antenna when all entries of $\boldsymbol{X}$ are nonzero.

Compared to (2.7), the data model (2.9) is more insightful in that it relates the received samples to the ST code matrix as a whole. It shows that an ST system transmits an ST code matrix over a matrix channel created by the use of multiple transmit- and receive-antennas.

As depicted in Figure 2.1, the receiver employs an ST decoder to recover $\boldsymbol{X}$ and thus $\boldsymbol{s}$ from $\boldsymbol{Y}$. Similar to (2.2), this decoding process can be conceptually viewed as a mapping

$$\bar{\mathcal{M}}_{\text{ST}} : \quad \boldsymbol{Y} \longrightarrow \boldsymbol{X} \text{ or } \boldsymbol{s}. \tag{2.12}$$

As we will discuss soon, the particular design of $\bar{\mathcal{M}}_{\text{ST}}$ depends on the selection of the ST encoder, the availability of channel state information (CSI), the decoding complexity that a system can afford, and the prescribed error performance.

Equation (2.9) provides a general system model for ST transmissions over flat fading multi-input multi-output (MIMO) channels. Under this model, designing an ST system amounts to selecting parameters N_s, N_x, N_t, N_r and constructing the two mappings (2.2) and (2.12) so that desirable trade-offs among error performance, spectral efficiency, and decoding complexity can be achieved.

2.2 ST CODING VIZ CHANNEL CODING

The usefulness of channel coding as an effective means of improving the reliability of communication links is well documented. However, we will soon argue that channel coding may not be as effective over fading channels as it has been over AWGN channels. This, in part, motivates the development and widespread applicability of ST coded multi-antenna systems. The purpose of this section is to highlight several advantages of ST coding over channel coding in the presence of fading. Our discussion will remain at a qualitative level without reference to any particular channel or ST coding scheme per se.

One major difference between channel coding and ST coding lies in the way data are transmitted. While ST coding relies on a two-dimensional ST transmission, channel coding was originally proposed for one-dimensional block transmissions with a single transmit-antenna. Since multiple receive-antennas are beneficial for both ST and block transmissions, let us consider in this section the simplest configuration with a single receive-antenna.

When one receive-antenna is used, the corresponding ST data model is a special form of (2.9) and is given by

$$\boldsymbol{y} = \sqrt{\frac{\bar{\gamma}}{N_t}}\boldsymbol{X}^T\boldsymbol{h} + \boldsymbol{w} = \sqrt{\frac{\bar{\gamma}}{N_t}}\mathcal{M}_{\text{ST}}(\boldsymbol{s})\boldsymbol{h} + \boldsymbol{w}, \tag{2.13}$$

where $\boldsymbol{y}^T \in \mathbb{C}^{1\times N_x}$ and $\boldsymbol{w}^T \in \mathbb{C}^{1\times N_x}$ denote a row of $\boldsymbol{Y}$ and $\boldsymbol{W}$, respectively; vector $\boldsymbol{h} \in \mathbb{C}^{N_t\times 1}$ collects channel coefficients between the N_t transmit-antennas and the single receive-antenna; and for the second equality we used (2.2).

By setting $N_t = 1$, (2.13) can be further reduced to the data model of a single-antenna block transmission; that is,

$$\boldsymbol{y} = \sqrt{\bar{\gamma}}\,\boldsymbol{x}h + \boldsymbol{w} = \sqrt{\bar{\gamma}}\,\mathcal{M}_{\text{non-ST}}(\boldsymbol{s})h + \boldsymbol{w}, \tag{2.14}$$

where h is the channel coefficient between the single transmit-antenna and the single receive-antenna; and $\boldsymbol{x} \in \mathbb{C}^{N_x\times 1}$ denotes the channel codeword obtained from $\boldsymbol{s}$ according to a certain channel coding scheme $\mathcal{M}_{\text{non-ST}}(\cdot)$.

Although $\boldsymbol{X}$ in (2.13) and $\boldsymbol{x}$ in (2.14) convey identical information (i.e., $\boldsymbol{s}$), they are obviously transmitted over different channels. Specifically, $\boldsymbol{X}$ is transmitted over a vector channel $\boldsymbol{h}$, whereas $\boldsymbol{x}$ goes through a scalar channel h. Because fading channels are random, the resulting performance clearly depends on channel conditions. Generally speaking, as channel gains increase, error performance improves. In the case of a block transmission, (2.14) reveals that the transmitted entries of $\boldsymbol{x}$ experience a single channel coefficient, h. Intuitively, if h happens to be small, those transmissions simultaneously become unreliable, due to bursty errors, often to an extent beyond the error-correcting capability of the underlying channel code. To improve reliability, one commonly used approach in practice is to incorporate interleaving in conjunction with channel coding so that different entries of $\boldsymbol{x}$ experience distinct channel coefficients. If a sufficiently long interleaver is designed properly, those channel coefficients are likely to be statistically independent. Therefore, it becomes unlikely that all of them are small simultaneously. Thus, transmission errors become less "bursty" and with the aid of channel coding, $\boldsymbol{x}$ can be recovered reliably. Unfortunately, the price paid for the use of possibly long interleaving is excess decoding delay, which may not be tolerable in a number of applications. On the other hand, instead of using an interleaver, ST transmission resorts to multiple transmit-antennas to create multiple channel coefficients that are statistically independent if multiple transmit-antennas are well separated from one another. Since $\boldsymbol{X}$ is transmitted as a whole over those channels, it is very likely that at least some of the channels are strong enough to ensure reliable recovery of $\boldsymbol{X}$. In short, ST transmissions function in a way similar to channel-coded block transmissions together with interleaving but without introducing undesirable decoding delay.

The potential advantages of ST transmissions over single-antenna block transmissions hinge on the assumption that different transmit-antennas create statistically independent (or at least uncorrelated) channels. To further justify this point, let us consider an extreme case where all multi-antenna transmissions experience identical channels with coefficient h, i.e., $\boldsymbol{h} = [h, \ldots, h]^T$. As a result, (2.13) can be rewritten as ($\mathbf{1}$ denotes the all-one column vector)

$$\boldsymbol{y} = \sqrt{\bar{\gamma}}\, h \left(\frac{\boldsymbol{X}^T \cdot \mathbf{1}}{\sqrt{N_t}} \right) + \boldsymbol{w}, \tag{2.15}$$

which is basically (2.14) with

$$\boldsymbol{x} := \frac{\boldsymbol{X}^T \cdot \mathbf{1}}{\sqrt{N_t}}.$$

The implication is that ST coding is not useful in this special case of identical channels, since channel coding is capable of achieving the same error performance as ST coding but without requiring additional transmit-antennas. More generally, use of ST coding is justified when transmit-antennas are separated enough so that transmissions from different antennas experience independent (or at least uncorrelated) fading channels.

So far, our intuitive analysis suggests that ST coded transmissions are potentially more robust against channel fading than is channel coding without interleaving. As we will see in Section 2.4, this improved robustness is due primarily to the diversity enabled by ST coding. On the other hand, as we discuss in Chapter 10, ST coding typically has smaller coding gain and may not be as effective as channel coding in coping with additive noise. Therefore, where fading and additive noise are both present, it makes a lot of sense in practice to combine ST coding with channel coding.

In addition to error performance advantages, ST coding has the potential to offer better bandwidth (a.k.a. spectral) efficiency relative to channel coding. ST coding maps a one-dimensional vector to a two-dimensional code matrix, while channel coding is a mapping between two one-dimensional vectors. Consequently, time-domain redundancy is typically a must in channel coding but may not be necessary in ST coding, thanks to the extra space dimension. In other words, spectral efficiency loss can be mitigated with ST coding. In addition, the extra space dimension brings about more flexibility in the design of ST coding, as we will see soon.

2.3 CAPACITY OF ST CHANNELS

An important figure of merit that benchmarks the maximum transmission rates for reliable communications is channel capacity. Channel capacity provides the rate limit determined by the characteristics of the underlying channel and is fundamental because it does not depend on any particular transceiver design. In this section the channel capacity of multi-antenna systems is investigated and compared to that of single-antenna systems. It will be shown that by creating a matrix channel through the use of multiple antennas, an ST system can potentially support much higher rates than a single-antenna system when operating in the presence of fading.

Let us take a closer look at the single-shot ST transmission model of (2.9). At any time interval we have the linear input-output relationship

$$\boldsymbol{y} = \sqrt{\frac{\bar{\gamma}}{N_t}} \boldsymbol{H}\boldsymbol{x} + \boldsymbol{w}, \tag{2.16}$$

where $\boldsymbol{y} \in \mathbb{C}^{N_r \times 1}$, $\boldsymbol{x} \in \mathbb{C}^{N_t \times 1}$, and $\boldsymbol{w} \in \mathbb{C}^{N_r \times 1}$ denote, respectively, a generic column of $\boldsymbol{Y}$, $\boldsymbol{X}$, and $\boldsymbol{W}$. The noise vector $\boldsymbol{w}$ is complex Gaussian distributed with zero mean and covariance matrix $\boldsymbol{I}_{N_r}$. We suppose that channel knowledge is available at the receiver but not at the transmitter. It has been shown [256, Theorem 1] that under the power constraint [cf. (2.1)]

$$\frac{1}{N_t} \operatorname{tr}(\boldsymbol{R}_{xx}) \leq 1,$$

the channel capacity of ST transmission in (2.16) is achieved when $\boldsymbol{x}$ is a circularly symmetric Gaussian vector with zero mean and covariance matrix $\boldsymbol{R}_{xx} = \boldsymbol{I}_{N_t}$; and the channel capacity, conditioned on $\boldsymbol{H}$, is given by

$$\begin{aligned} C(N_t, N_r|\boldsymbol{H}) &= I(\boldsymbol{x}; \boldsymbol{y}|\boldsymbol{H}, \boldsymbol{I}_{N_t}) \\ &= \log_2 \det \left[\boldsymbol{I}_{N_r} + \frac{\bar{\gamma}}{N_t} \boldsymbol{H}\boldsymbol{H}^{\mathcal{H}} \right] \\ &= \sum_{i=1}^{\operatorname{rank}(\boldsymbol{H})} \log_2 \left(1 + \lambda_i \frac{\bar{\gamma}}{N_t} \right), \end{aligned} \tag{2.17}$$

where $\det(\cdot)$ denotes matrix determinant, λ_i is the ith eigenvalue of $\boldsymbol{H}\boldsymbol{H}^{\mathcal{H}}$, and $\operatorname{rank}(\boldsymbol{H})$ denotes the rank of $\boldsymbol{H}$. The conditional channel capacity of a single-antenna system follows as a special case of (2.17) with $N_t = N_r = 1$ and can be expressed as

$$C(1, 1|h) = \log_2\left(1 + \bar{\gamma}|h|^2\right). \tag{2.18}$$

In *non-fading* channels, h is deterministic and can be normalized to 1. In this case, $C(1, 1|h)$ reduces to the well-known Shannon capacity for an AWGN channel [232]

$$C_{\text{AWGN}} = \log_2\left(1 + \bar{\gamma}\right), \tag{2.19}$$

which will be used as a benchmark to assess the effects of fading on channel capacity. Notice that $\bar{\gamma}$ in (2.19) represents in the non-fading case the standard receive-SNR of an AWGN channel.

In *fading* channels, (2.17) indicates that $C(N_t, N_r|\boldsymbol{H})$ is a function of the random channel matrix $\boldsymbol{H}$. As a result, $C(N_t, N_r|\boldsymbol{H})$ is a random variable, which can be characterized fully by its probability density function (pdf) and in part by its moments, e.g., its mean. Under assumptions A2.1) and A2.2) of Section 2.1, we consider in the following section two measures of $C(N_t, N_r|\boldsymbol{H})$: the outage rate [69] and the ergodic capacity [256]. The more general cases of Rician and correlated Rayleigh fading channels have been investigated in [133, 134], respectively.

2.3.1 Outage Capacity

Before we introduce the notion of outage capacity, let us first define the capacity outage probability $P_{\text{out}}(R; \bar{\gamma}; N_t, N_r)$, which is the probability that the conditional channel capacity $C(N_t, N_r|\boldsymbol{H})$ drops below a certain rate R:

$$P_{\text{out}}(R; \bar{\gamma}; N_t, N_r) := \Pr\left(C(N_t, N_r|\boldsymbol{H}) < R\right). \tag{2.20}$$

It is clear that the capacity outage probability is the cumulative distribution function (cdf) of the random variable $C(N_t, N_r|\boldsymbol{H})$. Using the matrix identity $\det(\boldsymbol{I}_{N_r} + \boldsymbol{AB}) = \det(\boldsymbol{I}_{N_t} + \boldsymbol{BA})$, where $\boldsymbol{A}$ and $\boldsymbol{B}$ are $N_r \times N_t$ and $N_t \times N_r$ matrices, respectively, it is easy to verify that under A2.2)

$$P_{\text{out}}(R; \bar{\gamma}; N_t, N_r) = P_{\text{out}}\left(R; \frac{N_r}{N_t}\bar{\gamma}; N_r, N_t\right). \tag{2.21}$$

This equation shows that P_{out} is asymmetric with respect to N_t and N_r. The reason is that the total receive-power grows as N_r increases since more receive-antennas are capable of collecting more transmit-power, whereas the total receive-power remains the same when N_t increases due to the transmit-power constraint.

An exact expression of $P_{\text{out}}(R; \bar{\gamma}; N_t, N_r)$ was derived in [290] and is summarized in the following proposition:

Proposition 2.1 *Let* $N_1 := \min(N_t, N_r)$ *and* $N_2 := \max(N_t, N_r)$*. Under assumptions A2.1) and A2.2), the moment generating function (MGF) of* $C(N_t, N_r|\boldsymbol{H})$*, that is,* $\Phi_C(s) := E_{\boldsymbol{H}}\{e^{sC(N_t, N_r|\boldsymbol{H})}\}$*, is given by*

$$\Phi_C(s) = B^{-1}\det[\boldsymbol{G}(s)], \tag{2.22}$$

where $B := \prod_{i=1}^{N_1} \Gamma(N_2 - N_1 + i)$ *and* $\boldsymbol{G}(s)$ *is an* $N_1 \times N_1$ *Hankel matrix whose* (i, j)*th entry is*

$$g_{ij}(s) = \int_0^\infty \left(1 + \frac{\bar{\gamma}}{N_t}\lambda\right)^{-s} \lambda^{i+j+N_2-N_1} e^{-\lambda}\, d\lambda,$$

$$i, j = 0, \ldots, N_1 - 1. \quad (2.23)$$

The pdf of $C(N_t, N_r|\boldsymbol{H})$ *is the inverse Laplace transform of* $\Phi_C(s)$ *and the outage probability* $P_{out}(R; \bar{\gamma}; N_t, N_r)$ *is given by the inverse Laplace transform of* $s^{-1}\Phi_C(s)$*:*

$$P_{out}(R; \bar{\gamma}; N_t, N_r) = \frac{1}{2\pi j} \int_{\sigma - j\infty}^{\sigma + j\infty} s^{-1} \Phi_C(s) e^{sR} ds, \quad (2.24)$$

where σ *is a fixed positive number.*

Notice that Proposition 2.1 neither assumes that N_t or N_r is large nor requires the SNR to be small or large. It offers an effective way of evaluating the outage probability for practical values of N_t and N_r. Figure 2.2 plots the pdf of $C(N_t, N_r|\boldsymbol{H})$ for $N_t = N_r = 3$ and $\bar{\gamma}$'s ranging from 0 to 20 dB.

A number of useful conclusions can be drawn from Figure 2.2:

1. The pdf of $C(N_t, N_r|\boldsymbol{H})$ resembles that of a Gaussian random variable.

2. The mean increases with SNR and the increase for high SNR is linearly proportional to the SNR increase in decibels.

3. The variance also increases with SNR and the rate of increase becomes smaller for higher SNR values.

4. The mass of the pdf stays mostly above a certain level (e.g., for $\bar{\gamma} = 20$ dB, the pdf is nearly zero for $C(N_t, N_r|\boldsymbol{H}) < 10$). This suggests that although capacity in the Shannon sense does not exist per channel realization, the outage probability can be quite small for reasonably high rates.

The fact that the pdf of $C(N_t, N_r|\boldsymbol{H})$ is Gaussian can be explained intuitively via a central limit theorem (CLT) argument: The conditional channel capacity can be written as a sum of $\log_2(\cdot)$ random variables that are correlated [cf. (2.17)]. Had they been uncorrelated, the classical CLT would apply directly. But here the eigenvalues are statistically dependent. The proof of Gaussianity for the asymptotic cases when N_t or N_r is large, or when both N_t and N_r are large and the SNR is either very small or very large, can be found in [117]. The proof for the general case can be found in [290] and the formal result can be summarized as follows:

Theorem 2.1 *Define the functions*

$$\tilde{\phi}_k(\lambda) := \left[\frac{k!}{(k + N_2 - N_1)!}\right]^{\frac{1}{2}} L_k^{N_2-N_1}(\lambda) \lambda^{\frac{N_2-N_1}{2}} e^{-\frac{\lambda}{2}},$$

where

$$L_k^{N_2-N_1}(\lambda) := \frac{1}{k!} e^{\lambda} \lambda^{N_1-N_2} \frac{d^k}{d\lambda^k} \left(e^{-\lambda} \lambda^{N_2-N_1+k} \right) \tag{2.25}$$

is the associated Laguerre polynomial of order k [221]. Also define

$$K(x,y) := \sum_{k=0}^{N_1-1} \tilde{\phi}_k(x) \tilde{\phi}_k(y).$$

The Gaussian approximation to the pdf of $C(N_t, N_r|\boldsymbol{H})$ is given by

$$p(C; N_t, N_r, \bar{\gamma}) \approx \frac{1}{\sqrt{2\pi}\,\sigma_C} \exp\left(-\frac{(C-\mu_C)^2}{2\sigma_C^2} \right), \tag{2.26}$$

where $\mu_C := \int_0^\infty \log_2(1 + \lambda\bar{\gamma}/N_t) K(\lambda,\lambda)\, d\lambda$ and

$$\sigma_C^2 = \int_0^\infty \log_2^2(1 + \lambda\bar{\gamma}/N_t) K(\lambda,\lambda)\, d\lambda$$
$$- \int_0^\infty \int_0^\infty \log_2(1 + \lambda_1\bar{\gamma}/N_t) \log_2(1 + \lambda_2\bar{\gamma}/N_t) K^2(\lambda_1, \lambda_2)\, d\lambda_1 d\lambda_2.$$

If $N_1 := \min(N_t, N_r)$ is sufficiently large, the mean μ_C is much larger than the square root of the variance σ_C^2 [117,290]. This is a sign of "channel hardening" [117] and indicates that when both N_t and N_r grow large, the capacity behaves more and more like a deterministic quantity. In this sense, the Shannon capacity exists and is infinite in the limit of infinitely many antennas deployed at the transmitter and the receiver.

For a fixed outage probability ϵ, the outage rate $R_{\text{out}}(\epsilon; \bar{\gamma}; N_t, N_r)$ is defined as the rate R for which the outage probability $P_{\text{out}}(R; \bar{\gamma}; N_t, N_r)$ is at the given level ϵ. In other words, $R_{\text{out}}(\epsilon; \bar{\gamma}; N_t, N_r)$ is the value of R satisfying

$$P_{\text{out}}(R; \bar{\gamma}; N_t, N_r) = \epsilon. \tag{2.27}$$

To find the outage rate for a given ϵ using Proposition 2.1, we can perform a bisection search over a starting interval, on whose left (right) boundary the outage probability takes values less (greater) than ϵ. Figure 2.3 plots the outage rate for $\epsilon = 10\%$ for different SNR values and numbers of antennas. It readily follows from this figure that:

1. The slope of the curves converges to a constant at high SNR (the curves tend to become straight lines).

2. The asymptotic slope increases when both N_t and N_r increase. The numerical values corresponding to the graph are listed in Table 2.1.

These observations are also corroborated by the following major result established in [290]:

Proposition 2.2 *At high SNR, the slope of the outage rate curve (number of bits versus SNR in dB) is* $\log_2(10^{0.1}) \min(N_t, N_r)$ *bits per dB, or* $\min(N_t, N_r)$ *bits every factor of 2 increase (about 3.01 dB) in SNR.*

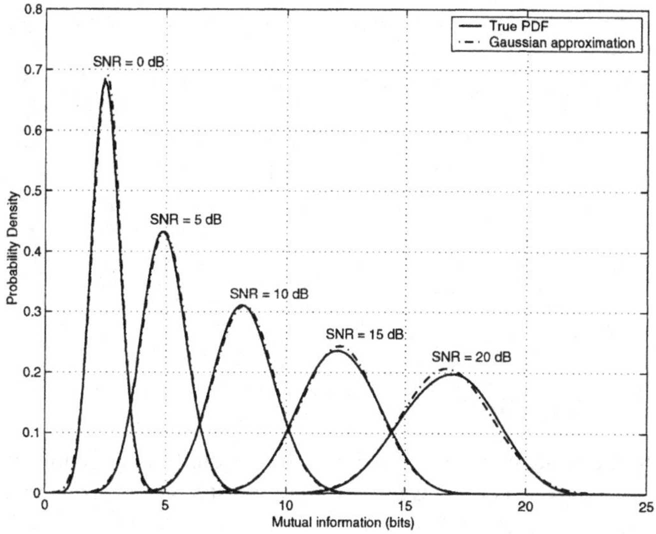

Figure 2.2 Pdfs of $C(N_t, N_r|\boldsymbol{H})$ at different SNR values with $N_t = N_r = 3$

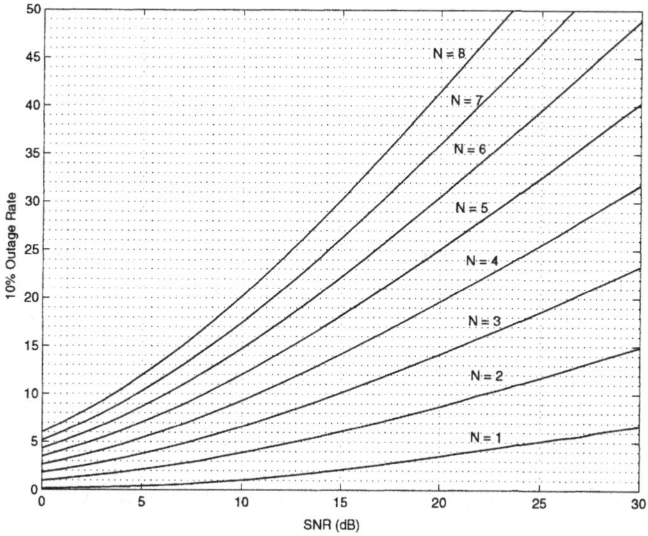

Figure 2.3 10% outage rate vs. SNR with $N = N_t = N_r = 1, 2, \ldots, 8$

Table 2.1 10% OUTAGE RATE FOR DIFFERENT SNR VALUES AND $N_t = N_r$.

$N_t = N_r$	0dB	5dB	10dB	15dB	20dB	25dB	30dB
1	0.15	0.41	1.04	2.11	3.53	5.10	6.73
2	0.97	2.13	3.89	6.10	8.72	11.71	14.91
3	1.80	3.76	6.60	10.15	14.16	18.57	23.30
4	2.64	5.41	9.32	14.16	19.63	25.53	31.78
5	3.48	7.04	12.04	18.19	25.10	32.52	40.28
6	4.31	8.67	14.76	22.21	30.57	39.52	48.87
7	5.15	10.31	17.48	26.24	36.04	46.53	57.45
8	5.99	11.94	20.21	30.25	41.52	53.52	66.05

2.3.2 Ergodic Capacity

In addition to the outage rate, another important measure of the conditional channel capacity $C(N_t, N_r|\boldsymbol{H})$ in (2.17) is the ergodic capacity. The latter is defined as

$$C(N_t, N_r) := E_{\boldsymbol{H}}\left[C(N_t, N_r|\boldsymbol{H})\right], \tag{2.28}$$

where the expectation is with respect to all possible channel realizations $\boldsymbol{H}$. Under assumption A2.2), the ergodic capacity was first evaluated in [256] and is given by

$$C(N_t, N_r) = \int_0^\infty \log_2\left(1 + \frac{\bar{\gamma}}{N_t}\lambda\right) \times \sum_{k=0}^{N_1-1} \frac{k!}{(k+N_2-N_1)!}\left[L_k^{N_2-N_1}(\lambda)\right]^2 \lambda^{N_2-N_1} e^{-\lambda}\, d\lambda. \tag{2.29}$$

Not surprisingly, $C(N_t, N_r)$ in (2.29) is the mean μ_C in Theorem 2.1.

The closed-form expression in (2.29) can be evaluated numerically. To this end, we fix the total number of antennas and consider different combinations of antennas at the transmitter and receiver. Figures 2.4 and 2.5 plot $C(N_t, N_r)$ versus the SNR for $N_t + N_r = 3$ and $N_t + N_r = 6$. From these figures it is easy to deduce that:

1. For a fixed total number of antennas, it is better to balance the number of transmit-antennas with the number of receive-antennas. In other words, it is always better to have $N_1 = \min(N_t, N_r)$ as large as possible.

2. For a fixed $N_1 = \min(N_t, N_r)$, it is always better to choose $N_r > N_t$.

The first conclusion can be justified after recognizing that deployment of N_t transmit-antennas and N_r receive-antennas generates a number of independent parallel channels equal to the rank of $\boldsymbol{H}$ [cf. (2.17)]; and this rank reaches its maximum value at N_1. The second conclusion is related to the asymmetry of the capacity outage probability that we alluded to in Section 2.3.1 and is due to the same fact, namely that

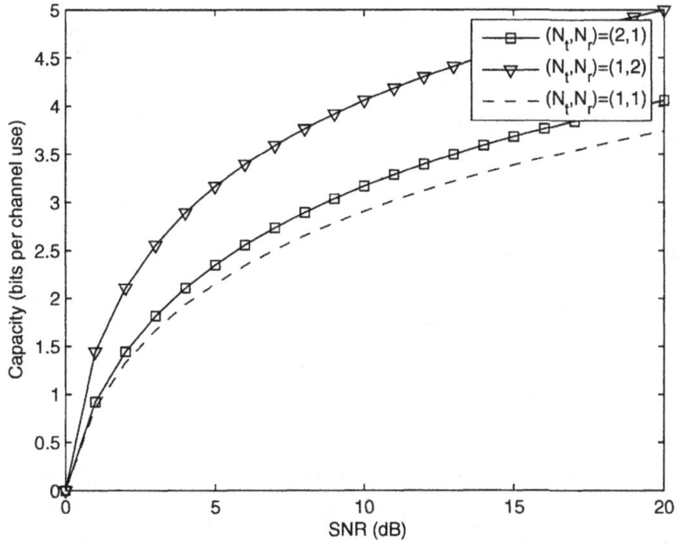

Figure 2.4 Ergodic capacity with variable allocation of transmit-receive antennas ($N_t + N_r = 3$)

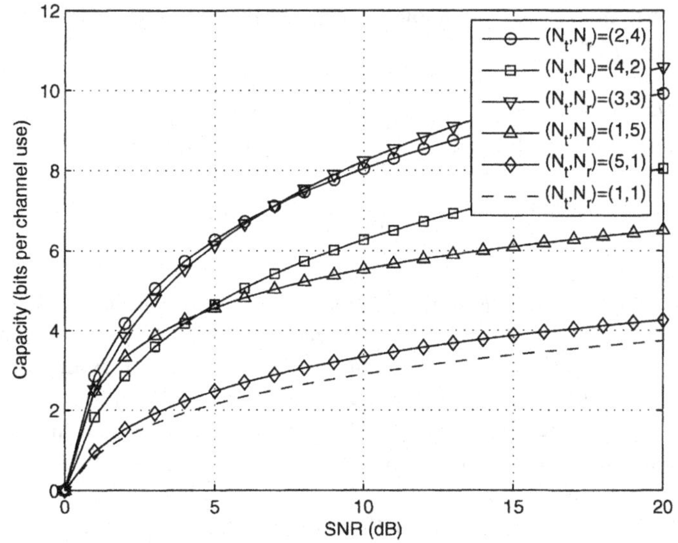

Figure 2.5 Ergodic capacity with variable allocation of transmit-receive antennas ($N_t + N_r = 6$)

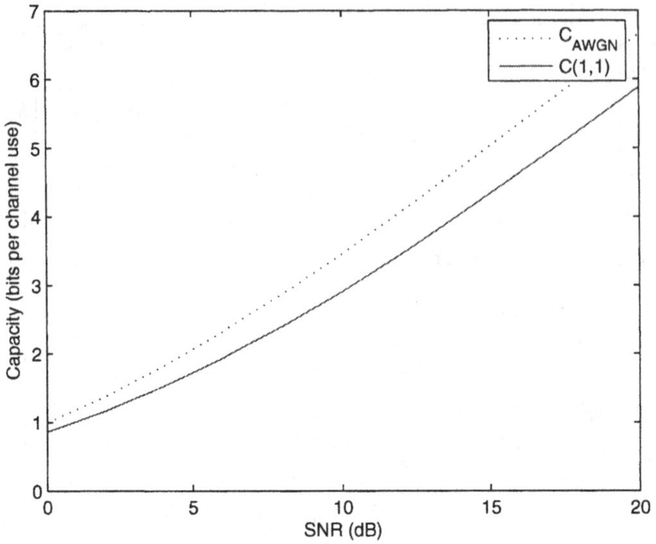

Figure 2.6 Effect of fading on ergodic capacity

more receive-antennas collect more power, while the total transmit-power remains constant even as the number of transmit-antennas grows.

In addition to numerical evaluation, further insights can be gained by considering the following important special cases:

Case 1: $N_t = N_r = 1$. This case corresponds to a single-antenna system. Substituting $N_1 = N_2 = 1$ and $L_0^{N_2-N_1}(\lambda) = 1$ into (2.29) yields

$$C(1,1) = \int_0^\infty \log_2 (1 + \bar{\gamma}\lambda)\, e^{-\lambda}\, d\lambda, \tag{2.30}$$

which is the capacity of a single-antenna system transmitting over flat Rayleigh fading channels. It is worth noting that (2.30) can be reached alternatively by averaging (2.18) with respect to the Gaussian distribution of h, in accordance with assumption A2.2). Figure 2.6 depicts $C(1,1)$ in comparison with C_{AWGN} and confirms that channel fading decreases the ergodic capacity. This provides an additional motivation for deploying multiple antennas when communicating over fading channels.

Case 2: $N_t > 1$ ***and*** $N_r = 1$. In this case we have $N_1 = 1$ and $N_2 = N_t$, for which (2.29) can be rewritten as

$$C(N_t, 1) = \frac{1}{(N_t - 1)!} \int_0^\infty \log_2 \left(1 + \frac{\bar{\gamma}}{N_t}\lambda\right) \lambda^{N_t - 1} e^{-\lambda}\, d\lambda. \tag{2.31}$$

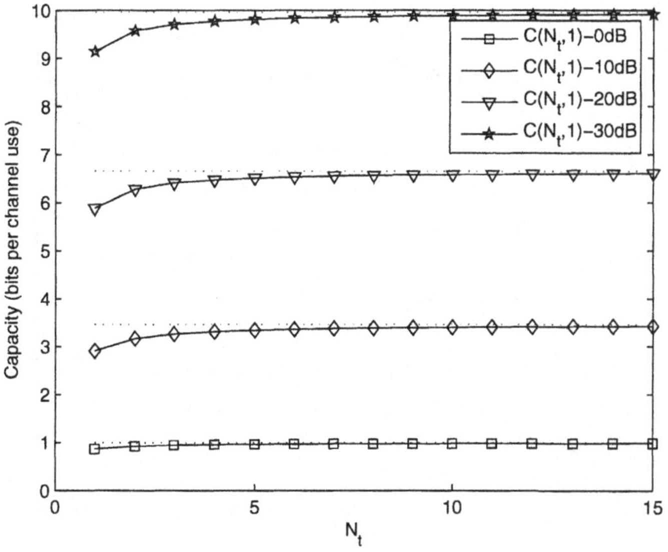

Figure 2.7 Ergodic capacity of an ST channel with $N_r = 1$

We are interested in the asymptotic value of $C(N_t, 1)$ as N_t increases. With $N_r = 1$, $\boldsymbol{H}$ is a row vector $\boldsymbol{h} := [h_1, \ldots, h_{N_t}]^T$. Accordingly, (2.17) can be expressed as

$$C(N_t, 1|\boldsymbol{h}) = \log_2 \left(1 + \frac{\bar{\gamma}}{N_t} \sum_{\mu=1}^{N_t} |h_\mu|^2 \right). \tag{2.32}$$

By the law of large numbers, we have

$$\frac{1}{N_t} \sum_{\mu=1}^{N_t} |h_\mu|^2 \to 1$$

almost surely as N_t grows. Thus, $C(N_t, 1)$ can be approximated for large N_t as

$$C(N_t, 1) \approx \log_2 (1 + \bar{\gamma}) = C_{\text{AWGN}}. \tag{2.33}$$

This suggests that as the number of transmit-antennas increases, the adverse effects of fading on channel capacity diminish asymptotically. Figure 2.7 depicts the values of $C(N_t, 1)$ versus N_t for different SNR values and compares them to the corresponding values of C_{AWGN} (dashed curves in Figure 2.7). We clearly see that $C(N_t, 1)$ increases as N_t grows. However, $C(N_t, 1)$ saturates quickly (as soon as $N_t \geq 4$). This observation implies that with a single receive-antenna, it makes little sense to use more than four transmit-antennas from the ergodic capacity point of view.

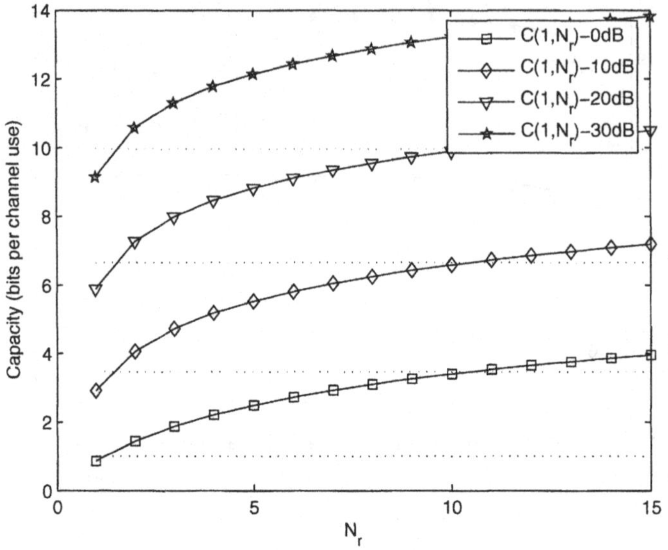

Figure 2.8 Ergodic capacity of an ST channel with $N_t = 1$

***Case 3:* $N_t = 1$ *and* $N_r > 1$.** In this case we have $N_1 = 1$ and $N_2 = N_r$, for which (2.29) yields

$$C(1, N_r) = \frac{1}{(N_r - 1)!} \int_0^\infty \log_2 (1 + \bar{\gamma}\lambda)\, \lambda^{N_r - 1} e^{-\lambda}\, d\lambda. \tag{2.34}$$

Following arguments similar to those used to arrive at (2.33), it can readily be shown that for large N_r we have

$$C(1, N_r) \approx \log_2 (1 + N_r \bar{\gamma}) > C_{\text{AWGN}}. \tag{2.35}$$

Therefore, it is possible to achieve channel capacity beyond C_{AWGN} if sufficiently many receive-antennas are deployed. This is not surprising since multiple receive-antennas as a matter of fact collect more transmit-power than does a single receive-antenna. Figure 2.8 shows the values of $C(1, N_r)$ versus N_r for different SNRs, together with C_{AWGN}.

***Case 4:* $N_t = N_r > 1$.** When $N_t = N_r := N$, we have $N_1 = N_2 = N$ and upon substituting into (2.29), we find that

$$C(N, N) = \int_0^\infty \log_2 \left(1 + \frac{\bar{\gamma}}{N}\lambda\right) \sum_{k=0}^{N-1} \left[L_k^0(\lambda)\right]^2 e^{-\lambda}\, d\lambda. \tag{2.36}$$

Under assumption A2.2), the law of large numbers implies that as N increases

$$\frac{1}{N} \boldsymbol{H}\boldsymbol{H}^{\mathcal{H}} \to \boldsymbol{I}_N.$$

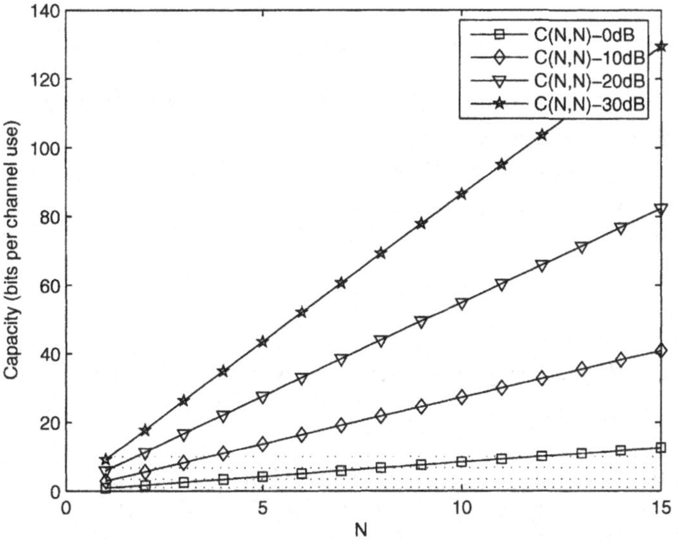

Figure 2.9 Ergodic capacity of an ST channel with $N_t = N_r = N$

It thus follows from (2.17) that for large N, we have

$$C(N, N) \approx N \log_2 (1 + \bar{\gamma}) = N C_{\text{AWGN}}. \tag{2.37}$$

Hence for large N, the ergodic capacity increases linearly with N, which is confirmed by Figure 2.9, where the capacity curve is very well approximated by a linear function of N. Note that this conclusion is also consistent with Proposition 2.2. Figure 2.9 also indicates that an ST system with a fairly large N is capable of supporting very high-rate transmissions. For example, when $N = 8$ and SNR = 20 dB, we have $C(N, N) = 70$ bits per channel use, which translates to a transmission rate of 1.4 gigabits per second (Gbps) for 20 MHz bandwidth.

From our discussion so far, we have seen that deploying multiple antennas boosts the capacity of fading channels. As in a single-antenna system, whether and what percentage of this ST capacity limit can be achieved is up to the design of ST codes. Capacity benchmarks the maximum rate that one can communicate reliably over a given channel. To measure the reliability of a communication system, another commonly used metric is error probability, which is investigated next.

2.4 ERROR PERFORMANCE OF ST CODING

In this section we derive error probability metrics pertaining to ST coded transmissions over flat fading MIMO channels. The ensuing derivations take into account possible spatial correlation among transmit- and receive-antennas. Specifically, we adopt the following channel model [23]:

$$\boldsymbol{H} = \boldsymbol{R}_r^{1/2}\boldsymbol{H}_w\boldsymbol{R}_t^{1/2}, \tag{2.38}$$

where $\boldsymbol{H}_w$ is an $N_r \times N_t$ matrix whose entries are i.i.d. complex Gaussian random variables with zero mean and unit variance, and the $N_t \times N_t$ matrix $\boldsymbol{R}_t = \boldsymbol{R}_t^{1/2}\boldsymbol{R}_t^{\mathcal{H}/2}$ and the $N_r \times N_r$ matrix $\boldsymbol{R}_r = \boldsymbol{R}_r^{1/2}\boldsymbol{R}_r^{\mathcal{H}/2}$ denote the transmit- and receive-correlation matrices. Clearly, when $\boldsymbol{R}_t = \boldsymbol{I}_{N_t}$ and $\boldsymbol{R}_r = \boldsymbol{I}_{N_r}$, (2.38) reduces to the channel model assumed in assumptions A2.1) and A2.2). In addition to (2.38), we further assume that:

A2.3) Maximum-likelihood (ML) detection is performed.

A2.4) Channel $\boldsymbol{H}$ is perfectly known at the receiver but not at the transmitter.

A2.5) SNR is sufficiently high; i.e., $\bar{\gamma} \gg 1$.

Because the entries of $\boldsymbol{W}$ are i.i.d. complex Gaussian with zero mean and unit variance, ML detection of $\boldsymbol{X}$ from $\boldsymbol{Y}$ can be carried out using [cf. (2.9)]

$$\hat{\boldsymbol{X}}_{\mathrm{ML}} = \arg\min_{\forall \boldsymbol{X} \in \mathbb{X}} d^2\left(\boldsymbol{Y}, \boldsymbol{X}|\boldsymbol{H}\right), \tag{2.39}$$

which amounts to finding the ST code matrix $\boldsymbol{X} \in \mathbb{X}$ for which the distance metric

$$d^2(\boldsymbol{Y},\boldsymbol{X}|\boldsymbol{H}) := \mathrm{tr}\left[\left(\boldsymbol{Y} - \sqrt{\frac{\bar{\gamma}}{N_t}}\boldsymbol{H}\boldsymbol{X}\right)\left(\boldsymbol{Y} - \sqrt{\frac{\bar{\gamma}}{N_t}}\boldsymbol{H}\boldsymbol{X}\right)^{\mathcal{H}}\right]$$

is minimized.

Since it is rarely possible to obtain and analyze the exact error probability of the ML detector in (2.39), we investigate instead its pairwise error probability (PEP), which provides a good approximation of the error probability at high SNR [120,253]. Let $\boldsymbol{X} \in \mathbb{X}$ and $\boldsymbol{X}' \in \mathbb{X}$ be two distinct ST code matrices. The PEP $P_2(\boldsymbol{X} \to \boldsymbol{X}')$ is defined as the probability that the ML detector erroneously decodes $\boldsymbol{X}$ as $\boldsymbol{X}'$ when $\boldsymbol{X}$ is actually sent. But this pairwise error occurs when $d^2(\boldsymbol{Y},\boldsymbol{X}'|\boldsymbol{H}) < d^2(\boldsymbol{Y},\boldsymbol{X}|\boldsymbol{H})$. Consequently, the conditional PEP can be expressed as [212]

$$P_2(\boldsymbol{X} \to \boldsymbol{X}'|\boldsymbol{H}) = \mathcal{Q}\left(\sqrt{\frac{\bar{\gamma}}{2N_t}\mathrm{tr}\left[(\boldsymbol{H}\boldsymbol{X} - \boldsymbol{H}\boldsymbol{X}')(\boldsymbol{H}\boldsymbol{X} - \boldsymbol{H}\boldsymbol{X}')^{\mathcal{H}}\right]}\right), \tag{2.40}$$

where $\mathcal{Q}(\cdot)$ is defined as $\mathcal{Q}(x) := \int_x^\infty \frac{1}{\sqrt{2\pi}}e^{-t^2/2}dt$. Using the Chernoff bound $\mathcal{Q}(x) \le e^{-x^2/2}$ [214], it is possible to upper bound the PEP in (2.40) by

$$P_2(\boldsymbol{X} \to \boldsymbol{X}'|\boldsymbol{H}) \le \exp\left(-\frac{\bar{\gamma}}{4N_t}\mathrm{tr}\left[(\boldsymbol{H}\boldsymbol{X} - \boldsymbol{H}\boldsymbol{X}')(\boldsymbol{H}\boldsymbol{X} - \boldsymbol{H}\boldsymbol{X}')^{\mathcal{H}}\right]\right). \tag{2.41}$$

Because the upper bound in (2.41) is fairly tight at high SNR, it will be used to approximate the PEP.

The conditional PEP in (2.41) clearly depends on the underlying channel realization. The unconditional PEP can be obtained by averaging $P_2(\boldsymbol{X} \to \boldsymbol{X}'|\boldsymbol{H})$ over all possible realizations $\boldsymbol{H}$ under the channel model (2.38). Let us now define the $N_tN_r \times N_tN_r$ matrix

$$\boldsymbol{C_y} = \left[(\boldsymbol{X} - \boldsymbol{X}')^T \boldsymbol{R}_t^T (\boldsymbol{X} - \boldsymbol{X}')^*\right] \otimes \boldsymbol{R}_r, \tag{2.42}$$

where $\otimes$ stands for the Kronecker product. If the $\{\lambda_i\}_{i=1}^{\text{rank}(\boldsymbol{C_y})}$ denote the positive eigenvalues of $\boldsymbol{C_y}$, the PEP can be well approximated at high SNR by [23]

$$P_2(\boldsymbol{X} \to \boldsymbol{X}') \approx \left[G_{e,c}\frac{\bar{\gamma}}{4N_t}\right]^{-G_{e,d}}, \tag{2.43}$$

where

$$G_{e,d} := \text{rank}(\boldsymbol{C_y}),$$
$$G_{e,c} := \left(\prod_{i=1}^{\text{rank}(\boldsymbol{C_y})} \lambda_i\right)^{1/\text{rank}(\boldsymbol{C_y})} \tag{2.44}$$

are, respectively, the pairwise diversity and coding gains that are commonly used to measure error performance in fading environments. Equation (2.43) implies that at high SNR, the PEP can be approximately quantified by $G_{e,c}$ and $G_{e,d}$. These two parameters affect the PEP in two different ways. With reference to Figure 2.10, the pairwise diversity gain determines the slope of the average PEP plotted in a log-log scale as a function of the SNR. On the same plot, the pairwise coding gain determines the parallel shift of the PEP curve relative to one without coding gain. At sufficiently high SNR, the diversity gain plays a more important role than the coding gain in terms of improving the error performance.

Noting that $G_{e,d}$ depends on a particular error event $\{\boldsymbol{X} \to \boldsymbol{X}'\}$, the need to account for all possible error events motivates defining the (overall) diversity gain as

$$G_d = \min_{\forall \boldsymbol{X} \neq \boldsymbol{X}' \in \mathbb{X}} G_{e,d}, \tag{2.45}$$

where the minimization is taken over all pairs of distinct ST code matrices. The (overall) coding gain depends on the (overall) diversity gain and is defined as

$$G_c = \min_{\forall \boldsymbol{X} \neq \boldsymbol{X}' \in \mathbb{X}, G_{e,d} = G_d} G_{e,c}, \tag{2.46}$$

where the minimization is taken among those pairs of distinct ST code matrices yielding $G_{e,d} = G_d$.

By checking the dimensionality of $\boldsymbol{C_y}$, we can deduce that the maximum possible diversity gain is

$$G_{d,\max} = N_tN_r, \tag{2.47}$$

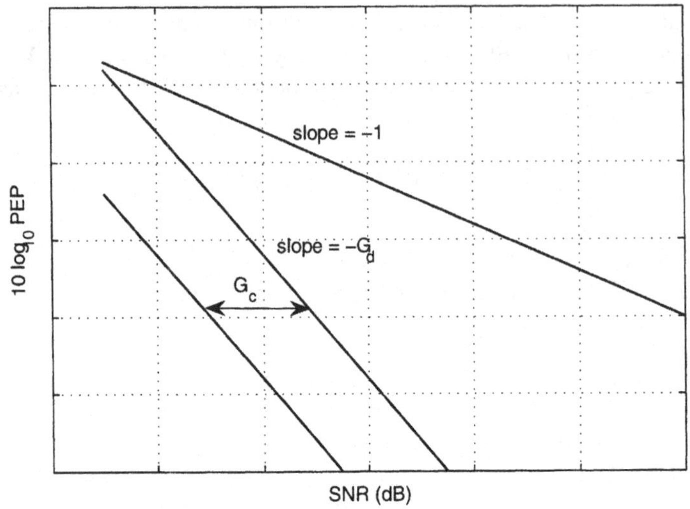

Figure 2.10 Diversity and coding gains

which is achieved if and only if $\boldsymbol{C}_{\boldsymbol{y}}$ has full rank for all pairs of distinct ST code matrices $\boldsymbol{X}$ and $\boldsymbol{X}'$.

The diversity gain clearly depends on the spatial correlation matrices $\boldsymbol{R}_t$ and $\boldsymbol{R}_r$ [cf. (2.44)]. In fact, G_d can be upper and lower bounded by [23]

$$\text{rank}(\boldsymbol{R}_r)[r_\Delta + \text{rank}(\boldsymbol{R}_t) - N_t] \le G_d \le \text{rank}(\boldsymbol{R}_r)\min[r_\Delta, \text{rank}(\boldsymbol{R}_t)], \qquad (2.48)$$

where r_Δ is the minimum rank of the $N_t \times N_t$ difference matrix

$$\boldsymbol{\Delta}_e := (\boldsymbol{X} - \boldsymbol{X}')(\boldsymbol{X} - \boldsymbol{X}')^{\mathcal{H}} \in \mathbb{C}^{N_t \times N_t} \qquad (2.49)$$

over all possible pairs of distinct ST code matrices $\boldsymbol{X}$ and $\boldsymbol{X}'$. It follows from (2.48) that if $\boldsymbol{\Delta}_e$ has full rank for any $\boldsymbol{X} \neq \boldsymbol{X}'$, i.e., $r_\Delta = N_t$, then

$$G_d = \text{rank}(\boldsymbol{R}_r)\,\text{rank}(\boldsymbol{R}_t). \qquad (2.50)$$

Equation (2.50) asserts that the maximum possible diversity gain can be achieved if and only if neither $\boldsymbol{R}_r$ nor $\boldsymbol{R}_t$ loses rank. On the other hand, when $\boldsymbol{\Delta}_e$ loses rank, i.e., $r_\Delta < N_t$, (2.48) implies that $G_d < N_t\text{rank}(\boldsymbol{R}_r)$. In this case, it is impossible to achieve the maximum possible diversity gain in (2.47) even if both $\boldsymbol{R}_r$ and $\boldsymbol{R}_t$ have full rank. In summary, to enable the maximum diversity gain necessitates designing $\mathbb{X}$ properly so that $r_\Delta = N_t$ for any $\boldsymbol{X} \neq \boldsymbol{X}'$.

The PEP analysis so far has established that the error probability associated with ST coded systems can be assessed at any finite (but sufficiently high) SNR using G_d and G_c, regardless of the underlying constellation used. Because G_d and G_c are in this sense universal indicators of error performance, ST code designs will aim at maximizing these two gains. Before we proceed to introduce the corresponding design criteria, it is important to point out that the error performance of ST transmissions is also affected by other factors, such as the kissing number [343], that is, the number of pairs $(\boldsymbol{X}, \boldsymbol{X}')$ yielding the same $\boldsymbol{\Delta}_e$. But since analytical forms of the latter are not available to optimize, G_c and primarily G_d are used throughout this book to evaluate error performance.

2.5 DESIGN CRITERIA FOR ST CODES

The expressions of $G_{e,d}$ and $G_{e,c}$ in (2.44) have important implications on the design of ST codes. Knowledge of the channel's spatial correlation is, however, not always available at the transmitter. But even if it is available, it may not remain invariant due to, e.g., mobility effects. For this reason, open-loop ST coded systems are typically designed without accounting for the MIMO channel's spatial correlation. In other words, most ST code designs assume that $\boldsymbol{R}_t = \boldsymbol{I}_{N_t}$ and $\boldsymbol{R}_r = \boldsymbol{I}_{N_r}$.

When $\boldsymbol{R}_t = \boldsymbol{I}_{N_t}$ and $\boldsymbol{R}_r = \boldsymbol{I}_{N_r}$, the diversity gain can be obtained from (2.48) as

$$G_d = N_r r_\Delta, \tag{2.51}$$

which implies that the maximum diversity can be achieved if and only if the error matrix $\boldsymbol{\Delta}_e$ has full rank for any two distinct ST code matrices. This suggests the following diversity-based criterion for designing ST codes:

C2.1) Design $\mathbb{X}$ such that $\boldsymbol{\Delta}_e$ has full rank for any $\boldsymbol{X} \neq \boldsymbol{X}' \in \mathbb{X}$.

To ensure that $\boldsymbol{\Delta}_e$ has full rank, one necessary condition is $N_x \geq N_t > 1$. The latter shows that time-domain processing is indispensable for ST systems if the maximum diversity gain is to be enabled.

As we mentioned earlier, diversity plays a critical role in improving error performance in fading channels. Therefore, it is not surprising that enabling the maximum diversity is typically a top priority task in the design of ST codes. When maximum diversity gain is ensured, i.e., $\text{rank}(\boldsymbol{\Delta}_e) = N_t$, the pairwise coding gain in (2.44) becomes

$$G_{e,c} = [\det(\boldsymbol{\Delta}_e)]^{\frac{1}{N_t}} .$$

To account for all possible pairwise errors, the following criterion is used to design ST transmissions with maximum coding gain:

C2.2) Design $\mathbb{X}$ such that the minimum value of $\det(\boldsymbol{\Delta}_e)$ for any $\boldsymbol{X} \neq \boldsymbol{X}' \in \mathbb{X}$ is maximized.

Notice that if C2.2) is satisfied, then C2.1) is satisfied automatically.

So far, we have evaluated error performance of ST systems in terms of diversity and coding gains. Because block transmissions can be viewed as a special case of ST transmissions, the diversity and coding gains of single-antenna block transmissions are obtained similarly as

$$G_{d,\text{block}} = 1 \quad \text{and} \quad G_{c,\text{block}} = d_{\min}^2, \tag{2.52}$$

where $d^2_{\min} := |\boldsymbol{x} - \boldsymbol{x}'|^2$ is the minimum Euclidean distance among two distinct channel codewords. Equation (2.52) asserts that channel coding can only affect the coding gain, which as we mentioned earlier is often less important than the diversity gain at high SNR. This further explains why channel coding is not as effective over flat fading environments when it is not combined with interleaving. If no channel coding is involved, i.e., $\boldsymbol{x}$ is uncoded, we can readily deduce that a single-antenna system can achieve

$$G_{d,\text{uncoded}} = 1 \quad \text{and} \quad G_{c,\text{uncoded}} = \Delta^2_{\min}, \tag{2.53}$$

where $\Delta_{\min}$ is the minimum distance among constellation points in $\mathbb{S}$.

2.6 DIVERSITY AND RATE: FINITE SNR VIZ ASYMPTOTICS

In practice, the performance of communication system designs at the physical layer is evaluated on the basis of their error probability and spectral efficiency for a given SNR or over a range of SNRs. For the ST coded systems dealt with in this book, error probability will be assessed primarily by the diversity gain G_d, which can be evaluated analytically as we discussed in Section 2.5. (The coding gain G_c will be used, too, but rarely, since it is possible to quantify in closed form only in a few cases.) As for spectral efficiency, we will rely on the transmission rate defined either in symbols per channel use or in bits per channel use, depending on the ST code adopted.

Both diversity and rate-related metrics have been reported in the literature based on either finite SNR or limiting SNR operations. Our goal in this section is to clarify these two viewpoints as well as their advantages, relationships, and limitations. Furthermore, we wish to justify the metrics chosen with an eye toward addressing this basic question: In what way do these metrics guide the design of ST coded multi-antenna systems and the choice of emerging trade-offs in practice?

Starting with the diversity gain in (2.45), we can easily rewrite it using a limiting SNR operation as [cf. (2.44)]

$$G_d := \lim_{\bar{\gamma}\to\infty} -\frac{\log_2 P_e(\bar{\gamma})}{\log_2 \bar{\gamma}}, \tag{2.54}$$

where $P_e(\bar{\gamma})$ denotes the average error probability at average SNR $\bar{\gamma}$. This limiting SNR definition as the slope of the $\log_2 P_e$ function of $\log_2 \bar{\gamma}$ depends not only on the underlying channel but also on the transmitter and receiver used. In fact, it expresses the *achievable system diversity*, which must be distinguished from the *channel diversity*, which is a fundamental characteristic of the channel.

Channel diversity can be defined based on the channel pdf without appealing to any limiting SNR operation [289]. Let us first consider for simplicity a single-antenna fading channel h with pdf $p(|h|)$ of its gain $|h|$. All reported pdfs modeling wireless channels (including Rayleigh, Rice, and general Nakagami-m ones) can be expressed as $p(|h|) = a|h|^t + o(|h|^{t+\epsilon})$ for $|h| \to 0^+$, where $a, t, \epsilon > 0$ are appropriate constants.[2] Based on this Taylor expansion, we can define the channel diversity as $(t/2)+1$, where the exponent t quantifies the degree of smoothness (number of nonzero derivatives) of the channel gain pdf sufficiently close to the origin. According to this definition, a Rayleigh pdf has exponent $t=0$, a Rice pdf has $t=0$, but Nakagami-m pdfs have exponents $t=2(m-1)$ which can be greater or less than zero and their diversity can even be fractional; e.g., if h is uniformly distributed, then $t=-1$ and its diversity is $1/2$.

Interestingly, it has been shown in [289] that the channel diversity coincides with the achievable diversity for sufficiently high SNR; that is, $G_d=(t/2)+1$ for $N_t=N_r=1$ and for general (N_t,N_r) systems [321]

$$G_d=\sum_{\mu=1}^{N_t}\sum_{\nu=1}^{N_r} t_{\nu\mu}/2+N_tN_r, \tag{2.55}$$

where $t_{\nu\mu}$ are the exponents in the Taylor expansion of the channel gain pdf $p(|h_{\nu\mu}|)$ corresponding to the channel between the μth transmit-antenna and the νth receive-antenna. Equation (2.55) explains why one may use the term diversity to describe system and channel characteristics at the same time. But we should always bear in mind that this only holds true for high SNR. And more important when dealing with system diversity, three verbs should be used carefully: the transmitter *enables*, the channel *provides*, and the receiver *collects* the diversity. Clearly, no matter what transceiver we use in practice, system diversity cannot exceed the diversity provided by the channel. On the other hand, even if the channel has diversity to provide, a suboptimum transmitter may not enable it (consider, e.g., the impractical case of transmitting an all-zero sequence); and even if the transmitter enables the full diversity provided by the channel, a suboptimum (non-ML) receiver may collect only part of it. In a nutshell, when the achievable system diversity is smaller than the channel diversity, we clearly have a suboptimum system.

The system diversity definition in (2.54) also leaves one to wonder how high the SNR should be in order for the asymptotics to "kick-in." The good news is that system diversity lends itself to practical interpretations since G_d at (perhaps impractically) high $\bar{\gamma}$ values can be brought down to reach the channel diversity at practical ranges of $\bar{\gamma}$ values using sufficiently powerful error control codes. Between two ST systems with identical spectral efficiency and a given $\bar{\gamma}$, this fact motivates the designer to select the one enabling the higher diversity order G_d.

Although finite and limiting SNR implications of diversity are clear, those of spectral efficiency are more intricate. A limiting SNR spectral efficiency metric was introduced by Zheng and Tse in [327] under the term *spatial multiplexing*. Somewhat analogous to (2.54), spatial multiplexing is defined as

$$G_{\text{sm}}:=\lim_{\bar{\gamma}\to\infty}\frac{R_{\text{out}}(\bar{\gamma})}{\log_2\bar{\gamma}}, \tag{2.56}$$

[2]A function $f(x)$ is $o(x)$ if and only if $\lim_{x\to 0} f(x)/x=0$.

where R_{out} denotes the outage capacity in bits per channel use. Two extreme values of G_{sm} are easy to identify: $G_{\text{sm,min}} = 0$ when transmitting with a fixed constellation at a constant transmission rate, and $G_{\text{sm,max}} = \min\{N_t, N_r\}$ when $\bar{\gamma} = \infty$. Since the latter represents the factor by which a multi-antenna ST system enhances the capacity of a single-antenna channel, values of G_{sm} between these two extremes represent the percentage of the MIMO channel capacity (degrees of freedom) that an ST design can attain. But the very definition of G_{sm} implies that for any ST design, $G_{\text{sm}} \neq 0$ is possible only if R_{out} increases as $\bar{\gamma}$ increases.

Clearly, for sufficiently high SNR, $2^{R_{\text{out}}}$ in (2.56) is proportional to $\bar{\gamma}^{G_{\text{sm}}}$ pretty much like P_e in (2.54) is proportional to $\bar{\gamma}^{-G_d}$. To improve rate and error performance, we would like to maximize both G_{sm} and G_d. However, [327] proved that there is a trade-off between G_{sm} and G_d which causes one to decrease when the other increases, as specified in the next proposition:

Proposition 2.3 *For an (N_t, N_r) multi-antenna system where the ST code matrix has time span $N_x \geq N_t + N_r - 1$ and the MIMO fading channel is i.i.d. Rayleigh distributed, the optimal trade-off curve $G_d^*(G_{\text{sm}})$ is given by the piecewise-linear function connecting the points $[G_{\text{sm}}, G_d^*(G_{\text{sm}})]$ as*

$$G_d^*(G_{\text{sm}}) = (N_t - G_{\text{sm}})(N_r - G_{\text{sm}}), \quad G_{\text{sm}} = 0, 1, \ldots, \min\{N_t, N_r\}, \quad (2.57)$$

where $G_d^*(G_{\text{sm}})$ denotes the supremum of the diversity orders achieved over all schemes for a given multiplexing gain G_{sm}.

To illustrate the optimal multiplexing-diversity trade-off curve, let us consider the following example.

Example 2.1 For an $(N_t, N_r) = (4, 4)$ system, suppose that $N_x \geq N_t + N_r - 1 = 3$ and the 4×4 channel matrix contains uncorrelated Rayleigh distributed entries with zero mean and unit variance. The maximum diversity is $G_{d,\max}^* = N_t N_r = 16$, and the maximum multiplexing gain is $G_{\text{sm,max}}^* = \min\{N_t, N_r\} = 4$. In accordance with Proposition 2.3, the optimal trade-off curve is plotted in Figure 2.11.

Extending the results in [289], the trade-off asserted by Proposition 2.3 for Rayleigh channels has been generalized to a wide class of channel distributions in [321]. But to gain intuition about what causes this trade-off, it suffices to consider even a single-antenna system with $N_t = N_r = 1$ signaling over a fading channel of any distribution (say, Nakagami-m). We know that if with a fixed constellation (hence, a constant transmission rate R_{ST} or outage rate R_{out}) we increase $\bar{\gamma}$, error probability P_e will improve and eventually the system diversity (slope of $\log_2 P_e$) will reach the channel diversity. On the other hand, if along with the $\bar{\gamma}$ we also increase the constellation size, as imposed *a fortiori* by the G_{sm} definition (2.56), $\log_2 P_e$ may not improve with SNR, and conceivably it could remain constant (a straight line as a function of the $\log_2 \bar{\gamma}$), never being able to reach the channel diversity.

Furthermore, since no channel provides zero diversity, the fact that for $G_{\text{sm,max}}^* = \min(N_t, N_r)$ the system diversity $G_{d,\min}^*$ can become zero means that the system

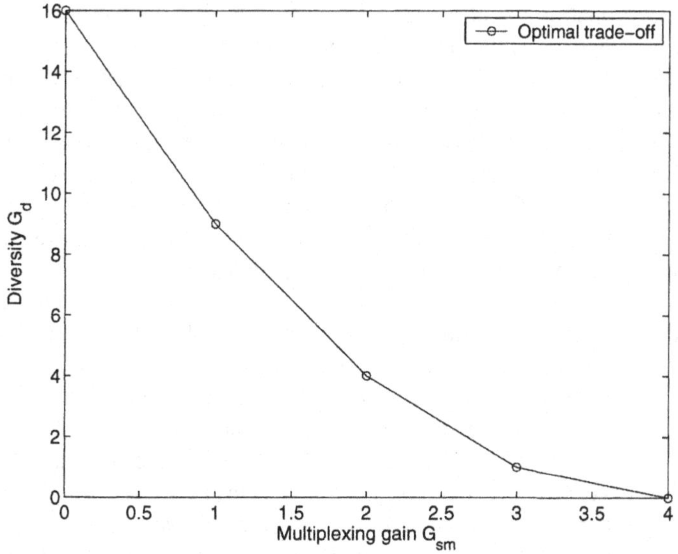

Figure 2.11 Diversity-multiplexing trade-off for $(N_t, N_r) = (4, 4)$

using rates adhering to (2.56) may be a badly suboptimum choice. At the other extreme, (2.57) implies that the system diversity attains its maximum $G^*_{d,\max} = N_t N_r$, if the spatial multiplexing gain reaches its minimum $G_{\text{sm},\min} = 0$. Notice, though, that zero multiplexing gain of a given ST configuration does not imply that the system has poor spectral efficiency since *any* transmission with fixed rate will have vanishing G_{sm} with increasing $\bar{\gamma}$, following (2.56). Hence, increasing the rate as the SNR increases is a must for G_{sm} to offer a potentially meaningful metric for spectral efficiency. But before exploring how relevant G_{sm} is in this role, two questions are in order: What is the physical meaning of G_{sm} in practice? And how can the designer pragmatically increase the rate with the SNR? Since rate and SNR are decoupled resources which in a number of system designs are chosen independently, providing a physical interpretation of G_{sm} over finite SNR values appears impossible. An answer to the second question is possible using adaptive modulation. However, in this case $\bar{\gamma}$ must be available at the transmitter, which requires more CSI knowledge than what an open-loop ST coded system assumes at the outset.

Recognizing these constraints, the trade-off between system diversity and spatial multiplexing is theoretically interesting but requires caution when interpreting its implications in practice. One case where this trade-off may provide useful guidelines is when the designer has to choose between systems $S1$ and $S2$ with equal diversity gains, say $G_{d,1} = G_{d,2}$, but different spatial multiplexing gains, say $G_{\text{sm},1} > G_{\text{sm},2}$. If the SNR is sufficiently high, system $S1$ should be preferred since it can afford a higher rate for the given SNR; i.e., in this case we will have $R_{\text{out},1} > R_{\text{out},2}$, or, if a larger constellation is used by $S2$ to equate rates ($R_{\text{out},1} = R_{\text{out},2}$), $S1$ will have a better error probability than $S2$ for the given SNR. But even in this comparison, the guideline provided will be safe if the SNR is sufficiently high, which as with the diversity, raises the issue of "how high" is "high SNR." Unfortunately, contrary to the case of system versus channel diversity, there are no means (such as channel coding) to bring impractically high SNRs closer to practical SNR ranges, precisely because rate and SNR are not fundamentally related in the finite SNR regime. Although in the literature the G_d versus G_{sm} trade-off has often been referred to as fundamental, it relates achievable quantities specific only to systems where rates increase with SNR and are not fundamentally related to the underlying MIMO channel.

Taking these considerations into account, one can infer that the spectral efficiency may not be well characterized through the limiting SNR definition of the spatial multiplexing gain. For this reason, we adopt the spectral efficiency η_{ST} defined in (2.4) to compare the performance of ST coded multi-antenna systems with equal diversity gains on the basis of their transmission rate (in symbols per channel use). Although this metric is well defined for any finite SNR, comparing the spectral efficiency of different systems based on η_{ST} can also be tricky. This is because ST code mappings $\mathcal{M}_{\text{ST}}$ can be either linear or nonlinear [cf. (2.2)], transmissions can be uncoded or channel coded; and different systems may adopt different constellations per SNR. To account for these cases, in this book, spectral efficiency-related comparisons will be carried out under the following conditions:

C2.3) If the systems under comparison entail ST code mappings $\mathcal{M}_{\text{ST}}$ which are linear, spectral efficiency will be characterized by η_{ST} in symbols per channel use, and systems will be compared using identical constellations per SNR.

C2.4) If a nonlinear (e.g., trellis based, differential, or noncoherent) ST code is involved, spectral efficiency will be quantified in terms of R_{ST} in bits per channel use [cf. (2.5)].

C2.5) If channel coding is used along with linear or nonlinear ST codes, spectral efficiency will be expressed as the product of η_{ST}, or correspondingly, R_{ST}, times the rate of the channel code.

Besides spectral efficiency, it is instructive to classify ST codes based on various criteria, a subject we consider in the next section.

2.7 CLASSIFICATION OF ST CODES

As with single-antenna systems, whether or how much of the MIMO channel's capacity can be achieved depends on the design of the ST codes used. Furthermore, error performance of a communication system also depends on the ST encoding and decoding modules. In the following chapters, we discuss many types of ST codes which are designed for different channels and with different trade-offs in mind. Before we start discussing those specific designs, it is useful to introduce categories of ST codes based on different criteria.

Similar to single-antenna systems, we can classify ST codes as linear or nonlinear. A *linear* ST block code is defined as [107]

$$\boldsymbol{X} = \sum_{n=1}^{N_s} [\Re(s_n)\boldsymbol{\Phi}_n + j\Im(s_n)\boldsymbol{\Psi}_n], \tag{2.58}$$

where $\Re(s_n)$ and $\Im(s_n)$ are the real and imaginary parts of the information symbol s_n, respectively, $\{\boldsymbol{\Phi}_n, \boldsymbol{\Psi}_n\}$ are $N_t \times N_x$ real matrices that depend on the specific code design, and N_s is the number of information symbols per code matrix. Equivalently, substituting $s_n = \Re(s_n) + j\Im(s_n)$, we can rewrite (2.58) as

$$\boldsymbol{X} = \sum_{n=1}^{N_s} (s_n \boldsymbol{A}_n + s_n^* \boldsymbol{B}_n), \tag{2.59}$$

where $\boldsymbol{A}_n := (\boldsymbol{\Phi}_n + \boldsymbol{\Psi}_n)/2$ and $\boldsymbol{B}_n := (\boldsymbol{\Phi}_n - \boldsymbol{\Psi}_n)/2$ are two $N_t \times N_x$ matrices with real entries. To normalize the transmission power per time slot, we usually set

$$\text{tr}(\boldsymbol{A}_n \boldsymbol{A}_n^T + \boldsymbol{B}_n \boldsymbol{B}_n^T) = N_t. \tag{2.60}$$

Later we will see that a number of ST block codes (e.g., [6, 107, 249, 313]) fall under this linear encoding category.

If an ST code cannot be written in the form (2.59), we say that it is generally a *nonlinear* ST code. Classical examples of nonlinear coding include trellis-coded modulation and differential encoding. The latter suggests another categorization of ST codes, depending on the amount of CSI required at the receiver. Specifically, ST codes requiring estimation of the MIMO channel at the receiver will belong to the *coherent* category, the subject of Chapter 3; while differential and *noncoherent* ST codes are covered in Chapter 6.

Besides linearity in encoding and coherence in decoding, ST codes can be categorized depending on the *type of MIMO channel* for which they are intended. Chapter 3 deals with ST codes designed for flat fading MIMO channels; Chapters 7 and 8 cover ST codes designed for frequency-selective MIMO channels; Chapter 9 presents the codes designed for time-varying MIMO channels; Chapters 14 and 15 describe ST codes in the presence of single- and multi-user interference.

Another categorization of ST coded systems is between *open-loop* and *closed-loop* ones. Division in this case is made depending on where CSI is assumed available. Historically, ST coded systems were first derived for coherent open-loop systems, where CSI is supposed available at the receiver. Closed-loop systems followed shortly after and require CSI also to be available at the multi-antenna transmitter, often via a feedback channel. Closed-loop ST coded systems are dealt with in Chapters 12 and 13.

The final classification we will use for ST codes is on the basis of error and rate performance. We call those ST codes designed to maximize diversity (and possibly coding) gains *error-performance-oriented* designs when the spectral efficiency they can attain does not exceed 1 symbol per channel use. On the other hand, those that target transmission rates as high as N_t symbols per channel use will henceforth be termed *rate-oriented* ST codes.

2.8 CLOSING COMMENTS

We have introduced basic notions and fundamental limits related to ST coding and argued why ST coded transmissions can be superior to single-antenna block-coded transmissions from the viewpoints of both channel capacity and error performance, which we quantified using the diversity and coding gains. We further delineated issues related to diversity and spectral efficiency of ST codes and presented various categories that can be used to classify ST coded multi-antenna systems.

In the next chapter we present several representative classes of ST codes that have been derived for flat fading channels. Those paradigm codes are carefully chosen to highlight pertinent trade-offs between the metrics we introduced in this chapter, rate and error performance, and also take complexity into account which certainly depends not only on the ST encoder adopted but also on the receiver processing that the system design can afford.

The basic topics we chose to cover in this chapter play an instrumental role in developing the ST coded systems presented in ensuing chapters, understanding their relative merits and evaluating their performance. However, the coverage is by no means exhaustive. Among the many important topics left uncovered, one pertains to the effects that mobility has on the capacity of ST channels [114, 257]. For this and additional fundamental topics, interested readers are referred to research monographs on related subjects [8, 16, 85, 97, 119, 126, 130, 142, 208, 263, 281].

3

Coherent ST Codes for Flat Fading Channels

In Chapter 2 we introduced fundamental figures of merit pertaining to rate and error performance of multi-antenna ST coded systems. The error performance metrics provided two design criteria which amount to selecting ST code matrices that maximize diversity and coding gains. Based on these criteria, in the present chapter we develop design paradigms representative of certain classes of ST codes, which are referred to as *coherent* because their decoding requires channel state information to be available at the receiver. The coherent designs outlined here for flat fading MIMO channels aim at ST codes optimizing high-SNR error performance at transmission rates not exceeding those achievable by single-antenna systems: 1 symbol per channel use. Nonetheless, the resulting options in these classes of ST codes will turn out to offer desirable trade-offs among error performance, bandwidth efficiency, and decoding complexity.

3.1 DELAY DIVERSITY ST CODES

The simplest class of coherent ST codes is the class of delay diversity (DD) codes first proposed in [231, 298]. Albeit simple, these codes turn out to be capable of effecting the maximum possible diversity gain for any number of transmit-antennas and for arbitrary signal constellations. Consider a block $\boldsymbol{s} := [s_1, \ldots, s_{N_s}]^T \in \mathbb{C}^{N_s \times 1}$ of N_s information symbols. As depicted in Figure 3.1, DD amounts to sending $\boldsymbol{s}$ through each of the N_t transmit-antennas but with different delays. Specifically, whereas there is no delay at the first transmit-antenna, which is used without loss of generality as a reference, $\boldsymbol{s}$ is delayed by one-symbol duration before transmission from the second

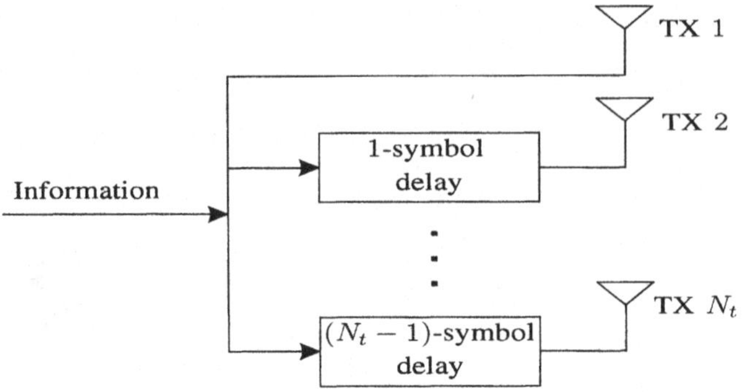

Figure 3.1 Delay diversity multi-antenna transmission

antenna, two-symbol durations at the third transmit-antenna, and so on. Accordingly, the ST code matrix for a DD transmission system is given by

$$\boldsymbol{X}_{\mathrm{DD}} = \begin{bmatrix} s_1 & s_2 & \cdots & s_{N_s} & 0 & \cdots & 0 \\ 0 & s_1 & s_2 & \cdots & s_{N_s} & \cdots & 0 \\ \vdots & \vdots & \ddots & \ddots & \cdots & \ddots & \vdots \\ 0 & 0 & \cdots & s_1 & s_2 & \cdots & s_{N_s} \end{bmatrix} \in \mathbb{C}^{N_t \times (N_s+N_t-1)}, \quad (3.1)$$

where the 0's correspond to no transmission. Upon substituting (3.1) into (2.7), the nth sample at the νth receive-antenna can be written as a linear convolution of $\boldsymbol{s}$ with the channel vector $\boldsymbol{h}_\nu := [h_{\nu 1}, \ldots, h_{\nu N_t}]^T$; that is,

$$y_{\nu n} = \sqrt{\frac{\bar{\gamma}}{N_t}}\, s_n \star h_{\nu n} + w_{\nu n}. \quad (3.2)$$

Therefore, DD can be thought of as a scheme transforming each set of N_t flat fading channels corresponding to a single receive-antenna (here the νth) to a single frequency-selective fading channel with N_t channel taps $h_{\nu\mu}$, $\mu = 1, \ldots, N_t$.

Since frequency-selective channels provide multipath diversity, as we explained in Chapter 1, one can intuitively expect that this ST equivalent channel will offer diversity of order N_t. And after accounting for all N_r receive-antennas, the overall diversity order will be $N_t N_r$. To establish this formally, consider two distinct information blocks $\boldsymbol{s}$ and $\boldsymbol{s}'$ and their corresponding code matrices $\boldsymbol{X}_{\mathrm{DD}}$ and $\boldsymbol{X}'_{\mathrm{DD}}$. The difference between $\boldsymbol{X}_{\mathrm{DD}}$ and $\boldsymbol{X}'_{\mathrm{DD}}$ is given by

$$\boldsymbol{X}_{\mathrm{DD}} - \boldsymbol{X}'_{\mathrm{DD}} = \begin{bmatrix} s_1-s'_1 & s_2-s'_2 & \cdots & s_{N_s}-s'_{N_s} & 0 & \cdots & 0 \\ 0 & s_1-s'_1 & s_2-s'_2 & \cdots & s_{N_s}-s'_{N_s} & \cdots & 0 \\ \vdots & \vdots & \ddots & \ddots & \cdots & \ddots & \vdots \\ 0 & 0 & \cdots & s_1-s'_1 & s_2-s'_2 & \cdots & s_{N_s}-s'_{N_s} \end{bmatrix}. \quad (3.3)$$

Because $s \neq s'$, there exists at least one entry n such that $s_n \neq s'_n$. Since the difference matrix in (3.3) is triangular, it has full row rank. It then follows that the error matrix in the DD scheme, defined as

$$\Delta_e := (X_{\text{DD}} - X'_{\text{DD}})(X_{\text{DD}} - X'_{\text{DD}})^{\mathcal{H}},$$

always has full rank N_t. Upon recalling (2.44), we deduce that DD coding enables the maximum possible diversity gain of order $N_t N_r$.

The time span (number of columns) of the ST code matrix in (3.1) is $N_x = N_s + N_t - 1$. Therefore, the code rate of the DD scheme is

$$\eta_{\text{DD}} = \frac{N_s}{N_s + N_t - 1} \quad \text{symbols per channel use},$$

which for a fixed number of transmit antennas approaches 1 asymptotically as N_s grows large. Hence, the bandwidth efficiency loss of a DD system relative to a single-antenna system becomes negligible for a sufficiently large N_s.

DD effects a banded structure in the corresponding ST code matrix. This structure not only ensures full diversity but also affords low-complexity decoding. As we will see soon, code matrices in the DD scheme can be generated by using a trellis. As a result, ST-DD decoding at the receiver can be implemented by using the computationally efficient Viterbi's algorithm. This is yet another manifestation of the ST-DD equivalence with a frequency-selective channel for which Viterbi decoding is applicable (see, e.g., [214]).

3.2 ST TRELLIS CODES

ST trellis codes (STTCs) constitute an important class of nonlinear ST codes aiming at maximum diversity and high coding gains at controllable bandwidth efficiency. As the term *trellis* suggests, a distinct feature of STTCs is that they can be generated and represented by a trellis structure, which also facilitates decoding via Viterbi's algorithm. For single-antenna systems, Viterbi decoding is known to incur complexity that depends on the number of states [214]. A number of ST codes, such as DD codes, can be viewed as special cases of STTCs. Among the available options of constructing STTCs, we present those introduced by [12, 101, 253].

3.2.1 Trellis Representation

STTCs can be derived using various approaches, all of which can be represented by a trellis. In the ST context, the latter is referred to as an ST trellis. Different from the conventional trellis commonly used to represent convolutional channel codes, each transition branch in an ST trellis entails N_t labels, each corresponding to the N_t symbols that are simultaneously transmitted over the N_t antennas. In particular, depending on the state of the ST trellis as well as the value of the input symbol at time n, one transition branch is chosen. If the chosen branch is labeled $l_{1n} l_{2n} \cdots l_{N_t n}$, the information symbol corresponding to $l_{\mu n}$ is transmitted over the μth transmit-antenna. This is equivalent to choosing $x_{\mu n}$ in the ST code matrix $\boldsymbol{X}$ as the information symbol corresponding to $l_{\mu n}$.

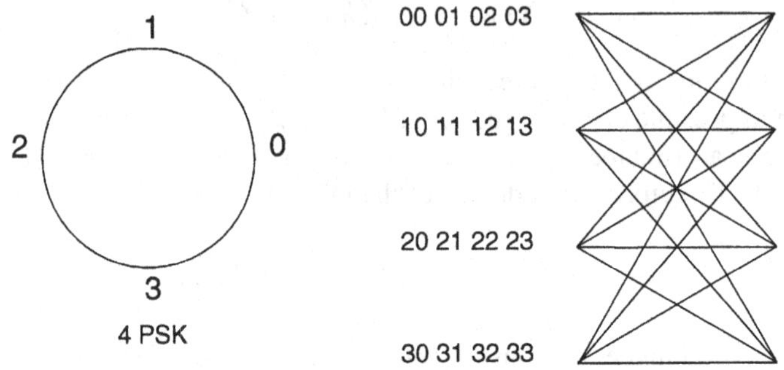

Figure 3.2 A 4-PSK, 4-state space-time trellis code

Let us consider as an example the 4-PSK, 4-state STTC introduced for two transmit-antennas by [253]. Figure 3.2 depicts the 4-PSK constellation, where the four possible 4-PSK symbols $\{1, j, -1, -j\}$ are respectively labeled as $\{0, 1, 2, 3\}$, and the corresponding ST trellis, which has four outgoing and four incoming branches per state. As in the conventional trellis, the ST trellis is initialized at the zero state. If, for instance, the input information block is $\boldsymbol{s} = [j, -j, -1, -j, 1, j, \ldots]^T$, the corresponding ST code matrix is generated from this ST trellis as

$$\boldsymbol{X}_{\text{TC}} = \begin{bmatrix} 0 & j & -j & -1 & -j & 1 & \cdots \\ j & -j & -1 & -j & 1 & j & \cdots \end{bmatrix}, \tag{3.4}$$

which is exactly the one generated when using the DD scheme. In this way, Figure 3.2 can be thought of as a trellis representation of DD. In other words, DD coding is just a special case of ST trellis coding.

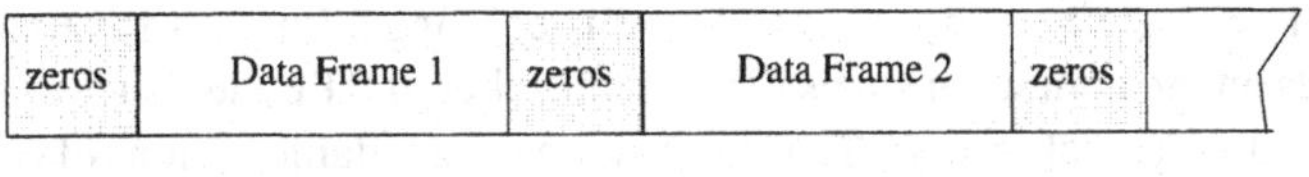

Figure 3.3 Signal structure for ST trellis coding

As mentioned earlier, it is common to have the ST trellis always starting and ending at the zero state. To accomplish this, information symbols are transmitted in a block-by-block fashion with a certain number of zeros periodically inserted in between, as illustrated in Figure 3.3. Since the number of zeros depends on the constraint length of the ST trellis regardless of the frame size, the bandwidth efficiency loss induced by those zeros could be made arbitrarily small if each frame is sufficiently large. In light of this, the transmission rate of ST trellis coding is

$$R_{\text{TC}} \approx \log_2 |\mathbb{S}| \quad \text{bits per channel use}$$

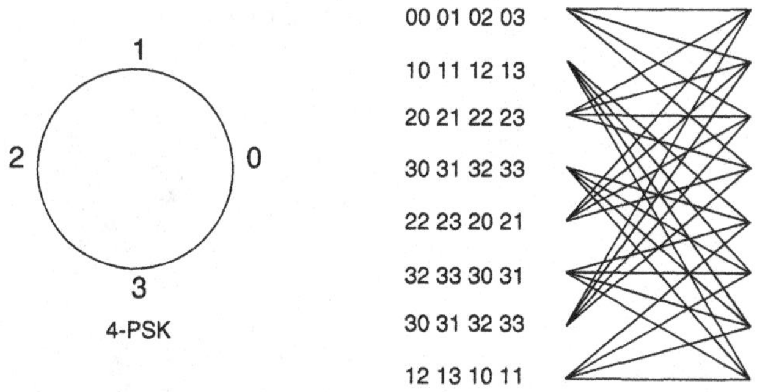

Figure 3.4 An 8-state 4-PSK STTC

for reasonably large data frames. Of course, the price paid for using larger frames is increased complexity and longer decoding delays.

Using the trellis representation, the design of STTCs amounts to constructing the ST trellis and labeling its branches properly with the objective of maximizing diversity and coding gains. A variety of approaches have been proposed to design STTCs. In what follows we present several representative STTC examples.

3.2.2 TSC ST Trellis Codes

In their pioneering work [253], Tarokh, Seshadri, and Calderbank (TSC) introduced the concept of ST coding, derived design criteria for ST codes, and proposed the following two simple rules for constructing STTCs when two transmit-antennas are used:

Rule 1) Transitions departing from the same state differ only in the second symbol.

Rule 2) Transitions merging at the same state differ only in the first symbol.

Following these two design rules, the difference between any two distinct code matrices $\boldsymbol{X}_{\text{TSC}}$ and $\boldsymbol{X}'_{\text{TSC}}$ becomes

$$\boldsymbol{X}_{\text{TSC}} - \boldsymbol{X}'_{\text{TSC}} = \begin{bmatrix} \cdots & 0 & \cdots & x_2 & \cdots \\ & x_1 & & 0 & \end{bmatrix}, \tag{3.5}$$

where x_1 and x_2 are two nonzero complex numbers. Thus, the error matrix

$$\Delta_e := (\boldsymbol{X}_{\text{TSC}} - \boldsymbol{X}'_{\text{TSC}})(\boldsymbol{X}_{\text{TSC}} - \boldsymbol{X}'_{\text{TSC}})^{\mathcal{H}}$$

has full rank 2, which confirms that maximum diversity of order 2 can be achieved. Figures 3.4, 3.5, 3.6, and 3.7 depict four design examples: 8-state 4-PSK, 16-state 4-PSK, 8-state 8-PSK, and 16-state 16-QAM STTCs [253].

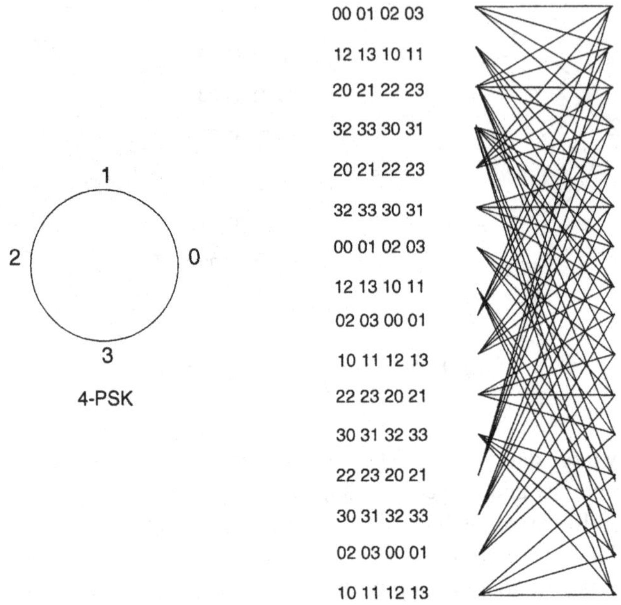

Figure 3.5 A 16-state 4-PSK STTC

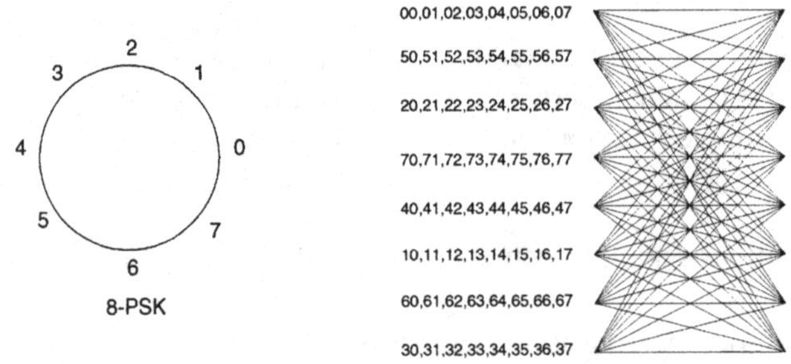

Figure 3.6 An 8-state 8-PSK STTC

3.2.3 BBH ST Trellis Codes

Although the two simple design rules of Section 3.2.2 effect the maximum diversity gain, they are suboptimal in terms of the coding gain they enable. In [12], Baro, Bauch, and Hansmann (BBH) represented the ST trellis in a more tractable generator matrix form, so that computer search can be used to construct STTC matrices systematically. To illustrate the basic idea behind the BBH representation, suppose that the information symbols are M-PSK modulated. Because each information symbol conveys

$$N_b = \log_2 M \tag{3.6}$$

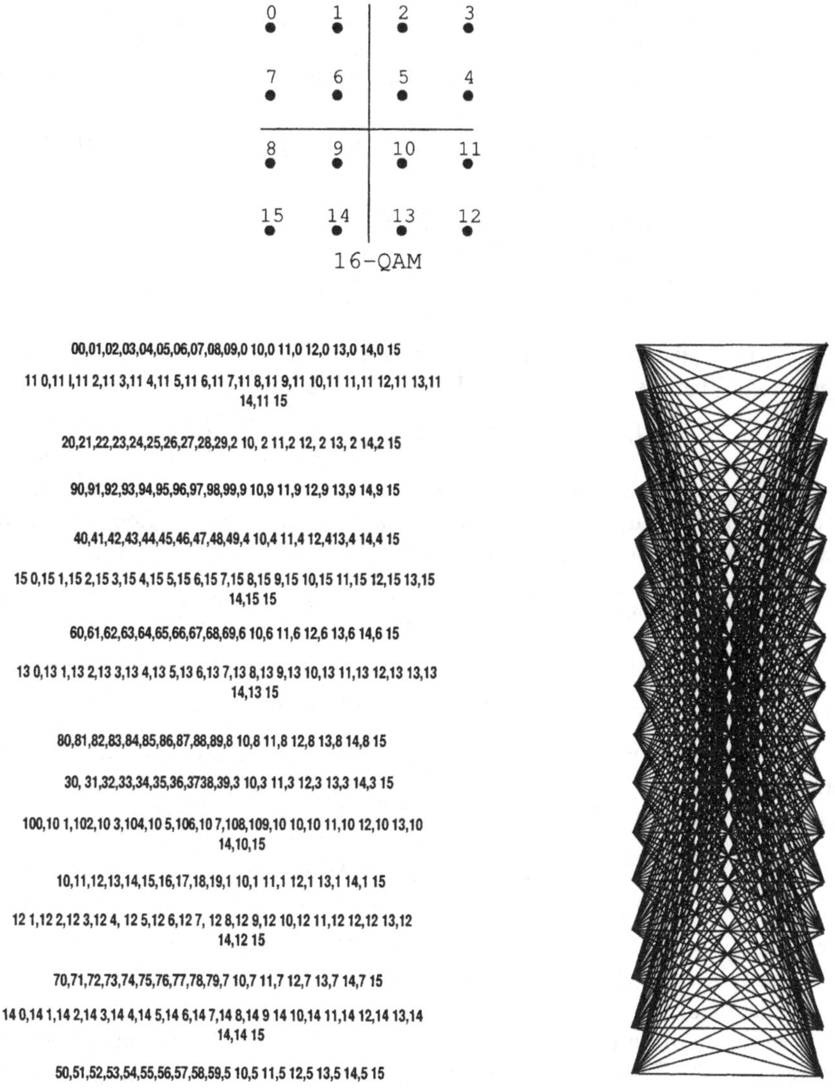

Figure 3.7 A 16-state 16-QAM STTC

bits, the entire information block $\boldsymbol{s} \in \mathbb{C}^{N_s \times 1}$ can be represented by a binary block $\boldsymbol{b} := [b_1, b_2, \ldots, b_{N_s N_b}]^T$ of length $N_s N_b$. Let us consider the $N_t \times (N_b + N_m)$ generator matrix

$$\boldsymbol{G} = \begin{bmatrix} g_{1,1} & g_{1,2} & \cdots & g_{1,N_b+N_m} \\ \vdots & \vdots & \ddots & \vdots \\ g_{N_t,1} & g_{N_t,2} & \cdots & g_{N_t,N_b+N_m} \end{bmatrix}, \tag{3.7}$$

Table 3.1 COMPARISON OF TSC WITH BBH STTCs (4-PSK)

States	G (TSC)	G_c	G (BBH)	G_c	Improvement
4	$\begin{bmatrix} 2 & 1 & 0 & 0 \\ 0 & 0 & 2 & 1 \end{bmatrix}$	2	$\begin{bmatrix} 2 & 0 & 1 & 3 \\ 2 & 2 & 0 & 1 \end{bmatrix}$	$\sqrt{8}$	1.50 dB
8	$\begin{bmatrix} 0 & 0 & 2 & 1 & 2 \\ 2 & 1 & 0 & 0 & 2 \end{bmatrix}$	$\sqrt{12}$	$\begin{bmatrix} 2 & 0 & 2 & 1 & 2 \\ 2 & 1 & 0 & 0 & 2 \end{bmatrix}$	4	0.62 dB

with N_m being the number of memory units in the encoder and entries g_{ij} taking integer values between 0 and $M-1$. Define also the function

$$\mathcal{M}(x) := e^{j\frac{2\pi}{M}x}, \quad x \in \{0, 1, \ldots, M-1\},$$

which maps the M integer values to the M possible M-PSK symbols. At time n, the ST encoder yields at its output an $N_t \times 1$ block $\boldsymbol{x}_n := [x_{1n}, x_{2n}, \ldots, x_{N_t n}]^T$. In this case, the nth column of $\boldsymbol{X}$ can be written as

$$\boldsymbol{x}_n = \mathcal{M}\left((\boldsymbol{G}\boldsymbol{b}_n) \text{mod } M\right), \quad n = 1, \ldots, N_x, \tag{3.8}$$

where

$$\boldsymbol{b}_n := [b_{N_b n}, \ldots, b_{N_b n - N_b + 1}, b_{N_b n - N_b}, \ldots, b_{N_b n - N_b - N_m + 1}]^T$$

comprises those bits in $\boldsymbol{b}$ that affect the output at time n. For example, the ST trellis corresponding to the DD scheme in Figure 3.2 can be represented in the form (3.8) by choosing parameters $M = 4$, $N_m = 2$, $N_b = 2$, and generator matrix

$$\boldsymbol{G} = \begin{bmatrix} 2 & 1 & 0 & 0 \\ 0 & 0 & 2 & 1 \end{bmatrix}. \tag{3.9}$$

Accordingly, the code matrix in (3.4) can be thought of as having been generated from (3.8) when the input is $\boldsymbol{b} = [0, 1, 1, 1, 1, 0, 1, 1, 0, 0, 0, 1, \ldots]^T$, which is the binary representation of $\boldsymbol{s} = [j, -j, -1, -j, 1, j, \ldots]^T$.

By using the form (3.8), the design of STTCs becomes equivalent to finding a generator matrix $\boldsymbol{G}$ so that the resultant STTCs are capable of optimizing both diversity and coding gains. Since each entry of $\boldsymbol{G}$ can take M possible values, there are a total of $M^{N_t(\log_2 M + N_m)}$ possible $\boldsymbol{G}$'s. Thus, a computer search is feasible when M, N_t, and N_m are relatively small. For 4-PSK with 4 and 8 states, Table 3.1 lists the generator matrices of optimal STTCs obtained using an exhaustive computer search [12]. Relative to TSC STTCs, the comparison in Table 3.1 shows that such a search does improve the coding gain G_c.

3.2.4 GFK ST Trellis Codes

Another generator matrix form leading to a systematic search for STTC matrices was put forth by Grimm, Fitz, and Krogmeier (GFK) in [101]. This form amounts to rewriting (3.8) in an equivalent form,

$$\boldsymbol{x}_n = \mathcal{M}\left(\left(\sum_{i=1}^{N_b} \boldsymbol{G}^{(i)}\boldsymbol{b}_n^{(i)}\right) \bmod M\right), \tag{3.10}$$

where the $N_t \times (1 + N_m^{(i)})$ matrix $\boldsymbol{G}^{(i)}$ and the $(1 + N_m^{(i)}) \times 1$ block $\boldsymbol{b}_n^{(i)}$ correspond to $\boldsymbol{G}$ and $\boldsymbol{b}_n$ through the relationships

$$\boldsymbol{G} = [\boldsymbol{G}^{(1)}, \ldots, \boldsymbol{G}^{(N_b)}],$$
$$\boldsymbol{b}_n = [\boldsymbol{b}_n^{(1)T}, \ldots, \boldsymbol{b}_n^{(N_b)T}]^T,$$

and $N_m^{(i)}$'s are chosen such that

$$N_m = \sum_{i=1}^{N_b} N_m^{(i)}.$$

Because (3.10) is mathematically identical to (3.8), exhaustive search for the optimal $\boldsymbol{G}$ or $\boldsymbol{G}^{(i)}$'s entails testing for $M^{N_t(\log_2 M + N_m)}$ possibilities if no additional constraints are imposed. This search is again feasible only if M, N_t, N_m are small.

To speed up the search process, [101] enforces a so-termed zero-symmetry constraint on the ST trellis code matrix. An ST trellis code matrix is said to satisfy this constraint if the difference matrix $\boldsymbol{X}_{\text{GFK}} - \boldsymbol{X}'_{\text{GFK}}$ between any two distinct code matrices $\boldsymbol{X}_{\text{GFK}}$ and $\boldsymbol{X}'_{\text{GFK}}$ is tridiagonal, as illustrated next for $N_t = 3$:

$$\boldsymbol{X}_{\text{GFK}} - \boldsymbol{X}'_{\text{GFK}} = \begin{bmatrix} x_1 & \cdots & \cdots & 0 & 0 \\ 0 & x_2 & \cdots & x_3 & 0 \\ 0 & 0 & \cdots & \cdots & x_4 \end{bmatrix}, \tag{3.11}$$

where the x_i's stand for some nonzero numbers. The zero-symmetry constraint has been proved sufficient to guarantee the maximum diversity gain and can be thought of as an extension of (3.5) to more than two transmit-antennas. Similarly, the generator matrix $\boldsymbol{G}^{(i)}$ is said to satisfy the zero-symmetry constraint if it is in a form similar to (3.11); that is,

$$\boldsymbol{G}^{(i)} = \begin{bmatrix} g_1^{(i)} & \cdots & \cdots & 0 & 0 \\ 0 & g_2^{(i)} & \cdots & g_3^{(i)} & 0 \\ 0 & 0 & \cdots & \cdots & g_4^{(i)} \end{bmatrix}. \tag{3.12}$$

The zero-symmetry constraint of an ST trellis code matrix is closely related to that of the corresponding generator matrix $\boldsymbol{G}^{(i)}$, as summarized next:

- When $M = 2$ ($N_b = 1$) and the generator matrix $\boldsymbol{G}$ has zero symmetry, the STTCs generated by (3.10) will also have zero symmetry, thereby guaranteeing maximum diversity gain.

Table 3.2 GFK STTCs FOR $M = 2$ (BPSK)

N_t	States	G	G_c
2	2	$\begin{bmatrix} 1 & 0 \\ 0 & 1 \end{bmatrix}$	2
2	8	$\begin{bmatrix} 1 & 1 & 1 & 0 \\ 0 & 1 & 0 & 1 \end{bmatrix}$	$\sqrt{20}$
3	16	$\begin{bmatrix} 1 & 1 & 1 & 0 & 0 \\ 0 & 1 & 0 & 1 & 0 \\ 0 & 0 & 1 & 1 & 1 \end{bmatrix}$	4.35
3	64	$\begin{bmatrix} 1 & 1 & 1 & 1 & 1 & 0 & 0 \\ 0 & 1 & 0 & 1 & 1 & 1 & 0 \\ 0 & 0 & 1 & 1 & 0 & 1 & 1 \end{bmatrix}$	7.54
4	64	$\begin{bmatrix} 1 & 0 & 1 & 1 & 0 & 0 & 0 \\ 0 & 1 & 1 & 0 & 1 & 0 & 0 \\ 0 & 0 & 1 & 1 & 1 & 1 & 0 \\ 0 & 0 & 0 & 1 & 1 & 0 & 1 \end{bmatrix}$	5.66

- When $M > 2$ ($N_b > 1$) and the generator matrices $G^{(i)}$ have zero symmetry, the STTCs generated by (3.10) will have zero symmetry if the following two extra conditions are satisfied $\forall\, j$:

$$g_j^{(i)} \neq g_j^{(i')}, \qquad \forall\, i \neq i'$$

$$\left(\sum_{i=1}^{N_b} g_j^{(i)} \right) \bmod M \neq 0.$$

Thanks to this last relationship, searching for the optimal generator matrices can be restricted over those satisfying the zero-symmetry constraint and the two additional conditions if $N_b > 1$. As a result, computational complexity can be reduced considerably. Table 3.2 lists the searched generator matrices for $M = 2$ (BPSK) with different numbers of transmit-antennas and states, together with the coding gain G_c they enable.

3.2.5 Viterbi Decoding of ST Trellis Codes

A major advantage of STTCs is that they can be decoded using Viterbi's algorithm. As we mentioned in Section 3.2.4, the only difference between an ST trellis and its counterpart in a single-antenna system is that each branch of an ST trellis is associated with N_t labels instead of one. As a result, Viterbi's algorithm originally proposed for decoding single-antenna trellis coded transmissions can be applied if the branch metric is chosen as

$$\sum_{\nu=1}^{N_r} \left| y_{\nu n} - \sum_{\mu=1}^{N_t} \sqrt{\frac{\bar{\gamma}}{N_t}} h_{\nu\mu} x_{\mu n} \right|^2 . \tag{3.13}$$

Using (3.13), the Viterbi algorithm returns the shortest path in terms of smallest accumulated metric as one traverses through the trellis; that is, the path with labels

$$(l_{11}, l_{21}, \ldots, l_{N_t 1}),\ (l_{12}, l_{22}, \ldots, l_{N_t 2}), \ldots, (l_{1N_x}, l_{2N_x}, \ldots, l_{N_t N_x})$$

minimizing over all possible paths $x_{\mu n}$ the metric

$$\sum_{n=1}^{N_x} \sum_{\nu=1}^{N_r} \left| y_{\nu n} - \sum_{\mu=1}^{N_t} \sqrt{\frac{\bar{\gamma}}{N_t}} h_{\nu\mu} x_{\mu n} \right|^2 . \tag{3.14}$$

At any time instant, the number of branches to be computed is

$$\text{number of states} \times \text{number of branches per state} = \text{number of states} \times 2^{R_{\text{ST}}},$$

where R_{ST} is the transmission rate defined in (2.5). Although this number is linear in N_x, unfortunately it remains exponential in the transmission rate. This constitutes a major limitation of STTCs.

3.3 ORTHOGONAL ST BLOCK CODES

Orthogonal ST block codes (OSTBCs) were first proposed for two transmit-antennas by Alamouti [6] and later were extended to an arbitrary number of transmit-antennas in the context of general orthogonal designs [249]. A signal processing interpretation of OSTBCs can be found in [80]. Compared with other ST codes, a distinct feature of OSTBCs is their low decoding complexity, which renders them attractive when receiver complexity is at a premium. Different from STTCs, the design of OSTBCs typically aims at optimizing the diversity gain only. As a result, OSTBCs often enable smaller coding gains compared to those of their trellis counterparts. More seriously, OSTBCs incur considerable loss in bandwidth efficiency when more than two transmit-antennas are used with generally complex constellations.

3.3.1 Encoding of OSTBCs

OSTBCs have been cast in the framework of general complex orthogonal designs of ST code matrices [249]. An OSTBC can be defined by specifying the mapping between an information block and the resulting ST code matrix, as follows:

Definition 3.1 *With* $\boldsymbol{s} := [s_1, \ldots, s_{N_s}]^T$ *and* * *denoting complex conjugation, the* $N_t \times N_x$ *matrix* $\boldsymbol{X}_O(\boldsymbol{s})$ *is an OSTBC matrix if and only if its nonzero entries are drawn from the set* $\{s_n, s_n^*\}_{n=1}^{N_s}$ *so that*

$$\boldsymbol{X}_O(\boldsymbol{s})\boldsymbol{X}_O^{\mathcal{H}}(\boldsymbol{s}) = \sum_{n=1}^{N_s} |s_n|^2 \boldsymbol{I}_{N_t}. \tag{3.15}$$

For $N_t = 2, 3, 4$, matrix $\boldsymbol{X}_O(\boldsymbol{s})$ takes correspondingly the form

$$\boldsymbol{O}_{2\times 2} = \begin{bmatrix} s_1 & -s_2^* \\ s_2 & s_1^* \end{bmatrix}, \tag{3.16}$$

$$\boldsymbol{O}_{3\times 4} = \begin{bmatrix} s_1 & -s_2^* & -s_3^* & 0 \\ s_2 & s_1^* & 0 & -s_3^* \\ s_3 & 0 & s_1^* & s_2^* \end{bmatrix}, \tag{3.17}$$

$$\boldsymbol{O}_{4\times 4} = \begin{bmatrix} s_1 & -s_2^* & -s_3^* & 0 \\ s_2 & s_1^* & 0 & -s_3^* \\ s_3 & 0 & s_1^* & s_2^* \\ 0 & s_3 & -s_2 & s_1 \end{bmatrix}. \tag{3.18}$$

Among them, $\boldsymbol{O}_{2\times 2}$ was first discovered by Alamouti in [6] and will be referred to as Alamouti's code. In its simplicity, the many attractive features that we will see it possesses render it the most powerful representative from the OSTBC class.

According to the classification in Section 2.7, OSTBCs are clearly linear ST codes. Indeed, since each entry of $\boldsymbol{X}_O(\boldsymbol{s})$ is either s_n or s_n^* or zero, $\boldsymbol{X}_O(\boldsymbol{s})$ can be written either in the form of (2.59) with $\{\boldsymbol{A}_n\}$ and $\{\boldsymbol{B}_n\}$ satisfying

$$\begin{aligned} \boldsymbol{A}_n\boldsymbol{A}_{n'}^T + \boldsymbol{B}_n\boldsymbol{B}_{n'}^T &= \boldsymbol{0}, & n &\neq n', \\ \boldsymbol{A}_n\boldsymbol{A}_n^T + \boldsymbol{B}_n\boldsymbol{B}_n^T &= \boldsymbol{I}_{N_t}, & \forall n&, \\ \boldsymbol{A}_n\boldsymbol{B}_{n'}^T &= \boldsymbol{0}, & \forall n&, n', \end{aligned} \tag{3.19}$$

or in the form of (2.58) with

$$\boldsymbol{\Phi}_n = \boldsymbol{A}_n + \boldsymbol{B}_n, \qquad \boldsymbol{\Psi}_n = \boldsymbol{A}_n - \boldsymbol{B}_n. \tag{3.20}$$

Using (3.19), we can readily verify that matrices $\boldsymbol{\Phi}_n$ and $\boldsymbol{\Psi}_n$ satisfy

$$\begin{aligned} &\boldsymbol{\Phi}_n\boldsymbol{\Phi}_n^T = \boldsymbol{I}_{N_t}, && \boldsymbol{\Psi}_n\boldsymbol{\Psi}_n^T = \boldsymbol{I}_{N_t}, && \forall n, \\ &\boldsymbol{\Phi}_n\boldsymbol{\Phi}_{n'}^T = -\boldsymbol{\Phi}_{n'}\boldsymbol{\Phi}_n, && \boldsymbol{\Psi}_n\boldsymbol{\Psi}_{n'}^T = -\boldsymbol{\Psi}_{n'}\boldsymbol{\Psi}_n, && \forall n \neq n', \\ &\boldsymbol{\Phi}_n\boldsymbol{\Psi}_{n'}^T = \boldsymbol{\Psi}_{n'}\boldsymbol{\Phi}_n, && && \forall n, n'. \end{aligned} \tag{3.21}$$

The general OSTBC representations in (2.59) and (2.58) will come in handy at the decoding stage and also when we analyze the error performance of OSTBCs.

To illustrate these representations, let us consider the Alamouti code. If (3.16) is expressed as in (2.59), the corresponding matrices $\boldsymbol{A}_n$ and $\boldsymbol{B}_n$ are

$$\boldsymbol{A}_1 = \begin{bmatrix} 1 & 0 \\ 0 & 0 \end{bmatrix}, \quad \boldsymbol{A}_2 = \begin{bmatrix} 0 & 0 \\ 1 & 0 \end{bmatrix}, \quad \boldsymbol{B}_1 = \begin{bmatrix} 0 & 0 \\ 0 & 1 \end{bmatrix}, \quad \boldsymbol{B}_2 = \begin{bmatrix} 0 & -1 \\ 0 & 0 \end{bmatrix}, \tag{3.22}$$

and clearly satisfy (3.19). If (2.58) is used, we have

$$\boldsymbol{\Phi}_1 = \begin{bmatrix} 1 & 0 \\ 0 & 1 \end{bmatrix}, \quad \boldsymbol{\Phi}_2 = \begin{bmatrix} 0 & -1 \\ 1 & 0 \end{bmatrix}, \quad \boldsymbol{\Psi}_1 = \begin{bmatrix} 1 & 0 \\ 0 & -1 \end{bmatrix}, \quad \boldsymbol{\Psi}_2 = \begin{bmatrix} 0 & 1 \\ 1 & 0 \end{bmatrix}, \tag{3.23}$$

which clearly satisfy (3.21).

Since N_s information symbols are transmitted over N_x time slots, the code rate of OSTBCs is

$$\eta_{\text{OSTBC}} = \frac{N_s}{N_x} \quad \text{symbols per channel use.} \tag{3.24}$$

With regard to the existence of OSTBCs and their rates for generally complex constellations, results in [80, 249] have established that:

- For $N_t = 2$, an OSTBC exists with rate $\eta_{\text{OSTBC}} = 1$ [cf. (3.16)].
- For $N_t = 3, 4$, OSTBCs exist but with rate $\eta_{\text{OSTBC}} = 3/4$.
- For $N_t > 4$, OSTBCs exist but with rate $\eta_{\text{OSTBC}} = 1/2$.

These results assert that when more than two transmit-antennas are used to transmit symbols drawn from a complex constellation, OSTBCs always induce considerable loss in bandwidth efficiency even relative to a single-antenna system. This constitutes a major disadvantage since as we have seen in Chapter 2 boosting data rates is one of the major reasons for deploying multi-antenna systems in the first place. Notwithstanding, the rate loss pertains to general OSTBCs with complex constellations. For the special case of real constellations, $\eta_{\text{OSTBC}} = 1$ is achievable for all $N_t \geq 2$.

3.3.2 Linear ML Decoding of OSTBCs

According to (2.9), an OSTBC system obeys the input-output relationship

$$\boldsymbol{Y} = \sqrt{\frac{\bar{\gamma}}{N_t}} \boldsymbol{H} \boldsymbol{X}_{\text{O}}(\boldsymbol{s}) + \boldsymbol{W}. \tag{3.25}$$

Since some entries of $\boldsymbol{X}_{\text{O}}$ could be zero, as pointed out in Section 2.1, $\bar{\gamma}$ in (3.25) no longer represents the average SNR per receive-antenna. Instead, it is related to the average SNR per receive-antenna in (2.11) via [cf. (3.15)]

$$\bar{\gamma} = \frac{\text{SNR}}{\eta_{\text{OSTBC}}}. \tag{3.26}$$

According to (3.24), $\bar{\gamma}$ can be also interpreted as the average symbol energy to noise ratio per receive-antenna. Recall that for a general ST code matrix, ML decoding of $\boldsymbol{s}$ from $\boldsymbol{Y}$ requires searching over all $|\mathbb{S}|^{N_s}$ possible information blocks $\boldsymbol{s}$. Fortunately, decoding complexity can be reduced if one exploits the rich structure of $\boldsymbol{X}_{\text{O}}(\boldsymbol{s})$ specified in Definition 3.1. To this end, we first use (2.59) to rewrite (3.25) as

$$\boldsymbol{Y} = \sqrt{\frac{\bar{\gamma}}{N_t}} \sum_{n=1}^{N_s} (\boldsymbol{H}\boldsymbol{A}_n s_n + \boldsymbol{H}\boldsymbol{B}_n s_n^*) + \boldsymbol{W}. \tag{3.27}$$

Then we form the $2N_rN_x \times 1$ vector

$$\bar{\boldsymbol{y}} = \begin{bmatrix} \text{vec}(\boldsymbol{Y}^T) \\ \text{vec}(\boldsymbol{Y}^{\mathcal{H}}) \end{bmatrix},$$

where vec($\boldsymbol{A}$) is a vector obtained after stacking all columns of $\boldsymbol{A}$ one on top of the other. From (3.27), $\bar{\boldsymbol{y}}$ can be written as

$$\bar{\boldsymbol{y}} = \sqrt{\frac{\bar{\gamma}}{N_t}}\bar{\boldsymbol{H}} \begin{bmatrix} \boldsymbol{s} \\ \boldsymbol{s}* \end{bmatrix} + \bar{\boldsymbol{w}}, \tag{3.28}$$

where the $2N_rN_x \times 2N_s$ matrix $\bar{\boldsymbol{H}}$ is given by

$$\bar{\boldsymbol{H}} = \begin{bmatrix} [\boldsymbol{A}_1^T\boldsymbol{H}^T]_1 & \cdots & [\boldsymbol{A}_{N_s}^T\boldsymbol{H}^T]_1 & [\boldsymbol{B}_1^T\boldsymbol{H}^T]_1 & \cdots & [\boldsymbol{B}_{N_s}^T\boldsymbol{H}^T]_1 \\ \vdots & & & & & \vdots \\ [\boldsymbol{A}_1^T\boldsymbol{H}^T]_{N_r} & \cdots & [\boldsymbol{A}_{N_s}^T\boldsymbol{H}^T]_{N_r} & [\boldsymbol{B}_1^T\boldsymbol{H}^T]_{N_r} & \cdots & [\boldsymbol{B}_{N_s}^T\boldsymbol{H}^T]_{N_r} \\ [\boldsymbol{B}_1^T\boldsymbol{H}^{\mathcal{H}}]_1 & \cdots & [\boldsymbol{B}_{N_s}^T\boldsymbol{H}^{\mathcal{H}}]_1 & [\boldsymbol{A}_1^T\boldsymbol{H}^{\mathcal{H}}]_1 & \cdots & [\boldsymbol{A}_{N_s}^T\boldsymbol{H}^{\mathcal{H}}]_1 \\ \vdots & & & & & \vdots \\ [\boldsymbol{B}_1^T\boldsymbol{H}^{\mathcal{H}}]_{N_r} & \cdots & [\boldsymbol{B}_{N_s}^T\boldsymbol{H}^{\mathcal{H}}]_{N_r} & [\boldsymbol{A}_1^T\boldsymbol{H}^{\mathcal{H}}]_{N_r} & \cdots & [\boldsymbol{A}_{N_s}^T\boldsymbol{H}^{\mathcal{H}}]_{N_r} \end{bmatrix}, \tag{3.29}$$

with $[\boldsymbol{A}]_\nu$ denoting the νth column of $\boldsymbol{A}$ and $\bar{\boldsymbol{w}}$ defined similar to $\bar{\boldsymbol{y}}$. Using the properties of $\boldsymbol{A}_n$ and $\boldsymbol{B}_n$ in (3.19), it can be readily verified that

$$\bar{\boldsymbol{H}}^{\mathcal{H}}\bar{\boldsymbol{H}} = ||\boldsymbol{H}||_F^2\boldsymbol{I}_{2N_s}, \tag{3.30}$$

with $||\boldsymbol{H}||_F^2 := \sum_{\nu=1}^{N_r}\sum_{\mu=1}^{N_t}|h_{\mu\nu}|^2$ standing for the squared Frobenius norm of $\boldsymbol{H}$. Therefore, we have

$$\tilde{\boldsymbol{y}} := \bar{\boldsymbol{H}}^{\mathcal{H}}\bar{\boldsymbol{y}} = \sqrt{\frac{\bar{\gamma}}{N_t}}||\boldsymbol{H}||_F^2 \begin{bmatrix} \boldsymbol{s} \\ \boldsymbol{s}* \end{bmatrix} + \bar{\boldsymbol{H}}^{\mathcal{H}}\bar{\boldsymbol{w}}, \tag{3.31}$$

from which we can form the decision statistics for s_n as

$$\tilde{y}_n = \sqrt{\frac{\bar{\gamma}}{N_t}}||\boldsymbol{H}||_F^2 s_n + \tilde{w}_n, \quad \forall\, n \in [1, N_s], \tag{3.32}$$

where $\tilde{y}_n$ and $\tilde{w}_n$ are the nth entries of $\tilde{\boldsymbol{y}}$ and $\bar{\boldsymbol{H}}^{\mathcal{H}}\bar{\boldsymbol{w}}$, respectively. Supposing that the entries of $\boldsymbol{W}$ are uncorrelated complex Gaussian with zero mean and unit variance, it follows from (3.30) that

$$E[\tilde{w}_n\tilde{w}_{n'}^*] = ||\boldsymbol{H}||_F^2\delta[n - n'], \tag{3.33}$$

which shows that $\tilde{w}_n$ remains white.

Taking (3.32) and (3.33) into account, we recognize that the linear processing in (3.31) has rendered the OSTBC transmission equivalent to a set of N_s independent serial transmissions each over a flat channel with coefficient $||\boldsymbol{H}||_F^2$ and noise variance

$||\boldsymbol{H}||_F^2$. As a result, ML decoding of $\boldsymbol{s}$ from $\boldsymbol{Y}$ has been transformed to ML decoding of s_n from $\tilde{y}_n$ for $n \in [1, N_s]$ without loss in error performance. The resulting decoding algorithm can be summarized in the following two steps:

S1) Given $\boldsymbol{H}$ and $\boldsymbol{Y}$, obtain $\tilde{y}_n$'s using (3.31).

S2) Given $\boldsymbol{H}$, decode s_n from $\tilde{y}_n$ via

$$\hat{s}_n = \underset{\forall s \in \mathbb{S}}{\arg\min} ||\tilde{y}_n - \sqrt{\frac{\bar{\gamma}}{N_t}} ||\boldsymbol{H}||_F^2 s||, \quad \forall\, n \in [1, N_s]. \tag{3.34}$$

Specializing this decoding algorithm to $N_t = 2$, we find that

$$\begin{bmatrix} \tilde{y}_1 \\ \tilde{y}_2 \end{bmatrix} = \sum_{\nu=1}^{N_r} \begin{bmatrix} h_{1\nu}^* & h_{2\nu} \\ h_{2\nu}^* & -h_{1\nu} \end{bmatrix} \begin{bmatrix} y_{\nu 1} \\ y_{\nu 2}^* \end{bmatrix}. \tag{3.35}$$

3.3.3 BER Performance with OSTBCs

Let us now analyze the diversity enabled by OSTBCs. Consider two distinct information blocks $\boldsymbol{s}$ and $\boldsymbol{s}'$, and their corresponding ST code matrices $\boldsymbol{X}_{\rm O}(\boldsymbol{s})$ and $\mathbf{X}_{\rm O}(\boldsymbol{s}')$. Using (2.59) and (3.19), the corresponding ST error matrix is given by [cf. (2.49)]

$$\begin{aligned} \boldsymbol{\Delta}_e &= [\boldsymbol{X}_{\rm O}(\boldsymbol{s}) - \boldsymbol{X}_{\rm O}(\boldsymbol{s}')][\boldsymbol{X}_{\rm O}(\boldsymbol{s}) - \boldsymbol{X}_{\rm O}(\boldsymbol{s}')]^{\mathcal{H}} \\ &= \left[\sum_{k=1}^{N_s} \boldsymbol{A}_k(s_k - s_k') + \boldsymbol{B}_k(s_k - s_k')\right] \left[\sum_{k=1}^{N_s} \boldsymbol{A}_k(s_k - s_k') + \boldsymbol{B}_k(s_k - s_k')\right]^{\mathcal{H}} \\ &= \sum_{k=1}^{N_s} |s_k - s_k'|^2 \boldsymbol{I}_{N_t}. \end{aligned} \tag{3.36}$$

Since $\boldsymbol{\Delta}_e$ has full rank, it follows from (2.45) that OSTBCs enable the maximum possible diversity gain of order $G_{d,\max} = N_t N_r$.

According to the equivalent symbol-by-symbol model in (3.32), the effective SNR at the decision point is

$$\gamma_{\text{effective}} = \frac{||\boldsymbol{H}||_F^2}{N_t}\bar{\gamma}, \tag{3.37}$$

based on which the exact BER of an OSTBC system can be derived once the underlying constellation is specified. For example, when QPSK modulation is used, the instantaneous BER can be expressed using Marcum's Q-function as

$$P_b = \mathcal{Q}\left(\sqrt{\frac{||\boldsymbol{H}||_F^2}{N_t}\bar{\gamma}}\right), \tag{3.38}$$

which clearly depends on the channel realization $\boldsymbol{H}$.

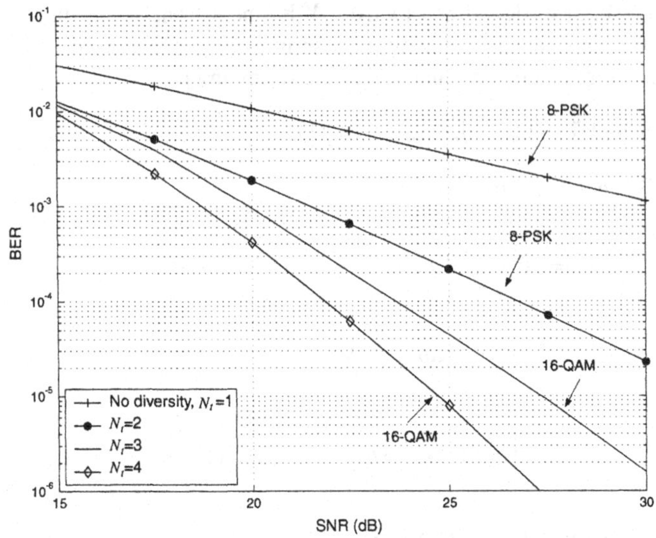

Figure 3.8 BER performance of OSTBCs

As an example, let us consider the simulated average BER of OSTBCs over flat Rayleigh fading channels with $N_t = 2, 3, 4$ transmit-antennas and one receive-antenna. For fairness, 8-PSK is used for $N_t = 2$ and 16-QAM is used for $N_t = 3, 4$, so that transmission rates for different N_t's are equal. Figure 3.8 depicts the average BER as a function of the SNR defined as in (3.26). It is clear that BER performance improves as N_t increases. However, the improvement tends to decrease as N_t becomes large.

3.3.4 Channel Capacity with OSTBCs

In orthogonal ST block coding, the maximum diversity gain is achieved by introducing redundancy among transmitted symbols at different time slots, e.g., $x_{11} = x_{22}^*$ in (3.16). As a consequence, the capacity achieved by an OSTBC system could be less than that predicted by (2.17). How much could this loss be? This has been investigated in [222], and basic results are summarized in this subsection.

As evidenced from (3.32), orthogonal ST coding decouples the transmission of each block $\boldsymbol{s}$ into N_s independent serial transmissions of its entries s_n with effective SNR given by (3.37). Accounting for the rate loss introduced by the OSTBC used, the capacity of an OSTBC system is [222]

$$C_{\text{OSTBC}}(N_t, N_r|\boldsymbol{H}) = \eta_{\text{OSTBC}} \log_2 \left(1 + \frac{\bar{\gamma}}{N_t}||\boldsymbol{H}||_F^2\right) \quad \text{bits per channel use,} \tag{3.39}$$

which is clearly different from $C(N_t, N_r|\boldsymbol{H})$ in (2.17).

Since $\boldsymbol{H}\boldsymbol{H}^{\mathcal{H}}$ is Hermitian, its eigendecomposition can be expressed as

$$\boldsymbol{H}\boldsymbol{H}^{\mathcal{H}} = \boldsymbol{U}\boldsymbol{\Lambda}\boldsymbol{U}^{\mathcal{H}}, \tag{3.40}$$

where $\boldsymbol{U} \in \mathbb{C}^{N_r \times N_r}$ is unitary and $\boldsymbol{\Lambda} \in \mathbb{R}^{N_r \times N_r}$ is a diagonal matrix containing on its diagonal the nonnegative real eigenvalues of $\boldsymbol{H}\boldsymbol{H}^{\mathcal{H}}$. Without loss of generality, these eigenvalues can be sorted in decreasing order,

$$\lambda_1 \geq \lambda_2 \geq \cdots \lambda_{R_H} > \lambda_{(R_H+1)} = \cdots = \lambda_{N_r} = 0,$$

where R_H is the rank of $\boldsymbol{H}\boldsymbol{H}^{\mathcal{H}}$. Substituting (3.40) into (2.17) yields

$$C(N_t, N_r|\boldsymbol{H}) = \log_2 \det\left(\boldsymbol{I}_{N_r} + \frac{\bar{\gamma}}{N_t}\boldsymbol{\Lambda}\right) = \log_2 \prod_{i=1}^{R_H}\left(1 + \frac{\bar{\gamma}}{N_t}\lambda_i\right), \tag{3.41}$$

and after expanding the product, we can rewrite (3.41) as

$$C(N_t, N_r|\boldsymbol{H}) = \log_2\left(1 + \frac{\bar{\gamma}}{N_t}\sum_{i=1}^{R_H}\lambda_i + S\right) = \log_2\left(1 + \frac{\bar{\gamma}}{N_t}||\boldsymbol{H}||_F^2 + S\right), \tag{3.42}$$

where

$$S := \frac{\bar{\gamma}^2}{N_t^2}\sum_{i_1 \neq i_2}^{i_1<i_2}\lambda_{i_1}\lambda_{i_2} + \frac{\bar{\gamma}^3}{N_t^3}\sum_{i_1\neq i_2\neq i_3}^{i_1<i_2<i_3}\lambda_{i_1}\lambda_{i_2}\lambda_{i_3} + \cdots + \frac{\bar{\gamma}^{R_H}}{N_t^{R_H}}\prod_{i=1}^{R_H}\lambda_i. \tag{3.43}$$

From (3.42) and (3.39), the difference $\Delta C := C(N_t, N_r|\boldsymbol{H}) - C_{\text{OSTBC}}(N_t, N_r|\boldsymbol{H})$, representing the potential loss in capacity induced by an OSTBC transmission, is

$$\Delta C = (1 - \eta_{\text{OSTBC}})\log_2\left(1 + \frac{\bar{\gamma}}{N_t}||\boldsymbol{H}||_F^2\right) + \log_2\left(1 + \frac{S}{1 + \frac{\bar{\gamma}}{N_t}||\boldsymbol{H}||_F^2}\right). \tag{3.44}$$

Clearly, ΔC is random since it depends on the underlying channel realization. However, we observe that ΔC is always nonnegative, which implies that an OSTBC system is not necessarily capacity achieving. It is thus of interest to identify the condition under which $\Delta C = 0$.

Let us ignore the trivial case $\boldsymbol{H} = \boldsymbol{0}$ and suppose that $||\boldsymbol{H}||_F^2 > 0$. It is clear from (3.44) that having $\Delta C = 0$ requires two conditions:

(i) $\eta_{\text{OSTBC}} = 1$.

(ii) $S = 0$.

Recalling our discussion by the end of Section 3.3.1, the first condition can be satisfied if and only if $N_t = 2$. Since $||\boldsymbol{H}||_F^2 > 0$, we must have $\lambda_1 > 0$. Because $\lambda_i \geq 0$, it follows that in order to meet the second condition, we must have $\lambda_i = 0$ for $i = 2, \ldots, N_r$. In other words, the rank of $\boldsymbol{H}\boldsymbol{H}^{\mathcal{H}}$ must be 1 for condition (ii) to be satisfied. On the other hand, under the assumption that the $h_{\mu\nu}$'s are uncorrelated, $\boldsymbol{H}\boldsymbol{H}^{\mathcal{H}}$ has rank equal to $\min(N_t, N_r)$ with probability one [222]. Thus, orthogonal ST block coding with more than one receive-antenna will almost surely incur a loss in capacity. Combining conditions (i) and (ii), we deduce that orthogonal ST block coding does not sacrifice channel capacity if and only if $N_t = 2$ and $N_r = 1$. Thus, within the class of OSTBCs, only Alamouti's code with two transmit-antennas and a single receive-antenna is optimal in the sense of achieving channel capacity.

3.4 QUASI-ORTHOGONAL ST BLOCK CODES

OSTBCs enable the maximum possible diversity gain with low-complexity linear ML decoding. Unfortunately, the achievable rate with OSTBCs is relatively low whenever more than two transmit-antennas are deployed with generally complex constellations. In a number of applications, however, achieving high transmission rate can be more important than achieving the maximum diversity gain. These considerations motivate the development of quasi-orthogonal ST block codes (QO-STBCs) which trade diversity gain and decoding simplicity for enhancing transmission rates [125, 259]. We next illustrate the basic idea behind quasi-orthogonal ST coding for the special case of $N_t = 4$ transmit-antennas.

With $\boldsymbol{s} := [s_1, s_2, s_3, s_4]^T$ and $N_t = 4$, the QO-STBC matrix is given by [125]

$$\boldsymbol{X}_{\text{QO}}(\boldsymbol{s}) = \begin{bmatrix} \boldsymbol{O}_{2\times2}(s_1, s_2) & -\boldsymbol{O}^*_{2\times2}(s_3, s_4) \\ \boldsymbol{O}_{2\times2}(s_3, s_4) & \boldsymbol{O}^*_{2\times2}(s_1, s_2) \end{bmatrix} = \begin{bmatrix} s_1 & -s_2^* & -s_3^* & s_4 \\ s_2 & s_1^* & -s_4^* & -s_3 \\ s_3 & -s_4^* & s_1^* & -s_2 \\ s_4 & s_3^* & s_2^* & s_1 \end{bmatrix}, \tag{3.45}$$

where we notice that $N_s = N_x = 4$. By computing the error matrix [cf. (2.49)]

$$\boldsymbol{\Delta}_e := [\boldsymbol{X}_{\text{QO}}(\boldsymbol{s}) - \boldsymbol{X}_{\text{QO}}(\boldsymbol{s}')][\boldsymbol{X}_{\text{QO}}(\boldsymbol{s}) - \boldsymbol{X}_{\text{QO}}(\boldsymbol{s}')]^{\mathcal{H}},$$

for any $\boldsymbol{s} \neq \boldsymbol{s}'$, it can be easily seen that the minimum rank of $\boldsymbol{\Delta}_e$ is only 2. Thus, the diversity gain offered by this code is $2N_r$, which is half of the maximum possible diversity gain $4N_r$. On the other hand, since four information symbols s_n, $n = 1, 2, 3, 4$, are transmitted in four time-slots, the rate of this code is one symbol per channel use. Recalling that when $N_t = 4$ the OSTBC achieves the maximum diversity gain at rate $3/4$, we recognize that QO-STBCs indeed trade-off diversity gain for improved code rate.

The input-output relationship of a QO-STBC system is [cf. (2.9)]

$$\boldsymbol{Y} = \sqrt{\frac{\bar{\gamma}}{N_t}} \boldsymbol{H} \boldsymbol{X}_{\text{QO}}(\boldsymbol{s}) + \boldsymbol{W}. \tag{3.46}$$

Because $\boldsymbol{X}_{\text{QO}}(\boldsymbol{s})$ is not orthogonal as in (3.15), the low-complexity linear ML decoding scheme of OSTBCs [cf. Section 3.3.2] is no longer applicable here. Again, let us suppose that the entries of $\boldsymbol{W}$ are zero mean, uncorrelated, and complex Gaussian distributed. Given $\boldsymbol{Y}$ and $\boldsymbol{H}$, ML decoding of $\boldsymbol{s}$ from $\boldsymbol{Y}$ in (3.46) amounts to

$$\hat{\boldsymbol{s}} = \arg\min_{\forall \boldsymbol{s}} d^2(\boldsymbol{s}), \tag{3.47}$$

where the decision metric is

$$d^2(\boldsymbol{s}) = \mathrm{tr}\left[\left(\boldsymbol{Y} - \sqrt{\frac{\bar{\gamma}}{N_t}}\boldsymbol{H}\boldsymbol{X}_{\mathrm{QO}}(\boldsymbol{s})\right)\left(\boldsymbol{Y} - \sqrt{\frac{\bar{\gamma}}{N_t}}\boldsymbol{H}\boldsymbol{X}_{QO}(\boldsymbol{s})\right)^{\mathcal{H}}\right]. \tag{3.48}$$

Recalling the code structure in (3.45), it is easy to verify the following properties:

$$\begin{aligned} &[\boldsymbol{X}_{\mathrm{QO}}]_1^{\mathcal{H}}[\boldsymbol{X}_{\mathrm{QO}}]_2 = 0, \quad [\boldsymbol{X}_{\mathrm{QO}}]_1^{\mathcal{H}}[\boldsymbol{X}_{\mathrm{QO}}]_3 = 0, \\ &[\boldsymbol{X}_{\mathrm{QO}}]_2^{\mathcal{H}}[\boldsymbol{X}_{\mathrm{QO}}]_4 = 0, \quad [\boldsymbol{X}_{\mathrm{QO}}]_3^{\mathcal{H}}[\boldsymbol{X}_{\mathrm{QO}}]_4 = 0, \end{aligned} \tag{3.49}$$

where $[\boldsymbol{X}_{\mathrm{QO}}]_i$ denotes the ith column of $\boldsymbol{X}_{\mathrm{QO}}$. Capitalizing on these properties, the decision metric in (3.48) can be decomposed in two summands: $d^2(\boldsymbol{s}) = d^2(s_1, s_4) + d^2(s_2, s_3)$, where [125]

$$\begin{aligned} d^2(s_1, s_4) = \sum_{\nu=1}^{N_r}\Bigg[&\left(\sum_{\mu=1}^{4}|h_{\mu\nu}|^2\right)(|s_1|^2 + |s_4|^2) \\ &+ 2\Re\Big\{(-h_{1\nu}y_{\nu1}^* - h_{2\nu}^* y_{\nu2} - h_{3\nu}^* y_{\nu3} - h_{4\nu}y_{\nu4}^*)s_1 \\ &+ (-h_{4\nu}y_{\nu1}^* + h_{3\nu}^* y_{\nu2} + h_{2\nu}^* y_{\nu3} - h_{1\nu}y_{\nu4}^*)s_4 \\ &+ (-h_{1\nu}h_{\nu4}^* - h_{2\nu}^* h_{3\nu} - h_{2\nu}h_{3\nu} + h_{1\nu}^* h_{4\nu})s_1 s_4^*\Big\}\Bigg], \end{aligned} \tag{3.50}$$

$$\begin{aligned} d^2(s_2, s_3) = \sum_{\nu=1}^{N_r}\Bigg[&\left(\sum_{\mu=1}^{4}|h_{\mu\nu}|^2\right)(|s_2|^2 + |s_3|^2) \\ &+ 2\Re\Big\{(-h_{2\nu}y_{\nu1}^* + h_{1\nu}^* y_{\nu2} - h_{4\nu}^* y_{\nu3} + h_{3\nu}y_{\nu4}^*)s_2 \\ &+ (-h_{3\nu}y_{\nu1}^* - h_{4\nu}^* y_{\nu2} + h_{1\nu}^* y_{\nu3} + h_{2\nu}y_{\nu4}^*)s_3 \\ &+ (h_{2\nu}h_{3\nu}^* - h_{1\nu}^* h_{4\nu} - h_{1\nu}h_{4\nu}^* + h_{2\nu}^* h_{3\nu})s_2 s_3^*\Big\}\Bigg]. \end{aligned} \tag{3.51}$$

Since $d^2(s_1, s_4)$ does not depend on s_2 and s_3 and $d^2(s_2, s_3)$ is not a function of s_1 and s_4, the minimization of $d^2(\boldsymbol{s})$ in (3.47) can be performed by minimizing $d^2(s_1, s_4)$ and $d^2(s_2, s_3)$ separately. As a result, the decoding complexity of QO-STBCs can be reduced considerably, although it is still higher than that of OSTBCs. Notwithstanding, this reduction does not sacrifice error performance.

For a QO-STBC transmission to enable the maximum diversity provided by the MIMO channel, a constellation rotation approach has been developed in [233, 243], where half of the information symbols use the original signal constellation and the rest rely on a rotated constellation. A general class of QO-STBC transmission systems is provided in [234].

3.5 ST LINEAR COMPLEX FIELD CODES

STTCs, OSTBCs, and QO-STBCs offer various trade-offs among error performance, bandwidth efficiency, and decoding complexity. One limitation common to all these codes is that their designs are not very flexible. Part of the reason is that their designs involve difference matrices between all possible distinct ST code matrices. In what follows, we describe a different class of ST codes, ST linear complex field codes (STLCFCs). We will see that STLCFCs entail systematically constructed code matrices and as such, they can be more flexible.

What we refer to as linear complex field (LCF) codes were first studied for single-antenna systems independently (up to 2001) by coding theorists (as early as 1992 [28]), and by researchers in the signal processing for communications community (starting with the filterbank precoders of [90] in 1997). Specifically, Belfiore and co-workers investigated lattice codes that rotate PAM or QAM constellations to gain what they termed *signal diversity*, when communicating over single-input single-output (SISO) flat fading channels [30,94,95,140]. In parallel, Giannakis and co-workers advocated (non-) redundant *linear precoding*, which amounts to multiplying a block of symbols with a (square) tall Vandermonde matrix $\boldsymbol{\Theta}$ to enable (even with linear FIR equalizers) what they termed *symbol detectability*, of block transmissions over SISO, MIMO, and multiuser frequency-selective finite impulse response (FIR) channels, regardless of the channel zero locations and irrespective of the underlying constellation (when $\boldsymbol{\Theta}$ is tall) [93, 158, 225, 286].

The link came in 2001, when it was recognized that these Vandermonde matrices can implement existing lattice codes and nicely complement Galois field (GF) codes [288]. They enable coding gains and the maximum diversity that can be provided not only by flat fading channels, but also by frequency- and time-selective SISO channels [163,180,183,288]. We will see in Chapters 7, 8, and 9 how LCF codes can be utilized in ST transmissions over these challenging propagation environments. Being square, these Vandermonde matrices constitute *bandwidth efficient* block coders, particularly attractive for fading channels; their entries are drawn from the complex field and are available in *closed form* that depends on their size and the constellation they precode. We will henceforth call them LCF codes, to delineate the features they differ and share with traditional GF codes.

Since LCF codes improve the reliability of SISO flat fading channels when transmissions are properly interleaved [30] and multiple transmit-antennas can implement interleaving (as we saw in Chapter 2), it is natural to explore the usefulness of LCF codes in ST transmissions over MIMO channels. For MIMO flat fading channels, STLCFCs were introduced by [314], and independently by [50], to enable the maximum diversity at 1 symbol per channel use, for *any* number of antennas (see also [313] reporting the first coding gain analysis of such LCFCs). For MIMO frequency- and time-selective channels, space, multipath, and Doppler diversity modes were enabled via LCF coding in [91, 162, 181, 184, 332], but again at transmission rates not exceeding 1 symbol per channel use. For $N_t = 2$ antennas, maximum diversity designs achieving 2 symbols per channel use were proposed in [51,53,260]. The first combination of LCF codes with a layered ST multiplexer trading off between space diversity and rate for any N_t, N_r appeared in [312] for flat MIMO channels, and with maximum diversity in [311] for frequency-selective MIMO channels. Shortly after, related combinations enabling full diversity and full rate (FDFR) for MIMO flat fading channels were reported in [74, 77, 196], and independently in [179] (see also [178, Section 4.2.1]).

In this section we outline error performance-oriented STLCFC designs attaining rates of 1 symbol per channel use. Rate-oriented STLCFC alternatives capable of reaching the MIMO capacity (which scales with the $\min(N_t, N_r)$ as we saw in Chapter 2) are discussed in Chapter 4 along with the emerging diversity-rate trade-offs.

3.5.1 Antenna Switching and Linear Precoding

The application of STLCFCs relies on the idea of antenna switching which was discussed early in [206]. Antenna switching plays a role similar to that of interleaving. When antenna switching is invoked at any time n, only the transmit-antenna with index $[n \pmod{N_t} + 1]$ is switched on, while the rest are turned off. Under the ST model in Section 2.1, antenna switching can be effected by choosing $N_x = N_t$ and setting to zero the nondiagonal entries of the STLCF code matrix $\boldsymbol{X}_{\text{LCF}}$. Because $\boldsymbol{X}_{\text{LCF}}$ is diagonal, it can be specified by its diagonal entries, which we denote as

$$\boldsymbol{X}_{\text{LCF}} = \text{diag}\left(\boldsymbol{x}_{\text{LCF}}\right). \tag{3.52}$$

Consequently, designing $\boldsymbol{X}_{\text{LCF}}$ amounts to selecting the vector-vector mapping

$$\mathcal{M}_{LC}: \quad \boldsymbol{s} \longleftrightarrow \boldsymbol{x}_{\text{LCF}}, \tag{3.53}$$

which will turn out to be much simpler than designing the vector-matrix mapping in (2.2). In ST linear complex field coding, the mapping in (3.53) will be further specialized to be a linear mapping. In this case, $\boldsymbol{x}_{\text{LCF}}$ is related to $\boldsymbol{s}$ via an $N_t \times N_t$ matrix that we term the linear precoding matrix $\boldsymbol{\Theta}_{\text{LCF}}$:

$$\boldsymbol{x}_{\text{LCF}} = \boldsymbol{\Theta}_{\text{LCF}}\boldsymbol{s}, \tag{3.54}$$

where $\boldsymbol{s}$ has length $N_s = N_t$. As a result, the ST linear complex field code matrix $\boldsymbol{X}_{\text{LCF}}$ can be written as

$$\boldsymbol{X}_{\text{LCF}}(\boldsymbol{s}) = \text{diag}(\boldsymbol{\Theta}_{\text{LCF}}\boldsymbol{s}) = \begin{bmatrix} \boldsymbol{\theta}_1^T\boldsymbol{s} & 0 & 0 & 0 \\ 0 & \boldsymbol{\theta}_2^T\boldsymbol{s} & 0 & 0 \\ \vdots & \vdots & \ddots & \vdots \\ 0 & \cdots & \cdots & \boldsymbol{\theta}_{N_t}^T\boldsymbol{s} \end{bmatrix}, \tag{3.55}$$

with $\boldsymbol{\theta}_i^T$ denoting the ith row of $\boldsymbol{\Theta}_{\text{LCF}}$. Because $\boldsymbol{X}_{\text{LCF}}$ is diagonal, the power constraint in (2.1) can be translated to a power constraint on $\boldsymbol{\Theta}_{\text{LCF}}$ as

$$\text{tr}(\boldsymbol{\Theta}_{\text{LCF}}\boldsymbol{\Theta}_{\text{LCF}}^{\mathcal{H}}) = N_t. \tag{3.56}$$

Due to the fact that $N_x = N_s$, the rate of STLCFCs is

$$\eta_{\text{STLCFC}} = 1 \quad \text{symbol per channel use}, \tag{3.57}$$

which coincides with that of a single-antenna system.

3.5.2 Designing Linear Precoding Matrices

An STLCF code is specified completely by the linear precoding matrix in (3.54). This section presents design criteria for selecting the precoding matrix $\boldsymbol{\Theta}_{\mathrm{LCF}}$ so that both diversity and coding gains are maximized. To this end, let us consider two distinct information blocks $\boldsymbol{s}$ and $\boldsymbol{s}'$. Substituting (3.55) into (2.49), the error matrix

$$\boldsymbol{\Delta}_e := [\boldsymbol{X}_{\mathrm{LCF}}(\boldsymbol{s}) - \boldsymbol{X}_{\mathrm{LCF}}(\boldsymbol{s}')]\,[\boldsymbol{X}_{\mathrm{LCF}}(\boldsymbol{s}) - \boldsymbol{X}_{\mathrm{LCF}}(\boldsymbol{s}')]^{\mathcal{H}}$$

can be written as

$$\boldsymbol{\Delta}_e = \begin{bmatrix} |\boldsymbol{\theta}_1^T(\boldsymbol{s}-\boldsymbol{s}')|^2 & 0 & 0 & 0 \\ 0 & |\boldsymbol{\theta}_2^T(\boldsymbol{s}-\boldsymbol{s}')|^2 & 0 & 0 \\ \vdots & \vdots & \ddots & \vdots \\ 0 & \cdots & \cdots & |\boldsymbol{\theta}_{N_t}^T(\boldsymbol{s}-\boldsymbol{s}')|^2 \end{bmatrix}. \tag{3.58}$$

With $\boldsymbol{\Delta}_e$ being diagonal, the precoding matrix $\boldsymbol{\Theta}_{\mathrm{LCF}}$ should obey the following criteria that can be readily obtained from c2.1) and c2.2) in Chapter 2:

c3.1) To maximize the diversity gain, the $N_t \times N_t$ matrix $\boldsymbol{\Theta}_{\mathrm{LCF}}$ should be designed so that

$$|\boldsymbol{\theta}_i^T(\boldsymbol{s}-\boldsymbol{s}')| \neq 0, \quad \forall i \in [1, N_t],\ \forall \boldsymbol{s} \neq \boldsymbol{s}' \in \mathbb{S}^{N_t \times 1}. \tag{3.59}$$

c3.2) To maximize the coding gain, the $N_t \times N_t$ matrix $\boldsymbol{\Theta}_{\mathrm{LCF}}$ should be designed to maximize the minimum product distance

$$d_{p,\min}^2 = \min_{\forall\, \boldsymbol{s} \neq \boldsymbol{s}' \in \mathbb{S}^{N_t \times 1}} \prod_{i=1}^{N_t} |\boldsymbol{\theta}_i^T(\boldsymbol{s}-\boldsymbol{s}')|^2 . \tag{3.60}$$

If c3.2) is met, then c3.1) will be satisfied automatically. Notice also that the resulting precoder $\boldsymbol{\Theta}_{\mathrm{LCF}}$ will generally depend on the underlying constellation $\mathbb{S}$.

3.5.3 Upper Bound on Coding Gain

In general, deriving the maximum coding gain of ST codes in closed form is difficult. However, this is not the case for STLCFCs, thanks mainly to the linearity present in their encoding process. As we discussed in Section 2.4, the maximum coding gain is instrumental in assessing performance of the code design.

With θ_{ij} denoting the (i,j)th entry of $\boldsymbol{\Theta}_{\mathrm{LCF}}$, it follows from (3.56) that

$$\sum_{j=1}^{N_t} \sum_{i=1}^{N_t} |\theta_{ij}|^2 = N_t,$$

from which we deduce that $\exists\, p \in [1, N_t]$, so that

$$\sum_{i=1}^{N_t} |\theta_{ip}|^2 \leq 1. \tag{3.61}$$

Consider two distinct information blocks $\boldsymbol{s}_1$, $\boldsymbol{s}_1'$ and let $\boldsymbol{s}_1 - \boldsymbol{s}_1' := \Delta_{\min}\boldsymbol{e}_p$, where $\Delta_{\min}$ is the minimum distance among constellation points in $\mathbb{S}$ and $\boldsymbol{e}_p$ is the p-th column of the identity matrix. Because the minimization in (3.60) is over all possible $\boldsymbol{s}$ and $\boldsymbol{s}'$, it follows that $d_{p,\min}^2$ can be upper bounded by

$$d_{p,\min}^2 \leq \prod_{i=1}^{N_t} |\boldsymbol{\theta}_i^T(\boldsymbol{s}_1 - \boldsymbol{s}_1')|^2 = \Delta_{\min}^{2N_t} \prod_{i=1}^{N_t} |\theta_{ip}|^2. \tag{3.62}$$

Using the arithmetic-geometric mean inequality and (3.61), $d_{p,\min}^2$ can be further bounded by

$$d_{p,\min}^2 \leq \Delta_{\min}^{2N_t} \left(\frac{\sum_{i=1}^{N_t} |\theta_{ip}|^2}{N_t} \right)^{N_t} \leq \left(\frac{\Delta_{\min}^2}{N_t} \right)^{N_t}. \tag{3.63}$$

Since the right hand side of (3.63) does not depend on $\boldsymbol{\Theta}_{\mathrm{LCF}}$, we deduce that the maximum coding gain achieved by STLCFCs is upper bounded as

$$G_{c,\mathrm{LCF}} \leq \frac{\Delta_{\min}^2}{N_t}. \tag{3.64}$$

Compared to (2.53), we infer that STLCFCs basically trade coding gain for diversity gain, which is more crucial when it comes to improving the reliability of wireless transmissions over the practical range of SNR values.

3.5.4 Construction Based on Parameterization

Besides the power constraint (3.56), the precoding matrix $\boldsymbol{\Theta}_{\mathrm{LCF}}$ can be further constrained to be unitary. Unitary precoding (UP) offers a distinct advantage over nonunitary alternatives: whereas a nonunitary $\boldsymbol{\Theta}_{\mathrm{LCF}}$ draws some pairs of constellation points closer and some farther, a unitary $\boldsymbol{\Theta}_{\mathrm{LCF}}$ corresponds to a rotation and thereby preserves the Euclidean distance among the N_t-dimensional signal points $\boldsymbol{s}$. This distance-preserving property is essential because it ensures that STLCF coding does not degrade error performance when the channels are AWGN (or near-AWGN). Since wireless channels vary between AWGN and Rayleigh fading channels, UP is particularly useful in practice.

Prior to designing unitary precoders, it is natural to ask whether the class of unitary precoders is rich enough to contain linear precoders achieving at least the maximum possible diversity gain. An affirmative answer is provided by the following proposition proved in [313]:

Proposition 3.1 *For a finite constellation* $\mathbb{S}$, *there always exists at least one unitary precoder* $\boldsymbol{\Theta}_{\mathrm{LCF}}$ *that satisfies* (3.59) *and thus achieves the maximum possible diversity gain of order* N_tN_r.

Ensured by Proposition 3.1, we can search for a unitary precoder $\boldsymbol{\Theta}_{\mathrm{LCF}}$ that yields maximum $d_{p,\min}$ in (3.60) among those diversity-maximizing unitary precoders. The reason that we look for complex unitary precoders instead of real orthogonal ones (which were considered in [54, 216]) is that unitary complex precoders have the potential to achieve coding gain at least as high as those of real orthogonal ones. As formulated in (3.60), searching for an optimal $\boldsymbol{\Theta}_{\text{LCF}}$ involves multidimensional nonlinear optimization over N_t^2 complex entries. To facilitate such a search, one can exploit the fact that any $N_t \times N_t$ unitary matrix can be parameterized by using $N_t(N_t - 1)$ parameters taken from finite intervals. Specifically, it is possible to express $\boldsymbol{\Theta}_{\text{LCF}}$ as a product of $N_t(N_t - 1)/2$ Givens matrices [99]; that is,

$$\boldsymbol{\Theta}_{\text{LCF}} = \prod_{p=1}^{N_t-1} \prod_{q=p+1}^{N_t} \boldsymbol{G}_{pq}(\psi_{pq}, \phi_{pq}), \tag{3.65}$$

where parameters ψ_{pq} and ϕ_{pq} take values over $[-\pi, \pi]$, and $\boldsymbol{G}_{pq}(\psi_{pq}, \phi_{pq})$ is the $N_t \times N_t$ Givens matrix which is the identity matrix $\boldsymbol{I}_{N_t}$ with the (p, p)th, (q, q)th, (p, q)th, and (q, p)th entries replaced by $\cos\psi_{pq}$, $\cos\psi_{pq}$, $e^{-j\phi_{pq}}\sin\psi_{pq}$, and $-e^{-j\phi_{pq}}\sin\psi_{pq}$, respectively. For example, when $N_t = 2$, $\boldsymbol{\Theta}_{\text{LCF}}$ can be parameterized as

$$\boldsymbol{\Theta}_{\text{LCF}} = \begin{bmatrix} \cos\psi & e^{-j\phi}\sin\psi \\ -e^{j\phi}\sin\psi & \cos\psi \end{bmatrix}, \tag{3.66}$$

which is a function of only two parameters, ϕ and ψ.

With $\boldsymbol{\Theta}_{\text{LCF}}$ expressed in (3.65), $d_{p,\min}^2$ in (3.60) needs to be maximized with respect to $N_t(N_t - 1)$ parameters over the finite interval $[-\pi, \pi]$. Exhaustive search then becomes computationally feasible for small N_t and $|\mathbb{S}|$. For example, when $N_t = 3$ and QPSK modulation is employed, the resulting optimal precoder is given by

$$\boldsymbol{\Theta}_{\text{LCF}} = \begin{bmatrix} 0.687 & 0.513 - 0.113j & -0.428 + 0.264j \\ -0.358 - 0.308j & 0.696 - 0.172j & -0.011 - 0.513j \\ 0.190 + 0.520j & 0.243 - 0.389j & 0.696 \end{bmatrix}. \tag{3.67}$$

3.5.5 Construction Based on Algebraic Tools

The exhaustive search following the parameterization of unitary precoding matrices incurs high computational complexity and thus is not feasible when either N_t or $|\mathbb{S}|$ is large. Considering this, a number of algebraic construction methods have been proposed to yield closed-form precoders with reasonably large coding gains even when N_t and/or $|\mathbb{S}|$ is large. In what follows we describe two construction methods proposed in [313], which we hereafter abbreviate as LCF-A and LCF-B.

3.5.5.1 Algebraic Construction: LCF-A This approach yields a matrix $\boldsymbol{\Theta}_{\text{LCF}}$ which applies to any number of transmit-antennas and coincides with those reported in [29, 95, 96] when N_t is a power of 2 and the resulting $\boldsymbol{\Theta}_{\text{LCF}}$ is unitary. When N_t is not a power of 2, this method yields nonunitary $\boldsymbol{\Theta}_{\text{LCF}}$ matrices.

Let $\mathbb{Z}[j]$ denote the ring of Gaussian integers whose elements are in the form of $p + jq$, with both p and q belonging to an integer ring, $\mathbb{Q}(j)$ the smallest subfield of complex number field including both $\mathbb{Q}$ and j, and $m_{\alpha,\mathbb{Q}(j)}(x)$ the minimal polynomial

Table 3.3 DESIGN EXAMPLES OF $\boldsymbol{\Theta}_{\text{LCF}}$ FOR $N_t = 2, 3, 4, 5, 6$

N_t	α_1	α_2	α_3	α_4	α_5	α_6
2	$e^{-j\frac{\pi}{4}}$	$e^{-j\frac{5\pi}{4}}$				
3	$\sqrt[3]{2}\,e^{-j\frac{\pi}{12}}$	$\sqrt[3]{2}\,e^{-j\frac{9\pi}{12}}$	$\sqrt[3]{2}\,e^{-j\frac{17\pi}{12}}$			
4	$e^{-j\frac{\pi}{8}}$	$e^{-j\frac{5\pi}{8}}$	$e^{-j\frac{9\pi}{8}}$	$e^{-j\frac{13\pi}{8}}$		
5	$\sqrt[5]{2}\,e^{-j\frac{\pi}{20}}$	$\sqrt[5]{2}\,e^{-j\frac{9\pi}{20}}$	$\sqrt[5]{2}\,e^{-j\frac{17\pi}{20}}$	$\sqrt[5]{2}\,e^{-j\frac{25\pi}{20}}$	$\sqrt[5]{2}\,e^{-j\frac{33\pi}{20}}$	
6	$e^{-j\frac{2\pi}{7}}$	$e^{-j\frac{4\pi}{7}}$	$e^{-j\frac{6\pi}{7}}$	$e^{-j\frac{8\pi}{7}}$	$e^{-j\frac{10\pi}{7}}$	$e^{-j\frac{12\pi}{7}}$

of α over $\mathbb{Q}(j)$ with degree $\deg(m_{\alpha,\mathbb{Q}(j)}(x))$. By choosing α as an integer over $\mathbb{Z}[j]$ such that $\deg(m_{\alpha,\mathbb{Q}(j)}(x)) = N_t$, LCF-A constructs $\boldsymbol{\Theta}_{\text{LCF}}$ as

$$\boldsymbol{\Theta}_{\text{LCF}} = \frac{1}{\lambda}\begin{bmatrix} 1 & \alpha_1 & \cdots & \alpha_1^{N_t-1} \\ 1 & \alpha_2 & \cdots & \alpha_2^{N_t-1} \\ \vdots & \vdots & & \vdots \\ 1 & \alpha_{N_t} & \cdots & \alpha_{N_t}^{N_t-1} \end{bmatrix}, \tag{3.68}$$

where $\{\alpha_k\}_{k=1}^{N_t}$ are the roots of $m_{\alpha,\mathbb{Q}(j)}(x)$ with $\alpha_1 = \alpha$, and λ is a normalization factor used to satisfy the power constraint (3.56).

Table 3.3 summarizes the designed α_k's for $N_t = 2, 3, 4, 5, 6$. For example, STLCF codes for $N_t = 2, 4$ are given by

$$\boldsymbol{\Theta}_{\text{LCF}} = \frac{1}{\sqrt{2}}\begin{bmatrix} 1 & e^{-j\frac{\pi}{4}} \\ 1 & e^{-j\frac{5\pi}{4}} \end{bmatrix}, \qquad N_t = 2,$$
$$\boldsymbol{\Theta}_{\text{LCF}} = \frac{1}{2}\begin{bmatrix} 1 & e^{-j\frac{\pi}{8}} & e^{-j\frac{2\pi}{8}} & e^{-j\frac{3\pi}{8}} \\ 1 & e^{-j\frac{5\pi}{8}} & e^{-j\frac{10\pi}{8}} & e^{-j\frac{15\pi}{8}} \\ 1 & e^{-j\frac{9\pi}{8}} & e^{-j\frac{18\pi}{8}} & e^{-j\frac{27\pi}{8}} \\ 1 & e^{-j\frac{13\pi}{8}} & e^{-j\frac{26\pi}{8}} & e^{-j\frac{39\pi}{8}} \end{bmatrix}, \qquad N_t = 4. \tag{3.69}$$

It is worth mentioning that since the constructed $\boldsymbol{\Theta}_{\text{LCF}}$ may not be unitary when N_t is not a power of 2, STLCF coding in (3.54) should not be interpreted as constellation rotation in general.

The constructed STLCF codes in (3.68) enable maximum diversity and coding gains. It is interesting to compare the coding gain achieved with its upper bound in (3.64). For QAM or PAM constellation, the following proposition quantifies the coding gain achieved by LCF-A [313]:

Proposition 3.2 *For QAM or PAM constellations, the minimum product distance $d_{p,\min}^2$ in (3.60) achieved with $\boldsymbol{\Theta}_{\text{LCF}}$ precoders constructed as in (3.68) is*

$$d_{p,\min}^2 = \left[\frac{\Delta_{\min}^2}{\lambda^2}\right]^{N_t}, \tag{3.70}$$

where

$$\lambda^2 = \begin{cases} N_t & N_t \in \mathbb{N}_1 \cup \mathbb{N}_2, \\ \frac{1}{2^{1/N_t}-1} & \text{otherwise}, \end{cases}$$

$\mathbb{N}_1$ *denotes the set of Euler numbers*[1] $\phi(M)$ *for all positive integers* $P(mod\ 4) \neq 0$, *and* $\mathbb{N}_2$ *stands for the set of integers that are positive powers of 2.*

Recalling (3.63), one can infer from Proposition 3.2 that for QAM (or PAM) constellations, STLCF codes obtained by LCF-A achieve the maximum possible diversity and coding gains when $N_t \in \mathbb{N}_1 \cup \mathbb{N}_2$, and they effect the maximum possible diversity gain but do not achieve the upper bound of the coding gain when $N_t \notin \mathbb{N}_1 \cup \mathbb{N}_2$. However, since $(2^{1/N_t} - 1) \geq (\ln 2)/N_t$, they can achieve at least 70% of the upper bound.

3.5.5.2 *Algebraic Construction: LCF-B* The method LCF-B constructs unitary $\boldsymbol{\Theta}_{\text{LCF}}$ precoders for any N_t. The STLCF codes designed in LCF-B are given by

$$\boldsymbol{\Theta}_{\text{LCF}} = \boldsymbol{F}_{N_t}^{\mathcal{H}} \text{diag}(1, \alpha, \ldots, \alpha^{N_t-1}), \tag{3.71}$$

where $\boldsymbol{F}_{N_t}$ is the normalized $N_t \times N_t$ discrete Fourier transform (DFT) matrix, and α is a complex number. Another form of $\boldsymbol{\Theta}_{\text{LCF}}$ is

$$\boldsymbol{\Theta}_{\text{LCF}} = \boldsymbol{F}_{N_t} \text{diag}(1, \alpha^*, \ldots, (\alpha^*)^{N_t-1}), \tag{3.72}$$

which is equivalent to (3.71) in the sense that both lead to the same BER. In what follows, we discuss how to choose α.

Let I denote the number of distinct minimal polynomials $p_i(x)$ of $\beta_m = \alpha e^{j2\pi(m-1)/N_t}$, $m = 1, \ldots, N_t$ over $\mathbb{Q}(j)$ and D_i the degree of $p_i(x)$. To improve the coding gain as much as possible, one should choose $\alpha = e^{j2\pi/P}$ such that I is small and D_i's are as low as possible, to make $\mathcal{X} := \sum_{i=1}^{I} D_i/N_t$ small. For the special case when N_t is a power of 2, we can select

$$\alpha = e^{j\frac{\pi}{2N_t}}, \tag{3.73}$$

such that all β_m's are roots of the minimum polynomial $x^{N_t} - j$ over $\mathbb{Q}(j)$. In this case, we have $I = 1$ and $\mathcal{X} = 1$. For the case when N_t is not a power of 2, one heuristic rule is to choose $\alpha = e^{j2\pi/P}$ with $P = lN_t \in \mathbb{N}_1 \cup \mathbb{N}_2$ for some positive integer l such that most β_m's are roots of $m_{\alpha,\mathbb{Q}(j)}(x)$. This will make $\mathcal{X}$ small and $D_i \geq N_t$. As an example, for $N_t = 3, 5$, one can choose $P = 6, 10$, respectively.

3.5.6 Decoding ST Linear Complex Field Codes

Design criteria c3.1) and c3.2) were derived based on the assumption that ML decoding is performed. Recalling (2.9), the input-output relationship of an STLCF coded

[1] Euler number $\phi(M)$ denotes the number of positive integers that are less than M and relatively prime to M. For example, $\phi(6) = 2$ and $\phi(7) = 6$.

system can be expressed as

$$Y = \sqrt{\frac{\bar{\gamma}}{N_t}} H X_{\mathrm{LCF}}(s) + W. \tag{3.74}$$

Clearly, ML decoding of s from Y requires testing $|\mathbb{S}|^{N_t}$ possible vectors s, which is computationally prohibitive if either N_t or/and $|\mathbb{S}|$ is large. In what follows we describe how it is possible to use the sphere decoding algorithm (SDA), which is detailed in Chapter 5, to decode s with ML or near-ML performance at decoding complexity which can be polynomial in N_t. The computational complexity of SDA certainly depends on N_t, but also on the constellation size $|\mathbb{S}|$, the SNR, and the rank of H. Nonetheless, its average complexity remains approximately cubic for ranges of these parameters encountered in a number of practical settings.

Exploiting the diagonal structure of $X_{\mathrm{LCF}}(s)$, we can first rewrite (3.74) in an equivalent form,

$$\bar{y} = \sqrt{\frac{\bar{\gamma}}{N_t}} \bar{H} \Theta_{\mathrm{LCF}} s + \bar{w}, \tag{3.75}$$

where the $N_r N_t \times 1$ blocks $\bar{y}$ and $\bar{w}$ are formed by respectively stacking N_r columns of Y^T and W^T into a single column, and the $N_r N_t \times N_t$ matrix $\bar{H}$ is a block diagonal matrix defined as

$$\bar{H} := \begin{bmatrix} h_{11} & 0 & \cdots & 0 \\ 0 & h_{12} & \cdots & 0 \\ \vdots & \vdots & \ddots & \vdots \\ 0 & 0 & \cdots & h_{1N_t} \\ \vdots & \vdots & \vdots & \vdots \\ h_{N_r 1} & 0 & \cdots & 0 \\ 0 & h_{N_r 2} & \cdots & 0 \\ \vdots & \vdots & \ddots & \vdots \\ 0 & 0 & \cdots & h_{N_r N_t} \end{bmatrix}. \tag{3.76}$$

The received vector in (3.75) is equivalent to a received block from N_t transmit-antennas to $N_t N_r$ receive-antennas, with the channel matrix $\bar{H}$ being almost always full rank. To decode s from $\bar{y}$, we can first apply MRC and then perform noise whitening to obtain $\tilde{y} := (\bar{H}^{\mathcal{H}} \bar{H})^{-1/2} \bar{H}^{\mathcal{H}} \bar{y}$ as

$$\tilde{y} = \sqrt{\frac{\bar{\gamma}}{N_t}} \tilde{H} \Theta_{\mathrm{LCF}} s + \tilde{w}, \tag{3.77}$$

where $\tilde{w} := (\bar{H}^{\mathcal{H}} \bar{H})^{-1/2} \bar{H}^{\mathcal{H}} \bar{w}$ is still AWGN, and

$$\tilde{H} := (\bar{H}^{\mathcal{H}} \bar{H})^{1/2} = \mathrm{diag}\left(\sqrt{\sum_{\nu=1}^{N_r} |h_{\nu 1}|^2}, \cdots, \sqrt{\sum_{\nu=1}^{N_r} |h_{\nu N_t}|^2} \right).$$

Since both $\tilde{\boldsymbol{y}}$ and $\boldsymbol{s}$ are generally complex, we separate them into their real and imaginary parts and rewrite (3.77) as

$$\begin{bmatrix} \Re(\tilde{\boldsymbol{y}}) \\ \Im(\tilde{\boldsymbol{y}}) \end{bmatrix} = \sqrt{\frac{\bar{\gamma}}{N_t}} \begin{bmatrix} \Re(\tilde{\boldsymbol{H}}\boldsymbol{\Theta}_{\mathrm{LCF}}) & -\Im(\tilde{\boldsymbol{H}}\boldsymbol{\Theta}_{\mathrm{LCF}}) \\ \Im(\tilde{\boldsymbol{H}}\boldsymbol{\Theta}_{\mathrm{LCF}}) & \Re(\tilde{\boldsymbol{H}}\boldsymbol{\Theta}_{\mathrm{LCF}}) \end{bmatrix} \begin{bmatrix} \Re(\boldsymbol{s}) \\ \Im(\boldsymbol{s}) \end{bmatrix} + \begin{bmatrix} \Re(\tilde{\boldsymbol{w}}) \\ \Im(\tilde{\boldsymbol{w}}) \end{bmatrix}. \quad (3.78)$$

Example 3.1 As we mentioned earlier, decoding $\boldsymbol{s}$ from (3.78) with ML or near-ML optimality at possibly polynomial complexity can be accomplished using the SDA [66, 276], which we describe in detail in Chapter 5. Certainly, linear equalizers or block decision-feedback equalizers are possible alternatives that can afford complexity which is cubic in N_t, but their error performance is often far from ML. The error performance of STLCFCs is compared in Figures 3.9 and 3.10 with that of OSTBCs in (3.16), (3.17), and (3.18) for $N_t = 2, 3, 4$ and $N_r = 1$. To ensure fairness in terms of the transmission rate, 4-QAM is used for both STLCFC and OSTBC when $N_t = 2$, while 256-QAM is used for OSTBC and 16-QAM is used in STLCFC when $N_t = 3, 4$. The performance of STLCFC is also compared in Figure 3.11 with that of QO-STBC when $N_t = 4$. From Figure 3.9 it can be observed that when $N_t = 2$, Alamouti's code outperforms the STLCFCs by offering a higher coding gain. This is not surprising since Alamouti's code maximizes the received SNR. Figure 3.10 shows that when $N_t > 2$, STLCFCs achieve better performance than OSTBCs, which is due to the fact that OSTBCs have to use a larger constellation than do STLCFCs to compensate for the rate loss. Figure 3.11 reveals that STLCFCs have a higher diversity gain than that of QO-STBCs.

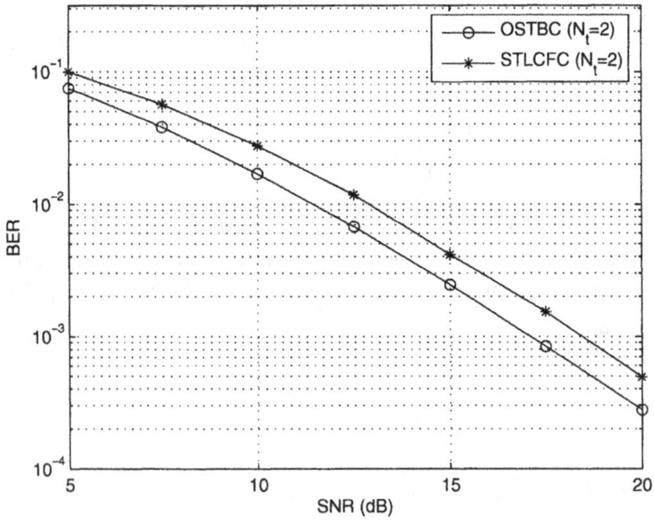

Figure 3.9 STLCFC vs. OSTBC ($N_t = 2$)

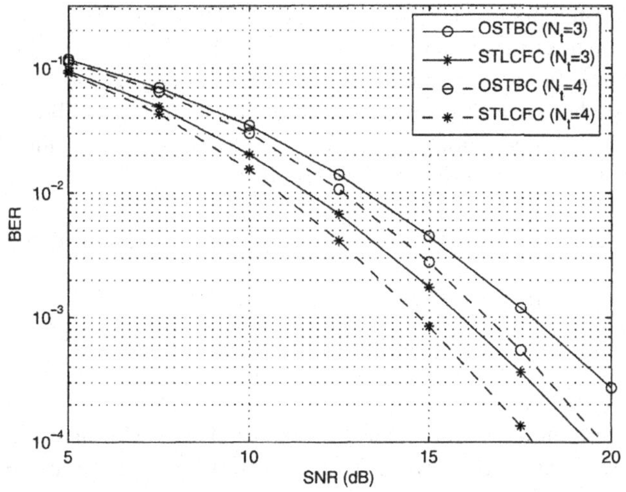

Figure 3.10 STLCFC vs. OSTBC ($N_t = 3, 4$)

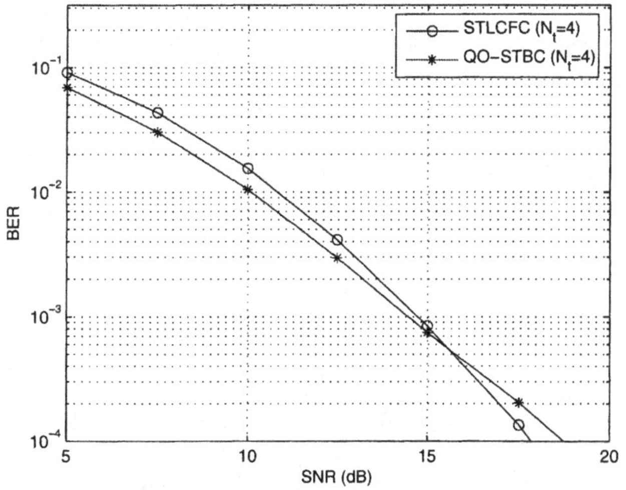

Figure 3.11 STLCFC vs. QO-STBC ($N_t = 4$)

3.5.7 Modulus-Preserving STLCFC

STLCFCs offer maximum diversity gain without rate loss relative to OSTBCs for any N_t. However, as one can verify by inspecting (3.55), STLCFC matrices do not preserve the modulus property of information symbols. As a result, they could potentially increase the peak-to-average power ratio (PAR) of the transmitted signal, which is undesirable at the power amplification stage. (PAR introduces nonlinear distortion effects in the transmitted waveform unless considerable backoff is implemented, thus reducing the efficiency of nonlinear power amplifiers.) To cope with PAR, an STLCFC variate has been developed in [338] which preserves symbol modulus. In the present subsection we outline the code construction of this modulus-preserving (MP) STLCFC scheme and reveal its relationship with the STLCFC design in Section 3.5.5.

With the parameters N_x and φ to be specified later, let us define two matrices,

$$\begin{aligned} \boldsymbol{\Lambda}(\varphi) &:= \mathrm{diag}(1, e^{-j\varphi}, \ldots, e^{-j\varphi(N_x-1)}), \\ \boldsymbol{J} &:= \begin{bmatrix} \mathbf{0}_{1\times(N_x-1)} & 1 \\ \boldsymbol{I}_{N_x-1} & \mathbf{0}_{(N_x-1)\times 1} \end{bmatrix}. \end{aligned} \tag{3.79}$$

When left-multiplying a column vector $\boldsymbol{a}$, the matrix $\boldsymbol{\Lambda}(\varphi)$ performs successive phase rotations on the elements of $\boldsymbol{a}$, while $\boldsymbol{J}$ (a downshift permutation matrix [99, Page 202]) performs a circular downshift on $\boldsymbol{a}$. Using these two matrices, the transmission matrix $\boldsymbol{X}_{\mathrm{MP}}$ corresponding to the MP-STLCFC scheme can be constructed in three steps:

S1) Collect N_x information symbols into a vector $\boldsymbol{s} := [s_1 \cdots s_{N_x}]^T$.

S2) Apply diagonal precoding (i.e., successive phase rotations) on $\boldsymbol{s}$ to obtain

$$\tilde{\boldsymbol{s}} = \boldsymbol{\Lambda}(\varphi)\boldsymbol{s}. \tag{3.80}$$

S3) Construct the $N_t \times N_x$ code matrix $\boldsymbol{X}_{\mathrm{MP}}$ as

$$\boldsymbol{X}_{\mathrm{MP}}(\boldsymbol{s}) = \begin{bmatrix} (\tilde{\boldsymbol{s}})^T \\ (\boldsymbol{J}\tilde{\boldsymbol{s}})^T \\ \vdots \\ (\boldsymbol{J}^{N_t-1}\tilde{\boldsymbol{s}})^T \end{bmatrix}. \tag{3.81}$$

Since neither diagonal precoding in (3.80) nor circular shifting in (3.81) affects symbol modulus, MP-STLCFC indeed preserves the modulus of information symbols. If $N_x = N_t = 3, 4$, for example, the square matrix $\boldsymbol{X}_{\mathrm{MP}}(\boldsymbol{s})$ takes the form

$$\boldsymbol{X}_{\mathrm{MP}}(\boldsymbol{s}) = \begin{bmatrix} s_1 & e^{-j\varphi}s_2 & e^{-j2\varphi}s_3 \\ e^{-j2\varphi}s_3 & s_1 & e^{-j\varphi}s_2 \\ e^{-j\varphi}s_2 & e^{-j2\varphi}s_3 & s_1 \end{bmatrix}, \tag{3.82}$$

$$\boldsymbol{X}_{\mathrm{MP}}(\boldsymbol{s}) = \begin{bmatrix} s_1 & e^{-j\varphi}s_2 & e^{-j2\varphi}s_3 & e^{-j3\varphi}s_4 \\ e^{-j3\varphi}s_4 & s_1 & e^{-j\varphi}s_2 & e^{-j2\varphi}s_3 \\ e^{-j2\varphi}s_3 & e^{-j3\varphi}s_4 & s_1 & e^{-j\varphi}s_2 \\ e^{-j\varphi}s_2 & e^{-j2\varphi}s_3 & e^{-j3\varphi}s_4 & s_1 \end{bmatrix}. \tag{3.83}$$

When $N_x > N_t$, the fat matrix $\boldsymbol{X}_{\mathrm{MP}}(\boldsymbol{s})$ can be obtained by keeping only the first N_t rows of a square matrix designed for N_x antennas.

We wish to show that the matrix in (3.81) is closely related to an STLCFC matrix. For simplicity in exposition, we assume that a single receive-antenna is used, bearing in mind that related claims can be readily established on a per receive-antenna basis when multiple antennas are deployed. In this case, the input-output relationship is

$$\boldsymbol{y}^T = \sqrt{\frac{\bar{\gamma}}{N_t}}\boldsymbol{h}^T\boldsymbol{X}_{\mathrm{MP}}(\boldsymbol{s}) + \boldsymbol{w}^T.$$

With $\boldsymbol{X}_{\mathrm{MP}}(\boldsymbol{s})$ as in (3.81), we obtain

$$\begin{aligned}\boldsymbol{y} &= \sqrt{\frac{\bar{\gamma}}{N_t}}\underbrace{\left(h_1\boldsymbol{I} + h_2\boldsymbol{J} + \cdots + h_{N_t}\boldsymbol{J}^{N_t-1}\right)}_{:=\tilde{\boldsymbol{H}}}\tilde{\boldsymbol{s}} + \boldsymbol{w} \\ &= \sqrt{\frac{\bar{\gamma}}{N_t}}\tilde{\boldsymbol{H}}\tilde{\boldsymbol{s}} + \boldsymbol{w}. \qquad (3.84)\end{aligned}$$

The $N_x \times N_x$ matrix $\tilde{\boldsymbol{H}}$ defined in (3.84) is circulant, which implies that it can be diagonalized using the N_x-point fast Fourier transform (FFT) matrix $\boldsymbol{F}_{N_x}$, as [99, Page 202]

$$\tilde{\boldsymbol{H}} = \boldsymbol{F}_{N_x}^{\mathcal{H}}\operatorname{diag}(\tilde{\boldsymbol{h}})\boldsymbol{F}_{N_x}, \qquad (3.85)$$

where $\tilde{\boldsymbol{h}} := [H(0), \cdots, H(N_x - 1)]^T$ with $H(p) := \sum_{\mu=1}^{N_t} h_\mu e^{-j\frac{2\pi}{N_x}(\mu-1)p}$. Upon taking the N_x-point FFT of $\boldsymbol{y}$, we can write $\tilde{\boldsymbol{y}} := \boldsymbol{F}_{N_x}\boldsymbol{y}$ as

$$\begin{aligned}\tilde{\boldsymbol{y}} &= \sqrt{\frac{\bar{\gamma}}{N_t}}\operatorname{diag}(\tilde{\boldsymbol{h}})\boldsymbol{F}_{N_x}\boldsymbol{\Lambda}(\varphi)\boldsymbol{s} + \boldsymbol{F}_{N_x}\boldsymbol{w} \\ &= \sqrt{\frac{\bar{\gamma}}{N_t}}\operatorname{diag}(\boldsymbol{F}_{N_x}\boldsymbol{\Lambda}(\varphi)\boldsymbol{s})\tilde{\boldsymbol{h}} + \tilde{\boldsymbol{w}}, \qquad (3.86)\end{aligned}$$

where $\tilde{\boldsymbol{w}} := \boldsymbol{F}_{N_x}\boldsymbol{w}$ has the same statistics as $\boldsymbol{w}$. Since $\operatorname{diag}(\boldsymbol{F}_{N_x}\boldsymbol{\Lambda}(\varphi)\boldsymbol{s})$ is diagonal, it can be treated as $\boldsymbol{X}_{\mathrm{LCF}}(\boldsymbol{s})$ in (3.55). Consequently, transmitting with the ST matrix in (3.81) over the channel $\boldsymbol{h}$ can be thought of equivalently as transmitting an STLCFC matrix $\operatorname{diag}(\boldsymbol{F}_{N_x}\boldsymbol{\Lambda}(\varphi)\boldsymbol{s})$ over the virtual channel $\tilde{\boldsymbol{h}}$.

In a nutshell, the key idea behind MP-STLCFC is to apply the LCFC in (3.71) in two steps: *explicit* diagonal precoding (implemented with $\boldsymbol{\Lambda}(\varphi)$) on information blocks and *implicit* FFT precoding (implemented with $\boldsymbol{F}_{N_x}$) via circularly shifted transmissions. As we will see in Section 8.4, this can also be viewed as a circular delay diversity transmission system. The MP-STLCFC system transmits with an STLCFC matrix over the virtual channel $\tilde{\boldsymbol{h}}$ in the frequency domain, whereas the original STLCFC system transmits its code matrix over $\boldsymbol{h}$ in the time domain.

What is left to specify are the choices for N_x and φ:

- If N_t is a positive power of 2 (i.e., $N_t \in \mathbb{N}_2$), we select $N_x = N_t$, and $\varphi = \pi/(2N_t)$ as in (3.73). Assuming that $\{h_\mu\}_{\mu=1}^{N_t}$ are Gaussian i.i.d., the entries of $\tilde{\boldsymbol{h}}$ have statistics identical to those of $\{h_\mu\}_{\mu=1}^{N_t}$. This implies that there is no difference in error performance between (3.86) and the transmission of STLCFCs in Section 3.5.5. In this case, the MP-STLCFC transmission preserves symbol modulus without any performance loss.

- When $N_t \notin \mathbb{N}_2$, a couple of options become available in designing the MP-STLCFC matrix. One choice is to set $N_x = N_t$ to minimize the decoding delay and complexity. The optimal φ can be found by, for example, a line search. In this case, the precoder $\boldsymbol{F}_{N_t}\boldsymbol{\Lambda}(\varphi)$ does not achieve the maximum coding gain. An alternative option is to choose $N_x > N_t$ as the smallest power of 2 which is larger than N_t, and then select φ as $\varphi = \pi/(2N_x)$. In this case, the precoder $\boldsymbol{F}_{N_x}\boldsymbol{\Lambda}(\varphi)$ follows the optimal design in (3.71)–(3.73), but the entries of $\tilde{\boldsymbol{h}}$ are correlated. Numerical simulations suggest that the second option exhibits better error performance than the first one [338]. Performance improvement comes at a slight increase in complexity, as $N_x > N_t$ symbols need to be decoded jointly.

3.6 LINKING OSTBC, QO-STBC, AND STLCFC DESIGNS

An interesting relationship has been established in [338] among the OSTBC class we presented in Section 3.3, the QO-STBC in Section 3.4, and the STLCFC in Section 3.5. Specifically, as we will see in the next subsection, it is possible to embed MP-STLCFCs into an OSTBC structure and obtain a QO-STBC matrix which can be modulus preserving (and will thus be called QO-MP-STLCFC). A general class of QO-STBCs along with a recursive scheme for constructing them can be found in [234]; see also [21] for a QO-STBC design based on linear Hadamard sequences. The QO-MP-STLCFC construction of the next subsection will offer an alternative construction of the general class of QO-STBCs in [234]. Although different QO-STBC matrices are equivalent as far as error performance is concerned, the QO-MP-STLCFC construction is unique in illuminating the relationship among STLCFC, OSTBC, and QO-STBC designs.

3.6.1 Embedding MP-STLCFCs into the Alamouti Code

Let us first demonstrate how MP-STLCFCs can be embedded in the 2×2 Alamouti code matrix in (3.16). This embedding will lead to a QO-STBC with rate 1 symbol per channel use. Without loss of generality, suppose that N_x is even, let $K = N_x/2$, and partition the symbol vector $\boldsymbol{s} = [\boldsymbol{s}_a^T, \boldsymbol{s}_b^T]^T$ into two subblocks $\boldsymbol{s}_a$ and $\boldsymbol{s}_b$ of equal length. Based on $\boldsymbol{s}_a$ and $\boldsymbol{s}_b$, construct the MP-STLCFC matrices $\boldsymbol{X}_a$ and $\boldsymbol{X}_b$ using (3.81). Embedding $\boldsymbol{X}_a$ and $\boldsymbol{X}_b$ into the Alamouti structure, the QO-MP-STLCFC without rate loss can be expressed as

$$\boldsymbol{X}_{\text{QO-MP}} = \begin{bmatrix} \boldsymbol{X}_a & -\boldsymbol{X}_b^* \\ \boldsymbol{X}_b & \boldsymbol{X}_a^* \end{bmatrix}. \tag{3.87}$$

For example, with $[\tilde{s}_1, \tilde{s}_2]^T = \boldsymbol{\Lambda}(\varphi)[s_1, s_2]^T$ and $[\tilde{s}_3, \tilde{s}_4]^T = \boldsymbol{\Lambda}(\varphi)[s_3, s_4]^T$, the corresponding $\boldsymbol{X}_{\text{QO-MP}}$ for $N_x = 4$ is

$$\boldsymbol{X}_4 = \begin{bmatrix} \tilde{s}_1 & \tilde{s}_2 & -\tilde{s}_3^* & -\tilde{s}_4^* \\ \tilde{s}_2 & \tilde{s}_1 & -\tilde{s}_4^* & -\tilde{s}_3^* \\ \tilde{s}_3 & \tilde{s}_4 & \tilde{s}_1^* & \tilde{s}_2^* \\ \tilde{s}_4 & \tilde{s}_3 & \tilde{s}_2^* & \tilde{s}_1^* \end{bmatrix}. \tag{3.88}$$

Matrix $\boldsymbol{X}_4$ in (3.88) can be obtained from the QO-STBC design (3.45) by row/column multiplication with -1, row/column permutation, symbol relabeling, and constellation rotation. For $N_x = 8$, the resulting QO-MP-STLCFC matrix is

$$\boldsymbol{X}_8 = \begin{bmatrix} \tilde{s}_1 & \tilde{s}_2 & \tilde{s}_3 & \tilde{s}_4 & -\tilde{s}_5^* & -\tilde{s}_6^* & -\tilde{s}_7^* & -\tilde{s}_8^* \\ \tilde{s}_4 & \tilde{s}_1 & \tilde{s}_2 & \tilde{s}_3 & -\tilde{s}_8^* & -\tilde{s}_5^* & -\tilde{s}_6^* & -\tilde{s}_7^* \\ \tilde{s}_3 & \tilde{s}_4 & \tilde{s}_1 & \tilde{s}_2 & -\tilde{s}_7^* & -\tilde{s}_8^* & -\tilde{s}_5^* & -\tilde{s}_6^* \\ \tilde{s}_2 & \tilde{s}_3 & \tilde{s}_4 & \tilde{s}_1 & -\tilde{s}_6^* & -\tilde{s}_7^* & -\tilde{s}_8^* & -\tilde{s}_5^* \\ \tilde{s}_5 & \tilde{s}_6 & \tilde{s}_7 & \tilde{s}_8 & \tilde{s}_1^* & \tilde{s}_2^* & \tilde{s}_3^* & \tilde{s}_4^* \\ \tilde{s}_8 & \tilde{s}_5 & \tilde{s}_6 & \tilde{s}_7 & \tilde{s}_4^* & \tilde{s}_1^* & \tilde{s}_2^* & \tilde{s}_3^* \\ \tilde{s}_7 & \tilde{s}_8 & \tilde{s}_5 & \tilde{s}_6 & \tilde{s}_3^* & \tilde{s}_4^* & \tilde{s}_1^* & \tilde{s}_2^* \\ \tilde{s}_6 & \tilde{s}_7 & \tilde{s}_8 & \tilde{s}_5 & \tilde{s}_2^* & \tilde{s}_3^* & \tilde{s}_4^* & \tilde{s}_1^* \end{bmatrix}, \tag{3.89}$$

where $[\tilde{s}_1, \ldots, \tilde{s}_4]^T = \boldsymbol{\Lambda}(\varphi)[s_1, \ldots, s_4]^T$ and $[\tilde{s}_5, \ldots, \tilde{s}_8]^T = \boldsymbol{\Lambda}(\varphi)[s_5, \ldots, s_8]^T$.

The QO-MP-STLCFC in (3.87) is applicable when $N_t = N_x$. With $N_t < N_x$, one can pick any N_t rows of $\boldsymbol{X}_{\text{QO-MP}}$ to form the corresponding ST code matrix.

3.6.2 Embedding 2×2 MP-STLCFCs into an OSTBC

Reduced-rate QO-MP-STLCFCs are also possible by embedding MP-STLCFCs into a general OSTBC other than the Alamouti code. For example, when the rate-3/4 OSTBC $\boldsymbol{O}_{4\times 4}$ with $N_s = 3$ is used [cf. (3.18)], we can pick $N_x = 2N_s = 6$ symbols and partition the vector $\boldsymbol{s} = [s_1, \ldots, s_6]^T$ into three 2×1 subvectors $\boldsymbol{s}_1$, $\boldsymbol{s}_2$, and $\boldsymbol{s}_3$. Generating the 2×2 MP-STLCFCs from $\{\boldsymbol{s}_i\}_{i=1}^3$ and then embedding them into $\boldsymbol{O}_{4\times 4}$, we obtain

$$\tilde{\boldsymbol{X}}_8 = \begin{bmatrix} \tilde{s}_1 & \tilde{s}_2 & -\tilde{s}_3^* & -\tilde{s}_4^* & -\tilde{s}_5^* & -\tilde{s}_6^* & 0 & 0 \\ \tilde{s}_2 & \tilde{s}_1 & -\tilde{s}_4^* & -\tilde{s}_3^* & -\tilde{s}_6^* & -\tilde{s}_5^* & 0 & 0 \\ \tilde{s}_3 & \tilde{s}_4 & \tilde{s}_1^* & \tilde{s}_2^* & 0 & 0 & -\tilde{s}_5^* & -\tilde{s}_6^* \\ \tilde{s}_4 & \tilde{s}_3 & \tilde{s}_2^* & \tilde{s}_1^* & 0 & 0 & -\tilde{s}_6^* & -\tilde{s}_5^* \\ \tilde{s}_5 & \tilde{s}_6 & 0 & 0 & \tilde{s}_1^* & \tilde{s}_2^* & \tilde{s}_3^* & \tilde{s}_4^* \\ \tilde{s}_6 & \tilde{s}_5 & 0 & 0 & \tilde{s}_2^* & \tilde{s}_1^* & \tilde{s}_4^* & \tilde{s}_3^* \\ 0 & 0 & \tilde{s}_5 & \tilde{s}_6 & -\tilde{s}_3 & -\tilde{s}_4 & \tilde{s}_1 & \tilde{s}_2 \\ 0 & 0 & \tilde{s}_6 & \tilde{s}_5 & -\tilde{s}_4 & -\tilde{s}_3 & \tilde{s}_2 & \tilde{s}_1 \end{bmatrix}. \tag{3.90}$$

It is easy to see that $\tilde{\boldsymbol{X}}_8$ is equivalent to the QO-STBC in [243, Equation (30)] up to row and column permutations. This is not surprising, since the quasi-orthogonal codes in [243] are the ABBA designs of [259] whose ST code matrices can be written as

$$\begin{bmatrix} A & B \\ B & A \end{bmatrix}, \tag{3.91}$$

where $\boldsymbol{A}$ and $\boldsymbol{B}$ are two OSTBCs based on two different sets of information symbols. Hence, the ABBA code embeds OSTBCs into a 2×2 STLCFC design. By row and column permutations, this is equivalent to embedding multiple 2×2 STLCFCs into the OSTBC presented here. Since different formats of QO-STBCs yield identical error performance, it suffices to use the ABBA type of construction and permuted versions of it.

3.6.3 Decoding QO-MP-STLCFC

An attractive feature of the QO-MP-STLCFC transmission we presented in Section 3.6.1 is that receiver processing can be simplified by decoding $\boldsymbol{s}_a$ and $\boldsymbol{s}_b$ separately. To demonstrate this, let $\boldsymbol{y}_a$ and $\boldsymbol{y}_b$ denote the received blocks corresponding to $\boldsymbol{s}_a$ and $\boldsymbol{s}_b$. Consider $N_t = N_x$ antennas with N_x even and collect the channel coefficients into two $K \times 1$ vectors, $\boldsymbol{h}_a$ and $\boldsymbol{h}_b$. With $\boldsymbol{w}_a$ and $\boldsymbol{w}_b$ denoting the corresponding additive noise terms, we have

$$\begin{bmatrix} \boldsymbol{y}_a \\ \boldsymbol{y}_b \end{bmatrix} = \sqrt{\frac{\bar{\gamma}}{N_t}} \begin{bmatrix} \boldsymbol{X}_a & -\boldsymbol{X}_b^* \\ \boldsymbol{X}_b & \boldsymbol{X}_a^* \end{bmatrix}^T \begin{bmatrix} \boldsymbol{h}_a \\ \boldsymbol{h}_b \end{bmatrix} + \begin{bmatrix} \boldsymbol{w}_a \\ \boldsymbol{w}_b \end{bmatrix}. \tag{3.92}$$

Taking the FFT of $\boldsymbol{y}_a$ and $\boldsymbol{y}_b$ and conjugating $\boldsymbol{F}_K \boldsymbol{y}_b$, we obtain

$$\begin{bmatrix} \boldsymbol{F}_K \boldsymbol{y}_a \\ (\boldsymbol{F}_K \boldsymbol{y}_b)^* \end{bmatrix} = \sqrt{\frac{\bar{\gamma}}{N_t}} \underbrace{\begin{bmatrix} \operatorname{diag}(\tilde{\boldsymbol{h}}_a) & \operatorname{diag}(\tilde{\boldsymbol{h}}_b) \\ -\operatorname{diag}(\tilde{\boldsymbol{h}}_b^*) & \operatorname{diag}(\tilde{\boldsymbol{h}}_a^*) \end{bmatrix}}_{:=\boldsymbol{D}_{ab}} \begin{bmatrix} \boldsymbol{\Theta}_K \boldsymbol{s}_a \\ \boldsymbol{\Theta}_K \boldsymbol{s}_b \end{bmatrix} + \begin{bmatrix} \boldsymbol{F}_K \boldsymbol{w}_a \\ (\boldsymbol{F}_K \boldsymbol{w}_b)^* \end{bmatrix}, \tag{3.93}$$

where $\boldsymbol{\Theta}_K := \boldsymbol{F}_K \boldsymbol{\Lambda}(\varphi)$. Matrix $\boldsymbol{D}_{ab}$ has orthogonal columns; that is, $\boldsymbol{D}_{ab}^{\mathcal{H}} \boldsymbol{D}_{ab} = \boldsymbol{I}_2 \otimes \boldsymbol{\Lambda}_{\text{equ}}^2$, where $\otimes$ denotes the Kronecker product and

$$\boldsymbol{\Lambda}_{\text{equ}} := \left[\operatorname{diag}^{\mathcal{H}}(\tilde{\boldsymbol{h}}_a)\operatorname{diag}(\tilde{\boldsymbol{h}}_a) + \operatorname{diag}^{\mathcal{H}}(\tilde{\boldsymbol{h}}_b)\operatorname{diag}(\tilde{\boldsymbol{h}}_b)\right]^{\frac{1}{2}}. \tag{3.94}$$

Left-multiplying (3.93) by a unitary matrix, we obtain

$$\begin{aligned} \begin{bmatrix} \boldsymbol{z}_a \\ \boldsymbol{z}_b \end{bmatrix} &= \sqrt{\frac{\bar{\gamma}}{N_t}} \underbrace{(\boldsymbol{D}_{ab}^{\mathcal{H}} \boldsymbol{D}_{ab})^{-\frac{1}{2}} \boldsymbol{D}_{ab}^{\mathcal{H}}}_{\text{unitary matrix}} \begin{bmatrix} \boldsymbol{F}_K \boldsymbol{y}_a \\ (\boldsymbol{F}_K \boldsymbol{y}_b)^* \end{bmatrix} \\ &= \sqrt{\frac{\bar{\gamma}}{N_t}} \begin{bmatrix} \boldsymbol{\Lambda}_{\text{equ}} & 0 \\ 0 & \boldsymbol{\Lambda}_{\text{equ}} \end{bmatrix} \begin{bmatrix} \boldsymbol{\Theta}_K \boldsymbol{s}_a \\ \boldsymbol{\Theta}_K \boldsymbol{s}_b \end{bmatrix} + \begin{bmatrix} \tilde{\boldsymbol{w}}_a \\ \tilde{\boldsymbol{w}}_b \end{bmatrix}, \end{aligned} \tag{3.95}$$

where the postprocessing noise $[\tilde{\boldsymbol{w}}_a^T, \tilde{\boldsymbol{w}}_b^T]^T$ remains white. Equation (3.95) shows that without loss of optimality, $\boldsymbol{s}_a$ and $\boldsymbol{s}_b$ can be decoded separately based on

$$\boldsymbol{z}_a = \sqrt{\frac{\bar{\gamma}}{N_t}} \boldsymbol{\Lambda}_{\text{equ}} \boldsymbol{\Theta}_K \boldsymbol{s}_a + \tilde{\boldsymbol{w}}_a, \quad \boldsymbol{z}_b = \sqrt{\frac{\bar{\gamma}}{N_t}} \boldsymbol{\Lambda}_{\text{equ}} \boldsymbol{\Theta}_K \boldsymbol{s}_b + \tilde{\boldsymbol{w}}_b. \tag{3.96}$$

The orthogonality of Alamouti's code matrix enables the receiver to perform two separate decoding steps with reduced size. Notice that MP-STLCFC is now applied on blocks of size $K = N_x/2$. When confined to the case where K is a power of 2, the optimal φ should be $\varphi = \pi/(2K)$.

In summary, embedding MP-STLCFC matrices into an OSTBC matrix yields a QO-STBC matrix. This construction links rather nicely seemingly unrelated code matrices. Furthermore, it shows clearly how decoding of individual QO-STBC blocks can be separated via linear processing. (Near-)optimal decoding of each block can be performed using the SDA at average complexity which has polynomial order (in the block size), as explained in Chapter 5.

3.7 CLOSING COMMENTS

In this chapter we dealt with representative classes of coherent ST codes for use in enhancing error performance of multi-antenna transmissions over flat fading MIMO channels. We saw that each of these codes has its own pros and cons in terms of rate and error performance as well as decoding complexity. For this reason, the designer's choice depends on application-specific trade-offs among error performance, decoding complexity, and bandwidth efficiency. When designing these ST codes, we focused whenever possible on optimizing the diversity gain as well as the coding gain. But as we also mentioned in Chapter 2, maximizing diversity and coding gains is necessary but not sufficient for minimizing error probability performance. Albeit challenging analytically, taking into account additional BER-critical factors in the design criteria can result in ST codes with improved error probability performance (see, e.g., [315]).

Being error performance (as opposed to rate-)oriented, the ST codes in this chapter achieve rates not exceeding those attained by single-antenna systems (up to 1 symbol per channel use). However, we saw in Chapter 2 that ST transmissions are capable of achieving rates higher than those of single-antenna systems by a factor approximately equal to $\min(N_t, N_r)$. Such rate-oriented ST codes and the associated rate-diversity trade-offs are the subject of Chapter 4.

A feature common to this chapter's codes is that their error performance relies on a certain structure of the associated codewords. In addition, their decoding requires channel state information at the receiver. If channel state information is not available or accurate enough, their code structure will be impaired and this may cause error performance to degrade markedly in terms of both diversity and coding gains. In Chapter 6 we describe several classes of what we will naturally term noncoherent ST codes whose decoding does not require channel state information.

4

Layered ST Codes

The ST block codes of Chapter 3 were designed with one main objective: to enable the highest possible diversity and coding advantages provided by MIMO channels while attaining rates no higher than 1 symbol per channel use, which are also achievable by single-antenna systems. Certainly, the rate of ST block codes can be enhanced by increasing the constellation size. Furthermore, ST trellis codes combine modulation with ST coding nonlinearly, which renders their design and decoding increasingly more complex as the rate increases. For these reasons, both ST block and trellis codes are more appropriate for applications where error performance is the major criterion to satisfy. On the other hand, we have seen in Chapter 2 that beyond improving error performance through increasing the diversity, multi-antenna systems can achieve higher rates than can their single-antenna counterparts. Indeed, even intuitively one can appreciate the fact that with, for example, three transmit antennas, it is possible to transmit three symbols simultaneously. This intuition has been analytically quantified in Chapter 2, where we saw that the ultimate information rates (ergodic and outage capacity) achievable by MIMO channels are higher than those of single-antenna systems by a factor as high as $\min(N_t, N_r)$. This is a manifestation of properly multiplexing symbol streams across antennas before transmission. The resulting rate advantage over single-antenna transmissions is known as ST multiplexing gain, and the ST codes enabling it belong to the class of what we term rate-oriented ST designs.

In this chapter we introduce such rate-oriented classes of ST designs, which are also known as layered ST codes because they effect ST multiplexing on groups of symbols which are called layers. We will see that layered ST codes can deliver rates higher than 1 symbol per channel use and in some cases they can reach (with proper error control coding) the fundamental capacity limits of MIMO channels. We also introduce hybrid

ST codes capable of trading off diversity for rate, thus bridging the classes of error performance-oriented with rate-oriented designs. Finally, we present ST codes that achieve the ultimate combination of full diversity with full transmission rate. These ST codes capitalize on LCF coding and layered ST multiplexing to offer designs that are universally applicable to *any* number of transmit-receive antennas. Moreover, they are flexible to achieve desirable trade-offs in rate, diversity, and complexity — a feature particularly attractive when it comes to practical multi-antenna systems.

As in Chapter 3, the MIMO channel here will be assumed known at the receiver to allow for coherent decoding. It can be obtained using the MIMO channel estimation algorithms we describe in Chapter 11. Furthermore, the coherent layered ST codes in this chapter are presented for flat fading MIMO channels, but as we will see in Chapters 7 through 9, the basic ideas behind their design carry over to frequency- and time-selective MIMO channels as well.

4.1 BLAST DESIGNS

Besides basic results on ergodic capacity of MIMO channels, G. J. Foschini and co-workers at Bell Laboratories pioneered two layered ST schemes which are known as D-BLAST (diagonal Bell Laboratories Layered Space-Time) and V-BLAST (vertical BLAST) [68–70]. These BLAST schemes were the first practical architectures to provably achieve transmission rates higher than 1 symbol per channel use (pcu), and demonstrate experimentally that error control coded BLAST can approach the fundamental limit provided by the ergodic capacity of a MIMO channel. Note that we assess transmission rate in symbols pcu, for the reason given in Section 2.6. Since 1996, BLAST designs have inspired most subsequent improvements on rate-oriented classes of ST codes.

4.1.1 D-BLAST

Historically, D-BLAST is the first layered architecture. Its ST code matrix is given by [68]

$$\boldsymbol{X} = \begin{bmatrix} x_1(1) & x_2(1) & \cdots & x_{N_l}(1) & & & \\ & x_1(2) & x_2(2) & \cdots & x_{N_l}(2) & & \\ & & \ddots & \ddots & & \ddots & \\ & & & x_1(N_t) & x_2(N_t) & \cdots & x_{N_l}(N_t) \end{bmatrix} \begin{matrix} \text{time} \\ \rightarrow \\ \downarrow \text{space}, \end{matrix} \qquad (4.1)$$

where $x_l(n)$ denotes the nth symbol of the lth layer transmitted from the nth antenna, N_l is the number of layers, and the blank spaces denote zeros. Tracing common subscripts in (4.1), we observe that each layer contains a group of (possibly coded) symbols placed on the corresponding diagonal of the ST code matrix $\boldsymbol{X}$. Although generalizations are possible, we suppose here for simplicity that the length of each layer equals the number of transmit-antennas N_t.

The ST code matrix in (4.1) is reminiscent of the matrix transmitted by a delay diversity (DD) multi-antenna system [cf. (3.1)]. However, instead of repeating the same symbol per DD layer, D-BLAST transmits a different symbol $x_l(n)$ from each antenna per time slot; and each $x_l(n)$ can be an uncoded symbol, the entry of a Galois-field (GF) codeword, or the entry of an LCF codeword. For example, we can employ the LCF matrix in (3.54) to encode each $N_t \times 1$ block of information symbols and use the resulting encoded block to form a corresponding layer (diagonal) of the ST code matrix in (4.1). In this case, too, D-BLAST transmits N_t symbols per layer whereas DD transmits only one symbol per layer. Besides this N_t-fold increase in transmission rate, every LCF- or GF-coded symbol $x_l(n)$ in D-BLAST contains information about all uncoded symbols in the lth layer. From this vantage point, D-BLAST can be viewed as a "smart delay diversity" ST code.

As far as spectral efficiency is concerned, simple inspection of (4.1) reveals that per ST code matrix D-BLAST transmits N_l layers carrying N_t symbols per layer over $N_l + N_t - 1$ time slots (columns of $\boldsymbol{X}$). Hence, based on condition *c2.3)* in Section 2.6, the (code) rate of uncoded or LCF-coded D-BLAST is

$$\eta_{\text{D-BLAST}} = \frac{N_t N_l}{N_l + N_t - 1} \quad \text{symbols pcu}, \tag{4.2}$$

which confirms that relative to the DD scheme in (3.1), D-BLAST offers an N_t times higher transmission rate. Clearly, as N_l grows large, this rate approaches N_t symbols pcu, which is higher than the 1 symbol pcu attained by single-antenna systems and by the ST codes of Chapter 3. If a GF-based error control code with rate $\eta_{\text{GF}} < 1$ bit pcu is used to generate each $x_l(n)$ entry in the D-BLAST code matrix, $\eta_{\text{D-BLAST}}$ in (4.2) must account for η_{GF} and be expressed in bits pcu.

Turning our attention to decoding, we recognize that the multiple symbols transmitted in parallel per time slot in D-BLAST interfere with each other at the receiver end. Nonetheless, the resulting input-output relationship is still given by (2.9) and ML decoding yields

$$\hat{\boldsymbol{X}} = \arg\min_{\boldsymbol{X}} \left\| \boldsymbol{Y} - \sqrt{\frac{\bar{\gamma}}{N_t}} \boldsymbol{H}\boldsymbol{X} \right\|_F^2 . \tag{4.3}$$

To assess the achievable diversity and required complexity, let us suppose as before that the MIMO channel is complex Gaussian distributed and that an LCF precoder is employed per layer; that is,

$$\boldsymbol{x}_l = \boldsymbol{\Theta} \boldsymbol{s}_l,$$

where $\boldsymbol{x}_l := [x_l(1), \ldots, x_l(N_t)]^T$, $\boldsymbol{\Theta}$ is the $N_t \times N_t$ LCF code matrix designed as in Section 3.5.5, and $\boldsymbol{s}_l$ is an $N_t \times 1$ vector containing N_t information symbols. Arguing as in the diversity analysis of Section 3.5.5, if the LCF precoded symbols $x_l(n)$ are transmitted through N_t antennas, then based on the structure of $\boldsymbol{X}$ in (4.1), it follows that the diversity order collected by the ML decoder in (4.3) is $N_t N_r$ (see [311] for detailed proof). In this case, the ML decoder incurs complexity $\mathcal{O}(|\mathbb{S}|^{N_t N_l})$, with

$|\mathbb{S}|$ denoting the number of bits per symbol. When either $|\mathbb{S}|$, N_t, or N_l is large, the computational burden becomes prohibitively high because all $N_t N_l$ symbols of all layers are decoded jointly.

Linear zero-forcing (ZF) or minimum mean-square error (MMSE) equalization can be alternatively employed to reduce the exponential complexity of ML decoding to cubic complexity (or linear complexity once the inverse of $\boldsymbol{H}$ is found), at the expense of possibly considerable degradation in error performance. A common performance-complexity compromise between linear equalization and nonlinear ML decoding is decision-feedback equalization (DFE). Block ZF or MMSE versions of DFE have been applied for decoding single-antenna block transmissions in single- and multi-user communication systems (see also Chapter 5). In the multi-antenna MIMO context, block DFE related algorithms, known as nulling-canceling (NC) decoders, have been derived to take advantage of the layered structure present in the D-BLAST code matrix. NC indeed offers a practical compromise because its error performance lies between ML and linear equalization, while its complexity is also in between (see, e.g., [288] for an example).

The NC decoder requires that $N_r \geq N_t$ and starts by applying QR-decomposition on the MIMO channel matrix to factor it as

$$\boldsymbol{H} = \boldsymbol{Q}\boldsymbol{R},$$

where $\boldsymbol{Q}$ is a unitary matrix and $\boldsymbol{R}$ is an upper triangular matrix. Using this decomposition and multiplying both sides of (2.9) by $\boldsymbol{Q}^{\mathcal{H}}$, we obtain

$$\boldsymbol{Q}^{\mathcal{H}}\boldsymbol{Y} = \sqrt{\frac{\bar{\gamma}}{N_t}}\boldsymbol{R}\boldsymbol{X} + \boldsymbol{Q}^{\mathcal{H}}\boldsymbol{W}. \tag{4.4}$$

With r_{mn} denoting the (m,n)th entry of the upper triangular factor $\boldsymbol{R}$, it follows that the matrix

$$\boldsymbol{R}\boldsymbol{X} = \begin{bmatrix} r_{11}x_1(1) & r_{11}x_2(1) + r_{12}x_1(2) & & \cdots & \\ & r_{22}x_1(2) & & r_{22}x_2(2) + r_{23}x_1(3) & \cdots \\ & & & \ddots & \\ & & & & r_{N_tN_t}x_{N_l}(N_t) \end{bmatrix} \tag{4.5}$$

is also upper triangular. The main diagonal of $\boldsymbol{Q}^{\mathcal{H}}\boldsymbol{Y}$ in (4.5) can be written as

$$\tilde{\boldsymbol{y}}_1 = \sqrt{\frac{\bar{\gamma}}{N_t}}\boldsymbol{D}_R\boldsymbol{x}_1 + \tilde{\boldsymbol{w}}_1, \tag{4.6}$$

where $\tilde{\boldsymbol{w}}_1$ is the main diagonal of $\boldsymbol{Q}^{\mathcal{H}}\boldsymbol{W}$ and the diagonal matrix $\boldsymbol{D}_R$ contains the diagonal entries of $\boldsymbol{R}$. Using (4.5), the NC algorithm detects the symbols in the first layer $\boldsymbol{x}_1$. Subsequently, these first-layer symbols are canceled from the second layer (second diagonal of $\boldsymbol{Q}^{\mathcal{H}}\boldsymbol{Y}$), and the second layer is recovered from the compensated second diagonal of $\boldsymbol{Q}^{\mathcal{H}}\boldsymbol{Y}$. Proceeding similarly, the NC decoder recovers all layers one after the other.

NC-based decoding of D-BLAST transmissions is computationally attractive and collects the receive diversity N_r even without GF or LCF coding, but not the full diversity $N_t N_r$. To guarantee this maximum diversity order, D-BLAST multiplexing must be applied on GF or LCF coded symbols at the transmitter, and joint ML decoding must be invoked at the receiver to collect it. But as we mentioned earlier, the latter comes at the price of high (and sometimes prohibitive) complexity. In a nutshell, D-BLAST attains rates higher than 1 symbol pcu and can either achieve the maximum possible diversity at high complexity, or settle for lower diversity at the affordable complexity of the NC decoder.

4.1.2 V-BLAST

Layers in the D-BLAST code matrix are arranged in a diagonal fashion to multiplex symbols across space and time. Unfortunately, this diagonal arrangement is also responsible for the large exponent ($N_l N_t$) in the complexity of the associated joint ML decoder [299]. Interestingly, transmitting layers in a vertical fashion leads to a simpler version of the BLAST architecture, which is naturally abbreviated as V-BLAST and can afford lower decoding complexity [299]. The pertinent V-BLAST code matrix has as many layers as the number of transmit-antennas ($N_l = N_t$) and is given by

$$\boldsymbol{X} = \begin{bmatrix} x_1(1) & \cdots & x_1(N_x) \\ x_2(1) & \cdots & x_2(N_x) \\ \vdots & & \vdots \\ x_{N_t}(1) & \cdots & x_{N_t}(N_x) \end{bmatrix}. \tag{4.7}$$

Unless coding has been applied across layers (which is not always advocated in practical V-BLAST systems), it suffices to consider $\boldsymbol{X}$ on a column-by-column basis. In other words, it is enough to focus on a single column of the ST matrix (or consider $\boldsymbol{X}$ in (4.7) with $N_x = 1$) and correspondingly, consider joint ML decoding of only one column of $\boldsymbol{X}$, comprising N_t symbols.

Because V-BLAST transmits N_t symbols per time slot (column of $\boldsymbol{X}$), the code rate in the uncoded and LCF coded cases is clearly

$$\eta_{\text{V-BLAST}} = N_t \quad \text{symbols pcu},$$

which should be properly scaled by η_{GF} if a GF code with rate η_{GF} is employed to code symbols per layer or/and across layers. Notice that the transmission rate of V-BLAST is at least as high as that of D-BLAST. Comparing the corresponding ST code matrices, this should not be surprising since V-BLAST fills the northeast and southwest triangles in the ST code matrix $\boldsymbol{X}$ used by D-BLAST, where symbols are not transmitted.

Using the corresponding received column, ML decoding each V-BLAST layer incurs complexity that is exponential in N_t. This is considerably lower than D-BLAST where the same exponent is N_l times larger. If lower complexity is desired, reduced-complexity V-BLAST receivers are possible using the NC decoder. Compared to the

NC scheme we described for D-BLAST, the NC decoder for V-BLAST has to operate only on column vectors $\boldsymbol{Rx}+\boldsymbol{Q}^{\mathcal{H}}\boldsymbol{w}$, which are obtained after QR decomposing $\boldsymbol{H}$ and multiplying each received vector by $\boldsymbol{Q}^{\mathcal{H}}$. Since $\boldsymbol{R}$ is an upper triangular matrix, the last layer (indexed by N_t) suffers no interference from other layers. After decoding it, the NC decoder cancels its effect from the received vector, proceeds backward to the (N_t-1)st layer, and so on, until the first layer is decoded. The order with which layers are decoded in the NC algorithm plays a performance-critical role. It turns out that the best NC scheme results when decoding the "strongest first" [299], which means that the NC algorithm exhibits best error performance if we start with the layer having the strongest channel magnitude and proceed toward the second strongest, down to the one with the weakest channel gain; see also Section 5.3.1 for more details on the decoding order.

Since each symbol entry in (4.7) is transmitted from one antenna and propagates through the N_r channels corresponding to this single transmit-antenna, the maximum diversity order cannot exceed N_r [263, Page 100]. This implies that even if symbols across layers are GF- or LCF-coded and if the MIMO channel remains invariant over the codeword length, V-BLAST does not enable any transmit diversity. This is in contrast with D-BLAST, where the enabled diversity is *at least* N_r and can be as high as N_tN_r.

Compared with D-BLAST, the V-BLAST architecture is easier to implement (especially for flat fading MIMO channels), it guarantees higher transmission rate, and can afford lower decoding complexity (only N_t symbols need to be decoded jointly) at the expense of reduced diversity.

4.1.3 Rate Performance with BLAST Codes

In Chapter 2 we presented the capacity of flat fading MIMO channels, against which in Chapter 3 we compared the maximum mutual information of an OSTBC multi-antenna system. Along this line of comparison, it is interesting to benchmark the rate performance of the D-BLAST and V-BLAST architectures. To this end, let us consider a single-shot (column-by-column) V-BLAST transmission, and notice that the pertinent input-output relationship is given by (2.16):

$$\boldsymbol{y} = \sqrt{\frac{\bar{\gamma}}{N_t}}\boldsymbol{Hx} + \boldsymbol{w}.$$

If each transmitted block $\boldsymbol{x}$ is circularly symmetric complex Gaussian distributed with zero mean and identity covariance matrix and the channel taps are i.i.d. complex Gaussian with zero mean and unit variance, the maximum mutual information associated with the V-BLAST system is given by [69]

$$C_{\text{BLAST}}(N_t, N_r|\boldsymbol{H}) = \log_2 \det\left[\boldsymbol{I}_{N_t} + \frac{\bar{\gamma}}{N_t}\boldsymbol{HH}^{\mathcal{H}}\right] \quad \text{bits pcu.} \tag{4.8}$$

A similar expression holds true for a D-BLAST system asymptotically in the number of columns. Compared with (2.17), equation (4.8) shows that BLAST transmissions incur no mutual information loss relative to the capacity of MIMO channels. Except for Alamouti's codes in the $(N_t, N_r) = (2, 1)$ configuration which we have already seen to be capacity achieving, all other OSTBC designs are lossy as far as mutual information is concerned. Even though BLAST schemes are capacity achieving, approaching the ergodic capacity limit of a MIMO channel in practice requires invoking along with the V-BLAST multiplexer a proper GF (e.g., Turbo or LDPC) encoder and a soft decoder at the receiver side. In Chapter 5 we elaborate further on the role of soft decoding in approaching the MIMO channel capacity.

4.2 ST CODES TRADING DIVERSITY FOR RATE

In Section 4.1 we saw that BLAST multiplexers can attain rates higher than 1 symbol pcu at the price of increasing decoding complexity (D-BLAST) or losing transmit-diversity (V-BLAST) relative to performance-oriented ST codes. On the other hand, the performance-oriented ST block codes of Chapter 3 take full advantage of the spatial diversity by sacrificing the transmission rate. In this section we explore hybrid ST coding-multiplexing schemes which allow for trade-offs between transmission rate and diversity.

4.2.1 Layered ST Codes with Antenna Grouping

A natural means of enabling the aforementioned rate-diversity trade-off is by combining any full-diversity ST code with the layered V-BLAST multiplexer. Specifically, we can partition the N_t transmit antennas into Q groups, each comprising n_j antennas so that $n_1 + \cdots + n_Q = N_t$ [252]. Different symbols directed to each group of antennas (say, the jth) are ST block or trellis coded to effect transmit diversity of order n_j; and the resulting ST encoded symbols across all groups are concatenated vertically and transmitted in parallel as one layer of the V-BLAST architecture. Although the transmission rate per group cannot exceed 1 symbol pcu, the overall transmission rate is higher because multiple distinct ST coded blocks are transmitted in parallel. And even if the symbols across the antenna groups are transmitted with V-BLAST, they inherit part of the transmit-diversity advantage, thanks to the ST block or trellis coding that has been applied per group.

To reduce receiver complexity, the NC decoder can be employed to separate the received subblocks corresponding to each group. After separation, each group's symbols can be recovered using the group-specific ST decoder. Compared with the layered codes, this hybrid architecture offers two advantages: (i) it collects higher spatial diversity ($n_j N_r$ for the jth group if ML decoding is used); and (ii) it relaxes the NC decoder requirement from $N_r \geq N_t$ to $N_r \geq Q$. The compromise for these advantages is twofold: (i) the transmission rate achieved depends on the ST codes used (e.g., if STLCFC is used per group, the rate is Q symbols pcu, which is less than V-BLAST's rate of N_t symbols pcu); and (ii) the encoding and decoding process is more complex than V-BLAST. Compared with the ST codes of Chapter

3, the combination of error performance-oriented ST block codes with rate-oriented layered codes leads to higher transmission rate but enables lower diversity order. If we suppose for simplicity that $n_j = N_t/Q, \forall j$, the transmission rate is at most Q symbols pcu, and the enabled diversity order is $N_t N_r/Q := q$. As Q increases, the rate increases while the diversity order decreases.

4.2.2 Layered High-Rate Codes

We saw in Section 4.2.1 how ST codes applied to groups of transmit antennas can be combined with layered ST multiplexers to trade diversity for transmission rate. However, because the diversity order (q) is inversely proportional to the rate ($N_t N_r/q$), when the rate increases, the achievable diversity order diminishes relatively quickly. In this subsection we describe briefly a different high-rate code which renders the relationship between rate and diversity linear [312].

Consider a layered multiplexer where each layer has length $N_d \in [1, N_t]$. Furthermore, suppose that each layer is LCF coded so that $\boldsymbol{x}_l = \boldsymbol{\Theta} \boldsymbol{s}_l$, where $\boldsymbol{\Theta}$ is designed as detailed in Section 3.5.5. Placing layers in a diagonal fashion and letting blanks denote zeros as before, let us consider the ST code matrix

$$\boldsymbol{X} = \begin{bmatrix} x_1(1) & & \\ x_2(1) & \ddots & \\ \vdots & \ddots & x_1(N_d) \\ \vdots & & x_2(N_d) \\ x_{N_l}(1) & & \vdots \\ & \ddots & \vdots \\ & & x_{N_l}(N_d) \end{bmatrix}_{N_t \times N_d} . \tag{4.9}$$

The number of layers here is $N_l = N_t - N_d + 1$. When $N_d = 1$, this layered ST code reduces to the original V-BLAST. When $N_d = N_t$, it becomes the STLCFC in Chapter 3. If ML decoding is used, the diversity order enabled by (4.9) is $N_d N_r$ and the transmission rate is $N_l = N_t - N_d + 1$ symbols pcu. Notice also that by tuning N_d, we can adjust the transmission rate and diversity. Different from the hybrid code of Section 4.2.1, this high-rate code design provides more flexibility to trade-off rate for diversity. An example is depicted in Figure 4.1 with $(N_t, N_r) = (8, 8)$ to compare these two designs. We observe that while achieving identical rates, the layered high-rate code in (4.9) enables higher diversity order (the maximum here is $N_t N_r = 64$).

4.3 FULL-DIVERSITY FULL-RATE ST CODES

Influenced by the ST designs of Section 4.2, which trade-off diversity for transmission rate, we might be tempted to believe that we always have to give up one figure of

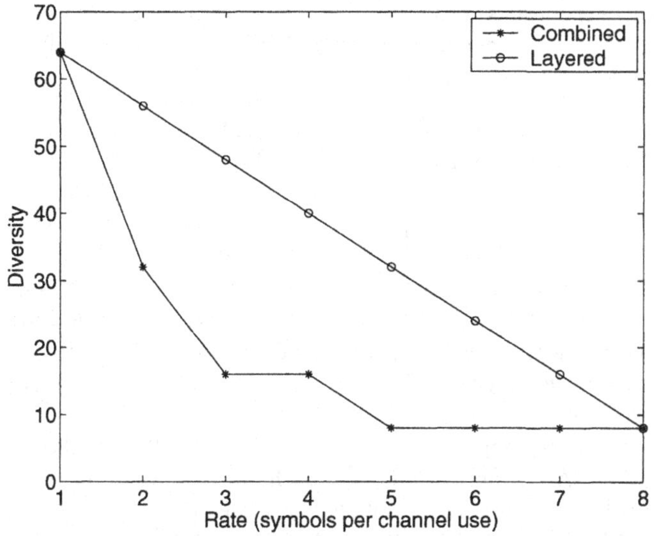

Figure 4.1 Rate-diversity trade-off

merit for the other. As we will see in this section, this is not necessarily the case. In fact, we will introduce a layered ST code capable of guaranteeing both full diversity (FD) as well as full transmission rate (FR) [74, 78, 182].

Based on average pairwise error probability analysis, we have already seen in Chapter 2 that the maximum diversity order provided by the $N_t \times N_r$ MIMO channel is $G_d^{\max} = N_t N_r$. Since it is possible to transmit up to one symbol per antenna per time slot (symbol period), the maximum possible transmission rate with N_t antennas is $\eta^{\max} = N_t$ symbols pcu. Parameters $G_d^{\max}$ and $\eta^{\max}$ quantify the full diversity and full rate, respectively. We have also seen in Section 4.1 that the V-BLAST architecture achieves full rate $\eta^{\max}$ but not full diversity, while the STLCFCs of Section 3.5 achieve full diversity $G_d^{\max}$ at rate $\eta = 1$ symbol pcu for any number of transmit antennas N_t. In the ensuing subsection, we present a design that we term naturally FDFR, because it guarantees full diversity and full rate for any (N_t, N_r) configuration.

4.3.1 FDFR Transceiver

To reach this ambitious FDFR objective, we need to design both transmitter and receiver judiciously. To this end, let us parse the information-bearing symbol stream into blocks $\boldsymbol{s}$ of length $N_s = N_t^2$, and divide each block into N_t subblocks $\{\boldsymbol{s}_g\}_{g=1}^{N_t}$ so that each subblock (that we will see soon forming a layer) has length N_t, for a total of $N_l = N_t$ layers. Each symbol block is LCF coded to form $\boldsymbol{u}$, which is accordingly split into $\{\boldsymbol{u}_g\}_{g=1}^{N_t}$, where $\boldsymbol{u}_g$ of the gth layer is given by $\boldsymbol{u}_g = \boldsymbol{\Theta}_g \boldsymbol{s}_g$. Matrix $\boldsymbol{\Theta}_g$ has entries drawn from $\mathbb{C}$ and implements LCF encoding as we detailed in Section 3.5.

The FDFR encoder consists of two modules. The inner module is a special layered ST multiplexer that is constructed from $\boldsymbol{u}_g$, $g \in [1, N_t]$, the N_t entries of $\boldsymbol{u}_g$ as follows [178, 182]:

$$\boldsymbol{X} = \begin{bmatrix} u_1(1) & u_{N_t}(2) & \cdots & u_2(N_t) \\ u_2(1) & u_1(2) & \cdots & u_3(N_t) \\ \vdots & \vdots & \cdots & \vdots \\ u_{N_t}(1) & u_{N_t-1}(2) & \cdots & u_1(N_t) \end{bmatrix} \quad \begin{matrix} \longrightarrow & \text{time} \\ \downarrow & \text{space.} \end{matrix} \tag{4.10}$$

Without LCF encoding, this ST multiplexer with "circularly wrapped around" layers was introduced by El Gamal and Hammons in [52, 79]. The main difference here is the rate-efficient fading-resilient precoding that is implemented in the outer LCF encoder module, which precedes the ST multiplexer and plays an instrumental role in realizing the FDFR goal.

It is worthwhile to compare the ST code matrix in (4.10) with those used by D-BLAST and V-BLAST. Since symbols in each entry of $\boldsymbol{X}$ are distinct, similar to V-BLAST, we have N_t^2 different symbols transmitted over N_t time slots, implying that the transmission rate is full, namely N_t. Because layers are wrapped around diagonals of $\boldsymbol{X}$ in a circular fashion, similar to D-BLAST, transmitted symbols are multiplexed across space and time, which offers the potential for full diversity. Both FR and FD properties are effected by the LCF encoder design. FR is ensured by having LCF matrices $\boldsymbol{\Theta}_g$ which are square and thus nonredundant. FD per layer is guaranteed by designing each $\boldsymbol{\Theta}_g$ as we have discussed in Section 3.5, while FD across layers is ensured by designing these LCF matrices to be layer specific. Using algebraic tools developed in [95, 313], such LCF encoders have the form

$$\boldsymbol{\Theta}_g = \beta^{g-1}\boldsymbol{\Theta}, \quad \forall g \in [1, N_t], \tag{4.11}$$

where $\boldsymbol{\Theta}$ is a unitary Vandermonde matrix and the scalar β is selected as we detail in Section 4.3.2.

With the inner and outer encoders designed respectively as in (4.10) and (4.11), and after concatenating receive vectors $\boldsymbol{y}(n)$ into a single vector, we can rewrite the resulting input-output relationship in a matrix-vector form as

$$\boldsymbol{y} = \sqrt{\frac{\bar{\gamma}}{N_t}}(\boldsymbol{I}_{N_t} \otimes \boldsymbol{H}) \begin{bmatrix} \boldsymbol{x}(1) \\ \vdots \\ \boldsymbol{x}(N_t) \end{bmatrix} + \boldsymbol{w}, \tag{4.12}$$

where $\boldsymbol{I}_{N_t}$ is the $N_t \times N_t$ identity matrix, $\otimes$ denotes the Kronecker product, and $\boldsymbol{x}(n)$ is the nth column of $\boldsymbol{X}$ in (4.10). The latter can be expressed as [cf. (4.11)]

$$\boldsymbol{x}(n) = [(\boldsymbol{P}_n \boldsymbol{D}_\beta) \otimes \boldsymbol{\theta}_n^T]\boldsymbol{s}, \tag{4.13}$$

where $\boldsymbol{\theta}_n^T$ denotes the nth row of $\boldsymbol{\Theta}$, while $\boldsymbol{P}_n$ is a permutation matrix and $\boldsymbol{D}_\beta$ is a diagonal matrix defined, respectively, as

$$\boldsymbol{P}_n := \begin{bmatrix} \boldsymbol{0} & \boldsymbol{I}_{n-1} \\ \boldsymbol{I}_{N_t-n+1} & \boldsymbol{0} \end{bmatrix} \quad \text{and} \quad \boldsymbol{D}_\beta := \text{diag}[1, \beta, \ldots, \beta^{N_t-1}]. \tag{4.14}$$

Upon defining $\mathcal{H} := I_{N_t} \otimes H$ and the unitary matrix [cf. (4.11)]

$$\Phi := \begin{bmatrix} (P_1 D_\beta) \otimes \theta_1^T \\ \vdots \\ (P_{N_t} D_\beta) \otimes \theta_{N_t}^T \end{bmatrix}, \tag{4.15}$$

we can rewrite (4.12) compactly as

$$y = \sqrt{\frac{\bar{\gamma}}{N_t}} \mathcal{H} \Phi s + w. \tag{4.16}$$

ML decoding can be employed to detect s from y optimally regardless of N_r, but possibly with high complexity. As we detail in Chapter 5, sphere decoding (SD) or semidefinite programming based decoding algorithms can also be used to achieve ML or near-ML performance. Among other factors, decoding complexity certainly depends on the length of s, which is $N = N_t^2$ here. When N_t is large, the decoding complexity is high even for near-ML decoders. To further reduce decoding complexity, one can resort to suboptimal (nulling-canceling based, or even linear) decoding. However, as with D-BLAST or V-BLAST, suboptimal decoders require that $N_r \geq N_t$.

We summarize our FDFR transceiver design in four steps:

S1) Given N_t and N_r, form information blocks s with length $N = N_t^2$.

S2) Design layer-specific LCF encoders according to (4.11), and use them to encode s to u.

S3) Multiplex the LCF-coded u using the ST code matrix (4.10), and transmit X from the N_t antennas.

S4) At the receiver, rely on (4.16) to decode s from y using SD- or NC-based algorithms.

If T_{coh} denotes channel coherence time and satisfies $T_{\text{coh}} \geq N_t$, and the channels satisfy A2.1), the main claim of the FDFR layered ST codes constructed as in (4.10) and (4.11) can be summarized as follows (see [182] for detailed proofs):

Proposition 4.1 *For information symbols s carved from $\mathbb{Z}[j]$ ($\mathbb{Z}[j]$ denotes the algebraic integer ring, with elements $p + jq$, where p, q are integers), and transmitted with the ST code matrix in* (4.10), *there exists at least one pair of (Θ, β) in* (4.11) *which enables full diversity $(N_t N_r)$ for the ST transmission in* (2.9), *at full-rate N_t symbols pcu.*

Proposition 4.1 reveals that selecting Θ and β is instrumental in enabling FDFR. Intuitively, as we commented earlier, Θ enables full diversity per layer (as shown in Chapter 3), while β "fully diversifies" transmissions across layers.

4.3.2 Algebraic FDFR Code Design

In this subsection we provide systematic design methods for selecting the pair $(\boldsymbol{\Theta}, \beta)$ for LCF encoding. The unitary Vandermonde matrix $\boldsymbol{\Theta}$ is chosen as in (3.71) to have the form

$$\boldsymbol{\Theta} = \boldsymbol{F}_{N_t}^{\mathcal{H}} \text{diag}[1, \alpha, \cdots, \alpha^{N_t-1}], \tag{4.17}$$

where $\boldsymbol{F}_{N_t}$ is the normalized $N_t \times N_t$ FFT matrix. Notice that $\boldsymbol{\Theta}$ in (4.17) is parameterized by a single parameter α. Also taking the scalar β in (4.11) into account, our ensuing design methodologies aim at (α, β) pairs that lead to $\boldsymbol{\Theta}_g$'s for which $\boldsymbol{X}$ in (4.10) ensures FDFR ST transmissions. To illustrate the design, we need the following notation: Let $\mathbb{Q}(j)$ be the smallest subfield of the set of complex numbers $\mathbb{C}$, including both $\mathbb{Q}$ and j; and let $\mathbb{Q}(j)(\alpha)$ denote the smallest subfield of $\mathbb{C}$, including both $\mathbb{Q}(j)$ and α, where α is algebraic over $\mathbb{Q}(j)$.

Design A: Select α such that the minimum polynomial of α over the field $\mathbb{Q}(j)$ has degree greater than or equal to N_t. Given α, choose β^{N_t} such that the minimum polynomial of β^{N_t} in the field $\mathbb{Q}(j)(e^{j2\pi/N_t})(\alpha)$ has degree greater than or equal to N_t.

Examples:

When $N_t = 2^k, k \in \mathbb{N}$, we select $\alpha = e^{j\pi/(2N_t)}$ and $\beta^{N_t} = e^{j\pi/(4N_t^2)}$.
When $N_t = 3$, we select $\alpha = e^{j\pi/9}$ and $\beta^{N_t} = e^{j\pi/54}$.
When $N_t = 5$, we select $\alpha = e^{j\pi/25}$ and $\beta^{N_t} = e^{j\pi/250}$.

Design B: Fixing $\beta^{N_t} = \alpha$, select α such that the minimum polynomial of α in the field $\mathbb{Q}(j)(e^{j2\pi/N_t})$ has degree greater than or equal to N_t^2.

Examples:

When $N_t = 2^k, k \in \mathbb{N}$, we select $\alpha = e^{j\pi/N_t^3}$.
When $N_t = 3$, we select $\alpha = e^{j\pi/54}$.
When $N_t = 5$, we select $\alpha = e^{j\pi/250}$.

Design C: First, select α such that the minimum polynomial of α in the field $\mathbb{Q}(j)$ has degree greater than or equal to N_t. Based on α, we can find one transcendental number in the field of $\mathbb{Q}(j)(e^{j2\pi/N_t})(\alpha)$. Alternatively, we can find a transcendental number α directly for the field $\mathbb{Q}(j)(e^{j2\pi/N_t})$.

Examples:

Given N_t, select $\boldsymbol{\Theta}$ as in Design A and let $\beta^{N_t} = e^{j/2}$.
Given N_t, select $\beta^{N_t} = \alpha$ and let $\alpha = e^{j/2}$.

Note that the transcendental number $e^{j/2}$ has also been used in [53]. According to Lindemann's theorem [84, Page 44], one can design transcendental numbers, e.g., $e^{jk}, \forall k \in \mathbb{Q}$. All three designs are capable of enabling full diversity. However, since we did not optimize the coding gain, theoretically, it is possible to find other FDFR encoders with improved coding gains. Later, we will test their relative performance through simulations.

4.3.3 Mutual Information Analysis

We have seen that the FDFR design implied by Proposition 4.1 enjoys full transmission rate. But how does it compare with the fundamental limits dictated on information rate by the ergodic capacity of MIMO channels? The answer is as favorable as V-BLAST and can be summarized as follows:

Proposition 4.2 *If the information symbols $\boldsymbol{s}$ are complex Gaussian distributed with zero mean and covariance matrix $(1/N_t)\boldsymbol{I}_N$, and the average signal-to-noise ratio (SNR) is $\bar{\gamma} := \mathcal{E}_s/N_0$, the mutual information of the LCF-coded layered FDFR design in Proposition 4.1 is given by*

$$\mathcal{C}_{\text{flat}} = \log_2 \det\left(\boldsymbol{I}_{N_r} + \frac{\bar{\gamma}}{N_t}\boldsymbol{H}\boldsymbol{H}^{\mathcal{H}}\right) \quad \text{bits pcu.} \tag{4.18}$$

Compared with the MIMO channel capacity provided in Chapter 2, the mutual information in (4.18) coincides with the instantaneous channel capacity. This means that our FDFR system incurs no mutual information loss. In contrast, although some other designs (e.g., OSTBC and STLCFC) can also achieve full diversity, they result in considerable mutual information loss, especially with complex constellations when $N_r > 1$ (see [178] for detailed comparisons).

4.3.4 Diversity-Rate-Performance Trade-offs

When the number of antennas is large, the diversity order $N_t N_r$ may be larger than what multi-antenna systems can exploit over the SNR range encountered in practice. At the same time, high performance and high rate come with high decoding complexity especially, when (near-)ML decoding is required. For this reason, with large (N_t, N_r) configurations, we may opt to give up performance gains (which may only show up for impractically high SNR) in order to reduce decoding complexity. The following two corollaries show that FDFR designs enjoy flexibility in trading off rate and performance with complexity. (As the complexity metric here, we take the block length that is to be decoded.)

Corollary 4.1 (Performance-complexity trade-off) *Keeping the same transmission rate (bits pcu), two performance-complexity trade-offs arise:*
(i) (diversity-complexity trade-off) *Using smaller LCF encoders $\boldsymbol{\Theta}_g$ with size $N_d < N_t$, we can lower the achieved diversity order to $N_d N_r < N_t N_r$ but also reduce the decoding block size to $N_d N_t$.*
(ii) (modulation-complexity trade-off) *We can combine several layers to one layer, provided that we increase the constellation size. If we eliminate N_z layers in the ST code matrix of the FDFR design by setting $\boldsymbol{u}_{g_1} = \cdots = \boldsymbol{u}_{g_{N_z}} = \boldsymbol{0}$, full diversity is maintained at reduced decoding block length $N_t(N_t - N_z)$, which obviously decreases as N_z increases.*

Regarding the diversity-complexity trade-off, it is worth noting that if we select $N_d = 1$, the resulting design reduces to V-BLAST. Instead of sacrificing performance, an alternative means of reducing decoding complexity is to decrease the transmission rate. This can be accomplished when $\min(N_t, N_r)$ is large, because we can then give up some rate to reduce decoding complexity. Similar to full diversity, full rate is not always necessary. For example, instead of having N_t layers in (4.10), we can design $\boldsymbol{X}$ with $N_t - 1$ or $N_t - 2$ layers. The following corollary quantifies this rate-complexity trade-off:

Corollary 4.2 (Rate-complexity trade-off) *If for the encoders in* (4.10) *and* (4.11), *we eliminate* N_z *layers by letting* $\boldsymbol{u}_{g_1} = \cdots = \boldsymbol{u}_{g_{N_z}} = \mathbf{0}$, *then ML (or near-ML) decoding collects the full diversity* $N_t N_r$ *with decoding block length* $N_t(N_t - N_z)$ *and transmission rate* $N_t - N_z$ *symbols pcu.*

Corollary 4.2 implies that when the entries of $\boldsymbol{s}$ are drawn from a fixed constellation, as the transmission rate increases (N_z decreases), the decoding complexity increases as well. Note that when the number of "null layers" $N_z > 0$, the condition $N_r \geq N_t$ for SD [109] or NC algorithms is relaxed to $N_r \geq N_t - N_z$. Here we have considered the trade-offs when the channel has fixed coherence time. If the latter varies, the trade-offs vary accordingly.

When the number of transmit antennas (N_t) is large, the block length (N_t^2) is also large. The affordable decoding complexity may not be enough to achieve FDFR. In this case, the diversity-rate trade-off we summarize next is well motivated.

Corollary 4.3 (Diversity-rate trade-off) *If the maximum affordable decoding block length* $N < N_t^2$, *FD and FR cannot be achieved simultaneously. By adjusting the size of* $\boldsymbol{\Theta}_g$, *it becomes possible to trade-off diversity for rate.*

To illustrate certain aspects of these desirable trade-offs enabled by the layered FDFR design, let us consider the following example.

Example 4.1 Consider an $(N_t, N_r) = (4, 1)$ system and suppose that the maximum block size our decoder's complexity can afford is $N = 12$. The following two designs (case 1 and case 2) in (4.19) illustrate the rate-diversity trade-off. Design case 1 contains four layers but each layer contains only three symbols. In this case, the transmission rate is 4, but the diversity order is only 3. Similarly for design case 2, we have three layers and each layer contains four symbols. So the transmission rate is 3, but the diversity order is 4.

Remark 4.1 The diversity-rate trade-offs described in Corollaries 4.1 to 4.3 are different from the diversity-multiplexing trade-off dealt with in [326]. As we clarified in Section 2.6, any fixed code has zero spatial multiplexing gain in the trade-off reported in [326]; while this section's trade-off refers to the transmission rate in symbols or bits pcu as detailed in Section 2.6. Notwithstanding, here the transmission rate does not vary with the SNR.

$$\text{case 1}: \quad \begin{bmatrix} u_1(1) & u_4(2) & u_3(3) \\ u_2(1) & u_1(2) & u_4(3) \\ u_3(1) & u_2(2) & u_1(3) \\ u_4(1) & u_3(2) & u_2(3) \end{bmatrix}$$

$$\text{case 2}: \quad \begin{bmatrix} u_1(1) & u_3(2) & 0 & u_2(4) \\ u_2(1) & u_1(2) & u_3(3) & 0 \\ 0 & u_2(2) & u_1(3) & u_3(4) \\ u_3(1) & 0 & u_2(3) & u_1(4) \end{bmatrix} \qquad (4.19)$$

	case 1	case 2
Rate (symbols pcu)	4	3
Diversity order	3	4

4.4 NUMERICAL EXAMPLES

In this section we rely on numerical examples to verify the analytical claims pertaining to the FDFR designs. The SNR is defined as the total transmitted signal power from N_t transmit-antennas versus noise power.

Example 4.2 (Flat fading channels with $N_t = 3$) Here, we use BPSK to signal over an $(N_t, N_r) = (3, 3)$ configuration at transmission rate 3 bits pcu. Channels are i.i.d. Gaussian distributed with zero mean and unit variance. The channel coherence time is greater than N_t. FDFR transmissions are decoded using the SD algorithm we detail in Chapter 5. First, we compare performance of the three different FDFR encoders given in Section 4.3. We deduce from Figure 4.2 that the three designs A to C lead to quite similar performance and exactly identical diversity order. To demonstrate the merits of FDFR, we next compare the layered FDFR system with one performance-oriented representative, namely the STLCFC of [313], and one rate-oriented representative, namely V-BLAST [299]. To maintain the same transmission rate, we employ 8-QAM for STLCFC and BPSK for V-BLAST. For V-BLAST we perform SD per time slot. Because the maximum achievable diversity for uncoded V-BLAST is only N_r, at high SNR it exhibits worse performance than STLCFCs. Figure 4.2 illustrates that the FDFR scheme outperforms both V-BLAST and STLCFC, because V-BLAST does not achieve full diversity, whereas STLCFC incurs rate loss. Also, we observe that the FDFR scheme achieves the full diversity order as STLCFC does.

Example 4.3 (Trade-offs with $N_t = 4$) In this example we use an $(N_t, N_r) = (4, 4)$ configuration over flat fading channels, in order to demonstrate the performance-rate-complexity trade-offs. The transmission power is fixed in all schemes tested. First, we confirm the performance-complexity trade-off in Corollary 4.1. Suppose that the rate is fixed at 4 bits pcu. We test three designs:

Case 1) BPSK per layer and four layers as in (4.10).

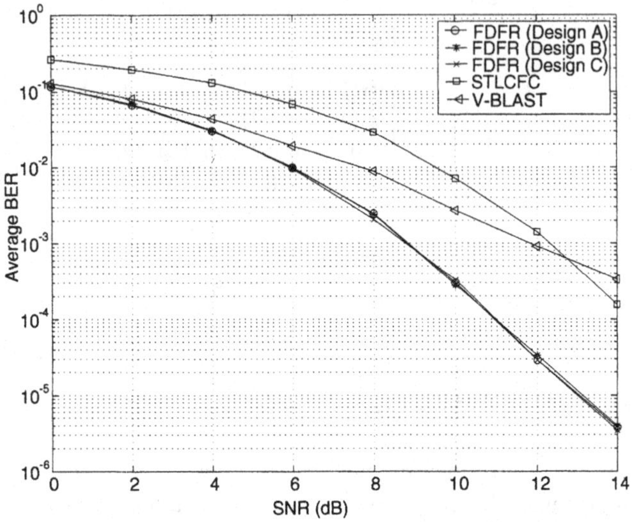

Figure 4.2 Complexity-performance trade-offs

Case 2) QPSK per layer and two layers.

Case 3) 16-QAM and one layer.

The ST code matrices for these three cases are given as follows:

$$\begin{aligned}
&Case\ 1): && \begin{bmatrix} u_1(1) & u_4(2) & u_3(3) & u_2(4) \\ u_2(1) & u_1(2) & u_4(3) & u_3(4) \\ u_3(1) & u_2(2) & u_1(3) & u_4(4) \\ u_4(1) & u_3(2) & u_2(3) & u_1(4) \end{bmatrix} \\
&Case\ 2): && \begin{bmatrix} u_1(1) & 0 & u_2(3) & 0 \\ 0 & u_1(2) & 0 & u_2(4) \\ u_2(1) & 0 & u_1(3) & 0 \\ 0 & u_2(2) & 0 & u_1(4) \end{bmatrix} \\
&Case\ 3): && \begin{bmatrix} u_1(1) & 0 & 0 & 0 \\ 0 & u_1(2) & 0 & 0 \\ 0 & 0 & u_1(3) & 0 \\ 0 & 0 & 0 & u_1(4) \end{bmatrix}
\end{aligned} \tag{4.20}$$

The BER performance for Cases 1) to 3) is depicted in Figure 4.3. We notice that all three designs achieve similar diversity order when SD is performed at the receiver. However, they exhibit considerably different coding gains, mainly due to their distinct

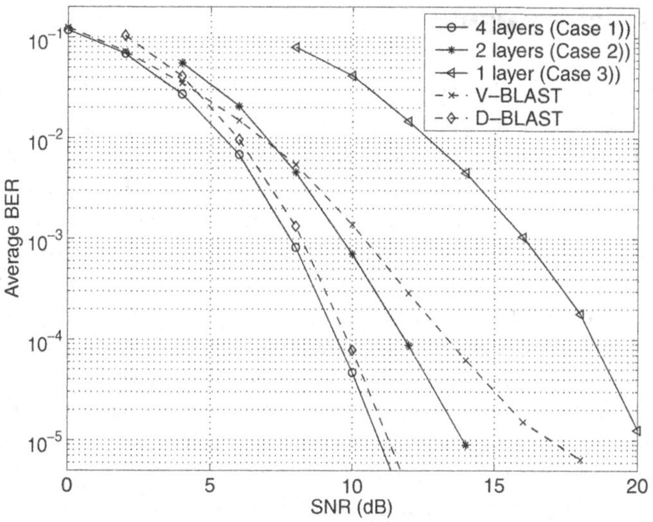

Figure 4.3 Complexity-performance trade-offs

constellation sizes. The decoding complexity for these three schemes is

$$\begin{array}{cccc} & \text{Case 1)} & \text{Case 2)} & \text{Case 3)} \\ \text{complexity} & \mathcal{O}((16)^3) & \mathcal{O}((8)^3) & \mathcal{O}((4)^3). \end{array} \tag{4.21}$$

Maintaining the same rate, we deduce from (4.21) that by paying a penalty in decoding complexity, we gain in performance. In Figure 4.3 we also compare FDFR designs with V-BLAST and D-BLAST [68]. For D-BLAST we design the ST matrix as

$$\begin{bmatrix} u_1(1) & u_2(1) & u_3(1) & & & \\ & \ddots & & \ddots & & \ddots & \\ & & & u_1(4) & u_2(4) & u_3(4) \end{bmatrix}_{4\times 6}. \tag{4.22}$$

To maintain the same transmission rate (bits pcu) while ensuring affordable decoding complexity, we use three layers in the D-BLAST scheme and select QPSK modulation to maintain the same information rate. Both performance and decoding complexity of D-BLAST in (4.22) lie between those of Cases 1) and 2), while D-BLAST has longer decoding delay. Note that V-BLAST provides a good compromise between complexity and performance. In this example, V-BLAST exhibits decoding complexity comparable to Case 3), while it outperforms Case 3) over a large range of SNR values.

4.5 CLOSING COMMENTS

In this chapter we dealt with layered ST codes for flat fading MIMO channels which exploit the multiplexing gain effected by multiple antennas to enhance transmission rates beyond those achieved by single-antenna systems. We also studied hybrid ST codes that enable trading off diversity for transmission rate. Furthermore, we introduced an LCF-coded layered architecture which guarantees both full diversity and full transmission rate for any number of transmit- and receive-antennas. We compared the various design alternatives analytically from both diversity and mutual information perspectives, and tested their error performance using simulated examples. Recently, full-diversity full-rate designs have been reported to optimize the coding advantage based on this chapter's algebraic LCF code designs [282, 319].

We further delineated the emerging performance-rate-complexity trade-offs which offer valuable flexibility in practical deployment of full-diversity full-rate designs. As we mentioned in both Chapters 3 and 4, although ML decoding provides optimal performance, it requires exponential complexity. In Chapter 5 we introduce the sphere decoding algorithm, which can provide ML or near-ML coherent decoding of ST coded transmissions at polynomial (and often cubic) average complexity. Its soft version will allow for near-capacity performance of the layered ST codes presented in this chapter. Reduced complexity suboptimal decoders are also mentioned in Chapter 5. Although the ST codes in Chapters 3 and 4 have been designed for flat fading channels, with proper modifications they can be generalized to frequency-selective channels. These generalizations are discussed in Chapters 7 and 8.

5

Sphere Decoding and (Near-)Optimal MIMO Demodulation

Multiple-input multiple-output (MIMO) models appear in a number of modern communication systems. In addition to multi-antenna ST systems (see, e.g., [107, 313]), which constitute the main theme of this book, MIMO decoding is also encountered in code division multiple access (CDMA) receivers performing multi-user detection [32, 269], as well as in certain point-to-point single-antenna links when (de)modulators operate on a block-by-block basis [286, 288]. Recently, fast MIMO demodulation algorithms have also found application to efficient decoding of linear block error control codes [322]. In all these applications, we are given received blocks of size $N_r \times 1$, and we wish to estimate transmitted blocks of size $N_t \times 1$, whose entries are integers drawn from a finite alphabet $\mathcal{A}$ (i.e., constellation) of size $|\mathcal{A}|$. This problem, which we referred to as MIMO demodulation or MIMO decoding, is also known as sequence estimation, closest point estimation, lattice decoding, or integer least-squares (ILS) problem. The last term is perhaps more encompassing since it goes beyond telecommunications to applications where the usual least-squares (LS) formulation constrains the LS solution to have integer entries.

Albeit optimal, the maximum likelihood (ML) block detector based on exhaustive search typically exhibits prohibitively high complexity, $\mathcal{O}(|\mathcal{A}|^{N_t})$, since it grows exponentially with the block size N_t (the problem dimension). Furthermore, dynamic programming implemented with Viterbi's algorithm is not suitable for general block, multi-user, or MIMO detection, since the number of trellis states also grows exponentially with the problem's dimension. Aiming for reduced complexity decoding, suboptimal receivers have been employed in practice. Those can be classified into linear and nonlinear receivers. Linear receivers include zero-forcing (ZF) and minimum mean-square error (MMSE) detectors; nonlinear receivers include decision feedback (DF), nulling-canceling (NC), and variants relying on successive interference cancellation (see, e.g., [68,214,269] and references therein). In general, these suboptimal receivers exhibit fixed complexity that is a polynomial function of the problem dimension (typically, cubic in N_t). On the other hand, suboptimal options may lead to considerable error performance degradation relative to ML decoding.

Recently, realistic ML and near-ML decoding algorithms have been developed based on the idea of sphere decoding. With reasonably low memory requirements, sphere decoding algorithms (SDAs) can achieve exact ML or near-ML error performance for certain N_t, alphabet sizes $|\mathcal{A}|$, and signal-to-noise ratio (SNR) values of practical interest, at *average* decoding complexity that is a polynomial function of N_t when $N_r \geq N_t$. One alternative to SDA with only approximate ML performance but guaranteed complexity $\mathcal{O}(N_t^4)$ across all SNR values is offered by the class of lattice reduction algorithms (LRAs) [147,297,303,304]. Another interesting recent alternative to SDA is the probabilistic data association algorithm (PDA), which also offers approximate ML performance at polynomial complexity [175,210]. By casting the ILS setup to a relaxed convex optimization problem, semidefinite programming (SDP) algorithms have also been employed to enable performance close to ML at complexity roughly $\mathcal{O}(N_t^{3.5})$ [177,239,245,246].

This chapter highlights recent advances in sphere decoding algorithms for use in exact (or near-)ML optimal demodulation of MIMO systems. We mentioned in Chapters 3 and 4 that SDA can be used to decode ST block and layered coded multi-antenna transmissions over flat fading MIMO channels. Subsequently, we will see in Chapters 7 to 9 that SDA is also useful for decoding ST coded transmissions over frequency- and/or time-selective MIMO channels. The reason we focus on SDA rather than PDA, LRA, and SDP alternatives is twofold: SDA is the only one capable of returning the exact ML solution in general MIMO settings and appears to be more flexible in terms of striking desirable error-complexity trade-offs when it comes to near-ML performance. After introducing the ILS problem in the communications context, we first present the original sphere decoding algorithm and its recent improvements. We then describe a reduced-complexity suboptimal algorithm based on sphere decoding and rely on Monte Carlo simulations to compare the average complexity of various SDA alternatives. We also outline basic ideas behind soft sphere decoding and iterative (a.k.a. turbo) MIMO (de)modulators which enable approaching the capacity of MIMO channels. Finally, we provide the pseudo-code of an efficient SDA implementation in the Appendix to this chapter.

5.1 SPHERE DECODING ALGORITHM

Consider the following *real* generic MIMO model:

$$\boldsymbol{y} = \boldsymbol{H}\,\boldsymbol{s}_o + \boldsymbol{v}, \tag{5.1}$$

where $\boldsymbol{y}, \boldsymbol{v} \in \mathbb{R}^{N_r}$, $\boldsymbol{s}_o \in \mathbb{Z}^{N_t}$, $\boldsymbol{H} \in \mathbb{R}^{N_r \times N_t}$, and $\mathbb{Z}$ and $\mathbb{R}$ denote the sets of integers and real numbers, respectively. Matrix $\boldsymbol{H}$ is known, tall (i.e., $N_r \geq N_t$), and is assumed to have full column rank. Operating on $\boldsymbol{s}$, matrix $\boldsymbol{H}$ generates a lattice denoted by $\Lambda(\boldsymbol{H}) := \{\boldsymbol{x} = \boldsymbol{H}\,\boldsymbol{s} \mid \boldsymbol{s} \in \mathbb{Z}^{N_t}\}$. The ILS problem is as follows: Given $\boldsymbol{y} \in \mathbb{R}^{N_r}$ and the lattice Λ with a known *generator* $\boldsymbol{H}$, find the lattice vector $\hat{\boldsymbol{x}} \in \Lambda$ that minimizes the Euclidean distance from $\boldsymbol{y}$ to $\hat{\boldsymbol{x}}$. Also known as the closest-point solution, the ILS estimate is $\hat{\boldsymbol{x}} = \arg\min_{\boldsymbol{x}\in\Lambda} \|\boldsymbol{y} - \boldsymbol{x}\|^2$, where $\|\cdot\|$ represents the Euclidean vector norm.

In the communications context, $\boldsymbol{s}_o$, $\boldsymbol{y}$, and $\boldsymbol{v}$ are the transmitted, received, and additive noise vectors; whereas $\boldsymbol{H}$ contains the MIMO channel coefficients. Vector $\boldsymbol{v}$ is typically modeled as additive white Gaussian noise (AWGN) with zero mean and known variance $\sigma^2\boldsymbol{I}$, henceforth denoted as $\boldsymbol{v} \sim \mathcal{N}(\mathbf{0}, \sigma^2\boldsymbol{I})$. In wireless applications, $\boldsymbol{H}$ is random with entries drawn from a known (e.g., Gaussian) distribution. It typically adheres to a block fading model, according to which every realization of $\boldsymbol{H}$ remains invariant for a number of (say, N) consecutively transmitted vectors $\boldsymbol{s}$. Training can be used to estimate the channel over the blocks it remains constant, which justifies our assumption that $\boldsymbol{H}$ in (5.1) is known. Furthermore, the transmitted vector $\boldsymbol{s}$ is confined to a finite subset $\mathcal{A}^{N_t} \subset \mathbb{Z}^{N_t}$ instead of the entire integer lattice $\mathbb{Z}^{N_t}$ (i.e., the entries of $\boldsymbol{s}$ are drawn from an M-ary PAM constellation).

We should mention at the outset that our focus on real constellations and real MIMO channels is only for notational brevity and entails no loss of generality. Indeed, by concatenating the real and imaginary parts of the received vector in the complex case, we can readily convert the complex counterpart of (5.1) to the following real one [cf. (3.78)]:

$$\begin{bmatrix}\Re(\boldsymbol{y})\\ \Im(\boldsymbol{y})\end{bmatrix} = \begin{bmatrix}\Re(\boldsymbol{H}) & -\Im(\boldsymbol{H})\\ \Im(\boldsymbol{H}) & \Re(\boldsymbol{H})\end{bmatrix}\begin{bmatrix}\Re(\boldsymbol{s}_o)\\ \Im(\boldsymbol{s}_o)\end{bmatrix} + \begin{bmatrix}\Re(\boldsymbol{v})\\ \Im(\boldsymbol{v})\end{bmatrix}, \tag{5.2}$$

where, as before, $\Re$ and $\Im$ denote real and imaginary parts of the corresponding vectors and matrices. Notice that dimensions in (5.2) are twice those of (5.1), but they have identical form, confirming that this chapter's results over the reals carry on to the complex case as well. It is also possible to tackle the complex case by working directly with the complex MIMO model, which treats PAM, QAM, and PSK modulations in a unifying manner [118, 210]. The two approaches have no essential difference in complexity, since instead of working with a real model of twice the dimension, the complex model involves symbols drawn from a two-dimensional constellation with size equal to the square of its real counterpart. Differences may appear, however, in their implementation stage when, for example, VLSI technology is used [33]. Furthermore, working directly with the complex model should be preferred for PSK transmissions whose real and imaginary parts are generally correlated and not integer valued.

Given $\boldsymbol{y}$, $\boldsymbol{H}$, and ignoring the finite alphabet (FA) constraints, the usual *unconstrained* LS (a.k.a. ZF) solution is (^{T} denotes transposition)

$$\hat{\boldsymbol{s}} = (\boldsymbol{H}^T\boldsymbol{H})^{-1}\boldsymbol{H}^T\boldsymbol{y}. \tag{5.3}$$

Computing $\hat{\boldsymbol{s}}$ incurs complexity $\mathcal{O}(N_t^3)$ to evaluate the pseudoinverse $(\boldsymbol{H}^T\boldsymbol{H})^{-1}\boldsymbol{H}^T$ and $\mathcal{O}(N_t^2)$ to obtain $\hat{\boldsymbol{s}}$. However, since the pseudoinverse can be reused for decoding a number (N) of transmitted blocks, the complexity for linear ZF decoding is basically quadratic. As far as error performance is concerned, ZF is clearly suboptimal

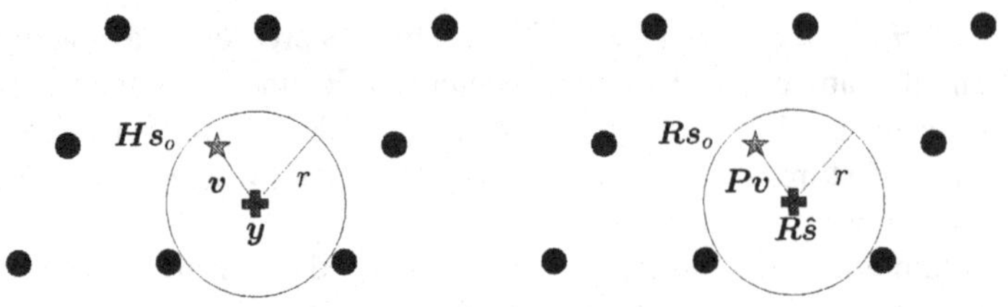

Figure 5.1 Sphere decoding

since it ignores the FA constraints. Adhering to the FA constraints requires "slicing" (quantizing) each entry of the ZF solution vector; but this two-step solution is generally suboptimal in the ML sense.

The optimal solution is provided by the ML demodulator, which for a given $\boldsymbol{H}$ realization under the AWGN model in (5.1) detects the transmitted vector as

$$\hat{\boldsymbol{s}}_{\text{ML}} = \arg\min_{\boldsymbol{s}\in\mathcal{A}^{N_t}} \|\boldsymbol{y} - \boldsymbol{H}\boldsymbol{s}\|^2. \tag{5.4}$$

Viewing $\boldsymbol{y}$ as a point in the N_r-dimensional space, (5.4) suggests searching exhaustively over all possible $|\mathcal{A}|^{N_t}$ candidate vectors $\boldsymbol{s}$ and selecting the one for which $\boldsymbol{H}\boldsymbol{s}$ lies closest to (has smallest Euclidean distance from) $\boldsymbol{y}$. Unfortunately, this exponential complexity is prohibitive in most practical scenarios where $|\mathcal{A}|$ can typically range from 2 to 16 and N_t from 4 to 32. The basic premise of the sphere decoding algorithm lies exactly at this point of paramount practical importance: it can provide an ML or near-ML optimal demodulator incurring polynomial complexity (preferably cubic or even quadratic in N_t) for a number of constellation sizes, SNR ranges, and block sizes encountered in MIMO decoding applications.

As its name indicates, SDA searches for transmitted candidates within a (hyper)sphere of radius r centered at the received vector. If the true vector transmitted is $\boldsymbol{s}_o$, SDA examines all $\boldsymbol{H}\boldsymbol{s}$ candidates in a hypersphere of radius r centered at $\boldsymbol{y}$. This is illustrated in Figure 5.1 (left), where the star represents the lattice point $\boldsymbol{H}\boldsymbol{s}_o$ and the cross denotes $\boldsymbol{y}$. Noise $\boldsymbol{v}$ carries $\boldsymbol{H}\boldsymbol{s}_o$ to $\boldsymbol{y}$. In its original form [66, 276], the SDA entails three basic steps: (i) selecting a search radius r, (ii) initializing the search with the unconstrained LS solution in (5.3), and (iii) searching exhaustively but efficiently within the chosen sphere of fixed radius r.

5.1.1 Selecting a Finite Search Radius

Since searching exhaustively with an infinite radius amounts to the exponentially complex ML approach, a finite (and preferably small) radius r clearly leads to a reduced complexity search. On the other hand, if r is too small, it becomes less likely to find the ML solution when searching around a small spherical neighborhood of $\boldsymbol{y}$. One safe option is to set the initial radius equal to $r_{\text{ZF}} := \|\boldsymbol{y} - \boldsymbol{H}\hat{\boldsymbol{s}}_{\text{ZF}}\|$, where $\hat{\boldsymbol{s}}_{\text{ZF}}$ is the ZF solution quantized according to the FA. If r_{ZF} is still large, the NC solution can be used instead to set the radius at $r_{\text{NC}} := \|\boldsymbol{y} - \boldsymbol{H}\hat{\boldsymbol{s}}_{\text{NC}}\|$. Since $r_{\text{ML}} \leq r_{\text{NC}} \leq r_{\text{ZF}}$

in general, the ML solution which lies on the sphere of radius r_{ML} is guaranteed to lie on or inside the sphere of radius r_{NC}. The issue is whether searching inside this sphere will yield the ML solution with polynomial complexity.

Although no specific finite radius was, suggested in the original SDA [66, 276], it is also possible to specify one based on the AWGN model [109], which asserts confidence in finding the ML solution. Consider the sphere of radius r centered at $\boldsymbol{y}$. The probability that $\boldsymbol{H}\boldsymbol{s}_o$ lies within this sphere is clearly given by $\Pr(\|\boldsymbol{y}-\boldsymbol{H}\boldsymbol{s}_o\|^2 \leq r^2) = \Pr(\|\boldsymbol{v}\|^2 \leq r^2)$. But under the AWGN model, $\|\boldsymbol{v}\|^2$ is central-chi-square distributed with N_r degrees of freedom and probability density function (pdf)

$$f_V(v) = \frac{1}{(2\sigma^2)^{N_t/2}\Gamma(N_t/2)} v^{N_t/2-1} \exp\left(-\frac{v}{2\sigma^2}\right), \quad v \geq 0,$$

where Γ denotes the gamma function. Using this pdf, the radius of the sphere centered at $\boldsymbol{y}$ containing $\boldsymbol{H}\boldsymbol{s}_o$ with probability p can be calculated by solving numerically (with respect to r_p) the equation

$$\Pr(\|\boldsymbol{v}\|^2 \leq r_p^2) = \int_0^{r_p^2} f_V(v)\, dv = p. \tag{5.5}$$

In addition to providing a systematic means of selecting the radius, (5.5) quantifies how likely it is for SDA to yield the ML solution when confining the search within a finite radius sphere. For instance, with a recommended $p = 0.99$, searching within the sphere of radius $r_{0.99}$ determined as in (5.5) ensures the designer that the ML solution will be found there with probability 0.99. We should clarify, though, that SDA is not guaranteed to yield the ML solution at polynomial complexity for any *fixed* radius of finite size. Despite this, being able to claim exact ML optimality with high probability after searching within a sphere of (preferably small) radius is an attractive feature not shared by alternative reduced-complexity algorithms (e.g., [175, 177]) that are only approximately ML over SNR values of practical concern.

In the next section we will see that either by using a sequence of search spheres with increasing radii, or, by starting with a sphere of large radius and successively reducing its size, it becomes possible to obtain exact ML or at least near-ML performance at complexity that remains polynomial for a number of practical constellation sizes, SNR values, and N_t dimensions. But before presenting these variates, let us see how SDA searches efficiently within a single sphere of a fixed radius, starting with an efficient initialization step.

5.1.2 Initializing with Unconstrained LS

The starting point of SDA is the unconstrained LS solution $\hat{\boldsymbol{s}}$ in (5.3) obtained using the QR-decomposition of the matrix $\boldsymbol{H}$. This first step enables a recursive search reminiscent of backward substitution within an equivalent sphere centered at $\boldsymbol{H}\hat{\boldsymbol{s}}$ instead of $\boldsymbol{y}$. Specifically, we will see that $\|\boldsymbol{y}-\boldsymbol{H}\boldsymbol{s}\|^2 = \|\boldsymbol{R}(\boldsymbol{s}-\hat{\boldsymbol{s}})\|^2 + c$, where $\boldsymbol{R}$

is the $N_t \times N_t$ upper triangular factor in the QR-decomposition[1] $\boldsymbol{H} = \boldsymbol{Q}\boldsymbol{R}$ and c is a constant that does not depend on $\boldsymbol{s}$.

To establish equivalence of the aforementioned spheres, let $\boldsymbol{P} := \boldsymbol{H}(\boldsymbol{H}^T\boldsymbol{H})^{-1}\boldsymbol{H}^T$ and $\boldsymbol{P}^{\perp} := \boldsymbol{I} - \boldsymbol{H}(\boldsymbol{H}^T\boldsymbol{H})^{-1}\boldsymbol{H}^T = \boldsymbol{I} - \boldsymbol{P}$ denote the orthonormal matrices projecting $\boldsymbol{y}$ to the column space of $\boldsymbol{H}$ and its orthogonal complement, respectively. Since $\boldsymbol{H}\boldsymbol{s}_o$ lies in the column space of $\boldsymbol{H}$, it follows that $\boldsymbol{P}^{\perp}\boldsymbol{y} = \boldsymbol{P}^{\perp}\boldsymbol{v}$. Using the latter, we can readily recognize that $||\boldsymbol{y} - \boldsymbol{H}\boldsymbol{s}||^2 = ||\boldsymbol{H}\boldsymbol{s} - (\boldsymbol{P} + \boldsymbol{P}^{\perp})\boldsymbol{y}||^2 = ||\boldsymbol{H}\boldsymbol{s} - \boldsymbol{P}\boldsymbol{y} - \boldsymbol{P}^{\perp}\boldsymbol{v}||^2 = ||\boldsymbol{H}(\boldsymbol{s} - \hat{\boldsymbol{s}})||^2 + ||\boldsymbol{P}^{\perp}\boldsymbol{v}||^2 = ||\boldsymbol{R}(\boldsymbol{s} - \hat{\boldsymbol{s}})||^2 + c$, where in deriving the third equality we used that $\boldsymbol{H}\hat{\boldsymbol{s}} = \boldsymbol{P}\boldsymbol{y}$ for the unconstrained LS solution $\hat{\boldsymbol{s}}$ and Pythagoras' theorem (recall that $\boldsymbol{P}^{\perp}\boldsymbol{v}$ is orthogonal to the column space of $\boldsymbol{H}$); while for the fourth equality we used the orthonormality of the $\boldsymbol{Q}$ factor and the definition $c := ||\boldsymbol{P}^{\perp}\boldsymbol{v}||^2$. Notice that c does not depend on $\boldsymbol{s}$; in fact, it vanishes in the square nonsingular case $(N_r = N_t)$, since $\boldsymbol{P}^{\perp} = \boldsymbol{0}$. This establishes the equivalence

$$\begin{aligned}\hat{\boldsymbol{s}}_{\mathrm{ML}} &= \arg\min_{\boldsymbol{s}\in\mathcal{A}^{N_t}} ||\boldsymbol{y} - \boldsymbol{H}\boldsymbol{s}||^2 = \arg\min_{\boldsymbol{s}\in\mathcal{A}^{N_t}} ||\boldsymbol{R}(\boldsymbol{s} - \hat{\boldsymbol{s}})||^2 \\ &= \arg\min_{\boldsymbol{s}\in\mathcal{A}^{N_t}} ||\hat{\boldsymbol{y}} - \boldsymbol{R}\boldsymbol{s}||^2, \end{aligned} \tag{5.6}$$

where $\hat{\boldsymbol{y}} := \boldsymbol{R}\hat{\boldsymbol{s}}$. Equation (5.6) confirms that instead of searching within a sphere to locate the $\boldsymbol{s}$ candidate for which the point $\boldsymbol{H}\boldsymbol{s}$ comes closest to the center $\boldsymbol{y}$, we can look for $\boldsymbol{s}$ candidates to find the point $\boldsymbol{R}\boldsymbol{s}$ (or equivalently, $\boldsymbol{H}\boldsymbol{s}$) closest to the center of the sphere located at $\boldsymbol{R}\hat{\boldsymbol{s}}$ (or equivalently $\boldsymbol{H}\hat{\boldsymbol{s}}$); see also Figure 5.1 (right). By changing the center of the sphere from $\boldsymbol{y}$ to $\boldsymbol{R}\hat{\boldsymbol{s}}$, we do not lose information because the LS estimate $\hat{\boldsymbol{s}}$ and the received data $\boldsymbol{y}$ are related through a one-to-one mapping under the full column rank assumption on $\boldsymbol{H}$. On the contrary, since $\hat{\boldsymbol{s}}$ is a projection of $\boldsymbol{y}$ to the "signal subspace," albeit suboptimal (since it ignores the FA constraints), this mapping mitigates noise effects by providing an initialization orthogonal to the "noise subspace." The robustness to noise offered by this initialization increases with the difference $N_r - N_t$. Moreover, the upper-triangular form of $\boldsymbol{R}$ enables an efficient recursive search within this sphere to find the (near-)ML optimal solution after imposing the FA constraints, as we detail next.

5.1.3 Searching Within the Fixed-Radius Sphere

Capitalizing on the upper-triangular structure of $\boldsymbol{R}$, we can express the norm $||\boldsymbol{R}(\boldsymbol{s} - \hat{\boldsymbol{s}})||^2$ componentwise, starting from the N_tth entry and proceeding backward to the

[1] To ensure uniqueness of the QR factors, we constrain the diagonal entries of $\boldsymbol{R}$ to be positive and set the lower-trianglar entries to zero (i.e., we select $\{R_{i,i} > 0\}_{i=1}^{N_t}$ and $R_{i,j} = 0, \forall\, i > j$).

first entry, as follows:

$$\begin{aligned}\|\boldsymbol{R}(\boldsymbol{s}-\hat{\boldsymbol{s}})\|^2 &= R^2_{N_t,N_t}\left[s_{N_t}-\hat{s}_{N_t}\right]^2\\ &\quad + R^2_{N_t-1,N_t-1}\left[s_{N_t-1}-\hat{s}_{N_t-1}+\frac{R_{N_t-1,N_t}}{R_{N_t-1,N_t-1}}(s_{N_t}-\hat{s}_{N_t})\right]^2\\ &\quad + \cdots + R^2_{1,1}\left[s_1-\hat{s}_1+\frac{R_{1,2}}{R_{1,1}}(s_2-\hat{s}_2)+\cdots+\frac{R_{1,N_t}}{R_{1,1}}(s_{N_t}-\hat{s}_{N_t})\right]^2\\ &= R^2_{N_t,N_t}(s_{N_t}-\rho_{N_t})^2 + R^2_{N_t-1,N_t-1}(s_{N_t-1}-\rho_{N_t-1})^2+\cdots\\ &\quad + R^2_{1,1}(s_1-\rho_1)^2, \end{aligned} \tag{5.7}$$

where $\rho_{N_t} := \hat{s}_{N_t}$ and $\rho_k := \hat{s}_k - \sum_{j=k+1}^{N_t}(s_j-\hat{s}_j)\,R_{k,j}/R_{k,k}$, for $1 \le k < N_t$. It is important to remember that ρ_k depends on $s_{k+1},\ldots,s_{N_t}$.

Confining our search within the sphere of radius r centered at $\boldsymbol{R}\hat{\boldsymbol{s}}$ dictates a set of $\boldsymbol{s}$ candidates satisfying the inequality $\|\boldsymbol{R}(\boldsymbol{s}-\hat{\boldsymbol{s}})\|^2 < r^2$, where for simplicity in exposition we have omitted the constant c that can be subtracted from the radius r. This inequality constrains the candidates for the last entry s_{N_t} of $\boldsymbol{s}$ to be such that $R^2_{N_t,N_t}(s_{N_t}-\rho_{N_t})^2 < r^2$; or equivalently, $\lceil \hat{s}_{N_t} - r/R_{N_t,N_t}\rceil \le s_{N_t} \le \lfloor \hat{s}_{N_t} + r/R_{N_t,N_t}\rfloor$, where $\lceil\cdot\rceil$ and $\lfloor\cdot\rfloor$ denote the ceiling and floor operators, respectively. If no s_{N_t} from the prescribed FA can be found to satisfy the latter with the $\hat{s}_{N_t}$ and R_{N_t,N_t} values obtained from the LS initialization, SDA with a fixed radius declares infeasibility. (This motivates increasing the chosen radius r, as we will see later.) On the other hand, suppose that a list of n_{N_t} candidates $\{s^{(n)}_{N_t}\}_{n=1}^{n_{N_t}}$ satisfy this inequality. Clearly, n_{N_t} can be as small as 0 (empty list) and as large as $|\mathcal{A}|$ (the FA size). For $n_{N_t} > 0$, we select one candidate, say the one closest to the lower bound $\lceil \hat{s}_{N_t} - r/R_{N_t,N_t}\rceil$ of the inequality, which we denote as $s^{(1)}_{N_t}$. We then proceed to the next (N_t-1)st dimension (entry of $\boldsymbol{s}$) that must satisfy the inequality $R^2_{N_t-1,N_t-1}(s_{N_t-1}-\rho_{N_t-1})^2 < r^2 - R^2_{N_t,N_t}(s_{N_t}-\rho_{N_t})^2$, which in turn constrains the s_{N_t-1} integer values to lie in an interval dictated by (a) the $s^{(1)}_{N_t}$ value chosen at the previous step, (b) the selected radius r, and (c) the entries of $\boldsymbol{R}$ and $\hat{\boldsymbol{s}}$ obtained from the initialization phase. Notice that with any fixed $s^{(n)}_{N_t}$ candidate, the radius constraining s_{N_t-1} candidates is smaller than r.

Similarly for each entry, we deduce that searching within a sphere of radius r imposes a *causal necessary condition*, which for the generic kth entry can be written as

$$R^2_{k,k}(s_k-\rho_k)^2 < r^2 - \sum_{i=k+1}^{N_t} R^2_{i,i}(s_i-\rho_i)^2. \tag{5.8}$$

Accordingly, given candidates of $\{s_i\}_{i=k+1}^{N_t}$, this condition constrains the admissible candidates of s_k to lie in the interval

$$\lceil \rho_k - \tau_k/R_{k,k}\rceil \le s_k \le \lfloor \rho_k + \tau_k/R_{k,k}\rfloor, \qquad k = N_t,\ldots,1, \tag{5.9}$$

where $\tau_{N_t} := r$, and $\tau_k := [r^2 - \sum_{i=k+1}^{N_t} R^2_{i,i}(s_i-\rho_i)^2]^{1/2}$ for $k = N_t-1,\ldots,1$ represents the radius reduction that is effected after fixing the candidates of $\{s_i\}_{i=k+1}^{N_t}$.

Clearly, if all possible lists of candidates per dimension are empty, the initially selected radius r must be increased. Supposing that this is not the case, the SDA search proceeds as follows.

As we move downward from dimension N_t to 1, we pick one candidate per dimension[2] (always the closest to the lower bound in (5.9)) until we reach a dimension for which the list is empty. If this does not happen, collecting each candidate entry we picked per dimension yields an admissible or valid candidate vector $\boldsymbol{s}^{(1)} = [s_{N_t}^{(1)}, s_{N_t-1}^{(1)}, \ldots, s_2^{(1)}, s_1^{(1)}]^T$. (Admissible here means that $\boldsymbol{s}^{(1)}$ obeys the FA constraints and the distance of $\boldsymbol{H}\boldsymbol{s}^{(1)}$ from $\boldsymbol{y}$ is smaller than r.) Of course, this does not imply that $\boldsymbol{s}^{(1)}$ is necessarily the ML solution, and one has to search more to ensure that. But even if $\boldsymbol{s}^{(1)}$ is not the ML solution, it is useful in two counts: first, we know that the exact ML solution can be found in a sphere of smaller radius $r^{(1)} := \|\boldsymbol{R}(\boldsymbol{s}^{(1)} - \hat{\boldsymbol{s}})\| < r$ that we can use in our subsequent search; and second, we can work our remaining search from dimension 1 to 2 up to N_t to improve it, while reusing already computed distances,

$$d_k^2 = \sum_{i=k}^{N_t} R_{i,i}^2 (s_i - \rho_i)^2 = d_{k+1}^2 + R_{k,k}^2 (s_k - \rho_k)^2.$$

Specifically, using the reduced radius $r^{(1)}$ we can now retain $s_1^{(1)}$ and consider, if available, the second candidate $s_2^{(2)}$ from the list of candidates for dimension 2. This candidate generates a new list of s_1 candidates. But in the corresponding distances d_1^2, the d_2^2 summand is already available.

In general, d_k^2 is basically what one needs to store per dimension and is updated *in place* for each candidate of s_k. This implies that online operation of SDA requires memory that is only linear in N_t. (Certainly, $\boldsymbol{R}$ (or $\boldsymbol{H}$) must be stored too as in any MIMO demodulation.) If an empty list is obtained when we reach dimension $k_0 \geq 1$, we go back to dimension $k_0 + 1$, try another candidate from its list, and proceed again with a depth-first search. If the radius r is sufficiently large, we are assured that after searching exhaustively, all these lists of candidates per dimension within a sphere (or spheres of shrinking radii) will yield the exact ML solution. Relative to the $\mathcal{O}(|\mathcal{A}|^{N_t})$ complexity of the exhaustive ML search, the reduction in average complexity brought by SDA (which we will see later to incur complexity approximately $\mathcal{O}({N_t}^3)$ in a number of practical settings) comes from three factors: (i) the AWGN statistics judiciously suggest the radius of a sphere that provides a search space of reduced size; (ii) the LS solution mitigates noise (when $N_r > N_t$) and guides the exhaustive search around a reasonably good initial candidate; (iii) the recursive search enabled by the QR-decomposition can be efficiently implemented through the causally dependent necessary conditions in (5.9), which restrict candidate entries satisfying the FA constraints to intervals of sizes that can even decrease over successive depth-first passes of the search process; and (iv) distances involved in the search are efficiently reused.

[2]This type of search is often referred to as depth-first search.

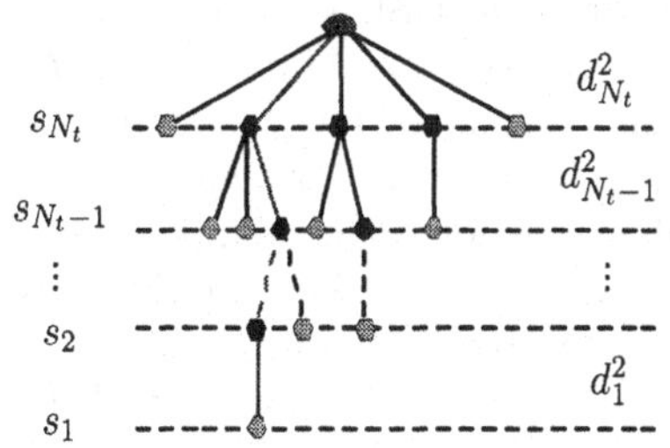

Figure 5.2 Tree associated with the SDA search

In fact, a search tree is implicitly constructed during the SDA search based on the necessary conditions in (5.9). Such a search tree is depicted in Figure 5.2, where nodes at the first level (indicated by the top dashed line) are possible candidates for dimension N_t; nodes at the second level correspond to admissible candidates for dimension $N_t - 1$ for each possible candidate of level 1; and the tree construction goes on in a similar fashion down to all plausible candidates of dimension 1. How intermediate computations are fully reused is also demonstrated by this search tree. A weight $R^2_{k,k}(s_k - \rho_k)^2$ is associated with the branch connecting a node at the kth level to its parent node at level $k-1$. Similarly, the weight of the partial path from this node to the root is $\sum_{i=k}^{N_t} R^2_{i,i}(s_i - \rho_i)^2$, which corresponds to a given set of candidates for $\{s_{N_t}, \ldots, s_k\}$. The key point here is that all admissible candidates of $s_{k-1}, \ldots, s_1$ share this distance. Expressed differently, memory savings and efficiency in the SDA search by reusing intermediate calculations come from the fact that children of the same node in the tree share the calculated weight of the partial path from this node to the root.

5.2 AVERAGE COMPLEXITY OF THE SDA IN PRACTICE

As SDA can achieve exact ML or near-ML optimal decoding of MIMO transmissions at reduced complexity relative to the exponentially complex ML enumeration, quantifying its complexity analytically is of paramount importance. For a problem of dimension N_t, SDA's complexity can be measured using the number of floating-point operations N_{flop} required for decoding a transmitted vector $\boldsymbol{s}$; or equivalently, through the complexity exponent $C := \log_{N_t}(N_{\text{flop}})$. Given the random nature of the MIMO model in (5.1), N_{flop} is clearly random and *fully* characterizing even its first- and second-order statistics remains a challenging open research problem. Intuition and corroborating simulations suggest that N_{flop} depends on the channel and noise statistics, the dimension N_t, and the FA size $|\mathcal{A}|$, which is intimately related to the operational SNR.

Before outlining existing results on SDA's average complexity, let us consider its asymptotic behavior as SNR and N_t grow arbitrarily large. As SNR $\rightarrow \infty$, the AWGN vector $\boldsymbol{v}$ in (5.1) can be ignored, and the model simplifies to $\boldsymbol{y} = \boldsymbol{H}\boldsymbol{s}$. In this case, $\boldsymbol{s}$ can be recovered after performing matrix inversion, which requires $\mathcal{O}(N_t^3)$ flops and can be used for decoding $\boldsymbol{s}$ vectors at quadratic complexity $\mathcal{O}(N_t^2)$ per vector, as long as $\boldsymbol{H}$ remains invariant under the block fading channel model. Furthermore, for *any fixed* dimension N_t there exists an SNR threshold above which SDA's complexity approaches that of linear (ZF or MMSE) equalization. This is an advantage of SDA over the deterministically suboptimal SDP [177], whose complexity is just above cubic but does not decrease as SNR increases.

Under specific assumptions on the randomness of the MIMO model, analytical expressions for the *average* decoding complexity of SDA have been derived in [110] for several finite PAM constellations. We summarize next only a special case from [110] for binary PAM.

Proposition 5.1 *For $N_t = N_r$ and given that the channel $\boldsymbol{H}$ comprises i.i.d. $\mathcal{N}(0,1)$ entries, the AWGN vector $\boldsymbol{v}$ follows $\mathcal{N}(\boldsymbol{0}, \sigma^2\boldsymbol{I})$, and $\boldsymbol{H}$ is independent of $\boldsymbol{v}$, the average complexity of SDA for a binary PAM constellation and a fixed radius r is given by*

$$\bar{N}_{\text{flop}}(N_t, \text{SNR}, r) = \sum_{k=1}^{N_t}(2k+17)\sum_{l=0}^{k}\binom{k}{l}\int_0^{\frac{\alpha}{2(1+lSNR)}} \frac{\lambda^{k/2-1}}{\Gamma(k/2)} e^{-\lambda}\, d\lambda, \quad (5.10)$$

where SNR $:= E[s^2]/\sigma^2$, $\int_0^x [\lambda^{k/2-1}/\Gamma(k/2)]e^{-\lambda}\, d\lambda$ *is the incomplete gamma function, and* $\alpha := r/\sigma^2$.

Numerically evaluating (5.10) and generalized versions of it in [110] reveals that for a wide range of dimensions and SNR values of practical concern, the expected complexity is polynomial, with exponent typically ranging from 3 to 6. Although this pertains only to average complexity, we should underscore that $\bar{N}_{\text{flop}}$ in Proposition 5.1 represents a pessimistic average, since the complexity reduction effected by radius shrinking is not accounted for (primarily because it is difficult to quantify its effect analytically). On the other hand, notice that if $\boldsymbol{H}$ has complex i.i.d. entries, its real counterpart in (5.2) will have non-i.i.d. entries and Proposition 5.1 will no longer be applicable.

A conclusion seemingly contradicting Proposition 5.1 was reached recently in [128, 129], where it was argued that SDA's average complexity scales exponentially with N_t for a *fixed* SNR value. Specifically, [128, 129] established the following *lower bound* on SDA's average complexity:

Proposition 5.2 *For $N_r \geq N_t$ and given that the channel matrix $\boldsymbol{H}$ comprises i.i.d. $\mathcal{N}(0,1)$ entries, the AWGN vector $\boldsymbol{v}$ follows $\mathcal{N}(\boldsymbol{0}, \sigma^2\boldsymbol{I})$, and $\boldsymbol{H}$ is independent of $\boldsymbol{v}$, the average complexity of SDA for any PAM constellation $\mathcal{A}$ and a fixed radius r is lower bounded by*

$$\bar{N}_{\text{flop}}(N_t, \text{SNR}) \geq \frac{|\mathcal{A}|^{\eta N_t} - 1}{|\mathcal{A}| - 1}, \quad \eta = \frac{1}{2(2N_r \cdot \text{SNR} + 1)}, \quad (5.11)$$

where, as before, $\text{SNR} := E[s^2]/\sigma^2$ *and* s *is drawn uniformly from any PAM constellation of size* $|\mathcal{A}|$.

To further study the implication of the lower bound in (5.11), we depict it versus N_t in Figure 5.3 and versus SNR in Figure 5.4. From Figure 5.3 we observe that the complexity exponent grows linearly with N_t, which confirms that the complexity bound increases exponentially for any *fixed* SNR value as $N_t \to \infty$. On the other hand, we can deduce from Figure 5.4 that the complexity bound[3] also decreases approximately exponentially with increasing SNR values for any *fixed dimension* N_t. Certainly, when SNR and N_t take values over finite intervals of practical interest ($[\text{SNR}_{\min}, \text{SNR}_{\max}]$ and $[N_{t,\min}, N_{t,\max}]$), exponential functions behave as polynomials and what is important is whether polynomial (often cubic) average complexity can be afforded for the SNR and N_t intervals encountered in MIMO applications. Simulations indicate that this holds true in a number of practical MIMO settings, especially when using the recent SDA improvements we outline in the next section.

As a word of caution, however, we have to clarify that there are cases, more typical of multi-user decoding scenarios, where SDA may fail to return the ML solution in polynomial time. Empirical evidence suggests that this may happen when (i) the problem dimension is high (say, $N_t > 32$); (ii) the constellation size is large (say, $|\mathcal{A}| > 4$; (iii) the SNR is low (say, < 5 dB); and (iv) the channel matrix $\boldsymbol{H}$ is fat ($N_r < N_t$) or when $\boldsymbol{H}$ is square or tall but has rank lower than the number of its columns. In each of these cases (and combinations thereof), suboptimum decoders such as PDA-, LRA-, or SDP-based schemes offer viable alternatives either to obtain quasi-ML solutions or to initialize SDA with a smaller radius and thus accelerate the ML search. Among the various quasi-ML variates, LRA offers guaranteed diversity across the SNR range [244] at worst complexity $\mathcal{O}(N_t^4)$ and can afford efficient implementation with complex constellations [320]; SDP has guaranteed complexity $\mathcal{O}(N_t^{3.5})$, remains operational with fat channel matrices, and has recently been shown to exhibit a constant and quantifiable performance gap from the ML solution for any fixed SNR value as N_t grows large [137].

Beyond asymptotics and average complexity indicators, simulations also reveal another important feature of SDA when the number of affordable flops is limited to the extent that SDA cannot attain the exact ML solution. If the SDA search must be terminated when only a partial path of the search tree is admissible (only some entries of $\boldsymbol{s}$ obey conditions (5.9)) and we complete the remaining entries by slicing the unconstrained LS estimates, for certain channels this suboptimal solution is closer to the ML solution than what PDA or SDP iterations return as their suboptimal solutions with the same limited number of flops; see, e.g., [145], where these decoders are tested for diagonally dominant nonfading channels, and also [174] for related "best first" tree strategies as well as [105, 175] for pertinent comparisons with PDA and SDP in the context of multi-user detection.

[3]Another issue yet to be quantified analytically is how tight the complexity bound is.

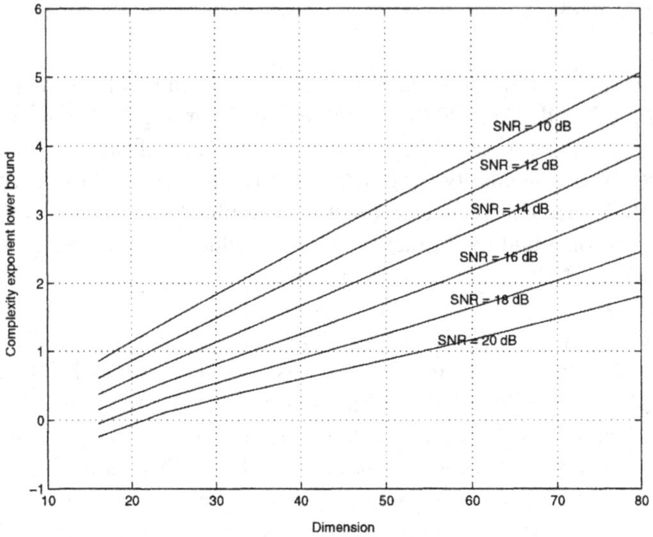

Figure 5.3 Exponent of the lower bound in (5.11) as a function of the dimension N_t

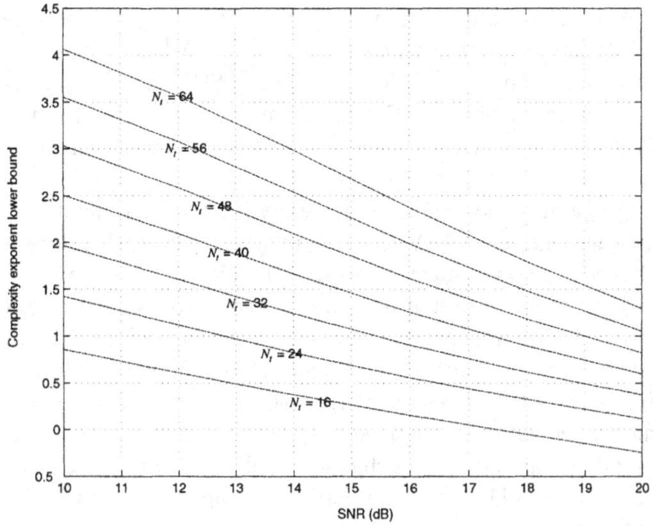

Figure 5.4 Exponent of the lower bound in (5.11) as a function of SNR

5.3 SDA IMPROVEMENTS

Several simple yet effective methods are available to improve error performance and/or reduce the complexity of the basic SDA search we presented so far for a fixed radius r. These techniques include (i) detection ordering across dimensions based on SNR [66], [323]; (ii) the Schnorr-Euchner ordering of candidates per dimension [4], [41]; and (iii) increasing radius search [110]. It is important to stress at the outset that all these methods can be conveniently incorporated in SDA without modifying the basic tree search structure.

5.3.1 SDA with Detection Ordering and Nulling-Canceling

Detection ordering (DO) is a preprocessing step leading to improved performance in early passes of the SDA search. DO-SDA was introduced in [66] as a useful heuristic, but it is possible to justify it statistically when the entries of $\boldsymbol{H}$ are independent identically distributed according to $\mathcal{N}(0,1)$ [323]. The motivation behind DO is to have the symbols we detect first in SDA be those with the largest SNR. We will illustrate this role of DO, in improving the nulling-canceling estimate we alluded to earlier.

Suppose that $N_r \geq N_t$ and rewrite (5.1) as $\boldsymbol{y} = \sum_{i=1}^{N_t} \boldsymbol{h}_i s_i + \boldsymbol{v}$, where $\boldsymbol{h}_i$ denotes the ith column of the channel matrix. Left multiplication with a unit-norm vector $\boldsymbol{u}_{N_t}^T$ from the null space of the submatrix $[\boldsymbol{h}_1 \cdots \boldsymbol{h}_{N_t-1}]$ isolates the N_tth symbol in the scalar equation $\zeta_{N_t} := \boldsymbol{u}_{N_t}^T \boldsymbol{y} = \boldsymbol{u}_{N_t}^T \boldsymbol{h}_{N_t} s_{N_t} + \boldsymbol{u}_{N_t}^T \boldsymbol{v}$. The latter shows that the receive SNR for detecting s_{N_t}, when all other symbols have been canceled, is determined by $\|\boldsymbol{h}_{N_t}\|^2$. As a result, the probability of correctly detecting s_{N_t} will increase in proportion to this norm. In NC without DO, we first detect s_{N_t} as $\hat{s}_{N_t}^{(\mathrm{NC})} = \zeta_{N_t}/(\boldsymbol{u}_{N_t}^T \boldsymbol{h}_{N_t})$, regardless of its receive SNR. Subsequently, we cancel $\boldsymbol{h}_{N_t} \hat{s}_{N_t}^{(\mathrm{NC})}$ from $\boldsymbol{y}$, null out all but the (N_t-1)st dimension, obtain $\hat{s}_{N_t-1}^{(\mathrm{NC})}$ as before, and proceed all the way down to dimension 1 until we obtain the NC solution $\hat{\boldsymbol{s}}_{\mathrm{NC}}$. This NC process implements a ZF or MMSE decision-feedback (DF) equalizer in a block-by-block fashion.

Clearly, the resulting NC solution entails error propagation, the effects of which will be less pronounced if early symbol detection steps can be made more reliable. Maintaining our notational convention of starting always with the N_tth dimension, this can be accomplished by rearranging the columns of $\boldsymbol{H}$ so that the N_tth column has the largest norm, the (N_t-1)st column the second largest, and so on. And this is precisely how NC with DO works: We identify a permutation matrix $\boldsymbol{\Pi}$ which yields a channel matrix $\boldsymbol{H}_\pi = \boldsymbol{H}\boldsymbol{\Pi}$ with its columns ordered so that the last column has the largest norm and the first column the smallest, i.e., $\boldsymbol{H}_\pi := [\boldsymbol{h}_{\pi(1)}, \boldsymbol{h}_{\pi(2)}, \ldots, \boldsymbol{h}_{\pi(N_t)}]$, such that $\|\boldsymbol{h}_{\pi(1)}\| \leq \|\boldsymbol{h}_{\pi(2)}\| \leq \cdots \leq \|\boldsymbol{h}_{\pi(N_t)}\|$. Accordingly, we permute the entries of $\hat{\boldsymbol{s}}$ to obtain $\hat{\boldsymbol{s}}_\pi = \boldsymbol{\Pi}^T \boldsymbol{s}$. This reordering conveniently increases the SNR in early detected symbols during the NC process.

DO preprocessing can also be applied to the basic SDA after reordering the MIMO matrix with $\boldsymbol{\Pi}$ and correspondingly, the initial LS estimate $\hat{\boldsymbol{s}}$ according to $\boldsymbol{\Pi}^T$. Notice that a different $\boldsymbol{\Pi}$ is required only when the channel realization changes in the block fading model, i.e., the extra complexity introduced by DO is minimal since the same permutation matrix can be re-used for decoding multiple symbol vectors as long as the MIMO channel remains invariant. As with the NC approach, where nulling relies on multiplying (5.1) with null-space vectors, reordering the columns of $\boldsymbol{H}$ based on their norm also enhances detection of symbols in the initialization of SDA where nulling is effected through the QR-decomposition. Indeed, the upper-triangular form of $\boldsymbol{R}_\pi$ in our MIMO system of linear equations allows decoding first the N_tth dimension (which now comes with the largest SNR) to obtain $\hat{s}_{N_t}^{(\mathrm{NC})} = \lceil \rho_{N_t} \rfloor = \lceil \hat{s}_{N_t} \rfloor$, where $\lceil x \rfloor$ denotes the closest integer to x. Canceling this estimate from the second scalar equation corresponding to dimension $N_t - 1$, selecting $\hat{s}_{N_t-1}^{(\mathrm{NC})} = \lceil \rho_{N_t-1} \rfloor$, and proceeding similarly down to dimension 1, we obtain the NC-DO estimate

$$\hat{\boldsymbol{s}}_{NC} := \left[\hat{s}_1^{(\mathrm{NC})} \cdots \hat{s}_{N_t}^{(\mathrm{NC})}\right] = [\lceil \rho_1 \rfloor \cdots \lceil \rho_{N_t} \rfloor]^T . \tag{5.12}$$

This NC-DO estimate offers a good initialization of the SDA search provided that it lies within the sphere of the selected radius r. Once more, the role of DO is to ensure that the first symbol estimates in the QR-based tree search are more reliable. Simulations have confirmed that DO preprocessing accelerates finding an admissible candidate within a prescribed radius, which in turn allows early shrinking of the search radius and thus reducing the overall complexity required to find an exact or near-ML solution.

5.3.2 Schnorr-Euchner Variate of the SDA

Whereas DO-SDA performs demodulation in an ordered fashion across dimensions, the Schnorr-Euchner (SE) SDA variate orders the list of admissible candidates per dimension and follows this order across the various depth-first passes of the SDA search [4, 41]. This is the main difference in SE-SDA relative to the basic SDA described earlier, where no order was necessarily imposed on the list of candidates dictated per dimension by the inequalities in (5.9). (Recall that we had arbitrarily chosen to consider as the first candidate of the list in the kth dimension the lower bound $\lceil \rho_k - \tau_k / R_{k,k} \rceil$.)

According to the SE ordering, the first candidate in the list of the kth dimension is basically the s_k value from the FA closest to ρ_k (i.e., $s_k^{(1)} = \lceil \rho_k \rfloor$). This choice selects $s_k^{(1)}$ as close as possible to the middle point of the interval defined by the admissible candidates in (5.9). Notice that this also leads to the smallest incremental square distance $R_{k,k}^2 (s_k - \rho_k)^2$ among all candidates in the list of dimension k [cf. (5.8)]. If the symbol estimates in all dimensions detected previously are perfect, $\lceil \rho_k \rfloor$ is the most likely candidate at dimension k, since it is the one contributing the smallest summand to the norm $\|\boldsymbol{R}(\boldsymbol{s} - \hat{\boldsymbol{s}})\|$, and under the AWGN model, minimum distance translates to ML.

With the first candidate always chosen to be $\lceil \rho_k \rfloor$, the second candidate in the list of the kth dimension is the second-closest point to the middle of the interval defined by (5.9). This is either $\lceil \rho_k \rfloor - 1$ if $\rho_k \leq \lceil \rho_k \rfloor$, or $\lceil \rho_k \rfloor + 1$ if $\rho_k \geq \lceil \rho_k \rfloor$. Ranking all n_k candidates of each dimension k similarly according to their incremental distance square, the SE-SDA examines candidates per dimension according to the following order:

$$[s_k^{(1)}, s_k^{(2)}, s_k^{(3)}, s_k^{(4)}, \ldots] = [\lceil \rho_k \rfloor, \lceil \rho_k \rfloor - 1, \lceil \rho_k \rfloor + 1, \lceil \rho_k \rfloor - 2, \ldots] \tag{5.13}$$

when $\rho_k \leq \lceil \rho_k \rfloor$, or a similar one with each candidate $\lceil \rho_k \rfloor - n$ exchanging order with the candidate $\lceil \rho_k \rfloor + n$ for all $n \in [1, n_k]$ if $\rho_k \geq \lceil \rho_k \rfloor$. The improvement SE brings to SDA comes from the fact that the ordering in (5.13) increases the likelihood to find early in the search a good admissible $\boldsymbol{s}$ for which $\boldsymbol{R}\boldsymbol{s}$ is very close to $\boldsymbol{R}\hat{\boldsymbol{s}}$ (and thus $\boldsymbol{H}\boldsymbol{s}$ is close to $\boldsymbol{y}$). Finding a good admissible solution early means that we can shrink our initial (and possibly large) radius early, which in turn implies that we can speed up our convergence to the exact (or at least near-)ML solution.

Checking the first entry in (5.13) across dimensions $k = N_t, N_t - 1, \ldots, 1$, reveals that if r is sufficiently large so that SE-SDA can return an admissible $\boldsymbol{s}$ solution (after traversing the search tree once from top to bottom), this first solution will be the NC estimate in (5.12). This observation immediately implies that DO can also be used as a preprocessing step in SE-SDA to form a DO-SE-SDA hybrid that yields a very reliable first admissible solution (the NC one with DO), which can be used to shrink considerably a possibly very large initial radius. In fact, DO-SE-SDA allows the designer to start with no radius constraint (i.e., with an infinitely large r), use the LS error of the DO-NC solution to select a radius: $r_{\mathrm{NC}}^2 = \|\boldsymbol{y} - \boldsymbol{H}\hat{\boldsymbol{s}}_{\mathrm{NC}}\|^2$, and proceed with the SE-SDA search within the sphere of radius r_{NC}. This approach guarantees that DO-SE-SDA will yield the exact ML solution, and for an r_{NC} of reasonable size it is the SDA variate of choice since it can drastically reduce complexity by successively shrinking the search radius. Its pseudocode with a fixed search radius r is provided in the Appendix.

Despite its attractive features, experience with DO-SE-SDA simulations suggests that especially as N_t grows large and/or the SNR drops to small values, it is possible (due to error propagation across dimensions) to start occasionally with a considerably suboptimal $\hat{\boldsymbol{s}}_{\mathrm{NC}}$ estimate, which in turn yields a large r_{NC} radius. Cases where DO-SE-SDA starts with a large radius will inevitably lead to an increased complexity search. Instead of running the risk of searching within a sphere of large radius, the next variate starts with a relatively small sphere but allows its radius to increase if necessary.

5.3.3 SDA with Increasing Radius Search

This extension capitalizes fully on the AWGN model to select a set of spheres with increasing radii which enable SDA to search for the exact (or a near-)ML solution at reduced complexity [110, 276]. The motivation behind SDA with increasing radius search (IRS) stems from the fact that for *any fixed* search radius, there is always a positive probability that neither the exact ML nor even an admissible candidate can be found with polynomial complexity [cf. (5.5)]. To overcome this probable failure with a single sphere of fixed radius, IRS-SDA searches within multiple spheres with increasing radii chosen from a suitable set r_{p_i}, where $p_i := 1 - \epsilon^i$ and $\epsilon \in (0, 1)$. The value $\epsilon = 0.1$ is recommended in [110].

Using this or any other value of ϵ, consider the set of radii $r_{p_1} < r_{p_2} < \cdots < r_{p_n}$, where n is determined by the maximum complexity that can be afforded. IRS-SDA starts the basic SDA search with fixed radius r_{p_1}. If this search successfully finds a winner within distance r_{p_1}, this will be the exact ML solution and the algorithm terminates; otherwise, we run SDA again with the next radius r_{p_2}. The search continues in this fashion until finding the ML solution or until the sphere with the maximum affordable radius r_{p_n} has been searched.

Of course, in every sphere of radius r_{p_i} used by IRS-SDA, one can run the DO-SE-SDA to take advantage of the reduced complexity search that the latter can afford. The resulting DO-SE-IRS-SDA hybrid is particularly useful when the initial NC-DO solution is very far from the ML optimum (large N_t and/or low SNR). In such cases, although the sphere with radius r_{NC} will deterministically contain the exact ML solution, the exhaustive search required can be prohibitively complex. In summary, DO-SE-IRS-SDA performs the following steps. Starting with r_{p_1}, perform s1) DO; followed by s2) SE ordering per dimension. If the NC-DO solution is available, shrink the radius and proceed all the way as in SE-SDA. But if SE-SDA fails to return an admissible $\boldsymbol{s}$ candidate, s3) increase the radius and loop back to s1) and s2). Notice that SE and NC computations performed with a certain radius can be reused in the next loop with the larger radius.

DO-SE-IRS-SDA can reach exact or near-ML error performance, with a corresponding increase in complexity relative to fixed-radius SDA alternatives. Let $\bar{N}_{p_i}$ denote the average complexity for the DO-SE-SDA search within the sphere of radius r_{p_i}, and recall that this search comes with probability p_i of finding the exact ML solution. Clearly, the probability of incurring the complexity $\bar{N}_{p_{i+1}}$ of the next DO-SE-SDA search is $1 - p_i$, which equals the probability that the search within the previous sphere of radius r_{p_i} fails. Based on these considerations, we deduce that the average complexity when n IRS spheres are involved is given by

$$\bar{N}_{\text{flop}} = \bar{N}_{p_1} + \sum_{i=1}^{n} (1 - p_i) \bar{N}_{p_{i+1}}. \tag{5.14}$$

As with Proposition 5.1 on average complexity, we should clarify that this is a conservative expression for the average complexity since it does not take into account the complexity reduction enabled by radius shrinking — an effect that is difficult to account for analytically.

5.3.4 Simulated Comparisons

Our experience with simulations suggests that the most efficient SDA to date capable of achieving exact ML error performance is the hybrid DO-SE-IRS-SDA we summarized in Section 5.3.3. To test performance and complexity as well as to compare the

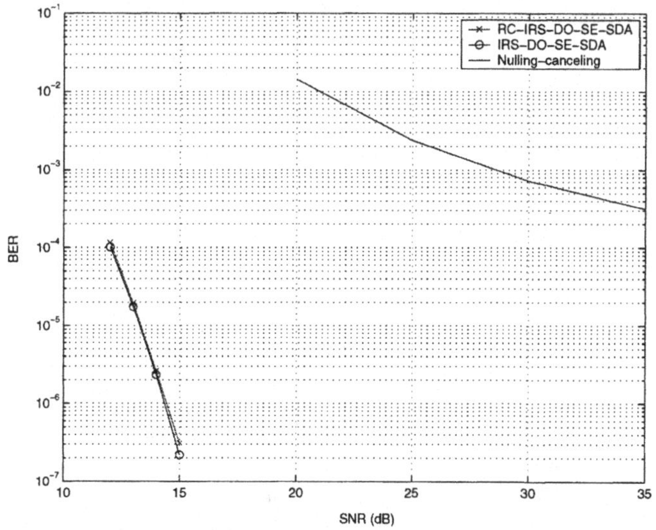

Figure 5.5 BER comparison among RC-IRS-DO-SE-SDA, IRS-DO-SE-SDA, and NC with DO for $N_t = N_r = 64$ and binary PAM

various SDA options, we rely on bit error rate (BER) and the complexity exponent $C := \log_{N_t}(N_{\text{flop}})$ averaged over multiple Monte Carlo realizations of the MIMO channel and the AWGN for various N_t sizes and SNR values encountered in practice. We focus on a block flat fading Rayleigh distributed MIMO channel, where many vectors $\boldsymbol{s}_o$ are transmitted through the same channel realization $\boldsymbol{H}$. These symbol blocks share the same preprocessing steps that include DO and QR-decomposition of $\boldsymbol{H}$. These common preprocessing computations are not included in the complexity plots.

Example 5.1 The MIMO system we simulate has $N_t = N_r = 64$ and uses 2-PAM constellation, which could correspond to a linearly precoded OFDM system or a symbol-synchronous CDMA setup. Here we compare four SDA variates: (i) the original SDA, (ii) IRS-SDA, (iii) SE-IRS-SDA, and (iv) DO-SE-IRS-SDA (the additional curve labeled RC-IRS-DO-SE-SDA corresponds to a reduced complexity suboptimal SDA presented in the next section). For the original version of SDA, we set the search radius to $r_{0.99}$, which in a number of realizations returns suboptimal solutions. For the IRS-SDA, we employ the set of radii as $\{r_{1-\epsilon}, r_{1-\epsilon^2}, r_{1-\epsilon^3}, \ldots\}$ with $\epsilon = 0.1$. To appreciate the importance of IRS-SDA in approaching ML optimality, we can observe the huge gap that it exhibits in BER relative to the suboptimal NC-DO solution in Figure 5.5. The BER of SDA with search radius $r_{0.99}$ is not plotted since it can be inferred from the block decoding failure rate, which is basically 1%. Also shown in the figure are the SE-IRS-SDA and the DO-SE-IRS-SDA, whose BER curves are indistinguishable from IRS-SDA since they too achieve ex-

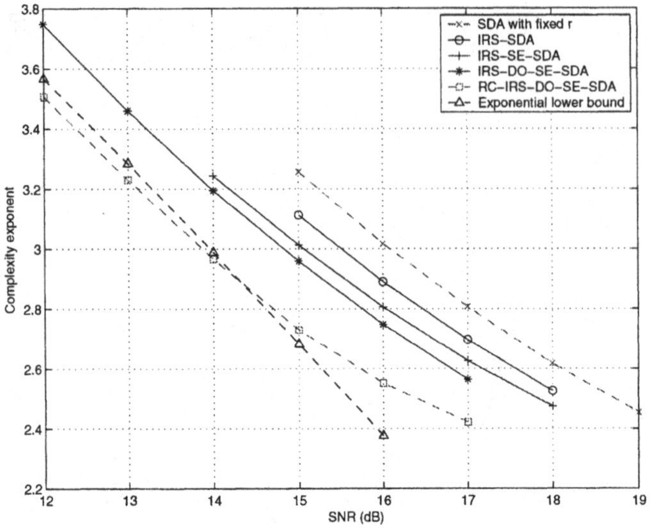

Figure 5.6 Average complexity exponent for $N_t = N_r = 64$ with binary PAM

act ML performance. Their average complexity exponents differ, however, as can be verified by checking the solid lines in Figure 5.6. These curves were averaged over 10000 randomly generated channel realizations and 100 noise realizations per channel. Notice that the decoding complexity for all these SDA improvements decreases with increasing SNR. Furthermore, for the number of antennas examined, their computational complexity appears to grow like N_t^2. It can also be observed from Figures 5.5 and 5.6 that SDA with $r_{0.99}$ not only incurs an error floor, but also exhibits higher decoding complexity than the ML algorithms. Notice also that DO-SE-IRS improvements systematically reduce the ML decoding complexity. For SNR values of practical concern, the average decoding complexity of these ML-optimal SDA versions remains polynomial in N_t. Finally, the expected complexity lower bound derived in [128, 129] is also plotted against SNR for $N_t = 64$ in Figure 5.6. Recall that this lower bound was used to argue that the average complexity of SDA grows exponentially with N_t for a fixed SNR. Nonetheless, at least for dimensions as large as 64, it can be verified that this lower bound does not prevent ML-optimal SDAs to achieve near cubic average complexity at practical SNR values. The implication here is that in practice, the usefulness of this lower bound may be limited. For a discussion on the reason behind and the delineation of practical scenarios where the true limiting behavior of the bound becomes effective, the interested reader is referred to [128].

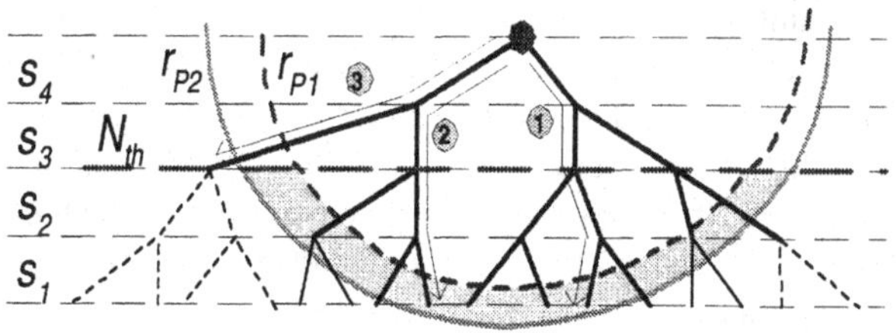

Figure 5.7 Reduced-complexity radius search

5.4 REDUCED-COMPLEXITY IRS-SDA

So far we have seen that SDAs with SE ordering or IRS search are capable of achieving exact ML optimality. In certain complexity-limited applications, however, suboptimal SDAs may be preferable if they can be flexible to trade-off ML optimality for reduced complexity (RC). In this section we outline such an IRS-SDA modification that is naturally termed RC-IRS-SDA [324]. Recall that when SDA searches within a sphere of fixed radius r_{p_i}, it is always possible that no admissible candidate can be found. When this happens, IRS-SDA increases the radius from r_{p_i} to $r_{p_{i+1}}$ and searches again. Computations in the search with radius r_{p_i} are discarded, but they are redone in the search with radius $r_{p_{i+1}}$. RC-IRS aims to reduce this waste and thereby lower the complexity of IRS-SDA.

Instead of a single radius, RC-IRS-SDA searches simultaneously within two spheres with radii $(r_{p_i}, r_{p_{i+1}})$. With r_{p_i} chosen to be very small, if the SDA search succeeds with this radius, it will do so very fast. Albeit this happens rarely since r_{p_i} is small, it provides one means of lowering complexity. But even when this search fails to find an admissible candidate beyond a certain dimension, it provides a set of partial paths, among which "promising" ones can be identified and stored for the $r_{p_{i+1}}$-based SDA search to have a head start. This is the second and more frequent reason why RC-IRS-SDA lowers complexity. The larger radius $r_{p_{i+1}}$ upper bounds the distance of those promising paths when the SDA search with r_{p_i} fails. As with IRS-SDA, if the $r_{p_{i+1}}$-based search fails to return an admissible solution, the radii of the pair of spheres are increased and RC-IRS-SDA is implemented with radii $(r_{p_{i+1}}, r_{p_{i+2}})$.

To outline how RC-IRS-SDA works with any pair of spheres, let $k_1 < N_t$ denote the dimension at which the r_{p_i}-based search fails. For this to happen, all partial paths (from the root dimension N_t down to dimension k_1) will have partial path metrics satisfying $d_{k_1}^2 > r_{p_i}^2$. Among those partial paths, a path is characterized as promising by the RC-IRS-SDA if two conditions are satisfied: (1) its metric must be less than the radii of the larger sphere (i.e., $d_{k_1}^2 < r_{p_{i+1}}^2$), and (2) the number of its branches should exceed a preselected threshold (i.e., $N_t - k_1 > N_{\text{th}}$). The threshold can be chosen to $N_{\text{th}} = N_t/2$, for example, and is set to make sure that the partial tree from the r_{p_i}-based search used by the $r_{p_{i+1}}$-based search (without repeating computations) is sufficiently informative. Notice that if N_{th} is set to be very small during the r_{p_i}-

based search, a prohibitively large list of promising paths must be stored. This not only increases memory requirements but also is less helpful to the $r_{p_{i+1}}$-based search since sorting and checking candidates from this large list can be as complex as starting the search within the larger sphere from scratch. Figure 5.7 illustrates a partial tree for $N_t = 4$ and $i = 1$. Starting from the root, each complete path corresponds to a candidate of $\boldsymbol{s}$, where the path metric is the sum of its branch metrics. In this example, the paths labeled as 1 and 2 from the r_{p_1}-based search are more promising than path 3, which being nonpromising will not be considered in the r_{p_2}-based search, i.e., only those paths falling in the shaded area of the figure will be checked when we search within the larger sphere.

To accelerate finding an admissible solution in the $r_{p_{i+1}}$-based search (which will allow us to shrink the radius quickly as in the basic SDA), the promising paths must be ranked. One approach to rank promising paths is to predict their full path metric and order them accordingly (first to be checked will be the one with the smallest full path metric). Using $l \in [1, L]$ to index promising paths, the full path metric can be predicted approximately as $d^{2(l)} = d^2_{k_1(l)} + [N_t - k_1(l)]\sigma^2$ [324]. Keeping the single most promising path with index $l_1 = \arg\min_{l\in[1,L]} d^{2(l)}$, and using it in the $r_{p_{i+1}}$-based search incurs minimal storage requirements. On the other hand, retaining the top two or three promising paths increases the chances to find the ML solution and thus improves error performance with a small increase in complexity. Although computationally more efficient than IRS-SDA, when RC-IRS-SDA is based on a single pair of spheres it is suboptimal. Its main asset is the flexibility to trade-off ML optimality for reduced complexity.

Simulation experience indicates that albeit suboptimal, RC-IRS-SDA comes close to ML performance with linear memory and modest complexity requirements. Recall that the average complexity of IRS-SDA is given by (5.14), and notice that most of $\bar{N}_{\text{flop}}$ is captured by $\bar{N}_{p_1} + (1 - p_1)\bar{N}_{p_2}$. With a suitably small r_{p_1}, RC-IRS-SDA can find an admissible candidate in the sphere of radius r_{p_2} with probability close to p_2, yet with complexity considerably lower than $\bar{N}_{p_1} + (1 - p_1)\bar{N}_{p_2}$, thanks to the reuse of the promising paths emerging from the search within radius r_{p_1} [324].

Example 5.2 Under the same system setting as in Example 5.1, we compare the error performance and decoding complexity of RC-IRS-SDA with the SDA improvements of Section 5.3 that are capable of exact ML performance. For the first RC radius search, we set $N_{\text{th}} = 34$ and $p_1 = 0.2$. The set of RC-IRS search radius pairs is $\{(r_{0.2}, r_{1-\epsilon^2}), (r_{1-\epsilon^2}, r_{1-\epsilon^4}), \ldots\}$ with $\epsilon = 0.1$. Notice that RC-IRS starts with a considerably smaller radius $r_{0.2}$ in the first search. The BER performance of RC-IRS combined with DO-SE-SDA is also depicted in Figure 5.5, corroborating that it comes very close to ML optimality. In fact, the probability of finding the exact ML can be found to be approximately 0.9876 [324], which is very close to $p_2 = 0.99$ (what a full IRS-SDA could guarantee) at lower complexity. Its complexity is plotted in Figure 5.6. For SNR values from 12 to 15 dB, RC-IRS is 2.7352, 2.5954, 2.5856, and 2.5988 times faster, respectively, than IRS. One key observation is that for sufficiently high SNR values, where current NC-based suboptimal receivers operate, both efficient ML and RC-IRS-based suboptimal receivers also achieve practical complexity, yet

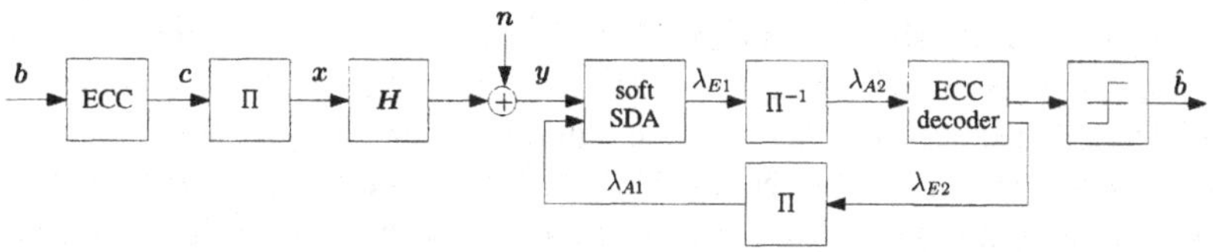

Figure 5.8 Block diagram of a MIMO system with turbo (de)coding

they offer considerable performance gain. For analytical parameter specifications and additional simulations, readers are referred to [324].

As a closing note of this overview of the main SDA options available for hard MIMO decoding, we should mention that the possible combinations are many. And since efficiency in the SDA search hinges heavily on the art of scientific programming, the possibilities are limited only by the designer's ingenuity. Furthermore, many alternatives can be envisioned toward combining SDA with other suboptimal but low-complexity MIMO decoders. Among those, we highlight the PDA-SDA hybrid in [144], where PDA is used first to determine symbols that can be decoded with high reliability, and SDA is applied subsequently to the reduced dimension problem after canceling detected symbols. With properly chosen parameters, this SDA-PDA hybrid approximates ML performance with reduced complexity and has practical merits for low to moderate SNR values.

5.5 SOFT-DECISION SPHERE DECODING

Error control coding (ECC) is universally present in practical MIMO systems. Turbo codes in particular, enable multi-antenna transmissions to approach the enhanced capacity of MIMO channels. However, turbo codes require iterative decoding at the receiver side, which relies on the exchange of *soft* information between the outer module of the turbo code and the inner MIMO decoding module. For such systems, the SDA variates we outlined so far are not directly applicable because the resulting MIMO decoders output *hard* decisions. In this section we outline two recent SDAs capable of generating bit-level soft information for use in capacity-approaching iterative MIMO decoders. The first performs list sphere decoding (LSD) and has been applied to turbo-coded layered ST (V-BLAST) systems [118]; see also [271] for a modified version of LSD. The second entails a soft-to-hard conversion which allows efficient implementation of a soft-decision SDA using multiple hard-decision SDAs [284].

Consider the block diagram of a MIMO system with turbo (de)coding depicted in Figure 5.8. Information bits $\boldsymbol{b}$ are first encoded by an ECC module to yield $\boldsymbol{c}$, which goes through a random interleaver Π. The interleaved bits $\boldsymbol{x}$ are then mapped to channel symbols and transmitted using a layered ST scheme, say V-BLAST [68]. (For simplicity, the scalar symbol (de)modulator and the ST (de)mapper following the interleaver are not shown in Figure 5.8.) At the receiver end, several iterations of soft information exchange are performed between the inner MIMO decoding module (labeled as a soft sphere decoder) and the outer ECC decoding module (labeled as an ECC decoder). The system under consideration falls under the general framework of bit-interleaved coded modulation with iterative detection and decoding. Specifically, extrinsic information denoted by λ_E is generated by one module and goes through either the interleaver Π or the deinterleaver Π^{-1} to serve as the new *a priori* information λ_A for the other module. The modules of this block diagram are common to the soft-decision SDAs we outline next.

5.5.1 List Sphere Decoding

Let $\boldsymbol{b} \in \{\pm 1\}^{N_t N_c}$ be the binary vector corresponding to the channel symbol vector $\boldsymbol{s}$ in (5.1), where each channel symbol is mapped to N_c binary symbols using, for example, Gray mapping. Thanks to the bit-level interleaver operating between the outer ECC and $\boldsymbol{H}$ (MIMO channel here replaces the inner ECC), we can safely assume that the entries of $\boldsymbol{b}$ are independent. The *a priori* information for the kth entry of $\boldsymbol{b}$ is thus captured by the log-likelihood ratio (LLR) $\lambda_A(b_k) := \ln[P(b_k = +1)/P(b_k = -1)]$. Aiming at a maximum *a posteriori* (MAP) decoder, application of Bayes' rule at the receiver allows one to decompose the LLR of the *a posteriori* probability of b_k given $\boldsymbol{y}$ as [118]

$$\lambda(b_k) := \ln \frac{P(b_k = +1|\boldsymbol{y})}{P(b_k = -1|\boldsymbol{y})} = \lambda_A(b_k) + \lambda_E(b_k),$$

where the difference $\lambda(b_k) - \lambda_A(b_k) := \lambda_E(b_k)$ represents the *extrinsic* information about b_k contained in $\boldsymbol{y}$. Letting $\boldsymbol{\lambda}_A := [\lambda_A(x_1), \ldots, \lambda_A(x_{N_t N_c})]^T$, this extrinsic information about b_k can be expressed as

$$\lambda_E(b_k) := \ln \frac{\sum_{\boldsymbol{b} \in \mathbb{B}_{k,+1}} P(\boldsymbol{y}|\boldsymbol{b}) \exp\{\frac{1}{2}\boldsymbol{b}_{[k]}^T \boldsymbol{\lambda}_{A,[k]}\}}{\sum_{\boldsymbol{b} \in \mathbb{B}_{k,-1}} P(\boldsymbol{y}|\boldsymbol{b}) \exp\{\frac{1}{2}\boldsymbol{b}_{[k]}^T \boldsymbol{\lambda}_{A,[k]}\}},$$

where $\mathbb{B}_{k,+1} := \{\boldsymbol{b}|b_k = +1\}$, $\mathbb{B}_{k,-1} := \{\boldsymbol{b}|b_k = -1\}$, $\boldsymbol{b}_{[k]}$ is the subvector of $\boldsymbol{b}$ obtained after omitting its kth entry b_k, and similarly for $\boldsymbol{\lambda}_{A,[k]}$, which comes from $\boldsymbol{\lambda}_A$ after removing $\lambda_A(b_k)$. Using the max-log approximation [220], $\lambda_E(b_k)$ can be approximated by

$$\begin{aligned}\lambda_E(b_k|\boldsymbol{y}) \approx & \frac{1}{2} \max_{\boldsymbol{b} \in \mathbb{B}_{k,+1}} \left\{-\frac{1}{\sigma^2}\|\boldsymbol{y} - \boldsymbol{H}\boldsymbol{s}\|^2 + \boldsymbol{b}_{[k]}^T \boldsymbol{\lambda}_{A,[k]}\right\} \\ & - \frac{1}{2} \max_{\boldsymbol{b} \in \mathbb{B}_{k,-1}} \left\{-\frac{1}{\sigma^2}\|\boldsymbol{y} - \boldsymbol{H}\boldsymbol{s}\|^2 + \boldsymbol{b}_{[k]}^T \boldsymbol{\lambda}_{A,[k]}\right\},\end{aligned} \tag{5.15}$$

where $\boldsymbol{b}$ coincides with $\boldsymbol{s}$ for the binary PAM assumed. The basic idea behind LSD, also explaining its acronym, is to carry out the maximization (5.15) over a constrained *list* of candidate bit vectors $\boldsymbol{b}$ giving rise to candidate symbol vectors for which the error norm $\|\boldsymbol{y} - \boldsymbol{H}\boldsymbol{s}\|$ remains within a *sphere* of preselected radius r. Specifically, letting $\mathcal{L} := \{\boldsymbol{b} \mid \|\boldsymbol{y} - \boldsymbol{H}\boldsymbol{s}\| < r\}$, the extrinsic information is determined by

$$\lambda_E(b_k|\boldsymbol{y}) \approx \frac{1}{2} \max_{\boldsymbol{b}\in\mathcal{L}\cap\mathbb{B}_{k,+1}} \left\{ -\frac{1}{\sigma^2}\|\boldsymbol{y} - \boldsymbol{H}\boldsymbol{s}\|^2 + \boldsymbol{b}_{[k]}^T \boldsymbol{\lambda}_{A,[k]} \right\} - \frac{1}{2} \max_{\boldsymbol{b}\in\mathcal{L}\cap\mathbb{B}_{k,-1}} \left\{ -\frac{1}{\sigma^2}\|\boldsymbol{y} - \boldsymbol{H}\boldsymbol{s}\|^2 + \boldsymbol{b}_{[k]}^T \boldsymbol{\lambda}_{A,[k]} \right\}.$$

With a suitable search radius r, LSD provides a satisfactory approximation to the extrinsic information $\lambda_E(b_k|\boldsymbol{y})$ in (5.15). When the list is incomplete, it is possible for one of the sets $\mathbb{B}_{k,\pm1}$ to be empty. In that case one has to assign a value to the maximum. When one works with log likelihoods, this is sometimes accomplished by clipping the log likelihood values [56, 118]. Nonetheless, a relatively large radius has to be chosen to generate sufficiently accurate soft information, which may incur high complexity. The soft-to-hard conversion that we outline next is motivated by the desire to reduce this complexity and also take advantage of the recent advances in hard-decision SDAs to obtain an alternative soft-decision SDA.

5.5.2 Soft SDA Using Hard SDAs

The key observation here is that we can combine the linear correcting term $\boldsymbol{b}_{[k]}^T \boldsymbol{\lambda}_{A,[k]}$ with $\|\boldsymbol{y} - \boldsymbol{H}\boldsymbol{s}\|^2$ in (5.15) to generate a modified quadratic form. For simplicity in exposition, let us consider binary channel symbols per real dimension (i.e., $\boldsymbol{s} = \boldsymbol{b}$). For more general constellations, the interested reader is referred to [284]. Notice that if the channel matrix $\boldsymbol{H}$ has full column rank, we can always find a vector $\tilde{\boldsymbol{y}}$ satisfying

$$2\boldsymbol{H}^T\tilde{\boldsymbol{y}} = \sigma^2\boldsymbol{\lambda}_A. \tag{5.16}$$

Plugging (5.16) and $\boldsymbol{s} = \boldsymbol{b}$ into (5.15), we obtain

$$\lambda_E(b_k|\boldsymbol{y}) \approx -\lambda_A(b_k) - \frac{1}{2\sigma^2} \min_{\boldsymbol{b}\in\mathbb{B}_{k,+1}} \|\boldsymbol{y} + \tilde{\boldsymbol{y}} - \boldsymbol{H}\boldsymbol{b}\|^2 + \frac{1}{2\sigma^2} \min_{\boldsymbol{b}\in\mathbb{B}_{k,-1}} \|\boldsymbol{y} + \tilde{\boldsymbol{y}} - \boldsymbol{H}\boldsymbol{b}\|^2. \tag{5.17}$$

This form of $\lambda_E(b_k|\boldsymbol{y})$ allows one to apply hard SDA twice for each b_k to compute the extrinsic information. For the given prior information $\boldsymbol{\lambda}_A$, the MAP estimate of $\boldsymbol{b}$ is found as $\hat{\boldsymbol{b}}_{\text{map}} = \min_{\boldsymbol{b}\in\mathbb{B}} \|\boldsymbol{y} + \tilde{\boldsymbol{y}} - \boldsymbol{H}\boldsymbol{b}\|^2$. Based on $\hat{\boldsymbol{b}}_{\text{map}}$, we can further simplify (5.17) to obtain

$$\lambda_E(b_k|\boldsymbol{y}) \approx -\lambda_A(b_k) - \frac{\hat{b}_{k,\text{map}}}{2\sigma^2}\|\boldsymbol{y}+\tilde{\boldsymbol{y}} - \boldsymbol{H}\hat{\boldsymbol{b}}_{\text{map}}\|^2 + \frac{\hat{b}_{k,\text{map}}}{2\sigma^2} \min_{\boldsymbol{b}\in\mathbb{B}_{k,-\hat{b}_{k,\text{map}}}} \|\boldsymbol{y} + \tilde{\boldsymbol{y}} - \boldsymbol{H}\boldsymbol{b}\|^2. \tag{5.18}$$

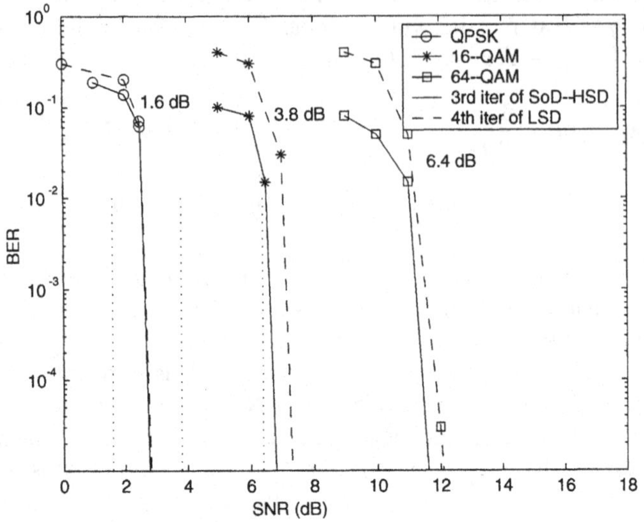

Figure 5.9 Average BER of an $(N_t, N_r) = (8, 8)$ MIMO system with turbo (de)coding based on soft-decision (SoD) via hard sphere decoding (HSD) versus LSD

Letting $\hat{\boldsymbol{b}}_k$ denote the "best vector" for which $b_k = -\hat{b}_{k,\text{map}}$, we have

$$\hat{\boldsymbol{b}}_k = \underset{\boldsymbol{b}\in\mathbb{B}_{k,-\hat{b}_{k,\text{map}}}}{\arg\min} \|\boldsymbol{y} + \tilde{\boldsymbol{y}} - \boldsymbol{H}\boldsymbol{b}\|^2, \quad k = 1, \ldots, N_t.$$

The latter shows that we can use hard SDA to find $\hat{\boldsymbol{b}}_{\text{map}}$ and $\{\hat{\boldsymbol{b}}_k\}_{k=1}^{N_t}$. Thus, to obtain the extrinsic information for the entire vector $\boldsymbol{b}$, we need one hard SDA step with block size N_t to find $\hat{\boldsymbol{b}}_{\text{map}}$, and N_t hard SDA steps with block size $N_t - 1$ to find $\{\hat{\boldsymbol{b}}_k\}_{k=1}^{N_t}$. Notice that unlike LSD, the extrinsic LLR values obtained from (5.18) are exact under the max-log approximation.

Example 5.3 To compare LSD against the soft-decision (SoD) via hard sphere decoding (HSD) approach, we simulated a MIMO system with $N_t = N_r = 8$ and turbo (de)coding modules as in Figure 5.8. The flat channel was simulated as block fading Rayleigh distributed and was supposed known to allow for coherent demodulation at the receiver. The outer code employed was a rate $1/2$ parallel concatenated turbo code. Each constituent convolutional code had memory 2 with feedback polynomial $1 + D + D^2$ and feedforward polynomial $1 + D^2$. The average BER for both SoD-HSD and LSD are depicted in Figure 5.9, where dotted lines at 1.6, 3.8, and 6.4 dB indicate the lowest possible SNR to achieve capacity for different constellations QPSK, 16-QAM, and 64-QAM. Three iterations are performed between the outer turbo module and the inner sphere decoding module. Curves with the same marker correspond to BER performance of the same constellation but at different iterations. It can be observed that SoD-HSD achieves essentially the same performance as LSD for QPSK signaling, whereas it outperforms LSD by about 0.5 dB for 16-QAM and 64-QAM. Further, this performance gain is expected to increase with increasing N_t.

5.6 CLOSING COMMENTS

The sphere decoding algorithm (SDA) and techniques to reduce its search complexity were outlined in this chapter. A relatively efficient version of SDA achieving exact ML error performance was identified to be the SE-SDA with increasing radius search (IRS) and detection ordering (DO). Three main factors contribute to its efficiency: (i) tree search, which effectively reuses intermediate computations; (ii) exploitation of the AWGN model with IRS; and (iii) examination of candidates with a likelihood order at each dimension (SE ordering). A reduced-complexity (RC) suboptimal SDA was also overviewed. Monte Carlo simulations confirmed that SE-SDA with RC-IRS is capable of achieving essentially ML error performance with considerably lower decoding complexity.

Two soft versions of sphere decoding were also briefly discussed. In particular, the generation of extrinsic information based on both list sphere decoding and soft-to-hard conversion was described. These soft SDAs enable multi-antenna systems to approach performance dictated by the capacity of MIMO channels.

Testament to the importance of SDA for hard or soft MIMO decoding is provided by the growing research efforts and by the recent development of SDA chips. Using only approximately 10 mm^2 in a 0.18 μm CMOS process, a soft-SDA architecture was reported in [83] implementing demodulation in a MIMO system with $N_t = N_r = 4$, capable of achieving 38.8 Mb/s data transmission with 16-QAM over a 5 MHz channel for use in the high-speed downlink packet access (HSDPA) extension of the Universal Mobile Telecommunications System (UMTS); see [33] for VLSI implementation of the complex SDA.

Topics of current and future research include analytical comparisons of SDA with PDA, LRA, and SDP approaches as well as delineation of the associated error performance versus complexity trade-offs. Simulation based evidence suggests that for cases where SDA terminates in polynomial (cubic on the average) complexity, it always outperforms competing alternatives, as it returns a solution that is either exactly or very close to ML. When SDA fails to terminate in polynomial time, PDA, LRA and especially SDP-based schemes offer feasible quasi-ML alternatives. A particularly promising but unexplored direction is using SDP-based decoding to initialize and accelerate the SDA. Such an SDP-SDA hybrid may lead to (near-)ML decoding at polynomial complexity for a very large class of application scenarios.

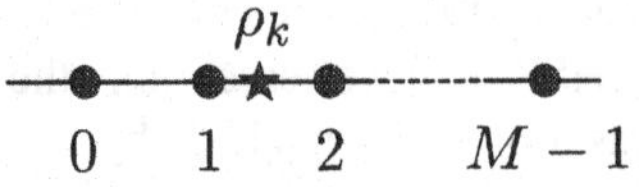

Figure 5.10 Subroutine *enum*

Appendix

The pseudocode for SE-SDA with FA constraints is provided here. Preprocessing steps such as DO and QR-decomposition applied to the channel matrix $\boldsymbol{H}$ are not included. The constellation employed is $\mathcal{A} := \{0, 1, \ldots, M-1\}$, where M is the constellation size. If the search within radius r fails, the return parameter d is set to -1; otherwise, $\boldsymbol{s}_{\mathrm{ML}}$ and its squared distance are returned. This algorithm naturally reduces the search radius whenever a candidate with a smaller LS error relative to $\hat{\boldsymbol{s}}$ is found.

$[\boldsymbol{s}_{\mathrm{ML}}, d] = \text{SE-SDA}(\boldsymbol{R}, \hat{\boldsymbol{s}}, r, M, n)$

$d := r^2;\ k := n;\ d_k^2 := 0;$ calculate ρ_k;
$[s_k, n_k] :=$ enum$(\rho_k, M, 0)$;
while(true) {
    $w := d_k^2 + R_{k,k}^2 (s_k - \rho_k)^2$;
    if $w < d$ {
        if $k \neq 1$ {
            $k := k - 1;\ d_k^2 := w;$ calculate ρ_k;
            $[s_k, n_k] :=$ enum$(\rho_k, M, 0)$;
        } else { $\boldsymbol{s}_{ML} := [s_1, \ldots, s_{N_t}];\ d := w;$ goto INC; }
    } else {
INC: while $k \leq n$ and $n_k = M$ { $k := k + 1$; }
        if $k > n$ { if $d < r^2$ return $[\boldsymbol{s}_{ML}, d]$; else return $[\mathbf{0}, -1]$; }
        $[s_k, n_k] :=$ enum(s_k, M, n_k);
    }
}

Given ρ_k, the constellation size M, and n_k (the number of candidates s_k that has been enumerated), the function of subroutine $[s_k, n_k] = \text{enum}(\rho_k, M, n_k)$ is to generate the next likely candidate and increase n_k by one. Both SE ordering and the FA constraints are incorporated in this subroutine. The enumerated s_k corresponds to increasing Euclidean distance from ρ_k. An example is illustrated in Figure 5.10, where the order of enumeration is $1, 2, 0, 3, \ldots, M-1$. For brevity, the pseudocode for this routine is not provided.

6

Noncoherent and Differential ST Codes for Flat Fading Channels

The coherent ST codes we described in Chapter 3 require at their decoding stage estimates of the MIMO channels. These estimates can be acquired using training-based or blind estimation algorithms. But since $N_t N_r$ channels are involved in ST transmissions, in addition to increasing receiver complexity, these estimation algorithms require the channels to remain invariant over an interval longer than that required by single-antenna transmissions. This however, may not be possible for a number of applications. Motivated by these considerations, a number of ST coding schemes have been developed for achieving the maximum diversity gain while circumventing the need for channel estimation. As with single-antenna communication systems, these schemes give rise to two classes of ST codes: noncoherent ST codes and differential ST codes. These two classes have their own pros and cons and one could be preferred over the other, depending on application-specific trade-offs among performance, complexity and bandwidth efficiency. The theme of this chapter is to present the basic ideas behind these schemes, analyze their performance, and compare them with their coherent counterparts.

6.1 NONCOHERENT ST CODES

A common means of bypassing channel knowledge in single-antenna (serial or block) transmissions is through the use of proper modulation formats which can be demodulated at the receiver using noncoherent detection. A natural question to ask is whether noncoherent modulation and detection is possible for ST transmissions. The first class of noncoherent ST codes that offer an affirmative answer to this question is the one introduced by [114] under the term unitary ST modulation.

Unitary ST modulation applies to channels obeying assumptions A2.1) and A2.2) in Chapter 2. Its constellation set $\mathbb{X}$ consists of L scaled $N_t \times N_x$ unitary ST code matrices $\boldsymbol{X} \in \mathbb{X}$ satisfying

$$\boldsymbol{X}\boldsymbol{X}^{\mathcal{H}} = N_x \boldsymbol{I}_{N_t}, \quad \forall \boldsymbol{X} \in \mathbb{X}. \tag{6.1}$$

Transmitting with such matrices is not only necessary for noncoherent ST coding to achieve channel capacity (under certain mild assumptions), but also facilitates noncoherent ST decoding in the absence of channel knowledge [114]. It is important to stress that unitary ST coded transmissions are capacity achieving only when the MIMO channel adheres to A2.1). For block flat fading channels with coherence time smaller than the duration of one ST code matrix, the use of nonunitary ST code matrices may offer better performance or lower decoding complexity [76].

The transmission of unitary ST code matrices $\boldsymbol{X}$ can be described by using the generic system model of (2.9). In noncoherent ST coding, neither the transmitter nor the receiver know the realization of the underlying MIMO channel $\boldsymbol{H}$. Instead, $\boldsymbol{H}$ is treated as a random matrix whose entries are i.i.d., complex Gaussian random variables each with zero mean and unit variance. As a result, conditioned on $\boldsymbol{X}$, the received matrix $\boldsymbol{Y}$ in (2.9) is a Gaussian random matrix with probability density function (pdf),

$$p(\boldsymbol{Y}|\boldsymbol{X}) = \frac{\exp\left[-\mathrm{tr}(\boldsymbol{\Sigma}^{-1}\boldsymbol{Y}^{\mathcal{H}}\boldsymbol{Y})\right]}{[\det(\pi\boldsymbol{\Sigma})]^{N_r}}, \tag{6.2}$$

where $\boldsymbol{\Sigma} = \boldsymbol{I} + (\bar{\gamma}/N_t)\boldsymbol{X}^{\mathcal{H}}\boldsymbol{X}$. Using (6.1), it can readily be verified that $\det(\boldsymbol{\Sigma})$ does not depend on $\boldsymbol{X}$ and $\boldsymbol{\Sigma}^{-1}$ is given by

$$\boldsymbol{\Sigma}^{-1} = \boldsymbol{I} - \frac{\bar{\gamma}}{N_x\bar{\gamma} + N_t}\boldsymbol{X}^{\mathcal{H}}\boldsymbol{X}.$$

It thus follows from (6.2) that ML detection can be performed optimally to recover the transmitted code matrix as [114]

$$\hat{\boldsymbol{X}} = \arg\max_{\boldsymbol{X}\in\mathbb{X}} ||\boldsymbol{X}\boldsymbol{Y}^{\mathcal{H}}||_F^2, \tag{6.3}$$

which clearly does not require channel knowledge and thus implements a noncoherent ST decoder. In fact, (6.3) can be viewed as an ST generalization of the classical energy (or envelope) detector which is used for demodulating scalar noncoherent transmissions modulated with, for example, frequency-shift keying (FSK). The same detector can also be interpreted as a generalized likelihood ratio test (GLRT). Indeed, it can easily be shown that (6.3) is equivalent to

$$\hat{\boldsymbol{X}} = \arg\max_{\boldsymbol{X}\in\mathbb{X}} \mathrm{tr}\left[-(\boldsymbol{Y} - \hat{\boldsymbol{H}}\boldsymbol{X})(\boldsymbol{Y} - \hat{\boldsymbol{H}}\boldsymbol{X})^{\mathcal{H}}\right],$$

which amounts to coherent ML correlation-detection with the channel matrix replaced by its estimate,

$$\hat{\boldsymbol{H}} = \frac{1}{N_x} \boldsymbol{Y} \boldsymbol{X}^{\mathcal{H}}.$$

Although the unitary ST codes adhering to (6.1) and the decoding rule in (6.3) allow for noncoherent ST system operation, the resulting error performance is expected to depend heavily on the choice of the ST constellation $\mathbb{X}$. The design options available for $\mathbb{X}$ can be divided in two categories: (i) search-based designs (see, e.g., [2, 113, 193, 194]); and (ii) training-based designs (see, e.g., [31, 55, 78, 250, 325]).

6.1.1 Search-Based Designs

Search-based designs rely on a proper distance metric that is used to search over and select among candidate matrices belonging to a parametric class of unitary ST constellations. Three distance metrics can be obtained by analyzing the error probability performance of the noncoherent detector in (6.3).

Specifically, this error probability can be union-bounded as

$$P_e \leq \frac{1}{L} \sum_{\boldsymbol{X}, \boldsymbol{X}' \in \mathbb{X}, \boldsymbol{X} \neq \boldsymbol{X}'} P_2(\boldsymbol{X} \to \boldsymbol{X}'), \tag{6.4}$$

where $P_2(\boldsymbol{X} \to \boldsymbol{X}')$ denotes the pairwise error probability (PEP) between two distinct unitary ST code matrices $\boldsymbol{X}$ and $\boldsymbol{X}'$ belonging to $\mathbb{X}$. The PEP can be further bounded using the Chernoff bound as [114, 120]

$$P_2(\boldsymbol{X} \to \boldsymbol{X}') \leq \frac{1}{2} \prod_{\mu=1}^{N_t} \left[1 + \frac{(\bar{\gamma} N_x / N_t)^2 \left[1 - d_\mu^2(\boldsymbol{X}, \boldsymbol{X}')\right]}{4(1 + \bar{\gamma} N_x / N_t)} \right]^{-N_r}, \tag{6.5}$$

where $1 \geq d_1(\boldsymbol{X}, \boldsymbol{X}') \geq \cdots \geq d_{N_t}(\boldsymbol{X}, \boldsymbol{X}') \geq 0$ are the N_t singular values of $\boldsymbol{X}\boldsymbol{X}'^{\mathcal{H}}/N_x$. By inspecting (6.5), it can readily be inferred that the maximum diversity gain is $N_t N_r$ and is attained when $d_\mu(\boldsymbol{X}, \boldsymbol{X}') \neq 1$ for all μ. A necessary condition for maximizing the diversity gain is [120]

$$N_x \geq 2N_t, \tag{6.6}$$

which implies that all unitary ST code matrices $\boldsymbol{X}$ must be fat.

The first distance metric is extracted from the Chernoff bound, which delineates error performance at high SNR. It is the diversity product distance, defined as

$$\delta_{\text{DP}} := \min_{\boldsymbol{X}, \boldsymbol{X}' \in \mathbb{X}, \boldsymbol{X} \neq \boldsymbol{X}'} \prod_{\mu=1}^{N_t} \left[1 - d_\mu^2(\boldsymbol{X}, \boldsymbol{X}')\right]. \tag{6.7}$$

This metric δ_{DP} is approximately equal to the second metric, known as the chordal distance, which can be obtained by further upper bounding the Chernoff bound in

(6.5) [113]. The chordal distance is defined as

$$\delta_C := 1 - \max_{\boldsymbol{X},\boldsymbol{X}' \in \mathbb{X}, \boldsymbol{X} \neq \boldsymbol{X}'} \sum_{\mu=1}^{N_t} d_\mu^2(\boldsymbol{X}, \boldsymbol{X}'). \tag{6.8}$$

Instead of the Chernoff bound, P_e can be also upper-bounded by the asymptotic union bound [194] as

$$P_e \leq \frac{1}{L}\binom{2N_tN_r - 1}{N_tN_r}\left(\frac{\bar{\gamma}}{N_t}\right)^{-N_tN_r} \sum_{\boldsymbol{X},\boldsymbol{X}' \in \mathbb{X}, \boldsymbol{X} \neq \boldsymbol{X}} \left[\det\left(\boldsymbol{I} - \frac{1}{N_x^2}\boldsymbol{X}\boldsymbol{X}'^{\mathcal{H}}\boldsymbol{X}'\boldsymbol{X}^{\mathcal{H}}\right)\right]^{-N_r},$$

from which the third distance metric, the union bound distance, can be obtained as

$$\delta_{\text{UB}} := \sum_{\boldsymbol{X},\boldsymbol{X}' \in \mathbb{X}, \boldsymbol{X} \neq \boldsymbol{X}} \left[\det\left(\boldsymbol{I} - \frac{1}{N_x^2}\boldsymbol{X}\boldsymbol{X}'^{\mathcal{H}}\boldsymbol{X}'\boldsymbol{X}^{\mathcal{H}}\right)\right]^{-N_r}. \tag{6.9}$$

To optimize error performance, one can design $\mathbb{X}$ so that any of the three distance metrics is maximized. One possible approach to this end is to use computer search. On the other hand, to support a transmission rate R, the number of distinct code matrices in $\mathbb{X}$ should be at least $L = 2^{RN_x}$. Since L is typically large (e.g., $L = 2^{20} \approx 10^6$ when $R = 2$ and $N_x = 10$), computer search could be computationally prohibitive. To facilitate this search, it is proposed in [113] to impose a block-circulant correlation structure on $\boldsymbol{X} \in \mathbb{X}$. With the factorization $L = \prod_{k=1}^{K} L_k$ for some positive integer K, let $\boldsymbol{U}'$ denote a $K \times (N_x - K)$ parity matrix with its entries drawn from the set $\{0, 1, \ldots, Q-1\}$, and let $\boldsymbol{U} = [\boldsymbol{I}_K, \boldsymbol{U}']$ be the corresponding systematic generator matrix. Also let $\boldsymbol{\Theta}_k$, $k = 1, \ldots, K$ denote an $N_x \times N_x$ diagonal complex matrix with diagonal entries given by

$$[\boldsymbol{\Theta}_k]_{i,i} = \frac{1}{\sqrt{N_x}} e^{j\frac{2\pi}{Q}[\boldsymbol{U}]_{k,i}}. \tag{6.10}$$

If the block-circulant correlation is imposed, any unitary ST code matrix $\boldsymbol{X} \in \mathbb{X}$ can be written in the form [113]

$$\boldsymbol{X} = \sqrt{N_x}\,\boldsymbol{\Theta}_1^{l_1}\boldsymbol{\Theta}_2^{l_2}\cdots\boldsymbol{\Theta}_K^{l_K}\boldsymbol{\Phi}_1, \quad 1 \leq l_1 \leq L_1, \ldots, 1 \leq l_K \leq L_K, \tag{6.11}$$

where the matrix $\boldsymbol{\Phi}_1$ is formed using N_t distinct rows of the $N_x \times N_x$ fast Fourier transform (FFT) matrix. Thanks to the parametrization in (6.11), the design of $\mathbb{X}$ reduces to searching over $Q^{K(N_x-K)}$ possible parity matrices $\boldsymbol{U}$. Table 6.1 lists noncoherent ST codes found based on the chordal distance metric δ_C along with their performance quantified also in terms of δ_C [113].

The imposed block-circulant correlation structure in [113] renders the search for high-performance unitary ST code matrices feasible and systematic. However, this correlation structure will generally render the resulting designs suboptimal [113]. Alternatively, it is possible to parameterize the code matrix $\boldsymbol{X}$ without constraining

Table 6.1 NONCOHERENT ST CODES FOR $N_t = 2$ AND $N_x = 8$

L	δ_C	K	Q	U'
4	0	2	2	$\begin{bmatrix} 0 & 1 & 1 & 0 & 0 & 1 \\ 0 & 1 & 0 & 1 & 0 & 1 \end{bmatrix}$
8	0.38	1	8	$\begin{bmatrix} 3 & 0 & 7 & 2 & 5 & 6 & 7 \end{bmatrix}$
17	0.47	1	17	$\begin{bmatrix} 12 & 11 & 9 & 14 & 6 & 10 & 0 \end{bmatrix}$
32	0.53	1	32	$\begin{bmatrix} 18 & 11 & 2 & 22 & 8 & 0 & 5 \end{bmatrix}$
67	0.58	1	67	$\begin{bmatrix} 7 & 31 & 15 & 3 & 29 & 20 & 0 \end{bmatrix}$
130	0.63	1	130	$\begin{bmatrix} 30 & 71 & 39 & 15 & 4 & 41 & 124 \end{bmatrix}$
257	0.67	1	257	$\begin{bmatrix} 7 & 60 & 79 & 187 & 125 & 198 & 154 \end{bmatrix}$
529	0.73	2	23	$\begin{bmatrix} 15 & 3 & 10 & 9 & 15 & 17 \\ 22 & 16 & 14 & 4 & 21 & 21 \end{bmatrix}$
1024	0.76	2	32	$\begin{bmatrix} 26 & 22 & 1 & 3 & 7 & 26 \\ 18 & 28 & 22 & 8 & 24 & 1 \end{bmatrix}$

the code structure as [194]

$$X = \sqrt{N_x}\, U \left(\prod_{k=1}^{N_x - 1} \prod_{l=N_x}^{k+1} G_{kl}(\phi_{kl}, \theta_{kl}) \right) V, \tag{6.12}$$

where U is a fixed $N_t \times N_x$ matrix, V is a fixed $N_x \times N_x$ diagonal matrix, and the complex-valued Givens matrix $G_{kl}(\phi_{kl}, \theta_{kl})$ is defined under (3.65) with two real parameters ϕ_{kl} and θ_{kl} taking values over the finite interval $[-\pi, \pi)$. According to (6.12), any $X \in \mathbb{X}$ is parameterized by $N_x N_t - N_t^2$ parameters over the interval $[-\pi, \pi)$. Since there are $L = 2^{RN_x}$ distinct code matrices in $\mathbb{X}$, any one of the three distance metrics can then be maximized with respect to about $2^{RN_x}(N_x N_t - N_t^2)$ parameters. This optimization problem was solved in [194] by using either a standard gradient-based approach or a suboptimal reduced-complexity variate based on successive updates. The resulting code designs were found to outperform their counterparts in [113].

6.1.2 Training-Based Designs

To bypass the need for computer search, training-based approaches to designing noncoherent ST codes were reported in [31,55,78,250,325]. In these designs, each ST code matrix consists of a part known to the receiver and a part containing unknown information symbols. To carry out noncoherent decoding, one approach is first to estimate the channel using the known part as training and then decode the unknown part coherently based on the estimated channel (see, e.g., [31,55,78,250]). Another approach is to still use (6.3) as in [325]. In what follows, we describe the training-based design derived in [325]. For simplicity, only the design for two transmit-antennas ($N_t = 2$) will be presented. The extension to more than two transmit-antennas can be found in [325].

Supposing that $L = M^2$ for some integer M, the ST constellation $\mathbb{X}$ in [325] is composed of L 2×4 ST code matrices satisfying (6.1) and can be expressed as $\mathbb{X} = \{\boldsymbol{X}_{k,l}, k, l \in [0, M-1]\}$. Each $\boldsymbol{X} \in \mathbb{X}$ takes the form

$$\boldsymbol{X} = [\boldsymbol{T},\ \boldsymbol{O}_{2\times 2}], \tag{6.13}$$

where

$$\boldsymbol{T} = \begin{bmatrix} 1 & 1 \\ -1 & 1 \end{bmatrix}$$

and $\boldsymbol{O}_{2\times 2}$ is the Alamouti code defined in (3.16). Matrix $\boldsymbol{T}$ is known at the receiver and $\boldsymbol{O}_{2\times 2}$ carries two unknown information symbols, s_1 and s_2. To satisfy (6.1), s_1 and s_2 are restricted to be M-PSK symbols. As a result, $\boldsymbol{X}_{k,l}$ is constructed as

$$\boldsymbol{X}_{k,l} := \begin{bmatrix} 1 & -1 & e^{j\frac{2\pi}{M}k} & -e^{-j\frac{2\pi}{M}l} \\ 1 & 1 & e^{j\frac{2\pi}{M}l} & e^{-j\frac{2\pi}{M}k} \end{bmatrix}, \quad k, l \in [0, M-1]. \tag{6.14}$$

It can easily be verified that $\boldsymbol{X}_{k,l}\boldsymbol{X}_{k,l}^{\mathcal{H}} = 4\boldsymbol{I}$. Furthermore, it can be shown that the design in (6.14) results in $\delta_{\text{DP}} = \sin(\pi/M)/\sqrt{2} > 0$, which suggests that the design in (6.14) guarantees the maximum diversity gain of order $N_t N_r$.

An issue of concern when decoding noncoherent ST coded transmissions is the relatively high complexity of the ML detector, especially when the constellation set has a large size. In what follows we show that decoding the noncoherent modulation in (6.14) incurs relatively low complexity.

Let us consider the noncoherent ST detection in (6.3) with the code matrix $\boldsymbol{X} = \boldsymbol{X}_{k,l}$. For each $\boldsymbol{X}_{k,l} \in \mathbb{X}$, it is possible to express $\|\boldsymbol{X}_{k,l}^{\mathcal{H}}\boldsymbol{Y}\|_F^2$ as a superposition of three terms:

$$\|\boldsymbol{X}_{k,l}\boldsymbol{Y}^{\mathcal{H}}\|^2 = \text{tr}\{\boldsymbol{Y}\boldsymbol{X}_{k,l}^{\mathcal{H}}\boldsymbol{X}_{k,l}\boldsymbol{Y}^{\mathcal{H}}\} = 2\,\|\boldsymbol{Y}\|^2 + \text{tr}\{\boldsymbol{Y}^{\mathcal{H}}\boldsymbol{Y}\boldsymbol{A}_k\} + \text{tr}\{\boldsymbol{Y}^{\mathcal{H}}\boldsymbol{Y}\boldsymbol{B}_l\},$$

where

$$A_k := \begin{bmatrix} 0 & e^{jk\frac{2\pi}{M}} & 0 & e^{-jk\frac{2\pi}{M}} \\ e^{-jk\frac{2\pi}{M}} & 0 & -e^{-jk\frac{2\pi}{M}} & 0 \\ 0 & -e^{jk\frac{2\pi}{M}} & 0 & e^{-jk\frac{2\pi}{M}} \\ e^{jk\frac{2\pi}{M}} & 0 & e^{jk\frac{2\pi}{M}} & 0 \end{bmatrix},$$

$$B_l := \begin{bmatrix} 0 & e^{jl\frac{2\pi}{M}} & 0 & -e^{-jl\frac{2\pi}{M}} \\ e^{-jl\frac{2\pi}{M}} & 0 & e^{-jl\frac{2\pi}{M}} & 0 \\ 0 & e^{jl\frac{2\pi}{M}} & 0 & e^{-jl\frac{2\pi}{M}} \\ -e^{jl\frac{2\pi}{M}} & 0 & e^{jl\frac{2\pi}{M}} & 0 \end{bmatrix}.$$

Notice that the first term in the decomposition does not depend on k or l, the second term is related only to k, while the third term is related only to l. Based on this decomposition, (6.3) simplifies to

$$\hat{X} = \arg\max_{X_{k,l} \in \mathbb{X}} \|X_{k,l} Y^{\mathcal{H}}\|_F^2 = X_{\hat{k}_{\mathrm{ML}}, \hat{l}_{\mathrm{ML}}}, \tag{6.15}$$

where

$$\hat{k}_{\mathrm{ML}} = \arg\max_{\forall\, k \in [0, M-1]} \operatorname{tr}\{Y^{\mathcal{H}} Y A_k\},$$

$$\hat{l}_{\mathrm{ML}} = \arg\max_{\forall\, l \in [0, M-1]} \operatorname{tr}\{Y^{\mathcal{H}} Y B_l\}.$$

Since $\hat{k}_{\mathrm{ML}}$ and $\hat{l}_{\mathrm{ML}}$ can be computed separately, this simplified noncoherent ST detector only needs to calculate $\operatorname{tr}\{Y^{\mathcal{H}} Y A_k\}$ M times and $\operatorname{tr}\{Y Y^{\mathcal{H}} B_l\}$ M times. This should be contrasted with the noncoherent ST detector in (6.3), where one needs to calculate $\|X_{k,l} Y^{\mathcal{H}}\|^2$ M^2 times. Clearly, the simplified detection scheme enjoys lower complexity, especially for larger M. Based on complexity considerations but also from the structure of its code matrix, we can view the training-based design in (6.14) as the noncoherent counterpart of the Alamouti code introduced in the coherent designs of Section 3.3.

Let us now compare the training-based design in (6.14) with the search-based design. For the search-based design, we will rely on the computer search algorithm of [113] but adopt the product distance metric δ_{DP} in (6.7) instead of the chordal distance. Two sets of ST constellations are obtained with cardinality $L = 16$ and $L = 64$. The simulated symbol error rate curves depicted in Figure 6.1 show that the two designs enjoy similar performance with the same L, but the training-based design incurs lower decoding complexity.

6.2 DIFFERENTIAL ST CODES

Similar to noncoherent designs, the development of differential ST coding schemes can be derived by generalizing scalar differential approaches used in single-antenna

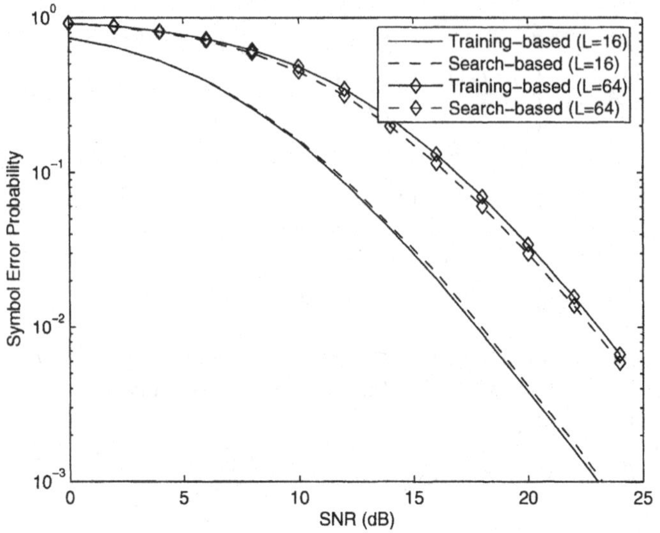

Figure 6.1 Performance comparison of training- with search-based noncoherent designs

systems to the matrix case. The basic idea behind matrix differential schemes is to introduce a proper encoding relationship between consecutive ST code matrices so that differential ST decoding can be accomplished at the receiver without knowledge of the underlying MIMO channel. Similar to scalar differential decoding, the price paid for differential ST coding is about 3 dB in the required SNR when compared to its coherent counterpart, which requires channel knowledge. Because differential ST coding closely resembles its scalar counterpart, it is instructive to begin our exposition with a brief review of scalar differential coding and decoding.

6.2.1 Scalar Differential Codes

The notion of scalar differential coding employed by single-antenna communication systems is traditionally presented in the context of PSK-modulated information-bearing symbols $s(n)$. Nonetheless, regardless of the underlying modulation, the idea is to relate successively transmitted symbols $x(n)$ and $x(n-1)$ through a multiplicative regression, as follows:

$$x(n) = \begin{cases} x(n-1)s(n), & n \geq 0, \\ 1, & n = -1. \end{cases} \tag{6.16}$$

Correspondingly, the differentially encoded symbols received after propagation through a flat fading channel can be expressed as

$$y(n) = \sqrt{\bar{\gamma}}\, hx(n) + w(n). \tag{6.17}$$

Under the assumption that the channel h remains time-invariant for at least two symbol intervals, combining (6.16) with (6.17) reveals that any two consecutively received samples are also related by a recursion,

$$\begin{aligned} y(n) &= \sqrt{\bar{\gamma}}\, h x(n-1) s(n) + w(n) \\ &= [y(n-1) - w(n-1)] s(n) + w(n) \\ &= y(n-1) s(n) + w(n) - w(n-1) s(n) \\ &= y(n-1) s(n) + \tilde{w}(n), \end{aligned} \tag{6.18}$$

where $\tilde{w}(n) := w(n) - w(n-1)s(n)$. Upon normalizing $s(n)$ to have unit energy $|s(n)| = 1$ and since $w(n)$'s are uncorrelated across time, we deduce that $\tilde{w}(n)$ is Gaussian with mean zero and variance 2. It thus follows from (6.18) that ML detection of $s(n)$ from $y(n)$ and $y(n-1)$ yields

$$\hat{s}(n) = \arg\min_{s(n)\in\mathbb{S}} ||y(n) - y(n-1)s(n)||^2. \tag{6.19}$$

Equation (6.19) implements scalar differential decoding, which clearly does not require knowledge of the channel coefficient h. The SNR at the output of the differential decoder can be found using (6.18) to be

$$\bar{\gamma}_{\text{dif}} \approx \frac{|h|^2 \bar{\gamma}}{2}, \tag{6.20}$$

which is about half of the output of a coherent decoder. This explains why differential decoding incurs 3 dB loss in SNR when compared to coherent ML decoding.

So far, we have seen that differential (de)coding is capable of forgoing channel estimation. However, (6.20) points out that $\bar{\gamma}_{\text{dif}}$ depends on $|h|$, which undergoes random fluctuations due to fading propagation and as a result, causes error performance of differential systems to degrade severely. This in part motivates well the development of differential ST coding schemes with diversity-enabling multi-antenna systems, which can mitigate this fading-induced degradation in error performance. In other words, differential ST coded systems can effect the diversity and coding gains without knowledge of the MIMO channel required by the coherent ST codes we dealt with in Chapter 3.

6.2.2 Differential Unitary ST Codes

Motivated by scalar differential coding, a differential ST coding scheme that applies to any number of transmit-antennas was introduced independently in [120] and [116]. Because ST code matrices of a differential ST system must be unitary, this scheme is often referred to as differential unitary ST coding.

Similar to scalar differential coding, differential unitary ST encoding amounts to relating two consecutive ST code matrices in a way that allows information symbols to be decoded without knowledge of the MIMO channel at the receiver. To describe the encoding process, we introduce the block index n and rewrite (2.9) as

$$\boldsymbol{Y}(n) = \sqrt{\bar{\gamma}}\, \boldsymbol{H}\boldsymbol{X}(n) + \boldsymbol{W}(n). \tag{6.21}$$

Each information-bearing symbol $s(n) \in \mathbb{S}$ drawn from the constellation set $\mathbb{S} := \{d_1, \ldots, d_Q\}$ is represented by a matrix $\boldsymbol{F}(n) \in \mathbb{F}$, where $\mathbb{F} := \{\boldsymbol{F}_0, \boldsymbol{F}_2, \ldots, \boldsymbol{F}_{Q-1}\}$ denotes a set of Q distinct $N_t \times N_t$ unitary matrices satisfying the group property

$$\boldsymbol{F}_p \boldsymbol{F}_q \in \mathbb{F}, \quad \forall\, \boldsymbol{F}_p, \boldsymbol{F}_q \in \mathbb{F}. \tag{6.22}$$

Matrix $\boldsymbol{F}_q$ is related to the qth element d_q of $\mathbb{S}$ with a one-to-one mapping

$$d_q \longleftrightarrow \boldsymbol{F}_q, \quad \forall\, q \in [0, Q-1], \tag{6.23}$$

where $Q := |\mathbb{S}|$. Because $\boldsymbol{F}(n)$ corresponds one-to-one with $s(n)$, transmitting or detecting $\boldsymbol{F}(n)$ is equivalent to transmitting or detecting $s(n)$. Having represented information symbols $s(n)$ by unitary matrices $\boldsymbol{F}(n)$, differential encoding of $\boldsymbol{F}(n)$ is carried out. The transmitted matrices obey the recursion

$$\begin{aligned} \boldsymbol{X}(n) &= \boldsymbol{X}(n-1)\boldsymbol{F}(n), \quad \forall\, n \geq 0, \\ \boldsymbol{X}(-1) &= \boldsymbol{I}. \end{aligned} \tag{6.24}$$

Relative to the system parameters introduced in Section 2.1, this encoding scheme has

$$N_x = N_t = N \quad \text{and} \quad N_s = 1.$$

According to (2.5), the transmission rate is

$$R_{\text{DUSTC}} = \frac{\log_2 |\mathbb{S}|}{N} = \frac{\log_2 Q}{N} \quad \text{bits per channel use.} \tag{6.25}$$

Turning our attention to the receiver end, we wish to show that $\boldsymbol{F}(n)$ can be differentially decoded from two consecutive received matrices $\boldsymbol{Y}(n)$ and $\boldsymbol{Y}(n-1)$ without knowing $\boldsymbol{H}$. Taking (6.21) and (6.24) into account, we have

$$\bar{\boldsymbol{Y}}(n) = \sqrt{\bar{\gamma}}\, \boldsymbol{H} \bar{\boldsymbol{X}}(n) + \bar{\boldsymbol{W}}(n), \tag{6.26}$$

where

$$\begin{aligned} \bar{\boldsymbol{Y}}(n) &:= [\boldsymbol{Y}(n-1)\ \boldsymbol{Y}(n)], \\ \bar{\boldsymbol{X}}(n) &:= [\boldsymbol{X}(n-1)\ \boldsymbol{X}(n)], \\ \bar{\boldsymbol{W}}(n) &:= [\boldsymbol{W}(n-1)\ \boldsymbol{W}(n)]. \end{aligned} \tag{6.27}$$

In writing (6.26), we have assumed that the channels remain invariant over two consecutive blocks. Note also that such an assumption is not required by noncoherent ST codes. Because $\boldsymbol{F}(n)$ is unitary, it follows from (6.24) that $\bar{\boldsymbol{X}}(n)$ satisfies

$$\bar{\boldsymbol{X}}(n)\bar{\boldsymbol{X}}^{\mathcal{H}}(n) = 2\boldsymbol{I}, \tag{6.28}$$

which implies that differential unitary ST coding can be thought of as a special form of noncoherent ST coding if we treat $\bar{\boldsymbol{X}}(n)$ as a noncoherent ST code matrix with

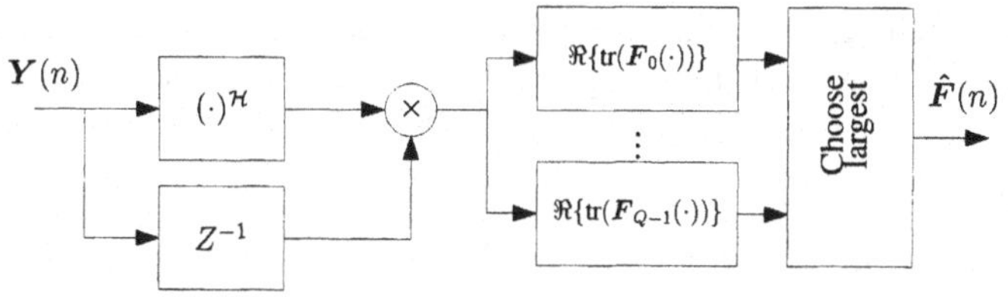

Figure 6.2 Differential unitary ST decoder

proper normalization. As a result, the ML decoder in (6.3) can be applied to detect $\bar{X}$ without channel knowledge as

$$\hat{\bar{X}} = \arg\max ||\bar{X}\bar{Y}^{\mathcal{H}}||_F^2. \tag{6.29}$$

Furthermore, using (6.24) and (6.27), we find that

$$\bar{X}^{\mathcal{H}}(n)\bar{X}(n) = \begin{bmatrix} I & F(n) \\ F^{\mathcal{H}}(n) & I \end{bmatrix} \tag{6.30}$$

is a function of $F(n)$. Substituting (6.30) into (6.29), we can simplify (6.29) to

$$\hat{F}(n) = \arg\max_{F \in \mathbb{F}} \Re\left\{\mathrm{tr}(FY^{\mathcal{H}}(n)Y(n-1))\right\}. \tag{6.31}$$

This differential unitary ST decoder is depicted in Figure 6.2, where Z^{-1} represents a one-block delay operator. Since (6.31) requires searching over Q candidate matrices F, it follows from (6.25) that decoding complexity grows exponentially with N_t and R_{DUSTC}.

In differential unitary ST coding, information symbols are conveyed through ST code matrices whose group property leads to an ML differential detector which does not require channel knowledge. Similar to coherent and noncoherent ST coding, a well-designed ST constellation plays an important role in optimizing error performance of a differential ST coded system. Because the PEP provides a good approximation of error probability performance, selecting ST constellation matrices for differential systems starts with analyzing the PEP of the differential decoder in (6.31).

Supposing that entries of the channel matrix H and the noise matrix $\bar{W}(n)$ in (6.26) are i.i.d. complex Gaussian random variables with zero mean and unit variance, the PEP of (6.31) can be upper-bounded by [212]

$$P_2(F \to F') \le \frac{1}{\left[\det(I + \frac{\bar{\gamma}}{8}(F - F')(F - F')^{\mathcal{H}})\right]^{N_r}}. \tag{6.32}$$

Comparing (6.32) with (2.43), we recognize that the bound in (6.32) is essentially identical to that in (2.43) when $R_t = I$ and $R_r = I$, except that the SNR is scaled by a factor $1/2$. This factor explains why error performance of differential unitary ST decoding degrades by 3 dB relative to its coherent counterpart. Furthermore, this link between coherent and differential ST systems suggests that to maximize both diversity and coding gains, the design of the unitary group $\mathbb{F}$ should follow criteria similar to C2.1) and C2.2) in Chapter 2. Specifically, one should design $\mathbb{F}$ according to the following criteria:

C6.1) Design the group $\mathbb{F}$ such that $\forall\, \boldsymbol{F}, \boldsymbol{F}' \in \mathbb{F}$, the matrix $\boldsymbol{\Delta}_F := (\boldsymbol{F} - \boldsymbol{F}')(\boldsymbol{F} - \boldsymbol{F}')^{\mathcal{H}}$ has full rank.

C6.2) Design the group $\mathbb{F}$ such that $\forall\, \boldsymbol{F}, \boldsymbol{F}' \in \mathbb{F}$, the minimum value of the determinant $\det(\boldsymbol{\Delta}_F)$ is maximized.

Note that C6.1) and C6.2) are identical to those used in designing coherent ST code matrices. However, the main difference here is that $\mathbb{F}$ must be restricted to be a unitary group satisfying (6.22).

For the special case corresponding to $N_t = 2$ and $Q = 2^p$, it was shown in [120] that all unitary groups with $|\mathbb{F}| = Q$ must form either a cyclic or a dicyclic group. If $< \boldsymbol{G}_1, \boldsymbol{G}_2, \ldots, \boldsymbol{G}_K >$ is the group comprising all distinct products of powers of $\boldsymbol{G}_1, \boldsymbol{G}_2, \ldots, \boldsymbol{G}_K$, a (Q, k) cyclic unitary group is defined as

$$\mathbb{F} := \left\langle \begin{bmatrix} \omega_Q & 0 \\ 0 & \omega_Q^k \end{bmatrix} \right\rangle,$$

where $\omega_Q := \exp(j2\pi/Q)$ and k is some odd number. For all $Q \geq 8$, a dicyclic unitary group is defined as

$$\mathbb{F} := \left\langle \begin{bmatrix} \omega_{Q/2} & 0 \\ 0 & \omega_{Q/2}^* \end{bmatrix}, \begin{bmatrix} 0 & -1 \\ 1 & 0 \end{bmatrix} \right\rangle.$$

Based on these definitions, the design of an optimal unitary group reduces to searching for a cyclic or dicyclic group with maximum diversity and coding gains. Table 6.2 lists the optimal unitary groups found using exhaustive search for $N_t = 2$.

6.2.3 Differential Alamouti Codes

For a two transmit- one receive-antenna system, Alamouti's coherent ST code presented in Chapter 3 achieves the maximum diversity gain and is capacity achieving while enjoying low-complexity linear ML decoding. These attractive features are, however, possible only when accurate channel estimates are available at the receiver. To obviate this requirement, Tarokh and Jafarkhani introduced in [248] a differential version of Alamouti's code which they proved capable of retaining most advantages of Alamouti's code without requiring channel knowledge. As compared to the differential unitary ST code matrix for $N_t = 2$, this differential Alamouti code can achieve higher transmission rate at lower decoding complexity.

To present the basic idea behind differential Alamouti coding we start with the input-output relationship presented in Section 2.1. But since the ST code matrix will change from slot to slot, we introduce the time index n and write the input-output

Table 6.2 OPTIMAL UNITARY GROUPS

R_{DUSTC}	$\mathbb{F}$	G_c
0.5	$\left\langle \begin{bmatrix} -1 & 0 \\ 0 & -1 \end{bmatrix} \right\rangle$	4
1.0	$\left\langle \begin{bmatrix} 0 & -1 \\ 1 & 0 \end{bmatrix} \right\rangle$	3
1.5	$\left\langle \begin{bmatrix} j & 0 \\ 0 & -j \end{bmatrix}, \begin{bmatrix} 0 & -1 \\ 1 & 0 \end{bmatrix} \right\rangle$	2
2.0	$\left\langle \begin{bmatrix} \omega_8 & 0 \\ 0 & \omega_8^* \end{bmatrix}, \begin{bmatrix} 0 & -1 \\ 1 & 0 \end{bmatrix} \right\rangle$	0.586
2.5	$\left\langle \begin{bmatrix} \omega_{32} & 0 \\ 0 & j\omega_{32} \end{bmatrix} \right\rangle$	0.249

relationship for the differentially coded Alamouti system as

$$\begin{bmatrix} y_1(n) \\ y_2(n) \end{bmatrix} = \sqrt{\frac{\bar{\gamma}}{2}} \begin{bmatrix} x_1(n) & x_2(n) \\ -x_2^*(n) & x_1^*(n) \end{bmatrix} \begin{bmatrix} h_1 \\ h_2 \end{bmatrix} + \begin{bmatrix} w_1(n) \\ w_2(n) \end{bmatrix}, \tag{6.33}$$

where $x_i(n)$ denotes the transmitted symbol in time slot n from antenna $i = 1, 2$. If $x_i(n)$ equals the information symbol $s_i(n)$, that is,

$$\begin{bmatrix} x_1(n) \\ x_2(n) \end{bmatrix} = \begin{bmatrix} s_1(n) \\ s_2(n) \end{bmatrix}, \tag{6.34}$$

then (6.33) is identical to the model used for Alamouti's coherent ST coded system. In this case, information symbols $s_1(n)$ and $s_2(n)$ can be detected from the decision statistics $\tilde{y}_1(n)$ and $\tilde{y}_2(n)$ [cf. (3.35)]:

$$\begin{bmatrix} \tilde{y}_1(n) \\ \tilde{y}_2(n) \end{bmatrix} = \begin{bmatrix} h_1^* & h_2 \\ h_2^* & -h_1 \end{bmatrix} \begin{bmatrix} y_1(n) \\ y_1^*(n) \end{bmatrix}, \tag{6.35}$$

provided that the channels can be estimated accurately at the receiver.

Similar to scalar differential coding, to bypass channel estimation, Alamouti's differential ST system relies on differentially encoding consecutive ST code matrices. To be specific, instead of (6.34), the transmitted symbols $x_1(n)$ and $x_2(n)$ are related to $s_1(n)$ and $s_2(n)$ according to the recursion

$$\begin{bmatrix} x_1(n) \\ x_2(n) \end{bmatrix} = \frac{1}{\sqrt{2}} \begin{bmatrix} x_1(n-1) & -x_2^*(n-1) \\ x_2(n-1) & x_1^*(n-1) \end{bmatrix} \begin{bmatrix} s_1(n) \\ s_2(n) \end{bmatrix}, \quad n \geq 0, \tag{6.36}$$

which is initialized with $x_1(-1) = x_2(-1) = 1$. Clearly, the pair of transmitted symbols per time slot depends not only on the corresponding pair of information symbols but also on the previous two transmitted symbols.

Replacing the index n in (6.33) by $n-1$, we find that

$$\begin{bmatrix} y_1(n-1) \\ y_2(n-1) \end{bmatrix} = \sqrt{\frac{\bar{\gamma}}{2}} \begin{bmatrix} x_1(n-1) & x_2(n-1) \\ -x_2^*(n-1) & x_1^*(n-1) \end{bmatrix} \begin{bmatrix} h_1 \\ h_2 \end{bmatrix} + \begin{bmatrix} w_1(n-1) \\ w_2(n-1) \end{bmatrix},$$

which can be rewritten as

$$\begin{aligned} &\sqrt{\frac{\bar{\gamma}}{2}} \begin{bmatrix} h_1 & h_2 \\ h_2^* & -h_1^* \end{bmatrix} \begin{bmatrix} x_1(n-1) & -x_2^*(n-1) \\ x_2(n-1) & x_1^*(n-1) \end{bmatrix} \\ &= \begin{bmatrix} y_1(n-1) & y_2(n-1) \\ y_2^*(n-1) & -y_1^*(n-1) \end{bmatrix} - \begin{bmatrix} w_1(n-1) & w_2(n-1) \\ w_2^*(n-1) & -w_1^*(n-1) \end{bmatrix}. \end{aligned} \tag{6.37}$$

To see how channel estimation is bypassed in the decoding stage, we rewrite (6.33) as

$$\begin{bmatrix} y_1(n) \\ y_2^*(n) \end{bmatrix} = \sqrt{\frac{\bar{\gamma}}{2}} \begin{bmatrix} h_1 & h_2 \\ h_2^* & -h_1^* \end{bmatrix} \begin{bmatrix} x_1(n) \\ x_2(n) \end{bmatrix} + \begin{bmatrix} w_1(n) \\ w_2^*(n) \end{bmatrix}, \tag{6.38}$$

and use (6.36) to obtain

$$\begin{bmatrix} y_1(n) \\ y_2^*(n) \end{bmatrix} = \sqrt{\frac{\bar{\gamma}}{2}} \begin{bmatrix} h_1 & h_2 \\ h_2^* & -h_1^* \end{bmatrix} \begin{bmatrix} x_1(n-1) & -x_2^*(n-1) \\ x_2(n-1) & x_1^*(n-1) \end{bmatrix} \begin{bmatrix} s_1(n) \\ s_2(n) \end{bmatrix} + \begin{bmatrix} w_1(n) \\ w_2^*(n) \end{bmatrix}. \tag{6.39}$$

Subsequent application of (6.37) yields

$$\begin{aligned} \begin{bmatrix} y_1(n) \\ y_2^*(n) \end{bmatrix} &= \frac{1}{\sqrt{2}} \begin{bmatrix} y_1(n-1) & y_2(n-1) \\ y_2^*(n-1) & -y_1^*(n-1) \end{bmatrix} \begin{bmatrix} s_1(n) \\ s_2(n) \end{bmatrix} \\ &\quad \underbrace{- \frac{1}{\sqrt{2}} \begin{bmatrix} w_1(n-1) & w_2(n-1) \\ w_2^*(n-1) & -w_1^*(n-1) \end{bmatrix} \begin{bmatrix} s_1(n) \\ s_2(n) \end{bmatrix} + \begin{bmatrix} w_1(n) \\ w_2^*(n) \end{bmatrix}}_{\tilde{\boldsymbol{w}}(n) := [\tilde{w}_1(n), \tilde{w}_2(n)]^T} \\ &= \frac{1}{\sqrt{2}} \begin{bmatrix} y_1(n-1) & y_2(n-1) \\ y_2^*(n-1) & -y_1^*(n-1) \end{bmatrix} \begin{bmatrix} s_1(n) \\ s_2(n) \end{bmatrix} + \begin{bmatrix} \tilde{w}_1(n) \\ \tilde{w}_2(n) \end{bmatrix}, \end{aligned} \tag{6.40}$$

where it can be verified that $\tilde{\boldsymbol{w}}(n)$ is a zero-mean Gaussian noise vector with correlation matrix $2\boldsymbol{I}_2$. Comparing (6.40) with (6.39), we see that the received samples $y_1(n-1)$ and $y_2(n-1)$ basically play the role that h_1 and h_2 have in the coherent ST code. Thus, decoding of the information symbols $s_1(n)$ and $s_2(n)$ can be accomplished without requiring estimates of h_1 and h_2. Specifically, the decision statistics $\tilde{y}_1(n)$ and $\tilde{y}_2(n)$ can be obtained as

$$\begin{bmatrix} \tilde{y}_1(n) \\ \tilde{y}_2(n) \end{bmatrix} = \sqrt{2} \underbrace{\begin{bmatrix} y_1^*(n-1) & y_2(n-1) \\ y_2^*(n-1) & -y_1(n-1) \end{bmatrix}}_{:=\boldsymbol{Y}(n-1)} \begin{bmatrix} s_1(n) \\ s_2(n) \end{bmatrix}. \tag{6.41}$$

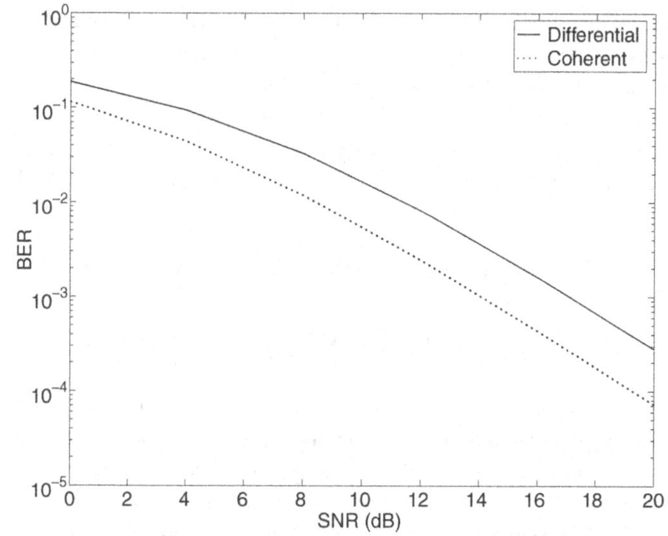

Figure 6.3 Comparison between Alamouti's coherent and differential ST coded systems

Because $\boldsymbol{Y}(n-1)$ defined under (6.41) is a scaled unitary matrix, we infer that decoding of $s_1(n)$ and $s_2(n)$ can be decoupled without sacrificing error performance, as in Alamouti's coherent ST decoder. In other words, the decoding simplicity of Alamouti's coherent ST code carries over to the differential setting. In Figure 6.3 we compare the performance of Alamouti's coherent code with its differential counterpart. We observe that while the diversity gains of these two schemes are equal, the differential Alamouti code incurs a 3 dB loss in SNR, as is the case with scalar differential single-antenna systems.

6.2.4 Differential OSTBCs

The differentially ST coded system we presented in Section 6.2.3 based on Alamouti's code was specifically tailored for two transmit-antennas. In this subsection we discuss its extension to an arbitrary number of transmit- and receive-antennas which uses differentially modulated OSTBCs [82].

In differential OSTBCs, the transmitted matrix $\boldsymbol{X}(n)$ in (6.21) is generated via the recursion

$$\boldsymbol{X}(n) = \boldsymbol{X}(n-1)\boldsymbol{X}_{\mathrm{O}}(\boldsymbol{s}), \tag{6.42}$$

where $\boldsymbol{X}_{\mathrm{O}}$ is the OSTBC matrix defined in Section 3.3.1. Clearly, differential OSTBCs exhibit code rates identical to their coherent counterparts in Section 3.3.1.

Because $\boldsymbol{X}_{\mathrm{O}}(\boldsymbol{s})$ satisfies (6.1), the differential decoder in (6.31) can be applied but with $\boldsymbol{F}$ replaced by $\boldsymbol{X}_{\mathrm{O}}(\boldsymbol{s})$, that is,

$$\hat{\boldsymbol{X}}_{\mathrm{O}}(\boldsymbol{s}) = \arg\max_{\boldsymbol{s}\in\mathbb{S}^{N_s\times 1}} \Re\left\{\mathrm{tr}(\boldsymbol{X}_{\mathrm{O}}(\boldsymbol{s})\boldsymbol{Y}^{\mathcal{H}}(n)\boldsymbol{Y}(n-1))\right\}. \tag{6.43}$$

The differential decoder in (6.43) involves an exhaustive search over $|\mathbb{S}|^{N_s}$ possible code matrices. Fortunately, the required computational complexity can be reduced considerably after taking into account the structure of $\boldsymbol{X}_{\mathrm{O}}(\boldsymbol{s})$.

To this end, let us replace $\boldsymbol{X}_{\mathrm{O}}(\boldsymbol{s})$ in (6.43) with its representation in (2.58). By using the properties in (3.21), it can easily be verified that decoding of s_i's in $\boldsymbol{s}$ can be performed on a symbol-by-symbol basis. Specifically, each s_i can be decoded separately by using the decoder

$$\hat{s}_i = \arg\max_{s\in\mathbb{S}} \Big\{ \Re\{\mathrm{tr}(\boldsymbol{\Phi}_i \boldsymbol{Y}^{\mathcal{H}}(n)\boldsymbol{Y}(n-1))\}\Re(s) + \Re\{\mathrm{tr}(j\boldsymbol{\Psi}_i \boldsymbol{Y}^{\mathcal{H}}(n)\boldsymbol{Y}(n-1))\}\Im(s) \Big\}, \tag{6.44}$$

whose computational complexity is much lower than that of (6.43).

Although complete channel knowledge does not need to be available at the transmitter or the receiver of a differential OSTBC system, partial channel knowledge, such as channel statistics, could possibly be estimated at the receiver and exploited by the transmitter to improve error performance. Such an idea was pursued in [34], where a differential OSTBC was combined with channel-adaptive transmit beamforming and power loading, both of which were optimized based on knowledge of the channel correlation at the transmitter. The resulting ST system was shown capable not only of improving error performance but also of increasing the data rate.

6.2.5 Cayley Differential Unitary ST Codes

Differential unitary ST coding in Section 6.2.2 relies on the use of matrix groups. However, the number of groups available is quite limited, and as the number of antennas increases, the groups do not attain rates as high as V-BLAST. The differential orthogonal designs in Section 6.2.3, on the other hand, have simple encoding and decoding modules but suffer from rate or error performance loss when the number of transmit-antennas is large. These considerations motivate yet another type of differential ST codes, which are known as Cayley differential unitary ST codes [106].

A challenge encountered with the design of unitary matrices $\boldsymbol{F}$ in (6.24) is the lack of simple parameterizations. This limitation can be alleviated by constructing $\boldsymbol{F}$ using the Cayley transform

$$\boldsymbol{F} = (\boldsymbol{I} + j\boldsymbol{A})^{-1}(\boldsymbol{I} - j\boldsymbol{A}), \tag{6.45}$$

where the Hermitian matrix $\boldsymbol{A}$ is given by

$$\boldsymbol{A} = \sum_{p=1}^{P} \alpha_p \boldsymbol{A}_p,$$

$\{\alpha_p\}_{p=1}^{P}$ are real scalars drawn from a finite alphabet $\mathcal{A}$ with cardinality M, and $\boldsymbol{A}_p$'s are fixed $N_t \times N_t$ complex Hermitian matrices. The differential unitary ST codes constructed based on (6.45) are called Cayley differential unitary ST codes.

To construct Cayley differential unitary ST codes, one needs to choose P, design the matrices $\{\boldsymbol{A}_p\}_{p=1}^P$ and select the alphabet $\mathcal{A}$ from which the scalars $\{\alpha_p\}_{p=1}^P$ are drawn. As detailed in [106], a Cayley ST code can be designed using the following steps:

S1) Choose $P \leq N_a(2N_t - N_a)$ with $N_a := \min(N_t, N_r)$.

S2) For a target transmission rate $R = \frac{P}{N_t}\log_2 M$, set $M = 2^{RN_t/P}$. Choose $\mathcal{A}$ to be the M-point discretization of the scalar Cauchy distribution obtained as the image of the function $\alpha = -\tan(\theta/2)$ applied to the set $\theta \in \{\pi/M, 3\pi/M, \ldots, (2M-1)\pi/M\}$.

S3) Choose matrices $\{\boldsymbol{A}_p\}_{p=1}^P$ so that the average distance in the matrix constellation is maximized.

To illustrate the Cayley differential ST code, let us consider an example with $N_t = 2$, $N_r = 1$, and $R = 1$. Choosing $P = 2$, the constellation size of $\mathcal{A}$ is $M = 2$ and the set $\mathcal{A} = \{1, -1\}$. The basis matrices are

$$\boldsymbol{A}_1 = \begin{bmatrix} \frac{1}{\sqrt{2}(\sqrt{3}+1)} & \frac{-i}{\sqrt{2}(\sqrt{3}-1)} \\ \frac{i}{\sqrt{2}(\sqrt{3}-1)} & \frac{-1}{\sqrt{2}(\sqrt{3}+1)} \end{bmatrix}, \text{ and } \boldsymbol{A}_2 = \begin{bmatrix} \frac{1}{\sqrt{2}(\sqrt{3}+1)} & \frac{i}{\sqrt{2}(\sqrt{3}-1)} \\ \frac{-i}{\sqrt{2}(\sqrt{3}-1)} & \frac{-1}{\sqrt{2}(\sqrt{3}+1)} \end{bmatrix}. \quad (6.46)$$

Turning our attention to decoding differentially coded ST transmissions based on the Cayley transform, we find that the ML decoder for the nth symbol $\boldsymbol{F}(n)$ represented by $\{\alpha_p(n)\}$ is given by

$$\{\hat{\alpha}_p\} = \arg\min_{\{\alpha_p\}} \|(\boldsymbol{I} + j\boldsymbol{A})^{-1}(\boldsymbol{Y}(n) - \boldsymbol{Y}(n-1) + j\boldsymbol{A}(\boldsymbol{Y}(n) + \boldsymbol{Y}(n-1)))\|^2.$$

Assuming that the noise remains white after differential decoding, the ML decoder can be simplified to

$$\{\hat{\alpha}_p\} = \arg\min_{\{\alpha_p\}} \|\boldsymbol{Y}(n) - \boldsymbol{Y}(n-1) + j\boldsymbol{A}(\boldsymbol{Y}(n) + \boldsymbol{Y}(n-1))\|^2. \quad (6.47)$$

For the simplified detector in (6.47), instead of exhaustive search, we can use the sphere-decoding or nulling-canceling algorithms of Chapter 5.

In summary, Cayley differential ST codes can be used for any number of transmit- and receive-antennas. They enable high transmission rates and can be decoded using near-ML or suboptimal algorithms. However, unlike alternative differential designs (e.g., [116]), Cayley codes offer no performance guarantees in terms of diversity or coding gains. Recently, LCF coded designs have been combined with the Cayley transform [78] to enable high diversity as well. Finally, the Cayley transform can also be used to design noncoherent ST codes as detailed in [131].

6.3 CLOSING COMMENTS

In this chapter we presented several representative noncoherent and differential ST codes for flat fading channels. Compared to coherent ST codes, these codes bypass the need for channel estimation at the price of moderate degradation in error performance. Whether this chapter's codes offer feasible ST designs for practical deployment depends on how fast the MIMO channel varies and how much complexity can be afforded. In addition, the design of these codes assumes flat fading channels. How to extend differential and noncoherent ST codes to frequency- and/or time-selective channels will be pointed out in Chapters 8 and 9.

7

ST Codes for Frequency-Selective Fading Channels: Single-Carrier Systems

In Chapters 2 to 4, we presented coherent ST codes for flat fading MIMO channels that are typically encountered in narrowband communication systems. Wireless broadband systems, on the other hand, aim at high data rates, which in our multi-antenna context necessitates dealing with frequency-selective (and hence time-dispersive) MIMO channels. Motivated by the widespread applications of broadband wireless systems, the present chapter and the next one lay out major classes of coherent ST codes for use over frequency-selective MIMO fading channels. We deal specifically with the design of ST codes for two popular transmission formats: block single-carrier in this chapter and multi-carrier in Chapter 8. We defer a comparison between the single- and multi-carrier approaches to the closing comments of Chapter 8.

We see first that the model for single-carrier multi-antenna transmissions over frequency-selective MIMO channels can be brought to a form equivalent to that of a flat fading MIMO channel created by a set of virtual transmit-antennas. The usefulness of this equivalence will turn out to be twofold: (i) it will help us assess rate and error performance when communicating over frequency-selective multi-antenna channels by resorting to the metrics we introduced for flat fading MIMO channels in Chapter 2; and (ii) it will guide the designs of ST codes for frequency-selective MIMO channels by properly modifying the corresponding designs of Chapter 3 for the flat fading virtual MIMO channel. We again adopt the diversity order as a metric for evaluating error performance at high SNR and also as a criterion for designing ST codes. Our figure of merit for rate performance and comparison of the various ST codes will be the rate outage probability introduced in Chapter 2. The ST codes we present here for single-carrier multi-antenna systems belong to the class of error-performance-oriented designs and generalize the STTCs and OSTBCs that we dealt with in Chapter 3 to frequency-selective channels. As with flat fading channels, we expose characteristics in their design and decoding complexity, diversity order, and spectral efficiency. Rate-oriented ST codes for frequency-selective MIMO channels are easier to design for multi-carrier systems and are presented in Chapter 8.

7.1 SYSTEM MODEL AND PERFORMANCE LIMITS

We consider a multi-antenna system with N_t transmit- and N_r receive-antennas. The fading channel between the μth transmit-antenna and the νth receive-antenna is assumed to be frequency selective but time flat, and is described by the discrete-time baseband equivalent impulse response vector

$$\boldsymbol{h}_{\nu\mu} := [h_{\nu\mu}(0), \ldots, h_{\nu\mu}(L)]^T \in \mathbb{C}^{(L+1)\times 1}, \tag{7.1}$$

where L stands for the channel order. The channel impulse response includes the effects of transmit-receive filters, physical multipath, and relative delays among antennas.

Let $x_\mu(n)$ denote the transmitted symbol from the μth antenna at time n and $y_\nu(n)$ the received symbol at the νth receive-antenna at time n. The channel input-output relationship is thus

$$y_\nu(n) = \sqrt{\frac{\overline{\gamma}}{N_t}} \sum_{\mu=1}^{N_t} \sum_{l=0}^{L} h_{\nu\mu}(l) x_\mu(n-l) + w_\nu(n), \tag{7.2}$$

where $w_\nu(n)$ denotes the AWGN having zero mean and unit variance, and $\overline{\gamma}$ stands for the average SNR per receive-antenna as we assume that the total variance of the channel taps $\sum_{l=0}^{L} E[|h_{\nu\mu}(l)|^2]$ is normalized to 1, for any antenna pair.

For future use, let us define the vectors

$$\boldsymbol{x}(n) := [x_1(n), \ldots, x_{N_t}(n)]^T \in \mathbb{C}^{N_t \times 1},$$
$$\boldsymbol{y}(n) := [y_1(n), \ldots, y_{N_r}(n)]^T \in \mathbb{C}^{N_r \times 1},$$
$$\boldsymbol{w}(n) := [w_1(n), \ldots, w_{N_r}(n)]^T \in \mathbb{C}^{N_r \times 1},$$

and for each tap l, the matrix

$$\boldsymbol{H}(l) := \begin{bmatrix} h_{11}(l) & \cdots & h_{1N_t}(l) \\ \vdots & \ddots & \vdots \\ h_{N_r 1}(l) & \cdots & h_{N_r N_t}(l) \end{bmatrix} \in \mathbb{C}^{N_r \times N_t}. \tag{7.3}$$

Using these definitions, we can rewrite (7.2) in vector-matrix form as

$$\boldsymbol{y}(n) = \sqrt{\frac{\overline{\gamma}}{N_t}} \sum_{l=0}^{L} \boldsymbol{H}(l) \boldsymbol{x}(n-l) + \boldsymbol{w}(n). \tag{7.4}$$

Notice that for $L = 0$, (7.4) reduces to the flat MIMO model in (2.9).

7.1.1 Flat Fading Equivalence and Diversity

Starting from (7.4), we will establish an equivalence that will allow us to view the model of a frequency-selective MIMO channel as an equivalent model of a flat fading MIMO channel. The same equivalence will also allow us to quantify the diversity provided by a frequency-selective MIMO channel using existing results on the diversity of the equivalent flat fading MIMO channel. Toward this objective, let us recall that ST coding in (2.2) was viewed as a one-to-one mapping between each information block $\boldsymbol{s} := [s_1, s_2, \ldots, s_{N_s}]^T$ and the corresponding ST code matrix

$$\boldsymbol{X} := [\boldsymbol{x}(0), \boldsymbol{x}(1), \ldots, \boldsymbol{x}(N_x - 1)] \in \mathbb{C}^{N_t \times N_x}. \tag{7.5}$$

Based on this mapping, N_s information symbols are ST coded and transmitted through N_t antennas in N_x time slots. To proceed, we define for the frequency-selective MIMO channel of order L, the set of matrices

$$\begin{aligned}
\boldsymbol{X}^{(0)} &:= [\boldsymbol{X}, \boldsymbol{0}_{N_t \times L}] \in \mathbb{C}^{N_t \times (N_x + L)}, \\
\boldsymbol{X}^{(1)} &:= [\boldsymbol{0}_{N_t \times 1}, \boldsymbol{X}, \boldsymbol{0}_{N_t \times (L-1)}] \in \mathbb{C}^{N_t \times (N_x + L)}, \\
&\vdots \\
\boldsymbol{X}^{(L)} &:= [\boldsymbol{0}_{N_t \times L}, \boldsymbol{X}] \in \mathbb{C}^{N_t \times (N_x + L)},
\end{aligned}$$

where $\boldsymbol{0}_{M \times N}$ stands for the all-zero matrix of size $M \times N$. With guard intervals between adjacent ST code matrices, the received $N_r \times (N_x + L)$ ST matrix is

$$\begin{aligned}
\boldsymbol{Y} &:= [\boldsymbol{y}(0), \ldots, \boldsymbol{y}(N_x + L)] \\
&= \sqrt{\frac{\bar{\gamma}}{N_t}} \sum_{l=0}^{L} \boldsymbol{H}(l) \boldsymbol{X}^{(l)} + \boldsymbol{W},
\end{aligned} \tag{7.6}$$

where $\boldsymbol{W}$ is defined similar to $\boldsymbol{Y}$. Defining two matrices

$$\boldsymbol{H}_{\text{eq}} := [\boldsymbol{H}(0), \ldots, \boldsymbol{H}(L)] \qquad \text{and} \qquad \boldsymbol{X}_{\text{eq}} := \begin{bmatrix} \boldsymbol{X}^{(0)} \\ \boldsymbol{X}^{(1)} \\ \vdots \\ \boldsymbol{X}^{(L)} \end{bmatrix}, \tag{7.7}$$

we can rewrite (7.6) as

$$\boldsymbol{Y} = \sqrt{\frac{\bar{\gamma}}{N_t}} \boldsymbol{H}_{\text{eq}} \boldsymbol{X}_{\text{eq}} + \boldsymbol{W}. \tag{7.8}$$

Upon comparing (7.8) with (2.9), we deduce deduce that the ST code matrix $\boldsymbol{X}_{\text{eq}}$ can be viewed as transmitted from $N_t(L+1)$ virtual transmit-antennas over a flat fading channel between each virtual transmit-antenna and each receive-antenna. However, it is worth stressing that the induced virtual antennas corresponding to multipath components are not equivalent to additional actual antennas since we can not control the signals transmitted through them. Indeed, signals transmitted over virtual antennas are simply shifted versions of the signals transmitted over the actual antennas.

The virtual antenna model in (7.8) suggests the following important result on the achievable diversity order provided by frequency-selective MIMO channels.

Proposition 7.1 *Define the augmented channel vector* $\boldsymbol{h}_{\text{eq}} = \text{vec}(\boldsymbol{H}_{\text{eq}}) \in \mathbb{C}^{N_t N_t (L+1)\times 1}$, *where the* vec(·) *operation stacks the columns of a matrix in a vector. Let* $\text{rank}(\boldsymbol{R}_{h_{\text{eq}}})$ *denote the rank of the correlation matrix of* $\boldsymbol{h}_{\text{eq}}$. *ST coded transmissions over frequency-selective MIMO channels enable diversity order no greater than*

$$G_{d,\max} = \text{rank}(\boldsymbol{R}_{h_{\text{eq}}}). \tag{7.9}$$

In rich scattering environments, where $\boldsymbol{R}_{h_{\text{eq}}}$ *has full rank equal to its dimension, the maximum diversity is*

$$G_{d,\max} = N_t N_r (L+1). \tag{7.10}$$

The maximum diversity is achieved if the error matrix

$$\boldsymbol{\Delta}_e := (\boldsymbol{X}_{\text{eq}} - \boldsymbol{X}'_{\text{eq}})(\boldsymbol{X}_{\text{eq}} - \boldsymbol{X}'_{\text{eq}})^{\mathcal{H}} \tag{7.11}$$

has full rank for any codeword $\boldsymbol{X} \neq \boldsymbol{X}'$.

Proof: We follow the steps in Section 2.4. One way to achieve the maximum diversity is through a low-rate repetition scheme, where each antenna transmits the same information symbol followed by L zeros in $L+1$ time slots, and different antennas occupy nonoverlapping time slots. As a result, only one information symbol is transmitted every $N_t(L+1)$ symbol periods with spectral efficiency $\eta = 1/[N_t(L+1)]$. Because the matrix $\boldsymbol{\Delta}_e$ in (7.11) is always proportional to an identity matrix, it has full rank. The arguments detailed in Section 2.4 complete the proof. □

Proposition 7.1 shows that relative to flat fading channels, frequency-selective MIMO channels provide $L+1$ times higher diversity. This diversity comes in two forms: space diversity through the deployment of multiple transmit- and receive-antennas, and multipath diversity arising due to multipath propagation effects. The resulting joint space-multipath diversity order is the product of the constituent forms of diversity. ST codes will be sought to enable the maximum space-multipath diversity available, with as high a transmission rate as possible.

7.1.2 Rate Outage Probability

Having established the diversity order as a performance indicator at high SNR, we consider here the outage mutual information of frequency-selective MIMO channels, which is related to the outage probability at a desirable rate. We will quantify this fundamental limit when the transmitter has no channel state information (CSI) while the receiver has perfect CSI. Conditioned on each realization of $\{\boldsymbol{H}(l)\}_{l=0}^{L}$, the mutual information between $\boldsymbol{x}(n)$ and $\boldsymbol{y}(n)$ in (7.4) is maximized when the input

$\boldsymbol{x}(n)$ is complex Gaussian with zero mean and identity covariance matrix [156]. The maximum mutual information is [156]

$$I(\boldsymbol{x};\boldsymbol{y}|\{\boldsymbol{H}(l)\}_{l=0}^{L}) = \int_0^1 \log_2 \det\left[\boldsymbol{I}_{N_r} + \frac{\bar{\gamma}}{N_t}\tilde{\boldsymbol{H}}(f)\tilde{\boldsymbol{H}}^{\mathcal{H}}(f)\right]df, \tag{7.12}$$

where

$$\tilde{\boldsymbol{H}}(f) := \sum_{l=0}^{L} \boldsymbol{H}(l)e^{-j2\pi fl}. \tag{7.13}$$

Since the maximum mutual information is a random variable, it can become arbitrarily small. Consequently, there is a nonzero probability for it to drop below a prescribed transmission rate R_0, no matter how small R_0 is. When this occurs, error-free transmission at this rate is impossible regardless of what coding scheme is adopted. Formally, one can define the outage probability corresponding to a transmission rate R_0 (in bits per channel use) as

$$P_{\text{out}}(R_0) = P\big(I(\boldsymbol{x};\boldsymbol{y}|\{\boldsymbol{H}(l)\}_{l=0}^{L}) < R_0\big). \tag{7.14}$$

The outage probability in (7.14) can be computed through Monte Carlo simulations, as we illustrate in the next example.

Outage probability provides a benchmark for various ST coding approaches. The performance of different ST codes (e.g., trellis codes or block codes) can readily be compared by measuring its distance from the outage probability.

Example 7.1 The outage probability in (7.14)can be evaluated numerically by Monte Carlo simulations after replacing the integral in (7.12) by a sum. If N is large enough, we can write

$$I(\boldsymbol{x};\boldsymbol{y}|\{\boldsymbol{H}(l)\}_{l=0}^{L}) = \frac{1}{N}\sum_{n=0}^{N-1} \log_2 \det\left[\boldsymbol{I}_{N_r} + \frac{\bar{\gamma}}{N_t}\tilde{\boldsymbol{H}}\Big(\frac{n}{N}\Big)\tilde{\boldsymbol{H}}^{\mathcal{H}}\Big(\frac{n}{N}\Big)\right]. \tag{7.15}$$

Using $N = 128$ suffices for this example since further increasing N brings no discernible improvement in the approximation.

Figure 7.1 depicts the outage probability for various $(N_t, N_r, L+1)$ triplets, where we assume that the entries of $\boldsymbol{H}_{\text{eq}}$ are i.i.d. complex Gaussian distributed with zero mean and variance $1/(L+1)$. Figure 7.1 verifies Proposition 7.1 and confirms that the diversity order for multi-antenna frequency-selective channels is $N_tN_r(L+1)$. For example, the system setup with $(N_t, N_r, L+1) = (2, 1, 2)$ exhibits in its outage probability curve the same slope as the system with $(N_t, N_r, L+1) = (4, 1, 1)$, which is known to have diversity order 4 from the results on flat fading MIMO channels.

As we mentioned earlier, outage probability is the ultimate performance limit for ST coding over frequency-selective MIMO channels and serves as a lower bound for the frame (or block) error rate of specific ST codes [156]. This bound can be evaluated for a fixed bit rate R_0 and each SNR value $\bar{\gamma}$ as follows. We generate multiple realizations of the MIMO channel and count how many times I in (7.15) is

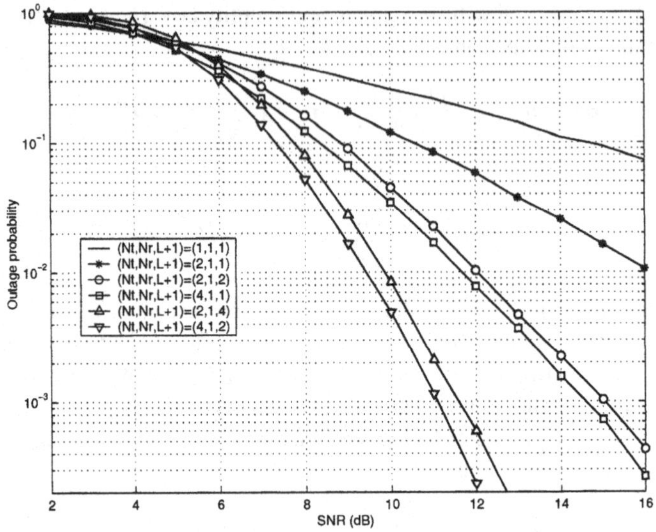

Figure 7.1 Outage probability at $R_0 = 2$ bits per channel use over frequency-selective MIMO channels

less than R_0. There are two causes generating frame errors: (i) outage due to fading since regardless of how strong error control codes are in place, decoding will yield frame errors in those realizations for which $I < R_0$; and (ii) errors due to "large" additive noise for those realizations in which $I \geq R_0$. The second source of error can be avoided asymptotically (in the frame length) by (de)coding the frame with sufficiently strong error control codes.

In the remainder of this chapter, we present ST code designs that can achieve the maximum diversity in (7.10) and compare them on the basis of the outage probability they exhibit at a given rate.

7.2 ST TRELLIS CODES

Following an exposition parallel to Chapter 3, we start our presentation with coherent ST trellis code (STTC) designs for frequency-selective MIMO channels. The simplest representative from this class is the design based on the notion of delay diversity.

7.2.1 Generalized Delay Diversity

As we saw in Section 3.1, delay diversity converts a flat fading MIMO channel with N_t transmit-antennas to an equivalent FIR channel with N_t taps, thus guaranteeing full space (or equivalently, N_t-path diversity). However, if we transmit the $\boldsymbol{X}_{\rm DD}$ in (3.1) directly over an $(L+1)$-tap frequency-selective MIMO channel, we end up with

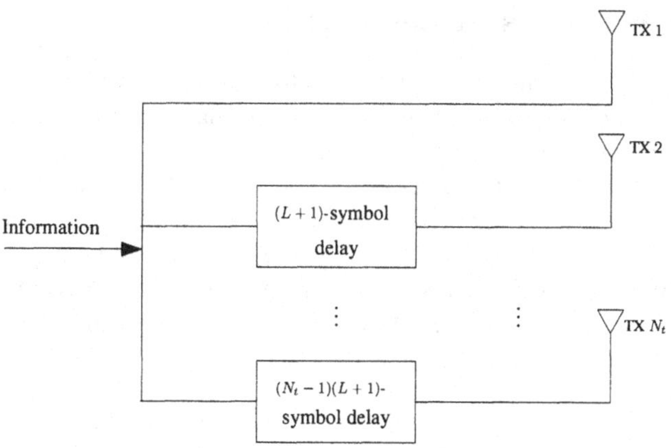

Figure 7.2 Multi-antenna transmitter enabling generalized delay diversity

an equivalent FIR channel having $N_t + L + 1$ taps. As a result, the minimum rank of $\boldsymbol{\Delta}_e$ in (7.11) will be $N_r(N_t + L + 1)$, and the maximum possible diversity of order $N_r N_t(L + 1)$ will not be achieved.

To enable the joint space-multipath diversity, the delay diversity scheme must be generalized as follows: Instead of delaying one symbol, each additional antenna should delay $(L + 1)$ symbols from the previous antenna [100]; see also Figure 7.2. In this way, each receive-antenna observes the output of an equivalent FIR channel with $N_t(L + 1)$ taps. Then $\boldsymbol{\Delta}_e$ has full row rank for any $\boldsymbol{X} \neq \boldsymbol{X}'$, and the maximum diversity order can be achieved.

To construct the ST code matrix $\boldsymbol{X}$, the channel order L must be known. For example, with $N_t = 2$ and $L = 1$, the ST code matrix $\boldsymbol{X}$ for the generalized delay diversity (GDD) is

$$\boldsymbol{X}_{\mathrm{GDD}} = \begin{bmatrix} s_1 & s_2 & s_3 & s_4 & \cdots & s_{N_s} & 0 & 0 \\ 0 & 0 & s_1 & s_2 & \cdots & s_{N_s-2} & s_{N_s-1} & s_{N_s} \end{bmatrix}. \tag{7.16}$$

Notice that if we use the virtual antenna representation from Section 7.1.1, there are $N_t(L + 1)$ virtual antennas in GDD, with each antenna transmitting a shifted version of the signal from the previous antenna. Hence, the spectral efficiency of a GDD system is

$$\eta_{\mathrm{GDD}} = \frac{N_s}{N_s + N_t(L + 1) - 1} \quad \text{symbols per channel use,} \tag{7.17}$$

which reduces to the efficiency of delay diversity in Section 3.1 when $L = 0$.

ML decoding of GDD coded transmissions can be carried out using Viterbi's algorithm, which has complexity on the order of $|\mathbb{S}|^{N_t(L+1)}$ per decoding step.

7.2.2 Search-Based STTC Construction

In Section 3.2 we designed ST trellis codes for flat fading MIMO channels. Good ST trellis codes effecting maximum diversity and large coding gains were found based on computer search. As a consequence of the equivalent flat fading MIMO channel in (7.8), good ST trellis codes for frequency-selective MIMO channels can be found following a similar search procedure. To this end, let us suppose without loss of generality that information symbols s_n are M-PSK modulated with $N_b = \log_2 M$ bits per symbol, and denote the information bits associated with s_n as $\{b_{N_b n+N_b-1}, \ldots, b_{N_b n+1}, b_{N_b n}\}$. As in Section 3.2.3, one can represent ST trellis codes using the $N_t \times (N_b + N_m)$ generator matrix

$$G = \begin{bmatrix} g_{1,1} & g_{1,2} & \cdots & g_{1,N_b+N_m} \\ \vdots & \vdots & \ddots & \vdots \\ g_{N_t,1} & g_{N_t,2} & \cdots & g_{N_t,N_b+N_m} \end{bmatrix}, \tag{7.18}$$

where N_m denotes the number of memory elements in the encoder and g_{ij}'s are integers taking values between 0 and $M-1$. At each time instant n, the ST encoder yields an output

$$\boldsymbol{x}_n = \mathcal{M}((\boldsymbol{G}\boldsymbol{b}_n) \bmod M),$$

where $\mathcal{M}(x) = e^{j2\pi x/M}$, and

$$\boldsymbol{b}_n := [b_{N_b n+N_b-1}, \ldots, b_{N_b n+1}, b_{N_b n}, \ldots, b_{N_b n-N_m}].$$

Based on the representation in (7.7), the block output at the nth time slot from all virtual antennas is

$$\begin{aligned} \boldsymbol{x}'_n = \begin{bmatrix} \boldsymbol{x}_n \\ \boldsymbol{x}_{n-1} \\ \vdots \\ \boldsymbol{x}_{n-L} \end{bmatrix} &= \mathcal{M}\left(\begin{bmatrix} \boldsymbol{G}\boldsymbol{b}_n \\ \boldsymbol{G}\boldsymbol{b}_{n-1} \\ \vdots \\ \boldsymbol{G}\boldsymbol{b}_{n-L} \end{bmatrix} \bmod M \right) \\ &:= f(\boldsymbol{G}, b_{N_b n+N_b-1}, \ldots, b_{N_b(n-L)-N_m}), \end{aligned}$$

where $f(\cdot)$ is a function of several variables. Therefore, for a trellis generated by a $\boldsymbol{G}$ having 2^{N_m} states, we obtain an enlarged trellis with $2^{N_m+LN_b}$ states. The computer search steps are identical to those used for flat fading MIMO channels, based on diversity and coding gains computed from $\boldsymbol{X}_{\text{eq}}$. Using the equivalent flat fading representation, ML decoding can be performed using the Viterbi algorithm on the enlarged trellis structure, which accounts for both the code trellis and the FIR channel.

ST trellis codes suitable for frequency-selective MIMO channels were first searched in [156, 157] using the flat MIMO equivalent representation (7.8). However, as no provision was made to optimize the coding gain, the example codes (with QPSK and $L = 1$) in [156, 157] only guarantee to achieve the maximum diversity. ST trellis codes with maximum diversity and coding gain optimized through computer search

Table 7.1 STTCs FOR $M = 2$ (BPSK) AND $L = 1$ (TWO-TAP CHANNELS)

N_t	States	$\boldsymbol{G}$	G_c
2	4	$\begin{bmatrix} 1 & 1 & 1 \\ 1 & 0 & 1 \end{bmatrix}$	4.000
2	8	$\begin{bmatrix} 1 & 1 & 1 & 1 \\ 1 & 0 & 0 & 1 \end{bmatrix}$	5.981
2	16	$\begin{bmatrix} 1 & 1 & 0 & 1 & 1 \\ 1 & 0 & 1 & 0 & 1 \end{bmatrix}$	7.445
2	32	$\begin{bmatrix} 1 & 1 & 1 & 0 & 0 & 1 \\ 0 & 0 & 1 & 1 & 0 & 1 \end{bmatrix}$	9.514

Table 7.2 STTCs FOR $M = 2$ (BPSK) AND $L = 2$ (THREE-TAP CHANNELS)

N_t	States	$\boldsymbol{G}$	G_c
2	8	$\begin{bmatrix} 1 & 1 & 1 & 1 \\ 1 & 1 & 0 & 1 \end{bmatrix}$	4.000
2	16	$\begin{bmatrix} 1 & 1 & 1 & 0 & 1 \\ 1 & 1 & 0 & 1 & 1 \end{bmatrix}$	5.532
2	32	$\begin{bmatrix} 1 & 1 & 1 & 1 & 1 & 1 \\ 1 & 1 & 0 & 0 & 0 & 1 \end{bmatrix}$	6.644

were reported for BPSK and $L = 1, 2$ in [215] and are also listed in Tables 7.1 and 7.2. The coding gain is defined as

$$G_c = \min_{\boldsymbol{X},\boldsymbol{X}'} \left\{ \det[(\boldsymbol{X}_{\text{eq}} - \boldsymbol{X}'_{\text{eq}})(\boldsymbol{X}_{\text{eq}} - \boldsymbol{X}'_{\text{eq}})^{\mathcal{H}}] \right\}^{1/N_t(L+1)}.$$

It turns out that there are many codes maximizing both diversity and coding gains. Among them, those with the largest average determinant $\boldsymbol{\Delta}_e$ in (7.11), averaged over all the shortest-path error events, are preferred because they typically lead to better error performance [215, 315]. Following the same procedure, ST trellis codes with maximum diversity and coding gains for other configurations can be found similarly, although no results have been reported so far.

Notice that GDD can be viewed as a special case of ST trellis codes (For delay diversity codes, this was noted in [253]). Within the framework of (7.18), the generator representation of a GDD-based ST code is listed in Table 7.3. For BPSK, the number of trellis states of the GDD generator is $2^{(N_t-1)(L+1)}$.

It can easily be verified that the GDD system has coding gain $G_c = 4$ for any N_t and L when BPSK is used. Through computer search, the coding gain of ST trellis codes can be further improved as the number of trellis states increases.

Example 7.2 We simulate the performance of the ST trellis codes in Table 7.1 for FIR channels with $L = 1$. The channel taps in $\boldsymbol{H}_{\text{eq}}$ are i.i.d. complex Gaussian

Table 7.3 TRELLIS REPRESENTATION FOR GDD, BPSK

N_t	L	G	G_c
2	1	$\begin{bmatrix} 1 & 0 & 0 \\ 0 & 0 & 1 \end{bmatrix}$	4.000
2	2	$\begin{bmatrix} 1 & 0 & 0 & 0 \\ 0 & 0 & 0 & 1 \end{bmatrix}$	4.000

distributed with zero mean and variance $1/(L+1) = 0.5$. We set the frame length N_x in (7.5) to be 130. The simulated frame error rates corresponding to different average SNRs $\overline{\gamma}$ are depicted in Figure 7.3.

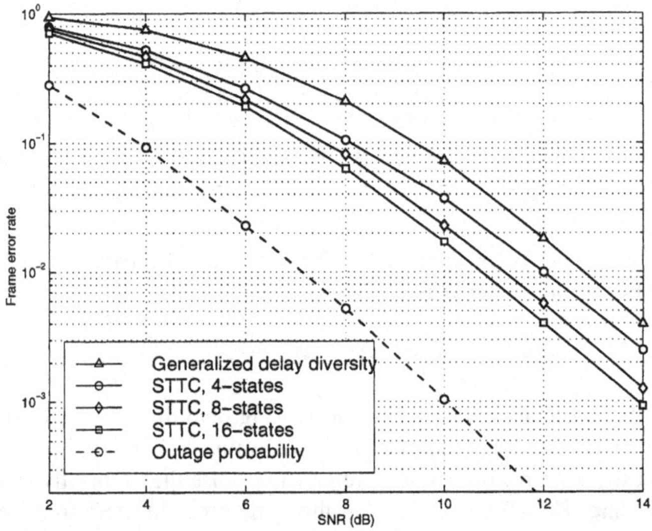

Figure 7.3 Frame error performance of ST trellis codes in FIR channels

Notice that the GDD exhibits much worse performance than the 4-state ST trellis code, although they both have identical coding gains and the same decoding complexity. This demonstrates that code optimization beyond the worst-case error event is necessary.

Among optimized ST trellis codes, the performance improves as the number of states increases. At frame error rate 10^{-2}, the 16-state ST trellis code is about 3.7 dB away from the outage probability at the rate $R_0 = 1$ bit per channel use.

7.3 ST BLOCK CODES

Let us now turn our attention to ST block codes tailored to single-carrier transmissions over frequency-selective MIMO channels. For simplicity in exposition, we focus first on the simple configuration with two transmit-antennas and a single receive-antenna. Later, we pursue extensions to multiple receive-antennas and more than two transmit-antennas.

7.3.1 Block Coding with Two Transmit-Antennas

With $N_t = 2$ transmit-antennas, we consider first generalizing the Alamouti code in (3.16) to frequency-selective MIMO channels. Since frequency-selective channels are time dispersive, such a generalization must rely on properly defined symbol blocks rather than individual information symbols. This intuition prompts us to consider two symbol blocks $\boldsymbol{s}_1$ and $\boldsymbol{s}_2$, each of length J, and construct the $2J \times 2$ matrix

$$\boldsymbol{S} = \begin{bmatrix} \boldsymbol{s}_1 & -\boldsymbol{P}\boldsymbol{s}_2^* \\ \boldsymbol{s}_2 & \boldsymbol{P}\boldsymbol{s}_1^* \end{bmatrix}, \tag{7.19}$$

where $\boldsymbol{P}$ is a permutation matrix. This matrix is drawn from a set of permutation matrices $\{\boldsymbol{P}_J^{(n)}\}_{n=0}^{J-1}$, where the subscript J signifies the dimensionality $J \times J$. Each $\boldsymbol{P}_J^{(n)}$ performs a reverse cyclic shift (that depends on n) when applied to a $J \times 1$ vector $\boldsymbol{a} := [a(0), a(1), \ldots, a(J-1)]^T$. Specifically, the $(p+1)$st entry of $\boldsymbol{P}_J^{(n)}\boldsymbol{a}$ is

$$\left[\boldsymbol{P}_J^{(n)}\boldsymbol{a}\right]_p = a\big((J - p + n - 1) \bmod J\big). \tag{7.20}$$

Two important special cases are $\boldsymbol{P}_J^{(0)}$ and $\boldsymbol{P}_J^{(1)}$, where

$$\boldsymbol{P}_J^{(0)}\boldsymbol{a} = \begin{bmatrix} a(J-1) \\ a(J-2) \\ \vdots \\ a(0) \end{bmatrix}, \qquad \boldsymbol{P}_J^{(1)}\boldsymbol{a} = \begin{bmatrix} a(0) \\ a(J-1) \\ \vdots \\ a(1) \end{bmatrix}. \tag{7.21}$$

We observe from (7.21) that $\boldsymbol{P}_J^{(0)}$ performs a *time reversal* on $\boldsymbol{a}$, while $\boldsymbol{P}_J^{(1)}$ corresponds to taking the J-point FFT or IFFT twice on the vector $\boldsymbol{a}$; that is,

$$\boldsymbol{P}_J^{(1)}\boldsymbol{a} = \boldsymbol{F}_J\boldsymbol{F}_J\boldsymbol{a} = \boldsymbol{F}_J^{\mathcal{H}}\boldsymbol{F}_J^{\mathcal{H}}\boldsymbol{a}. \tag{7.22}$$

The ST code matrix in (7.19) was considered first in [154] with $\boldsymbol{P} = \boldsymbol{P}_J^{(0)}$; and later in [5, 9] with $\boldsymbol{P} = \boldsymbol{P}_J^{(1)}$; while [280] considered both $\boldsymbol{P} = \boldsymbol{P}_J^{(0)}$ and $\boldsymbol{P} = \boldsymbol{P}_J^{(1)}$. The unifying view presented here is taken from [332] and applies to any $\boldsymbol{P}$ from the set $\{\boldsymbol{P}_J^{(n)}\}_{n=0}^{J-1}$.

Based on $\boldsymbol{S}$ in (7.19), the transmitter forwards $\boldsymbol{s}_1$ and $\boldsymbol{s}_2$ to the first and second antennas, respectively, during the first block time interval. During the second block

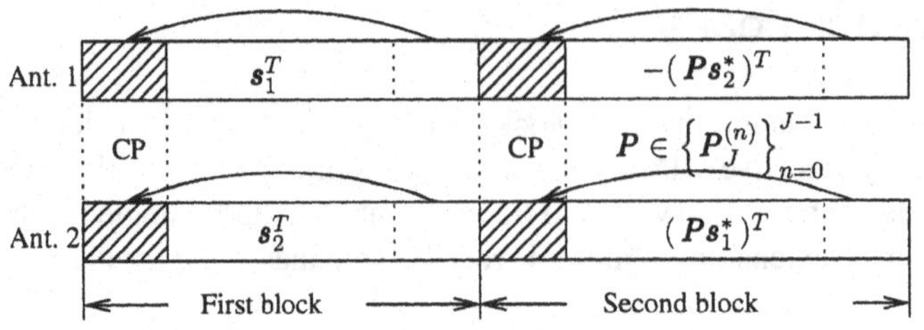

Figure 7.4 CP-based orthogonal STBC

time interval, $-Ps_2^*$ and Ps_1^* are forwarded to the first and second antennas, respectively. Since each block is transmitted over a time-dispersive channel, interblock interference (IBI) occurs between adjacent blocks. One popular approach to avoiding IBI is to insert a cyclic prefix (CP) per block transmission. Specifically, to transmit the block s_i of length J, the transmitter forms a block $T_{\text{cp}}s_i$ of length $P = J + L$, by replicating the last L entries of s_i to its front. The CP matrix implementing this operation is

$$T_{\text{cp}} = \begin{bmatrix} 0_{L\times(J-L)} \mid I_L \\ I_J \end{bmatrix}_{P\times J}. \tag{7.23}$$

With the matrix in (7.19) and CP-based block transmissions, the overall $2 \times 2P$ transmitted ST code matrix is thus

$$X = \begin{bmatrix} (T_{\text{cp}}s_1)^T & -(T_{\text{cp}}Ps_2^*)^T \\ (T_{\text{cp}}s_2)^T & (T_{\text{cp}}Ps_1^*)^T \end{bmatrix}. \tag{7.24}$$

As in [332], we term the design in (7.24) a CP-only system, where "only" is used to contrast the single-carrier system considered here with the multi-carrier CP-OFDM system in Chapter 8. The blocks transmitted for CP-only are depicted in Figure 7.4. For flat fading channels, symbol blocking is unnecessary (i.e., $J = 1, P = 1, P = 1$), and the design of (7.24) reduces to the well-known Alamouti code matrix in (3.16).

Example 7.3 For illustration purposes, consider the case with $L = 1$, $J = 4$, $P = P^{(0)}$, and let $s_1 := [a_0, a_1, a_2, a_3]^T$ and $s_2 := [b_0, b_1, b_2, b_3]^T$. Upon substituting s_1 and s_2 into (7.24), we obtain the ST code matrix

$$X = \begin{bmatrix} a_3 & a_0 & a_1 & a_2 & a_3 & -b_0^* & -b_3^* & -b_2^* & -b_1^* & -b_0^* \\ b_3 & b_0 & b_1 & b_2 & b_3 & a_0^* & a_3^* & a_2^* & a_1^* & a_0^* \end{bmatrix}. \tag{7.25}$$

We next detail two important variates of the CP-only design, affine precoded CP-only and ZP-only, where the latter derives its name because instead of CP it implements zero padding (ZP) per block for IBI suppression.

7.3.1.1 Affine Precoded CP-Only As will be clear soon, embedding L known symbols in s_i facilitates ML decoding because it enables the use of Viterbi's algorithm.

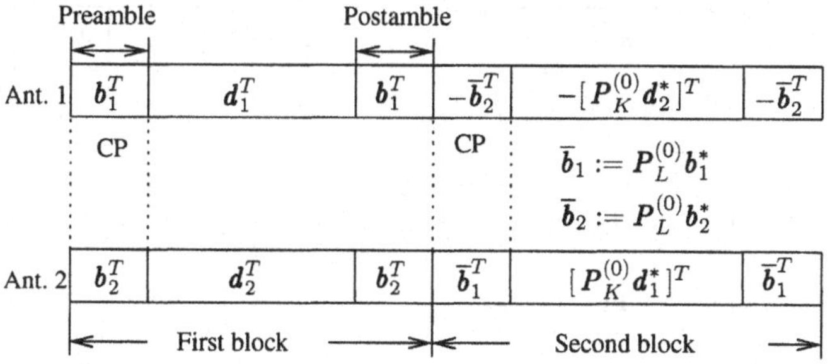

Figure 7.5 Affine precoded CP-only, with $s_i = [d_i^T, b_i^T]^T$ and $P = P_J^{(K)}$

We can effect this embedding by using symbol blocks of the form

$$s_i = T_1 d_i + T_2 b_i = \begin{bmatrix} d_i \\ b_i \end{bmatrix}, \tag{7.26}$$

where the precoder T_1 is formed by the first $K = J - L$ columns of I_J, the precoder T_2 comprises the last L columns of I_J, and the known symbol vector b has size $L \times 1$ with entries drawn from the same alphabet $\mathbb{S}$ as the $K \times 1$ information blocks d_i. Since (7.26) is a special form of affine precoding (AP), we abbreviate the transmission format in (7.24) with s_i in (7.26) as AP-CP-only . Notice that in this scheme, $J = K + L$ and $P = J + L$.

Although b_i here is placed at the end of the block s_i for convenience, we could also place b_i at any position within s_i. As long as L consecutive symbols are known in s_i, all decoding schemes we will see in Section 7.3.2 are applicable.

In the CP-based schemes depicted in Figure 7.4, the CP portion of the transmitted sequence is generally unknown, because it is replicated from the unknown data blocks. However, with AP-CP-only in (7.26) and with the specific choice of $P = P_J^{(K)}$, we have

$$P_J^{(K)} s_i^* = \begin{bmatrix} P_K^{(0)} d_i^* \\ P_L^{(0)} b_i^* \end{bmatrix}, \tag{7.27}$$

which implies that *both* the data block and the known symbol block are time reversed, but without interchanging their positions. The last L entries of $P_J^{(K)} s_i$ are again known and are then replicated as cyclic prefixes. For this special case, we depict the transmitted sequences in Figure 7.5. In this format, the data block d_i is surrounded by two known blocks, which correspond to the preamble and postamble segments in [154]. The general design based on the CP structure includes this known preamble and postamble transmission format as a special case.

7.3.1.2 ZP-Only Suppose now that in AP-CP-only we let $b_i = \mathbf{0}$ instead of having known symbols drawn from the constellation alphabet, and we fix $P = P_J^{(K)}$. Adjacent data blocks are now guarded by two zero blocks, each having length L, as depicted in Figure 7.6. Since the channel has order no greater than L, the presence

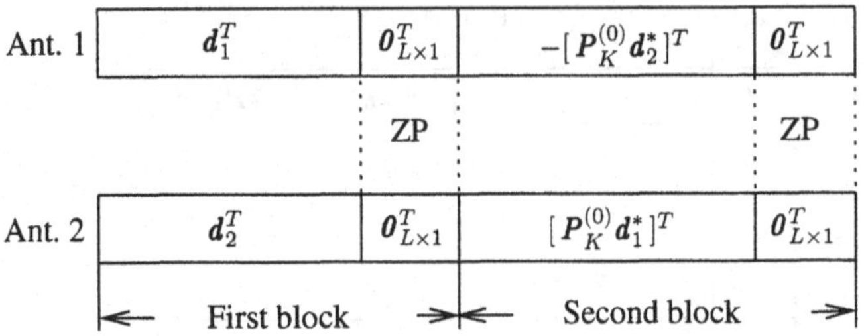

Figure 7.6 Transmitted sequence for ZP-only

of $2L$ zeros in the middle of two adjacent data blocks is not necessary. Keeping only a single block of L zeros corresponds to removing the CP-insertion operation at the transmitter, while IBI is avoided due to the ZP operation. The resulting transmission format achieves higher bandwidth efficiency than AP-CP-only. We call this scheme ZP-only and note that $J = K + L$ and $P = J$.

Considering the zero block in the previous block as the CP for the current block, all derivations and claims about AP-CP-only are still valid for ZP-only. Hence, ZP-only shares the same properties and decoding options as AP-CP-only; a detailed derivation tailored for ZP-only is available in [329].

7.3.2 Receiver Processing

At the receiver side, the first step is to discard the CP inserted at the transmitter in order to remove the IBI. To proceed, we let $\boldsymbol{y}_1$ denote the block received in the first block interval, and $\boldsymbol{y}_2$ the one in the second block interval, after CP removal. We further let $\{h_1(l)\}_{l=0}^{L}$ and $\{h_2(l)\}_{l=0}^{L}$ denote the channel coefficients corresponding to the two transmit antennas. It is well-known that CP insertion at the transmitter and CP removal at the receiver converts a linear convolution into a circular convolution [332]. To describe the resulting model in vector-matrix form, we need to introduce the $J \times J$ circulant matrix

$$\tilde{\boldsymbol{H}}_\mu = \begin{bmatrix} h_\mu(0) & 0 & \cdots & h_\mu(L) & \cdots & h_\mu(1) \\ \vdots & h_\mu(0) & \cdots & \cdots & \ddots & \vdots \\ \vdots & \cdots & \ddots & \cdots & \cdots & h_\mu(L) \\ h_\mu(L) & \cdots & \cdots & \ddots & \cdots & 0 \\ \vdots & \ddots & \cdots & \cdots & \ddots & \vdots \\ 0 & \cdots & h_\mu(L) & \cdots & \cdots & h_\mu(0) \end{bmatrix}_{J\times J}, \quad \mu = 1, 2, \tag{7.28}$$

with the $(p+1, q+1)$st entry

$$[\tilde{\boldsymbol{H}}_\mu]_{p,q} := h_\mu((p-q) \mod J). \tag{7.29}$$

Using (7.28), the input-output relationship after CP removal can be written as

$$\boldsymbol{y}_1 = \sqrt{\frac{\bar{\gamma}}{2}}\tilde{\boldsymbol{H}}_1\boldsymbol{s}_1 + \sqrt{\frac{\bar{\gamma}}{2}}\tilde{\boldsymbol{H}}_2\boldsymbol{s}_2 + \boldsymbol{w}_1, \tag{7.30}$$

$$\boldsymbol{y}_2 = -\sqrt{\frac{\bar{\gamma}}{2}}\tilde{\boldsymbol{H}}_1\boldsymbol{P}\boldsymbol{s}_2^* + \sqrt{\frac{\bar{\gamma}}{2}}\tilde{\boldsymbol{H}}_2\boldsymbol{P}\boldsymbol{s}_1^* + \boldsymbol{w}_2, \tag{7.31}$$

where $\boldsymbol{w}_1$ and $\boldsymbol{w}_2$ denote additive noise blocks.

Receiver processing to recover $\boldsymbol{s}_1$, $\boldsymbol{s}_2$ from $\boldsymbol{y}_1$, $\boldsymbol{y}_2$ will turn out to be efficient if one exploits the following two properties of circulant matrices:

P1) Circulant matrices can be diagonalized by FFT operations [99, Page 202]

$$\tilde{\boldsymbol{H}}_\mu = \boldsymbol{F}_J^{\mathcal{H}}\boldsymbol{\mathcal{D}}_\mu\boldsymbol{F}_J \quad \text{and} \quad \tilde{\boldsymbol{H}}_\mu^{\mathcal{H}} = \boldsymbol{F}_J^{\mathcal{H}}\boldsymbol{\mathcal{D}}_\mu^*\boldsymbol{F}_J, \tag{7.32}$$

where the diagonal matrix

$$\boldsymbol{\mathcal{D}}_\mu := \text{diag}\big(H_\mu(e^{j0}), \ldots, H_\mu(e^{j2\pi(J-1)/J})\big), \tag{7.33}$$

has the $(p+1)$st entry given by the channel response $H_\mu(z) := \sum_{l=0}^{L} h_\mu(l)z^{-l}$ evaluated at the frequency $z = e^{j2\pi p/J}$.

P2) Pre- and post-multiplying $\tilde{\boldsymbol{H}}_\mu$ by $\boldsymbol{P}$ yields $\tilde{\boldsymbol{H}}_\mu^T$ [332, Appendix]

$$\boldsymbol{P}\tilde{\boldsymbol{H}}_\mu\boldsymbol{P} = \tilde{\boldsymbol{H}}_\mu^T \quad \text{and} \quad \boldsymbol{P}\tilde{\boldsymbol{H}}_\mu^*\boldsymbol{P} = \tilde{\boldsymbol{H}}_\mu^{\mathcal{H}}. \tag{7.34}$$

Left-multiplying (7.31) by $\boldsymbol{P}$, conjugating, and then using (7.34), we obtain

$$\boldsymbol{P}\boldsymbol{y}_2^* = -\sqrt{\frac{\bar{\gamma}}{2}}\tilde{\boldsymbol{H}}_1^{\mathcal{H}}\boldsymbol{s}_2 + \sqrt{\frac{\bar{\gamma}}{2}}\tilde{\boldsymbol{H}}_2^{\mathcal{H}}\boldsymbol{s}_1 + \boldsymbol{P}\boldsymbol{w}_2^*. \tag{7.35}$$

Concatenating (7.30) and (7.35) leads to

$$\begin{bmatrix}\boldsymbol{y}_1 \\ \boldsymbol{P}\boldsymbol{y}_2^*\end{bmatrix} = \sqrt{\frac{\bar{\gamma}}{2}}\begin{bmatrix}\tilde{\boldsymbol{H}}_1 & \tilde{\boldsymbol{H}}_2 \\ \tilde{\boldsymbol{H}}_2^{\mathcal{H}} & -\tilde{\boldsymbol{H}}_1^{\mathcal{H}}\end{bmatrix}\begin{bmatrix}\boldsymbol{s}_1 \\ \boldsymbol{s}_2\end{bmatrix} + \begin{bmatrix}\boldsymbol{w}_1 \\ \boldsymbol{P}\boldsymbol{w}_2^*\end{bmatrix}. \tag{7.36}$$

Based on (7.32), we can rewrite (7.36) as

$$\begin{bmatrix}\boldsymbol{y}_1 \\ \boldsymbol{P}\boldsymbol{y}_2^*\end{bmatrix} = \sqrt{\frac{\bar{\gamma}}{2}}\begin{bmatrix}\boldsymbol{F}_J^{\mathcal{H}} & \boldsymbol{0} \\ \boldsymbol{0} & \boldsymbol{F}_J^{\mathcal{H}}\end{bmatrix}\begin{bmatrix}\boldsymbol{\mathcal{D}}_1 & \boldsymbol{\mathcal{D}}_2 \\ \boldsymbol{\mathcal{D}}_2^* & -\boldsymbol{\mathcal{D}}_1^*\end{bmatrix}\begin{bmatrix}\boldsymbol{F}_J & \boldsymbol{0} \\ \boldsymbol{0} & \boldsymbol{F}_J\end{bmatrix}\begin{bmatrix}\boldsymbol{s}_1 \\ \boldsymbol{s}_2\end{bmatrix} + \begin{bmatrix}\boldsymbol{w}_1 \\ \boldsymbol{P}\boldsymbol{w}_2^*\end{bmatrix}. \tag{7.37}$$

We wish to show that $\boldsymbol{s}_1$ and $\boldsymbol{s}_2$ in (7.37) can be demodulated separately without sacrificing ML optimality. Toward this objective, we consider a $J \times J$ diagonal matrix $\overline{\boldsymbol{\mathcal{D}}}_{12}$ with nonnegative diagonal entries as

$$\overline{\boldsymbol{\mathcal{D}}}_{12} = [\boldsymbol{\mathcal{D}}_1^*\boldsymbol{\mathcal{D}}_1 + \boldsymbol{\mathcal{D}}_2^*\boldsymbol{\mathcal{D}}_2]^{1/2}, \tag{7.38}$$

and construct the matrix

$$U = \begin{bmatrix} \overline{\mathcal{D}}_{12}^{-1} & 0 \\ 0 & \overline{\mathcal{D}}_{12}^{-1} \end{bmatrix} \begin{bmatrix} \mathcal{D}_1^* & \mathcal{D}_2 \\ \mathcal{D}_2^* & -\mathcal{D}_1 \end{bmatrix} \begin{bmatrix} F_J & 0 \\ 0 & F_J \end{bmatrix}, \tag{7.39}$$

which can easily be verified to be unitary (i.e., $UU^{\mathcal{H}} = I_{2J}$). A unitary matrix U can be constructed even when $\overline{\mathcal{D}}_{12}$ loses rank, as detailed in [332]. Multiplying (7.37) by the unitary matrix U yields

$$\begin{bmatrix} z_1 \\ z_2 \end{bmatrix} = U \begin{bmatrix} y_1 \\ P y_2^* \end{bmatrix} = \sqrt{\frac{\overline{\gamma}}{2}} \begin{bmatrix} \overline{\mathcal{D}}_{12} F_J s_1 \\ \overline{\mathcal{D}}_{12} F_J s_2 \end{bmatrix} + \begin{bmatrix} \eta_1 \\ \eta_2 \end{bmatrix}, \tag{7.40}$$

where the resulting noise $[\eta_1^T, \eta_2^T]^T := U[w_1^T, (P w_2^*)^T]^T$ is still white with unit variance per entry.

Since the noise in (7.40) is white, we infer that the blocks s_1 and s_2 can be demodulated separately, without compromising the ML optimality. Specifically, we only need to demodulate each information block s_i separately from the blocks

$$\begin{aligned} z_i &= \sqrt{\frac{\overline{\gamma}}{2}} \overline{\mathcal{D}}_{12} F_J s_i + \eta_i \\ &= \sqrt{\frac{\overline{\gamma}}{2}} A s_i + \eta_i, \quad i = 1, 2, \end{aligned} \tag{7.41}$$

where $A := \overline{\mathcal{D}}_{12} F_J$ is the mixing matrix for the symbol block s_i.

So far, we have performed at the receiver three linear unitary operations after the CP removal: (i) permutation (via P), (ii) conjugation, and (iii) unitary combining (via U). Linear receiver processing enables separate decoding for individual symbol blocks, thanks to the structure of the ST coding in (7.19). Notice that this would have been impossible without the permutation matrix P inserted at the transmitter.

Let us now focus on the detection of each individual block from (7.41), which is a well-studied block equalization problem. All detectors described in Chapter 5 apply. In a number of practical settings, the sphere decoding algorithm (SDA) can offer superior performance at average complexity, which is only about cubic in the block size.

One point worth stressing here is that the MMSE equalizer has very low complexity in this setup. Indeed, if the symbol vectors are white with covariance matrix $R_s = E\{s_i s_i^{\mathcal{H}}\} = I_J$, the block MMSE equalizer is given by

$$\begin{aligned} \Gamma_{\text{MMSE}} &= \sqrt{\frac{\overline{\gamma}}{2}} \left(\frac{\overline{\gamma}}{2} A^{\mathcal{H}} A + I_J \right)^{-1} A^{\mathcal{H}}, \\ &= \sqrt{\frac{\overline{\gamma}}{2}} F_J^{\mathcal{H}} \left(\frac{\overline{\gamma}}{2} \overline{\mathcal{D}}_{12}^2 + I_J \right)^{-1} \overline{\mathcal{D}}_{12}, \end{aligned} \tag{7.42}$$

which requires only a diagonal matrix inversion followed by an IFFT operation.

More interestingly, it turns out that Viterbi's algorithm is exactly applicable to AP-CP-only and ZP-only, as we detail next.

7.3.3 ML Decoding Based on the Viterbi Algorithm

For simplicity, we drop the block index i, and simplify (7.41) to

$$z = \sqrt{\frac{\overline{\gamma}}{2}}\, \overline{\mathcal{D}}_{12} F_J s + \eta. \tag{7.43}$$

Recall also that in AP-CP-only and ZP-only, the last L entries of s are known.

In the presence of white noise, ML decoding yields

$$\begin{aligned} \hat{s}_{\mathrm{ML}} &= \arg\max_{s} \ \ln P(z|s) \\ &= \arg\max_{s} \left\{ - \left\| z - \sqrt{\frac{\overline{\gamma}}{2}}\, \overline{\mathcal{D}}_{12} F_J s \right\|^2 \right\}. \end{aligned} \tag{7.44}$$

We next simplify (7.44), starting with

$$\begin{aligned} -\left\| z - \sqrt{\frac{\overline{\gamma}}{2}}\, \overline{\mathcal{D}}_{12} F_J s \right\|^2 &= 2\sqrt{\frac{\overline{\gamma}}{2}}\, \Re\{s^{\mathcal{H}} F_J^{\mathcal{H}} \overline{\mathcal{D}}_{12} z\} - \frac{\overline{\gamma}}{2} s^{\mathcal{H}} F_J^{\mathcal{H}} \overline{\mathcal{D}}_{12}^2 F_J s - z^{\mathcal{H}} z \\ &= 2\sqrt{\frac{\overline{\gamma}}{2}}\, \Re\{s^{\mathcal{H}} r\} - \frac{\overline{\gamma}}{2} \sum_{\mu=1}^{2} \|\tilde{H}_\mu s\|^2 - z^{\mathcal{H}} z, \end{aligned} \tag{7.45}$$

where $r := F_J^{\mathcal{H}} \overline{\mathcal{D}}_{12} z$. We let $r_n := [r]_n$ and $s_n := [s]_n$. Recognizing that $\tilde{H}_\mu s$ is a matrix-vector description of the circular convolution of the channel h with s, we have $[\tilde{H}_\mu s]_n = \sum_{l=0}^{L} h_\mu(l) s_{(n-l) \bmod J}$. Ignoring the common term $z^{\mathcal{H}} z$, we obtain

$$\hat{s}_{\mathrm{ML}} = \arg\max_{s} \sum_{n=0}^{J-1} \left[2\sqrt{\frac{\overline{\gamma}}{2}} \Re\{s_n^* r_n\} - \frac{\overline{\gamma}}{2} \sum_{\mu=1}^{2} \left| \sum_{l=0}^{L} h_\mu(l) s_{(n-l) \bmod J} \right|^2 \right]. \tag{7.46}$$

For each $n = 0, 1, \ldots, J$, let us define a sequence of state vectors as $\zeta_n = [s_{(n-1) \bmod J}, \ldots, s_{(n-L) \bmod J}]^T$, where the first and last states are identical and known: $\zeta_0 = \zeta_J = [s_{J-1}, \ldots, s_{J-L}]^T$. The symbol sequence $s_0, \ldots, s_{J-1}$ determines a unique path evolving from the known initial state ζ_0 to the known final state ζ_J. Thus, Viterbi's algorithm is applicable. Specifically, we have

$$\hat{s}_{\mathrm{ML}} = \arg\max_{s} \sum_{n=0}^{J-1} f(\zeta_n, \zeta_{n+1}), \tag{7.47}$$

where $f(\zeta_n, \zeta_{n+1})$ denotes the branch metric, that is, the expression inside the brackets in (7.46). The explicit recursion performed by Viterbi's algorithm is well-known and can be found in e.g., [275, Equation (7)].

Aiming at further simplification of the branch metric, we start with

$$\sum_{\mu=1}^{2} \|\tilde{H}_\mu s\|^2 = s^{\mathcal{H}} \sum_{\mu=1}^{2} (\tilde{H}_\mu^{\mathcal{H}} \tilde{H}_\mu) s, \tag{7.48}$$

where the matrix $\overline{\boldsymbol{H}} := \sum_{\mu=1}^{2}(\tilde{\boldsymbol{H}}_{\mu}^{\mathcal{H}}\tilde{\boldsymbol{H}}_{\mu})$ has the $(p+1, q+1)$st entry

$$[\overline{\boldsymbol{H}}]_{p,q} = \sum_{\mu=1}^{2}\sum_{n=0}^{J-1} h_{\mu}^{*}((n-p) \bmod J) h_{\mu}((n-q) \bmod J). \tag{7.49}$$

Let us now select $J > 2L$, and define

$$\beta_n = \sum_{\mu=1}^{2}\sum_{l=0}^{L} h_{\mu}^{*}(l) h_{\mu}(n+l), \quad \forall n \in [0, L]. \tag{7.50}$$

We can verify that the first column of $\overline{\boldsymbol{H}}$ is $[\beta_0, \beta_1, \ldots, \beta_L, 0, \ldots, 0, \beta_L^*, \ldots, \beta_1^*]^T$. If $\breve{\boldsymbol{H}}$ denotes a circulant matrix with first column $[\beta_0/2, \beta_1, \ldots, \beta_L, 0, \ldots, 0]^T$, then since $\overline{\boldsymbol{H}}$ is circulant and Hermitian, it can be decomposed as $\overline{\boldsymbol{H}} = \breve{\boldsymbol{H}} + \breve{\boldsymbol{H}}^{\mathcal{H}}$. We thus obtain $\boldsymbol{s}^{\mathcal{H}}\overline{\boldsymbol{H}}\boldsymbol{s} = 2\Re\{\boldsymbol{s}^{\mathcal{H}}\breve{\boldsymbol{H}}\boldsymbol{s}\}$. Recognizing that $[\breve{\boldsymbol{H}}\boldsymbol{s}]_n = (1/2)\beta_0 s_n + \sum_{l=1}^{L}\beta_l s_{(n-l) \bmod J}$, and combining with (7.46), we obtain the simplified metric

$$f(\zeta_n, \zeta_{n+1}) = 2\Re\left\{\sqrt{\frac{\overline{\gamma}}{2}} s_n^* \left[r_n - \sqrt{\frac{\overline{\gamma}}{2}}\left(\frac{1}{2}\beta_0 s_n + \sum_{l=1}^{L}\beta_l s_{(n-l) \bmod J}\right)\right]\right\}. \tag{7.51}$$

The branch metric in (7.51) has a format analogous to the one proposed by Ungerboeck for ML sequence estimation (MLSE) in a system involving *single-antenna serial transmissions* [27,266]. (Note also that the derivation here deals with a normalized signal constellation.)

For ML decoding in AP-CP-only, the known symbol vector $\boldsymbol{b}$ can be placed in an arbitrary position within the vector $\boldsymbol{s}$. If the known symbols occupy positions $B-L, \ldots, B-1$, we just need to redefine the states as $\zeta_n := [s_{(n+B-1) \bmod J}, \ldots, s_{(n+B-L) \bmod J}]^T$.

Notice that for channels with order L, the complexity of Viterbi's algorithm is $\mathcal{O}(|\mathbb{S}|^{L+1})$ per symbol; thus, ML decoding with exact application of Viterbi's algorithm is particularly attractive for transmissions with small constellation size, over relatively short channels.

7.3.4 Turbo Equalization

So far, we have only considered uncoded information streams in ST block coded transmissions over frequency-selective MIMO channels. As with single-antenna systems, error performance of multi-antenna systems can be further improved through the incorporation of channel coding. For example, outer convolutional codes can be used in AP-CP-only and ZP-only, as depicted in Figure 7.7. Other options include trellis coded modulation (TCM) and parallel or serially concatenated turbo codes.

For frequency-selective channels, channel coded transmissions can be decoded at the receiver using iterative (turbo) equalization which is known to improve error performance, at least for single-antenna systems [62]. This motivates consideration of turbo equalizers for the coded AP-CP-only and ZP-only multi-antenna systems.

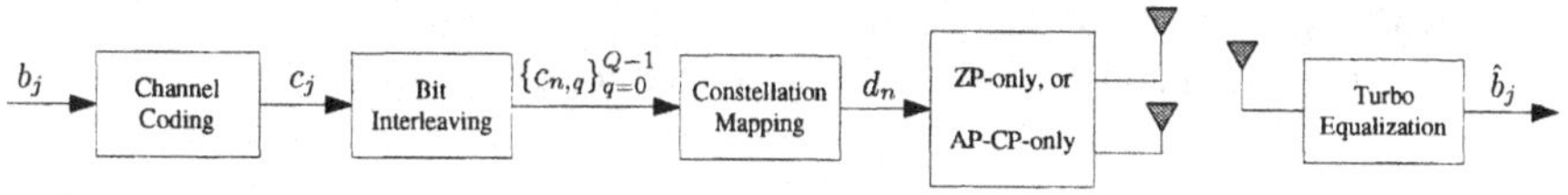

Figure 7.7 Coded AP-CP-only or ZP-only with turbo equalization

To enable turbo equalization, one needs to find the a posteriori probability of the transmitted symbols s_n based on the received vector $\boldsymbol{z}$. Suppose that each constellation point s_n is determined by $Q = \log_2 |\mathbb{S}|$ bits $\{c_{n,0}, \ldots, c_{n,Q-1}\}$. Let us consider the log-likelihood ratio (LLR)

$$\mathcal{L}_{n,q} = \ln \frac{P(c_{n,q} = +1|\boldsymbol{z})}{P(c_{n,q} = -1|\boldsymbol{z})}, \quad \forall n \in [0, J-1], \; q \in [0, Q-1]. \tag{7.52}$$

As detailed in [275], the LLR in (7.52) can be obtained by running two generalized Viterbi recursions: one in the forward direction evolving from $\boldsymbol{\zeta}_0$ to $\boldsymbol{\zeta}_J$, and the other in the backward direction going from $\boldsymbol{\zeta}_J$ to $\boldsymbol{\zeta}_0$. We refer readers to [275, Equations (7'),(8'),(10')] for explicit expressions. The only change required is to modify our branch metric as follows:

$$g(\boldsymbol{\zeta}_n, \boldsymbol{\zeta}_{n+1}) = f(\boldsymbol{\zeta}_n, \boldsymbol{\zeta}_{n+1}) + \ln P(\boldsymbol{\zeta}_{n+1}|\boldsymbol{\zeta}_n). \tag{7.53}$$

This modification is needed to take into account the a priori probability $P(\boldsymbol{\zeta}_{n+1}|\boldsymbol{\zeta}_n)$, determined by the extrinsic information from the convolutional channel decoders during the turbo iteration. When the transition from $\boldsymbol{\zeta}_n$ to $\boldsymbol{\zeta}_{n+1}$ is caused by the input symbol s_n, we have $\ln P(\boldsymbol{\zeta}_{n+1}|\boldsymbol{\zeta}_n) = \ln P(s_n)$. We assume that the bit interleaver in Figure 7.7 renders the symbols s_n independent and equally likely, such that $\ln P(s_n) = \sum_{q=0}^{Q-1} \ln P(c_{n,q})$, which in turn can be determined by the LLRs for the bits $\{c_{n,q}\}_{q=0}^{Q-1}$.

Finally, we remark that one could also adopt the turbo decoder described in Section 5.5, which relies on the soft SDA for equalization, or, the iterative algorithm in [285], which is based on soft MMSE equalization. These two iterative receivers are applicable not only to AP-CP-only and ZP-only, but also to the CP-only systems; see Chapter 10 for more details on iterative decoding.

7.3.5 Multi-Antenna Extensions

So far, we focused on $N_t = 2$ transmit-antennas and $N_r = 1$ receive-antenna. In this subsection we generalize the system design to $N_t > 2$ and/or $N_r > 1$ antennas. The channel between the μth transmit- and νth receive- antennas is defined in (7.1) as $\boldsymbol{h}_{\nu\mu} := [h_{\nu\mu}(0), \ldots, h_{\nu\mu}(L)]^T$, where $\mu = 1, \ldots, N_t$ and $\nu = 1, \ldots, N_r$.

Orthogonal STBCs for flat fading channels were presented in Section 3.3. As in (2.59), the $N_t \times N_x$ matrix $\boldsymbol{X}_{\text{OSTBC}}$ can be represented through a set of constant matrices $\{\boldsymbol{A}_n, \boldsymbol{B}_n\}_{n=1}^{N_s}$ associated with N_s information symbols. To extend OSTBCs to frequency-selective MIMO channels, we rely on N_s symbol blocks $\boldsymbol{s}_1, \boldsymbol{s}_2, \ldots, \boldsymbol{s}_{N_s}$, each of length J, and construct the $N_t J \times N_x$ ST code matrix

$$\boldsymbol{S} = \sum_{n=1}^{N_s} \left[\boldsymbol{A}_n \otimes \boldsymbol{s}_n + \boldsymbol{B}_n \otimes (\boldsymbol{P}\boldsymbol{s}_n^*) \right], \tag{7.54}$$

where $\boldsymbol{P}$ is defined as in Section 7.3.1 and $\otimes$ denotes the Kronecker product. The construction of $\boldsymbol{S}$ can be explained intuitively in the following three steps:

S1) With N_t antennas, pick the matrix $\boldsymbol{X}_{\text{OSTBC}}$ constructed from the symbols $s_1, \ldots, s_{N_s}$.

S2) Replace $s_1, \ldots, s_{N_s}$ in $\boldsymbol{X}_{\text{OSTBC}}$ by $\boldsymbol{s}_1, \ldots, \boldsymbol{s}_{N_s}$.

S3) Replace $s_1^*, \ldots, s_{N_s}^*$ in $\boldsymbol{X}_{\text{OSTBC}}$ by $\boldsymbol{P}\boldsymbol{s}_1^*, \ldots, \boldsymbol{P}\boldsymbol{s}_{N_s}^*$.

Matrix $\boldsymbol{S}$ is then transmitted from the N_t antennas block-by-block with either CP or ZP, as depicted in Figures 7.4 and 7.6. Note that with $N_t = 2$ transmit antennas, steps S1-S3 have been applied to the 2×2 matrix in (3.16) to obtain (7.19).

All possible matrices $\boldsymbol{S}$ can be divided in two categories. Category 1 contains all matrices whose elements in each column are either all unconjugated or all conjugated; and category 2 contains matrices with at least one column having both conjugated and unconjugated entries.

All codes with rate $1/2$ belong to category 1. With $N_t = 4$, we have, for instance,

$$\boldsymbol{S} = \begin{bmatrix} \boldsymbol{s}_1 & -\boldsymbol{s}_2 & -\boldsymbol{s}_3 & -\boldsymbol{s}_4 & \boldsymbol{P}\boldsymbol{s}_1^* & -\boldsymbol{P}\boldsymbol{s}_2^* & -\boldsymbol{P}\boldsymbol{s}_3^* & -\boldsymbol{P}\boldsymbol{s}_4^* \\ \boldsymbol{s}_2 & \boldsymbol{s}_1 & \boldsymbol{s}_4 & -\boldsymbol{s}_3 & \boldsymbol{P}\boldsymbol{s}_2^* & \boldsymbol{P}\boldsymbol{s}_1^* & \boldsymbol{P}\boldsymbol{s}_4^* & -\boldsymbol{P}\boldsymbol{s}_3^* \\ \boldsymbol{s}_3 & -\boldsymbol{s}_4 & \boldsymbol{s}_1 & \boldsymbol{s}_2 & \boldsymbol{P}\boldsymbol{s}_3^* & -\boldsymbol{P}\boldsymbol{s}_4^* & \boldsymbol{P}\boldsymbol{s}_1^* & \boldsymbol{P}\boldsymbol{s}_2^* \\ \boldsymbol{s}_4 & \boldsymbol{s}_3 & -\boldsymbol{s}_2 & \boldsymbol{s}_1 & \boldsymbol{P}\boldsymbol{s}_4^* & \boldsymbol{P}\boldsymbol{s}_3^* & -\boldsymbol{P}\boldsymbol{s}_2^* & \boldsymbol{P}\boldsymbol{s}_1^* \end{bmatrix}. \tag{7.55}$$

With $N_t = 3$, we can construct a rate-$3/4$ code in category 1 as

$$\boldsymbol{S} = \begin{bmatrix} \boldsymbol{s}_1 & \boldsymbol{0} & -\boldsymbol{P}\boldsymbol{s}_2^* & \boldsymbol{P}\boldsymbol{s}_3^* \\ \boldsymbol{s}_2 & \boldsymbol{P}\boldsymbol{s}_3^* & \boldsymbol{P}\boldsymbol{s}_1^* & \boldsymbol{0} \\ -\boldsymbol{s}_3 & \boldsymbol{P}\boldsymbol{s}_2^* & \boldsymbol{0} & \boldsymbol{P}\boldsymbol{s}_1^* \end{bmatrix}. \tag{7.56}$$

However, the rate-$3/4$ code for $N_t = 4$ belongs to category 2 and leads to the matrix

$$\boldsymbol{S} = \begin{bmatrix} \boldsymbol{s}_1 & \boldsymbol{0} & -\boldsymbol{P}\boldsymbol{s}_2^* & \boldsymbol{P}\boldsymbol{s}_3^* \\ \boldsymbol{0} & \boldsymbol{s}_1 & -\boldsymbol{s}_3 & -\boldsymbol{s}_2 \\ \boldsymbol{s}_2 & \boldsymbol{P}\boldsymbol{s}_3^* & \boldsymbol{P}\boldsymbol{s}_1^* & \boldsymbol{0} \\ -\boldsymbol{s}_3 & \boldsymbol{P}\boldsymbol{s}_2^* & \boldsymbol{0} & \boldsymbol{P}\boldsymbol{s}_1^* \end{bmatrix}. \tag{7.57}$$

Let us first focus on receiver processing for the codes belonging to category 1. The processing can be described in three steps:

S1) At each receive-antenna ν, we collect blocks $\boldsymbol{y}_{\nu,1}, \ldots, \boldsymbol{y}_{\nu,N_x}$ in N_x block transmission slots.

S2) We use the following rules to process $\boldsymbol{y}_{\nu,i}$: If $\boldsymbol{y}_{\nu,i}$ contains only unconjugated symbol blocks, we evaluate $\tilde{\boldsymbol{y}}_{\nu,i} = \boldsymbol{F}_J\boldsymbol{y}_{\nu,i}$; otherwise, we obtain $\tilde{\boldsymbol{y}}_{\nu,i} = \boldsymbol{F}_J^*\boldsymbol{P}\boldsymbol{y}_{\nu,i}$.

S3) ST decoding is performed on $\{\tilde{\boldsymbol{y}}_{\nu,i}\}_{\nu=1,i=1}^{N_r,N_x}$.

Example 7.4 Let us illustrate step S2 for the rate-3/4 code in (7.56), where $N_x = 4$. Collecting the processed blocks, we have

$$[\boldsymbol{F}_J\boldsymbol{y}_{\nu,1}, \boldsymbol{F}_J^*\boldsymbol{y}_{\nu,2}, \boldsymbol{F}_J^*\boldsymbol{P}\boldsymbol{y}_{\nu,3}, \boldsymbol{F}_J^*\boldsymbol{P}\boldsymbol{y}_{\nu,4}]$$
$$= \sqrt{\frac{\bar{\gamma}}{N_t}}\,[\boldsymbol{\mathcal{D}}_{\nu,1}, \boldsymbol{\mathcal{D}}_{\nu,2}, \boldsymbol{\mathcal{D}}_{\nu,3}] \begin{bmatrix} \boldsymbol{F}_J\boldsymbol{s}_1 & 0 & -(\boldsymbol{F}_J\boldsymbol{s}_2)^* & (\boldsymbol{F}_J\boldsymbol{s}_3)^* \\ \boldsymbol{F}_J\boldsymbol{s}_2 & (\boldsymbol{F}_J\boldsymbol{s}_3)^* & (\boldsymbol{F}_J\boldsymbol{s}_1)^* & 0 \\ -\boldsymbol{F}_J\boldsymbol{s}_3 & (\boldsymbol{F}_J\boldsymbol{s}_2)^* & 0 & (\boldsymbol{F}_J\boldsymbol{s}_1)^* \end{bmatrix} + \text{white noise}, \quad (7.58)$$

where $\boldsymbol{\mathcal{D}}_{\nu\mu}$ is defined similar to $\boldsymbol{\mathcal{D}}_\mu$ in Section 7.3.1. Hence, we have J parallel OSTBCs, each going through a flat fading MIMO channel corresponding to the nth frequency $\{H_{\nu\mu}(e^{j2\pi n/J})\}$.

For the codes in category 2, step S2 cannot be applied directly since both unconjugated and conjugated symbols are mixed in one column of $\boldsymbol{S}$. In fact, S2 is valid only when $\boldsymbol{P} = \boldsymbol{P}_J^{(1)}$ is used. This is due to the nice property of $\boldsymbol{F}_J^*\boldsymbol{P}_J^{(1)} = \boldsymbol{F}_J$ which implies that the operation $\boldsymbol{F}_J$ for unconjugated blocks and the operation $\boldsymbol{F}_J^*\boldsymbol{P}_J^{(1)}$ for conjugated blocks coincide. As a result, the same FFT processing $\boldsymbol{F}_J$ can be applied to all received vectors corresponding to the codes in category 2.

Example 7.5 For the rate-3/4 code in (7.57), we have $N_x = 4$. After FFT processing, we obtain

$$[\boldsymbol{F}_J\boldsymbol{y}_{\nu,1}, \boldsymbol{F}_J\boldsymbol{y}_{\nu,2}, \boldsymbol{F}_J\boldsymbol{y}_{\nu,3}, \boldsymbol{F}_J\boldsymbol{y}_{\nu,4}]$$
$$= \sqrt{\frac{\bar{\gamma}}{N_t}}\,[\boldsymbol{\mathcal{D}}_{\nu,1}, \boldsymbol{\mathcal{D}}_{\nu,2}, \boldsymbol{\mathcal{D}}_{\nu,3}, \boldsymbol{\mathcal{D}}_{\nu,4}] \begin{bmatrix} \boldsymbol{F}_J\boldsymbol{s}_1 & 0 & -(\boldsymbol{F}_J\boldsymbol{s}_2)^* & (\boldsymbol{F}_J\boldsymbol{s}_3)^* \\ 0 & \boldsymbol{F}_J\boldsymbol{s}_1 & -\boldsymbol{F}_J\boldsymbol{s}_3 & -\boldsymbol{F}_J\boldsymbol{s}_2 \\ \boldsymbol{F}_J\boldsymbol{s}_2 & (\boldsymbol{F}_J\boldsymbol{s}_3)^* & (\boldsymbol{F}_J\boldsymbol{s}_1)^* & 0 \\ -\boldsymbol{F}_J\boldsymbol{s}_3 & (\boldsymbol{F}_J\boldsymbol{s}_2)^* & 0 & (\boldsymbol{F}_J\boldsymbol{s}_1)^* \end{bmatrix} + \text{white noise}, \quad (7.59)$$

which has a format identical to the one in Example 7.4.

In summary, for the codes in category 1, all $\boldsymbol{P}_J^{(n)}$ can be used in the ST block codes, while only $\boldsymbol{P}_J^{(1)}$ is allowed for the codes in category 2 [e.g., (7.57)]. As a consequence, the codes in category 2 do not allow for a ZP based transmission, since ZP is applicable only when $\boldsymbol{P} = \boldsymbol{P}_J^{(K)}$.

For all codes in both categories 1 and 2, we have J parallel OSTBCs after step S2, each going through a flat fading MIMO channel. In step S3, ST decoding is applied on each flat fading MIMO subchannel. After noise normalization, we have

$$z_i = \sqrt{\frac{\bar{\gamma}}{N_t}} \,\overline{\mathcal{D}} F_J s_i + \eta_i, \quad i = 1, 2, \ldots, N_s, \tag{7.60}$$

where the noise η_i is white with each entry having unit variance, and

$$\overline{\mathcal{D}} := \left[\sum_{\nu=1}^{N_r} \sum_{\mu=1}^{N_t} \mathcal{D}_{\nu\mu}^* \mathcal{D}_{\nu\mu} \right]^{1/2} = \operatorname{diag}\left(\begin{bmatrix} \|\tilde{H}(0)\|_F \\ \vdots \\ \|\tilde{H}(\frac{J-1}{J})\|_F \end{bmatrix} \right), \tag{7.61}$$

where $\tilde{H}(f)$ is defined in (7.13). Subsequently, decoding of each symbol block is performed separately.

After STBC decoding, the equalization methods described in Section 7.3.2 can be applied to (7.60), with the same complexity as in single-antenna transmissions. For AP-CP-only and ZP-only, the ML estimate (after dropping the block index i for brevity)

$$\hat{s}_{\mathrm{ML}} = \arg\max_{s} \left(- \left\| z - \sqrt{\frac{\bar{\gamma}}{N_t}} \,\overline{\mathcal{D}} F_J s \right\|^2 \right) \tag{7.62}$$

can be obtained via exact application of Viterbi's algorithm. Relative to the two antenna case detailed in Section 7.3.3, we can use the same expression for the branch metric of (7.51), with the following modifications:

- Replace $\sqrt{\bar{\gamma}/2}$ by $\sqrt{\bar{\gamma}/N_t}$;
- Redefine $r = F_J^{\mathcal{H}} \overline{\mathcal{D}} z$, which leads to $r_n = [r]_n$.
- Recalculate $\{\beta_n\}$ as

$$\beta_n = \sum_{\nu=1}^{N_r} \sum_{\mu=1}^{N_t} \sum_{l=0}^{L} h_{\nu\mu}^*(l) h_{\nu\mu}(n+l), \quad \forall n \in [0, L]. \tag{7.63}$$

7.3.6 OSTBC Properties

In this subsection, we present properties pertaining to OSTBCs for frequency-selective MIMO channels from three perspectives: diversity, complexity, and capacity.

7.3.6.1 Diversity Order As established in Proposition 7.1, the maximum achievable diversity order in the presence of frequency-selective channels is the rank of the correlation matrix of h_{eq}, which equals $N_t N_r (L+1)$ when the correlation matrix has full rank. For OSTBCs, we have a related result [332]:

Proposition 7.2 *If the correlation matrix of the virtual channel $\boldsymbol{h}_{\text{eq}}$ in Proposition 7.1 has full rank, then*

1. *CP-only achieves only space diversity of order $N_t N_r$ irrespective of the underlying signal constellation.*

2. *AP-CP-only and ZP-only achieve the maximum diversity gain irrespective of the underlying signal constellation.*

The worst case in CP-only leading to diversity loss corresponds to having $\boldsymbol{s} = a\mathbf{1}_{J\times1}$ transmitted but decoded erroneously as $\boldsymbol{s}' = a'\mathbf{1}_{J\times1}$, where $a, a' \in \mathcal{A}$ [332]. This type of error event is rare, since it requires all symbols to be decoded erroneously at the same time. As verified by simulations, the performance of CP-only is almost identical to those achieved by AP-CP-only and ZP-only, over the practical SNR range. On the other hand, AP-CP-only and ZP-only are attractive since they guarantee full space-multipath diversity and enable exact ML decoding through the use of Viterbi's algorithm. Detailed comparisons among CP-only, AP-CP-only, and ZP-only can be found in [332].

7.3.6.2 Receiver Complexity For OSTBC transmissions over frequency-selective MIMO channels, receiver complexity should take into account the linear processing step needed to separate the ST coded blocks and the block equalization step required to detect the original information symbols. Omitting the complexity of permutation and diagonal matrix multiplication, the linear processing to reach (7.41) requires only one J-point FFT per block, which amounts to $\mathcal{O}(\log_2 J)$ per information symbol. On the other hand, the linear ML processing to reach (7.60) requires a total of $N_x N_r \leq 2N_s N_r$ FFTs corresponding to each ST coded block in (7.54), which amounts to at most $2N_r$ FFTs per information block. The block equalization step based on (7.41) or (7.60), entails complexity identical to that required to equalize single-antenna block transmissions over FIR channels [288]. For coded AP-CP-only and ZP-only, the complexity of turbo equalization is again the same as that of single-antenna transmissions [62].

Assessment of the overall complexity can be summarized as follows:

Proposition 7.3 *The CP-only, AP-CP-only and ZP-only ST block coded systems with $N_t > 2$ ($N_t = 2$) transmit-antennas and N_r receive-antennas require extra complexity on the order of $\mathcal{O}(2N_r \log_2 J)$ [respectively, $\mathcal{O}(N_r \log_2 J)$] per information symbol relative to the corresponding block transmission systems equipped with a single transmit- and a single receive-antenna. (J is the FFT size.)*

Increasing the number of antennas incurs only minor increase in complexity. This nice property comes from the OSTBC design (cf. Section 3.3), which enables linear ML processing to collect the available space diversity. In fact, the block codes of this section can be considered as OSBTCs for frequency-selective MIMO channels.

7.3.6.3 Capacity Turning our attention to the capacity of the ST block coding format in (7.54), the equivalent channel input-output relationship, after receiver processing, is described by (7.60) as: $\boldsymbol{z} = \sqrt{\overline{\gamma}/N_t}\,\overline{\boldsymbol{\mathcal{D}}}\boldsymbol{F}_J\boldsymbol{s} + \boldsymbol{\eta}$, where we dropped the block index for brevity. Let $\mathcal{I}(\boldsymbol{z};\boldsymbol{s})$ denote the mutual information between $\boldsymbol{z}$ and $\boldsymbol{s}$, and recall that $\mathcal{I}(\boldsymbol{z};\boldsymbol{s})$ is maximized when $\boldsymbol{s}$ is Gaussian distributed [49]. Due to the lack of channel knowledge at the transmitter, the transmit-power is equally distributed among symbols that have correlation matrix $\boldsymbol{R}_s := E\{\boldsymbol{s}\boldsymbol{s}^{\mathcal{H}}\} = \boldsymbol{I}_J$. Taking into account the CP of length L and the OSTBC spectral efficiency (see η_{OSTBC} in Section 3.3.1), the channel capacity for a given channel realization is

$$
\begin{aligned}
C_J &= \eta_{\text{OSTBC}}\frac{1}{J+L}\max \mathcal{I}(\boldsymbol{z};\boldsymbol{s}) \\
&= \eta_{\text{OSTBC}}\frac{1}{J+L}\log_2\det\left(\boldsymbol{I}_J + \frac{\overline{\gamma}}{N_t}\overline{\boldsymbol{\mathcal{D}}}\boldsymbol{F}_J\boldsymbol{F}_J^{\mathcal{H}}\overline{\boldsymbol{\mathcal{D}}}^{\mathcal{H}}\right) \\
&= \eta_{\text{OSTBC}}\frac{1}{J+L}\sum_{n=0}^{J-1}\log_2\left(1 + \frac{\overline{\gamma}}{N_t}\|\tilde{\boldsymbol{H}}(n/J)\|_F^2\right).
\end{aligned}
\tag{7.64}
$$

As the block size J increases, we obtain

$$
C_{J\to\infty} = \eta_{\text{OSTBC}}\int_0^1 \log_2\left(1 + \frac{\overline{\gamma}}{N_t}\|\tilde{\boldsymbol{H}}(f)\|_F^2\right)df. \tag{7.65}
$$

With a single receive-antenna, $\tilde{\boldsymbol{H}}(f)$ becomes a row vector, and thus $\|\tilde{\boldsymbol{H}}(f)\|_F^2 = \tilde{\boldsymbol{H}}(f)\tilde{\boldsymbol{H}}^{\mathcal{H}}(f)$. On the other hand, $\eta_{\text{OSTBC}} = 1$ when $N_t = 2$. Therefore, in the special case of two transmit- and one receive-antenna, (7.65) coincides with (7.12) which implies the following result:

Proposition 7.4 *The STBC transmission in* (7.54) *does not incur capacity loss only in the special case with* $(N_t, N_r) = (2, 1)$.

Proposition 7.4 is consistent with the results in Section 3.3, where the Alamouti code [6] was shown to achieve capacity for frequency-flat fading channels, only when $(N_t, N_r) = (2, 1)$. To approach the capacity limit with two transmit-antennas and a single receive-antenna, it suffices to equip the OSTBC system with sufficiently strong one-dimensional (i.e., scalar) channel codes [81].

7.3.7 Numerical Examples

In this section we present simulations for systems with two transmit- and one receive-antenna. To facilitate FFT processing, we choose the block size J to be a power of 2. In all figures, SNR denotes the average symbol energy-to-noise-ratio at the receive-antenna. For reference, we also depict the ideal error performance limit that can be evaluated through the rate outage probability metric (7.14), as we explained after Example 7.1.

Example 7.6 (Comparisons with different equalizers) We first set $L = 2$ and assume that all channel taps in $\boldsymbol{H}_{\text{eq}}$ are i.i.d. Gaussian with covariance $1/(L+1)$. We

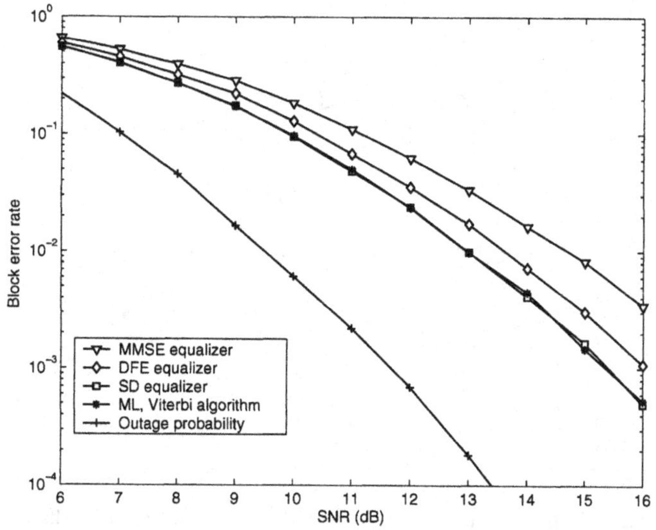

Figure 7.8 Comparisons of various equalizers for uncoded ZP-only

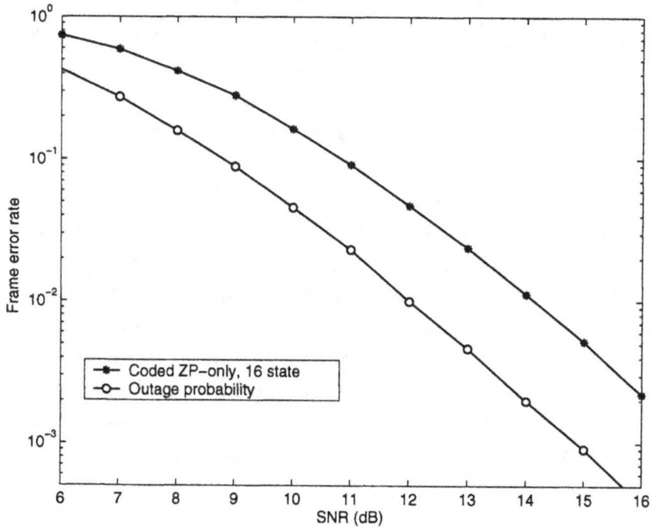

Figure 7.9 Convolutionally coded ZP-only in two-tap channels

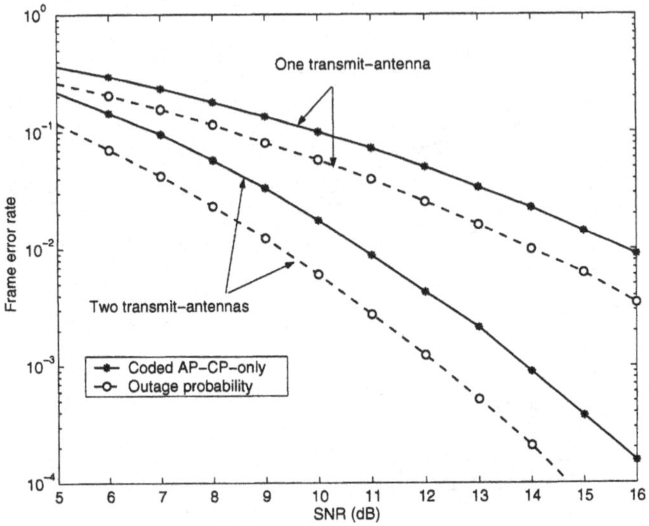

Figure 7.10 Convolutionally coded AP-CP-only over EDGE channels

test the performance of ZP-only with QPSK and block sizes $K = 14$ and $P = J = 16$. Figure 7.8 depicts the block error rate performance corresponding to MMSE, DFE, SDA-based, and ML equalizers. We observe that the SDA-based equalizer indeed achieves near-ML performance and outperforms the suboptimal block DFE as well as the block MMSE alternatives. Without channel coding, the performance of ZP-only is far away from the outage probability at rate $2K/(K + L) = 1.75$ bits per channel use.

Example 7.7 (Convolutionally coded ZP-only) We suppose here that each FIR channel consists of two i.i.d. channel taps (i.e., $L = 1$). We choose $K = 127$, $P = J = 128$ for the ZP-only system, and use 8-PSK constellation. For convenience, we view each block of length $P = 128$ as one data frame, with the ST codes applied to two adjacent frames. Within each frame, the information bits are convolutionally coded (CC) with a 16-state rate-$2/3$ encoder taken from [13, Table 11.6]. Omitting the trailing bits to terminate the CC trellis and ignoring the rate loss induced by the CP since $L \ll K$, we obtain a transmission rate of 2 bits per channel use.

A total of five turbo decoding iterations are performed. With the 16-state convolutional code, the frame error rate for ZP-only is within 2.3 dB away from the outage probability, as shown in Figure 7.9.

Example 7.8 (Convolutionally coded AP-CP-only over EDGE channels) We test the typical urban (TU) channel with a linearized Gaussian-minimum-shift-keying (GMSK) transmit pulse shape and symbol duration $T = 3.69\ \mu s$, as prescribed in the proposed third-generation TDMA cellular standard EDGE (Enhanced Date Rates for

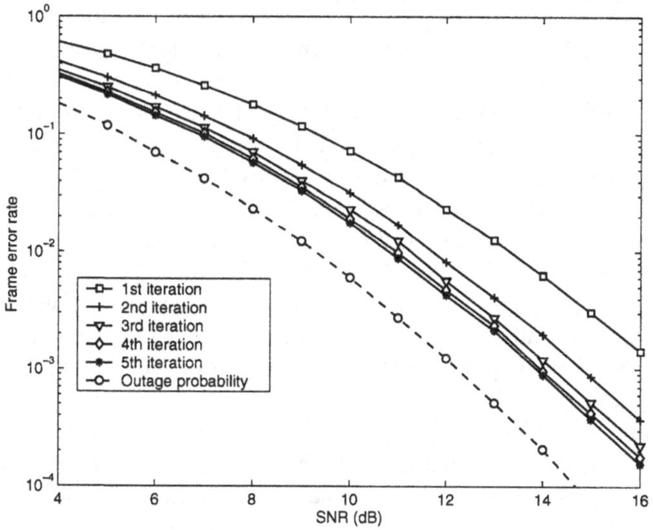

Figure 7.11 Turbo iterations for coded AP-CP-only with two transmit-antennas

GSM Evolution) [73]. The channels have order $L = 3$ and correlated taps [5]. We use QPSK constellations and set the block size $J = 128$. We adopt AP-CP-only where the last 3 symbols per frame (of 128 symbols) are known. Information bits are coded using a 16-state rate-$1/2$ convolutional code taken from [13, Table 11.2]. Taking into account the known symbols, the CP, and zero bits to terminate the CC trellis, the overall transmission rate of the AP-CP-only system is $(128-3-4)/(128+3) = 0.924$ bit per channel use or 250.4 kbps.

Figure 7.10 confirms that the system with two transmit-antennas significantly outperforms its counterpart with one transmit-antenna. At frame error rate 10^{-2}, about 5 dB SNR gain is achieved. Figure 7.11 depicts the performance improvement with turbo iterations. We notice that a large portion of the performance gain is achieved with three iterations only.

7.4 CLOSING COMMENTS

In this chapter we first represented a finite impulse response frequency-selective MIMO channel as an equivalent flat fading MIMO channel. Based on this equivalence, we found that the space-multipath diversity provided by a frequency-selective (tapped-delay line) MIMO channel can be as high as the product of the number of transmit-receive antennas times the number of the finite impulse response taps. The same equivalence led us to design ST codes for frequency-selective MIMO channels along the lines of coherent STTCs and OSTBCs for flat MIMO channels. Albeit more complex to design and decode than their flat fading counterparts, STTCs for frequency-selective MIMO channels can be devised to enable the maximum possible space-multipath diversity gain and in certain cases maximize the coding gain as well. OSTBCs, on the other hand, entail rate loss when used with more than two transmit-antennas (as with flat fading MIMO channels) but can also afford the lowest receiver complexity that involves linear ST decoding (as with flat MIMO channels) followed by equalization of the frequency-selective channel effects. Remarkably, it is possible to generalize Alamouti's ST coding scheme to frequency-selective MIMO channels, modify it to allow for maximum likelihood decoding using Viterbi's algorithm, and show its optimality in terms of requiring relatively low complexity and achieving the ergodic capacity of a system with two-transmit antennas and a single receive-antenna. Finally, we saw how the rate outage probability can be used to benchmark and compare performance of ST codes designed for frequency-selective MIMO channels.

In the next chapter we develop ST codes for multi-carrier transmissions over frequency-selective MIMO channels. At the closing comments of Chapter 8, we compare briefly single-carrier with multi-carrier transmission modalities.

8

ST Codes for Frequency-Selective Fading Channels: Multi-Carrier Systems

As we mentioned in Chapter 7, designing ST codes for frequency-selective MIMO channels is well motivated by broadband applications where multi-antenna (e.g., fixed wireless) systems have to deliver multimedia information content at high data rates. Relative to flat MIMO channels, frequency-selective MIMO channels are more challenging to estimate, as we will see in Chapter 11. But as we explained in Chapter 7, they offer more degrees of freedom which in return provide a higher order of diversity. To enable the latter, in Chapter 7 we designed ST codes for single-carrier systems. In this chapter, we pursue the same objective for multi-carrier systems.

The cornerstone of multi-carrier systems is orthogonal frequency division multiplexing (OFDM) — the "workhorse" modulation present in a number of practical wireless systems. Multi-antenna systems relying on OFDM are typically referred to as MIMO OFDM systems. OFDM converts single- (or multi-) antenna transmissions over frequency-selective (MIMO) channels to an equivalent set of flat fading (MIMO) channels. Using this equivalence, ST codes for frequency-selective MIMO channels will be sought to maximize diversity along the lines of their flat fading counterparts that we designed in Chapter 3. Their performance can be evaluated using the rate outage probability elaborated in Section 7.1.2. Recall that we also established a somewhat similar equivalence with a *single* flat fading MIMO channel using single-carrier multi-antenna transmissions. The OFDM-based equivalence is, however with a set of *multiple* flat fading MIMO channels of smaller dimension. This difference leads to a major advantage of OFDM over single-carrier alternatives since it reduces considerably the receiver complexity in equalization and decoding.

Besides space and time dimensions which are also present in the ST code matrices designed for single-carrier block transmissions, MIMO OFDM introduces a third dimension across frequency, which spans the set of subcarriers per block. This extra dimension will allow design of code matrices across space-time, space-frequency, or, space-time-frequency dimensions to achieve desirable trade-offs in complexity, diversity, and rate performance. Error performance oriented designs will include OSTBCs, STTCs, and digital-phase sweeping (DPS) based designs capable of effecting the maximum possible space-multipath diversity at complexity lower than their single-carrier counterparts and a transmission rate approaching one symbol per channel use. Unique to MIMO OFDM transmissions will be their ability to generalize to frequency-selective MIMO channels the rate-oriented layered and full-diversity full-rate (FDFR) designs we saw in Chapter 4.

8.1 GENERAL MIMO OFDM FRAMEWORK

8.1.1 OFDM Basics

A number of benefits OFDM brings to multi-antenna systems originate from basic features that OFDM possesses when it is applied even to a system with a single transmit-antenna and a single receive-antenna. To appreciate those, we first outline OFDM's operation using the discrete-time baseband equivalent block model of a single-antenna system depicted in Figure 8.1. For a thorough exposition of single- and multiuser OFDM and general multi-carrier as well as block single-carrier wireless systems, we refer the reader to [104, 286, 288, 291, 307] and references therein.

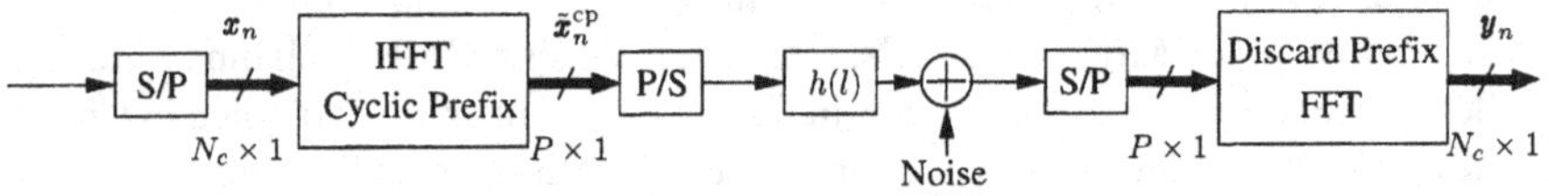

Figure 8.1 OFDM transceiver model

Different from a serial transmission, OFDM is a multi-carrier block transmission where, as the name suggests, information-bearing symbols are processed in blocks at both the transmitter and the receiver. Let N_c denote the number of subcarriers. At the transmitter, each symbol (say, the pth) in a block is conveyed on what is referred to as the pth subcarrier, which is a digital complex exponential sequence $\{1/\sqrt{N_c}\exp(j2\pi(p-1)n/N_c)\}_{n=0}^{N_c-1}$ of finite duration N_c corresponding to the pth column of the unitary IFFT matrix $\boldsymbol{F}_{N_c}^{\mathcal{H}}$. Specifically, the information symbols coming out of the modulator are parsed into blocks of length N_c. The N_c-point IFFT of each block (say, the nth)

$$\boldsymbol{x}_n := [x_n(0), x_n(1), \ldots, x_n(N_c-1)]^T, \tag{8.1}$$

is taken to obtain the $N_c \times 1$ block $\boldsymbol{F}_{N_c}^{\mathcal{H}}\boldsymbol{x}_n$. Block $\boldsymbol{x}_n$ is sometimes called an OFDM symbol in the frequency domain, whereas $\boldsymbol{F}_{N_c}^{\mathcal{H}}\boldsymbol{x}_n$ is referred to as an OFDM symbol in the time domain. Applying the triangle inequality to the N_c-point IFFT definition shows that the entries of $\boldsymbol{F}_{N_c}^{\mathcal{H}}\boldsymbol{x}_n$ have magnitudes that can exceed those of $\boldsymbol{x}_n$ by a factor as high as N_c. In other words, IFFT processing can increase the peak to average power ratio (PAR) by a factor as high as the number of subcarriers (which in certain applications can exceed $1,000$).

Following IFFT processing and using the matrix operator $\boldsymbol{T}_{\mathrm{cp}}$ defined in (7.23), we can replicate a number of (say, $P - N_c$) entries from the "tail" to the "head" of the vector $\boldsymbol{F}_{N_c}^{\mathcal{H}}\boldsymbol{x}_n$, to obtain a longer vector $\tilde{\boldsymbol{x}}_n^{\mathrm{cp}} := [\tilde{x}_n^{\mathrm{cp}}(0), \ldots, \tilde{x}_n^{\mathrm{cp}}(P-1)]^T = \boldsymbol{T}_{\mathrm{cp}}\boldsymbol{F}_{N_c}^{\mathcal{H}}\boldsymbol{x}_n$ of size $P > N_c$. The segment comprising the replicated entries is the cyclic prefix (CP).

The block $\tilde{\boldsymbol{x}}_n^{\mathrm{cp}}$ is transmitted over the channel, which is modeled as frequency selective with a finite impulse response (FIR) vector $\boldsymbol{h} := [h(0), \ldots, h(L)]^T$ of size $(L+1) \times 1$, where L is the channel order. When the CP length is selected larger than or equal to the channel order (i.e., $P - N_c \geq L$), the interblock interference (IBI) among consecutive $\tilde{\boldsymbol{x}}_n^{\mathrm{cp}}$ blocks (arising due to their convolution with $\boldsymbol{h}$) can be avoided by discarding the CP segment of each received block. (Notice that IBI is confined in the CP segment.)

Let $\overline{\boldsymbol{y}}_n$ denote the nth received block, which is free of IBI after the CP removal. It turns out that CP insertion at the transmitter together with CP removal at the receiver converts linear convolution to circular convolution [286]. As a result, the input-output relationship after CP removal can be written as

$$\overline{\boldsymbol{y}}_n = \sqrt{\overline{\gamma}}\,\tilde{\boldsymbol{H}}\boldsymbol{F}_{N_c}^{\mathcal{H}}\boldsymbol{x}_n + \overline{\boldsymbol{w}}_n, \tag{8.2}$$

where $\overline{\gamma}$ is the average SNR, $\tilde{\boldsymbol{H}}$ is the circulant channel matrix

$$\tilde{\boldsymbol{H}} := \begin{bmatrix} h(0) & 0 & \cdots & h(L) & \cdots & h(1) \\ \vdots & h(0) & \cdots & \cdots & \ddots & \vdots \\ \vdots & \cdots & \ddots & \cdots & \cdots & h(L) \\ h(L) & \cdots & \cdots & \ddots & \cdots & 0 \\ \vdots & \ddots & \cdots & \cdots & \ddots & \vdots \\ 0 & \cdots & h(L) & \cdots & \cdots & h(0) \end{bmatrix}_{N_c \times N_c}, \tag{8.3}$$

and $\overline{\boldsymbol{w}}_n$ denotes the additive noise vector whose entries are uncorrelated, circularly symmetric, and complex Gaussian distributed with zero mean and unit variance.

Using the property of circulant matrices in (7.32), $\tilde{\boldsymbol{H}}$ can be diagonalized by (post- and) premultiplication with (I)FFT matrices; that is,

$$\boldsymbol{F}_{N_c}\tilde{\boldsymbol{H}}\boldsymbol{F}_{N_c}^{\mathcal{H}} = \underbrace{\mathrm{diag}[H(0), \ldots, H(N_c - 1)]}_{:=\boldsymbol{D}_H}, \tag{8.4}$$

where $H(p) := \sum_{l=0}^{L} h(l)e^{-j2\pi lp/N_c}$. This nice property suggests FFT processing of $\overline{\boldsymbol{y}}_n$ at the receiver to obtain

$$\boldsymbol{y}_n := \boldsymbol{F}_{N_c}\overline{\boldsymbol{y}}_n = \sqrt{\overline{\gamma}}\,\boldsymbol{D}_H\boldsymbol{x}_n + \boldsymbol{w}_n, \tag{8.5}$$

where $\boldsymbol{w}_n := \boldsymbol{F}_{N_c}\overline{\boldsymbol{w}}_n$ is the FFT processed noise, which remains white since $\boldsymbol{F}_{N_c}$ is a unitary matrix.

If $y_n(p)$ and $w_n(p)$ denote the $(p+1)$st entry of $\boldsymbol{y}_n$ and $\boldsymbol{w}_n$, respectively, the scalar version of (8.5) is then

$$y_n(p) = \sqrt{\bar{\gamma}}\, H(p)x_n(p) + w_n(p), \quad p = 0, 1, \ldots, N_c - 1. \tag{8.6}$$

Equation (8.6) shows that an OFDM system which relies on N_c subcarriers to transmit the symbols of each block $\boldsymbol{x}_n$ converts an FIR frequency-selective channel to an equivalent set of N_c flat fading subchannels [carrying one symbol $x_n(p)$ per subcarrier] as depicted in Figure 8.2. This is intuitively reasonable since each narrowband subcarrier that is used to convey each information-bearing symbol per OFDM block "sees" a narrow portion of the broadband frequency-selective channel which can be considered frequency flat.

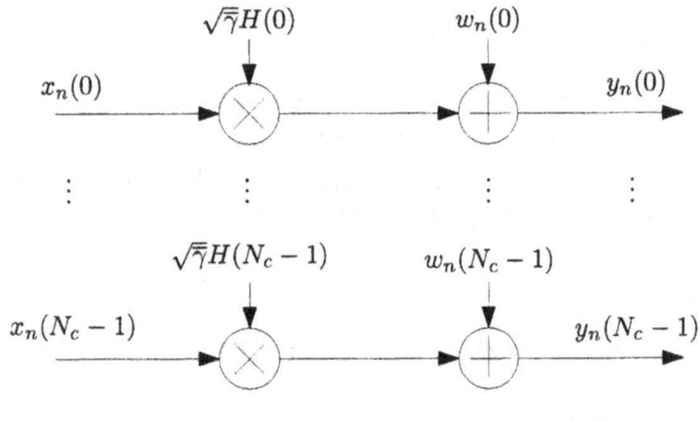

Figure 8.2 OFDM equivalent model in the frequency domain

Whereas subscript n in (8.6) denotes the time dimension, the argument p is used to index the frequency dimension inherent to an OFDM system. In fact, the simplified scalar model in (8.6) represents OFDM's operation in the frequency domain and constitutes the main reason behind its popularity in practical systems. This scalar model enables simple equalization of the FIR channel [by dividing (8.6) with the corresponding scalar subchannel $H(p)$] as well as low-complexity decoding across subchannels using, for example, Viterbi's algorithm if symbols $x_p(n)$ comprise convolutionally coded source bits.

Transmission of symbols over subcarriers also allows for flexible allocation of the available bandwidth to multiple users operating with possibly different rate requirements imposed by multimedia applications, which may include communication of data, audio, or video. Users with lower rate requirements will be allocated fewer subcarriers; while on a per user basis, more subcarriers can be designated for applications demanding higher rates. When channel state information (CSI) is available at the transmitter side, power and bits can be adaptively loaded per OFDM subcarrier, depending on the strength of the intended subchannel. When more power and bits are loaded to stronger subchannels (to obey the water-filling principle) and sufficiently long channel coding is incorporated, the resulting coded-OFDM system attains optimality in the sense of approaching the capacity of the underlying frequency-selective channel (see, e.g., [18]). Even though only partial CSI can be pragmatically available at the transmitter when communicating over wireless links, adaptive OFDM is the modulation used by high-speed digital subscriber lines (DSLs), where it is referred to as discrete-multitone (DMT) modulation [18, 46]. In our multi-antenna systems, use of partial CSI at the transmitter will be considered in Chapters 12 and 13 to combine STC with transmit beamforming. But for this chapter, no form of CSI will be assumed available at the transmitter.

The price paid for OFDM's attractive features in equalization, decoding, and (possibly adaptive power and) bandwidth allocation is its sensitivity to subcarrier drifts and the high PAR that IFFT processing introduces to the entries of each block transmitted. Subcarrier drifts come either from the carrier-frequency and phase offsets between transmit-receive oscillators or from mobility-induced Doppler effects, with the latter causing a spectrum of frequency drifts. Subcarrier drifts cause intercarrier interference, which renders (8.6) invalid. On the other hand, high PAR necessitates backing-off transmit-power amplifiers to avoid nonlinear distortion effects [224].

The advantages and limitations of single-antenna OFDM systems are basically present in the multi-antenna scenario we present next under the name MIMO OFDM.

8.1.2 MIMO OFDM

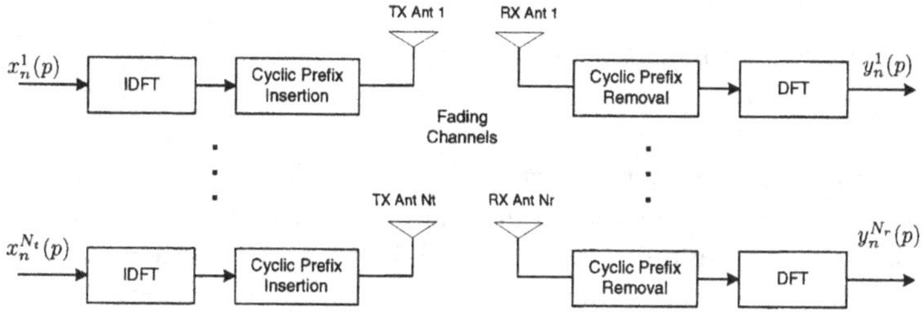

Figure 8.3 A multi-antenna space-time OFDM system

Let us now consider a multi-antenna system with N_t transmit-antennas and N_r receive-antennas, signaling over a frequency-selective MIMO channel using OFDM (de)modulation per antenna, as depicted in Figure 8.3. The model from any transmit-antenna element to any receive-antenna element in Figure 8.3 is a single-antenna OFDM transceiver model such as the one in Figure 8.1. Each symbol $x_n^\mu(p)$ at the OFDM modulator input is indexed by three variables: (i) the space variable $\mu \in [1, N_t]$ specifying the transmit-antenna, (ii) the time variable n denoting the OFDM symbol (block index), and (iii) the frequency variable $p \in [0, N_c - 1]$ indexing the subcarrier onto which the corresponding symbol in the generic nth block will "ride on" after OFDM modulation.

Notice that p can also be viewed as the time index within each block. In a single-carrier serial transmission, only one variable is used to index time; whereas in a single-carrier block transmission one variable is used to index blocks of symbols and a second one to index symbols within each block. OFDM basically views this second time variable as a subcarrier (or frequency-bin) variable. This three-dimensional (3D) representation of symbols in space-time-frequency (STF) is characteristic of MIMO OFDM and will be used in the general framework of designing STF coded multi-carrier multi-antenna transmissions over frequency-selective MIMO channels.

As in (7.1), the fading channel between the μth transmit-antenna and the νth receive-antenna is denoted by the corresponding FIR vector $\boldsymbol{h}_{\nu\mu} := [h_{\nu\mu}(0), \ldots, h_{\nu\mu}(L)]^T \in \mathbb{C}^{(L+1)\times 1}$, where L denotes the maximum channel order.

After CP removal and FFT processing per receive-antenna (here the νth), the received symbol $y_n^\nu(p)$ comprises the superposition of OFDM symbols transmitted from all N_t antennas. Using (8.6) for each one, we obtain the scalar input-output relationship

$$y_n^\nu(p) = \sqrt{\frac{\overline{\gamma}}{N_t}} \sum_{\mu=1}^{N_t} H_{\nu\mu}(p) x_n^\mu(p) + w_n^\nu(p), \quad \nu = 1, \ldots, N_r, \tag{8.7}$$

where $H_{\nu\mu}(p)$ is the subchannel gain from the μth transmit-antenna to the νth receive-antenna evaluated on the pth subcarrier:

$$H_{\nu\mu}(p) := \sum_{l=0}^{L} h_{\nu\mu}(l) e^{-j2\pi lp/N_c}, \tag{8.8}$$

and the additive noise $w_n^\nu(p)$ is circularly symmetric, zero mean, complex Gaussian with unit variance, that is also assumed uncorrelated across all three dimensions: ν, n, and p.

Equation (8.7) describes the operation of MIMO OFDM systems whether symbols $x_n^\mu(p)$ are uncoded or coded. Note that coding can be performed within each OFDM block but also across multiple (say, $N_x \geq 1$) OFDM blocks. In the most general case, coding amounts to mapping information symbols $s(n)$ to symbols $x_n^\mu(p)$, which can be designated to (μ, n, p) points in the 3D parallelepiped defined by the Cartesian product $[1, N_t] \times [0, N_x - 1] \times [0, N_c - 1]$. As usual, the designer's selection of a particular mapping (STF code) for MIMO OFDM operation will be guided by application-specific trade-offs among error performance, decoding complexity, and transmission rate metrics. Before delving into possible code designs, we will formulate analytically the STF coding framework and outline the pertinent metrics.

8.1.3 STF Framework

Suppose that information symbols are parsed into blocks of size N_I and consider one such block denoted generically (without block index) as $\boldsymbol{s} := [s_0, \ldots, s_{N_I-1}]^T \in \mathbb{C}^{N_I\times 1}$. STF coding is then defined as the one-to-one mapping

$$\Psi : \quad \boldsymbol{s} \rightarrow \boldsymbol{X}, \tag{8.9}$$

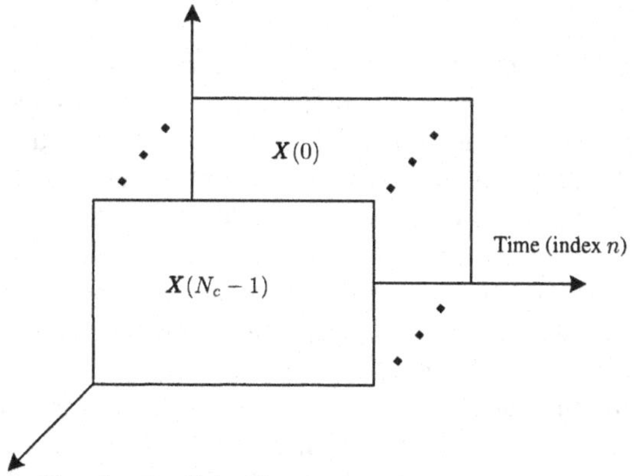

Figure 8.4 Space-time-frequency coded transmissions

where the STF codeword X comprises symbols $x_n^\mu(p)$ with $\mu \in [1, N_t]$, $n \in [0, N_x - 1]$ and $p \in [0, N_c - 1]$. Since the number of (μ, n, p) triplets (points) in the 3D parallelepiped spanned by N_t transmit-antennas, N_c subcarriers, and N_x blocks (OFDM symbols) is $N_t N_c N_x$, each STF codeword contains $N_t N_x N_c$ symbols $x_n^\mu(p)$. Organizing these symbols in an $N_t \times N_c N_x$ block matrix, we henceforth represent each codeword with the STF matrix

$$\boldsymbol{X} := [\boldsymbol{X}(0)\, \boldsymbol{X}(1) \cdots \boldsymbol{X}(N_c - 1)] \ \in \mathbb{C}^{N_t \times N_c N_x}, \tag{8.10}$$

where for each subcarrier $p \in [0, N_c - 1]$

$$\boldsymbol{X}(p) := \begin{bmatrix} x_0^1(p) & \cdots & x_{N_x-1}^1(p) \\ \vdots & & \vdots \\ x_0^{N_t}(p) & \cdots & x_{N_x-1}^{N_t}(p) \end{bmatrix} \in \mathbb{C}^{N_t \times N_x}. \tag{8.11}$$

Entries in each row of $\boldsymbol{X}(p)$ are transmitted from the same antenna; whereas each column contains symbols in a snapshot across transmit-antennas. Corresponding to each subcarrier p, we can also define a flat fading MIMO channel matrix $\boldsymbol{H}(p)$ as

$$\boldsymbol{H}(p) := \begin{bmatrix} H_{11}(p) & \cdots & H_{1N_t}(p) \\ \vdots & & \vdots \\ H_{N_r 1}(p) & \cdots & H_{N_r N_t}(p) \end{bmatrix} \in \mathbb{C}^{N_r \times N_t}. \tag{8.12}$$

Using these definitions and $y_n^\nu(p)$ from (8.7) to create the receive matrix $\boldsymbol{Y}(p) \in \mathbb{C}^{N_r \times N_x}$ with entries $[\boldsymbol{Y}(p)]_{\nu n} := y_n^\nu(p)$, we can rewrite the per subcarrier scalar MIMO OFDM model (8.7) in matrix form as

$$\boldsymbol{Y}(p) = \sqrt{\frac{\overline{\gamma}}{N_t}} \boldsymbol{H}(p) \boldsymbol{X}(p) + \boldsymbol{W}(p), \qquad p = 0, 1, \ldots, N_c - 1, \tag{8.13}$$

where the noise matrix $\boldsymbol{W}(p) \in \mathbb{C}^{N_r \times N_x}$ is formed similarly with entries $[\boldsymbol{W}(p)]_{\nu n} := w_n^{\nu}(p)$.

Equation (8.13) shows that for each subcarrier p, the input-output relationship of a MIMO OFDM transceiver signaling over a frequency-selective MIMO channel is equivalent to that of a flat fading MIMO channel. As with the scalar case, the matrix model (8.13) applies to both uncoded symbols (where the mapping Ψ implements the identity operator) as well as to STF coded ones. Picturing each ST matrix $\boldsymbol{X}(p)$ in (8.13) as a slab in Figure 8.4, we can view the N_c slabs obeying (8.13) as a 3D STF codeword which we represent by the STF code matrix $\boldsymbol{X}$ in (8.10). From this viewpoint, we can visualize graphically how STF coding encodes information simultaneously over the space, time and frequency (subcarrier) dimensions.

To quantify the transmission rate of STF coded transmissions, let us suppose that the information symbols s_n are drawn from a finite alphabet $\mathbb{S}$ which has cardinality $|\mathbb{S}|$. Since $\boldsymbol{X}$ is uniquely mapped from $\boldsymbol{s}$, the number of possible STF codewords $\boldsymbol{X}$ is $|\mathbb{S}|^{N_I}$, which we collect in a finite set $\mathbb{X}$ with cardinality $|\mathbb{X}| = |\mathbb{S}|^{N_I}$. Constructing an STF code amounts to specifying the finite set $\mathbb{X}$ and the mapping Ψ. By the definition of $\boldsymbol{X}$ in (8.10) and (8.11), it is clear that in each snapshot (channel use) across N_t antennas, N_I information symbols are transmitted during N_x blocks (OFDM symbols) over N_c subcarriers. Therefore, the STF *code rate* is

$$\eta_{\text{STF}} = \frac{N_I}{N_c N_x}. \tag{8.14}$$

Accounting for the rate loss due to the CP which contains L symbols, the resulting transmission rate for a constellation of size $|\mathbb{S}|$ is

$$R_{\text{STF}} = \frac{N_I}{(N_c + L)N_x} \log_2 |\mathbb{S}| \quad \text{bits per channel use.} \tag{8.15}$$

Turning our attention to error performance metrics, we already know from Proposition 7.1 that regardless of the modulation used, a frequency-selective MIMO channel with $L + 1$ taps can provide diversity as high as the product of the number of transmit-receive antennas times the number of taps. Thus, the maximum possible diversity order that a general STF coded MIMO OFDM transmission can enable is $N_t N_r (L + 1)$. In later sections we design STF codes capable of achieving this maximum diversity.

To assess frame-error-rate performance and compare STF codes that achieve the maximum diversity provided by the channel, one can rely on the rate outage probability defined in (7.14). The outage probability at rate R_{STF} serves as the ultimate frame error rate bound for STF codes used by MIMO OFDM and can also be used to compare the latter with the ST codes designed in Chapter 7 for single-carrier systems.

As we mentioned in Section 8.1.2, the frequency dimension we alluded to so far is basically a time dimension within each block (OFDM symbol); and such an interpretation of time as frequency is unique to OFDM since each symbol in a block is transmitted over a digital subcarrier (frequency bin). In reality, though, the physical dimensions over which an STF coded MIMO OFDM transmission takes place are space and time. To stress this point and illustrate how each STF code matrix $\boldsymbol{X}$ is transmitted, it is instructive to look at an example.

Example 8.1 Consider a MIMO OFDM system with $N_t = 2$, $N_x = 2$, $N_c = 4$, and $L = 1$. For these parameters, the STF code matrix in (8.10) becomes

$$\boldsymbol{X} = \Big[\underbrace{\begin{matrix} a_0 & c_0 \\ b_0 & d_0 \end{matrix}}_{\boldsymbol{X}(0)} \quad \underbrace{\begin{matrix} a_1 & c_1 \\ b_1 & d_1 \end{matrix}}_{\boldsymbol{X}(1)} \quad \underbrace{\begin{matrix} a_2 & c_2 \\ b_2 & d_2 \end{matrix}}_{\boldsymbol{X}(2)} \quad \underbrace{\begin{matrix} a_3 & c_3 \\ b_3 & d_3 \end{matrix}}_{\boldsymbol{X}(3)}\Big], \tag{8.16}$$

where we have two rows since there are two transmit-antennas and eight columns organized in four submatrices since we have two time blocks (two OFDM symbols) and each block consists of four symbols (four subcarriers). Since we wish to specify what is actually transmitted from the $N_t = 2$ antennas, we need to OFDM-modulate each block. To this end, we first perform IFFT processing per block of size 4. Using matrix representation, the 4-point IFFT processing of the first block of the first antenna yields

$$\begin{bmatrix} a'_0 \\ a'_1 \\ a'_2 \\ a'_3 \end{bmatrix} := \boldsymbol{F}_4^{\mathcal{H}} \begin{bmatrix} a_0 \\ a_1 \\ a_2 \\ a_3 \end{bmatrix},$$

and we can obtain $\{c_i\}_{i=0}^3$ similarly for the second block of the first antenna, and $\{b_i\}_{i=0}^3$ and $\{d_i\}_{i=0}^3$ for the corresponding blocks of the second antenna. The next step of the OFDM modulator is to replicate the last symbol of each IFFT-processed block to the beginning of the block in order to insert the CP, which here has size $L = 1$. After CP insertion, the ST matrix transmitted over the MIMO frequency-selective channel is

$$\boldsymbol{X}_{\mathrm{st}} = \begin{bmatrix} a'_3 & a'_0 & a'_1 & a'_2 & a'_3 & c'_3 & c'_0 & c'_1 & c'_2 & c'_3 \\ b'_3 & b'_0 & b'_1 & b'_2 & b'_3 & d'_3 & d'_0 & d'_1 & d'_2 & d'_3 \end{bmatrix}, \tag{8.17}$$

where we notice that the first (respectively, sixth) column is a replica of the fifth (tenth) column since it manifests the CP of the first (second) block in each antenna. At the receiver end, $\boldsymbol{X}_{\mathrm{st}}$ is OFDM demodulated (CP is removed from each block, which is subsequently FFT processed). The demodulator outputs in matrix form obey (8.13) per subcarrier.

Example 8.1 illustrates that STF coding in the MIMO OFDM context is one realization of the general ST coding for block multi-antenna transmissions over frequency-selective MIMO channels. In fact, Example 8.1 could have been constructed for a single-carrier block transmission after omitting the (I)FFT steps. Differences between multi- and single-carrier systems do not stem from STF encoding per se, since an STF code matrix can be viewed as an ST code matrix of a precoded single-carrier block transmission when the IFFT matrix is used as a precoding matrix. In other words, what matters more is the distinction between serial and block transmissions rather than the difference between block single- and multi-carrier systems. No matter whether one views OFDM as a multi-carrier transmission or as block precoded single-carrier transmission, the attractive features we outlined in the single-antenna case carry over to the multi-antenna setting.

8.2 ST AND SF CODED MIMO OFDM

Reflecting further on Example 8.1, we recognize that the STF code matrix can be viewed as an ST code matrix but also as an SF code matrix when $N_x = 1$. Various designs inspired by these ST and SF viewpoints have appeared in the literature. In what follows, we outline these through representative examples.

A class of ST coded OFDM systems considered early was one where a different block of information symbols was mapped to construct each $\boldsymbol{X}(p)$ separately without coding across subcarriers. Unfortunately, using ST codes of Chapter 3 to design each $\boldsymbol{X}(p)$ separately for the per subcarrier flat fading model in (8.13), enables only the space diversity $N_t N_r$. Among the first such ST coded OFDM systems were those in [151, 161], where OSTBCs were adopted to design each $\boldsymbol{X}(p)$ separately. With $N_t = 2$, we can, for example, use Alamouti's ST code per subcarrier to construct $\boldsymbol{X}(p)$ as

$$\boldsymbol{X}(p) = \begin{bmatrix} s_1(p) & -s_2^*(p) \\ s_2(p) & s_1^*(p) \end{bmatrix}, \quad p = 0, 1, \ldots, N_c - 1. \tag{8.18}$$

The merit of such a separable ST coded MIMO OFDM system is simplicity: not only during the encoding phase, since simple ST codes from Chapter 3 can be applied on a per subcarrier basis, but also at the receiver side, where reduced complexity equalization and demodulation can be afforded per subcarrier. The price paid for not coding across subcarriers is the loss of the available multipath diversity in (7.10).

Another approach to designing codes for MIMO OFDM is to focus on a single block ($N_x = 1$) and encode only across space and frequency (i.e., across space and time within one block, but not across blocks). The first space-frequency (SF) coded OFDM system was introduced in [3]; the term SF-OFDM was coined in [24]. Existing STTCs (specifically the 16-state STTC in [253]) were employed in both [3] and [24] simply by viewing the time variable within each block as the subcarrier variable. Heuristic designs of STTCs for SF-OFDM were also suggested in [170]. It was argued in [24, 170] that SF coded MIMO OFDM can achieve the maximum diversity of order $N_t N_r (L + 1)$, but no SF code designs were offered to enable this maximum joint space-multipath diversity order.

The first SF coded MIMO OFDM system effecting the maximum possible diversity was put forth in [184] using the notion of digital-phase sweeping (DPS). Because the DPS-based MIMO OFDM system outperforms existing ST, SF, and STF coded multi-carrier systems at rates approaching 1 symbol per channel use, we present it separately in Section 8.4. Interestingly, even a MIMO OFDM system attaining full diversity and full transmission rate (N_t symbols per channel use) falls under the class of SF coded OFDM systems that we present in Section 8.5.

8.3 STF CODED OFDM

In this section we present the STF designs (OSTBCs and STTCs) of [162], which were the first to offer a MIMO OFDM system with full diversity $N_tN_r(L+1)$. The idea in [162] was to use LCFCs and OSTBCs or STTCs to encode not only across space and time but also across subcarriers. A critical step in reducing the high dimensionality of the resulting STF code matrix $\boldsymbol{X}$ in (8.10) that facilitates the code design and decoding, is a subcarrier grouping step.

8.3.1 Subcarrier Grouping

Consider a MIMO OFDM system where the number of subcarriers is selected to be an integer multiple of the channel length; that is,

$$N_c = N_g(L+1), \tag{8.19}$$

where N_g is a positive integer denoting the number of groups. Notice that each group contains $L+1$ subcarriers, as many as the channel taps, a number also equal to the maximum multipath diversity offered by the frequency-selective channel. In accordance with this grouping, let us also divide each $N_I \times 1$ information block $\boldsymbol{s}$ used in (8.9) into N_g subblocks $\{\boldsymbol{s}_g\}_{g=0}^{N_g-1}$ each of size N_{sub} so that $N_I = N_gN_{\text{sub}}$. Block size N_I can be selected so that N_g and N_{sub} are integers. Furthermore, we divide the $N_t \times N_cN_x$ STF code matrix $\boldsymbol{X}$ in (8.10) into N_g submatrices $\{\boldsymbol{X}_g\}_{g=0}^{N_g-1}$ each formed by $N_x(L+1)$ columns of $\boldsymbol{X}$, so that

$$\boldsymbol{X}_g = [\boldsymbol{X}_g(0), \boldsymbol{X}_g(1), \ldots, \boldsymbol{X}_g(L)] \in \mathbb{C}^{N_t \times N_x(L+1)}, \tag{8.20}$$

where

$$\boldsymbol{X}_g(l) := \boldsymbol{X}(N_gl+g), \quad l = 0, \ldots, L.$$

Instead of (8.9), we will pursue mappings from each subblock $\boldsymbol{s}_g$ to each submatrix $\boldsymbol{X}_g$; that is, we will design a group STF (GSTF) codeword for each group according to

$$\Psi_g : \quad \boldsymbol{s}_g \rightarrow \boldsymbol{X}_g. \tag{8.21}$$

To appreciate the usefulness of this grouping, let us proceed to divide the STF system described by (8.13) into N_g GSTF (sub) systems, each of which obeys the input-output relationship

$$\boldsymbol{Y}_g(l) = \sqrt{\frac{\overline{\gamma}}{N_t}}\boldsymbol{H}_g(l)\boldsymbol{X}_g(l) + \boldsymbol{W}_g(l), \tag{8.22}$$

where $l = 0, 1, \ldots, L$, and

$$\begin{aligned}
\boldsymbol{Y}_g(l) &:= \boldsymbol{Y}(N_gl+g), \\
\boldsymbol{H}_g(l) &:= \boldsymbol{H}(N_gl+g), \\
\boldsymbol{W}_g(l) &:= \boldsymbol{W}(N_gl+g).
\end{aligned}$$

Relative to the STF system described by (8.13), each GSTF subsystem is a simpler STF system with much smaller size in the frequency dimension ($L+1$ instead of N_c). Subchannel grouping reduces the complexity and allows the design of code matrices per GSTF subsystem since STF coding constructs $\boldsymbol{X}_g$ separately as in (8.21) instead of constructing the entire $\boldsymbol{X}$ as in (8.10).

Because each subgroup contains $L+1$ well-spaced subcarriers, it turns out that such decomposition of the N_c subcarriers does not lose the multipath diversity factor while it reduces complexity at the code design and decoding stages[1]. Specifically, the multipath diversity factor can be effected by using LCFC matrices to precode each subblock $\boldsymbol{s}_g$ as we described in Chapter 3; and together with the ST diversity factor $N_t N_r$ enabled by OSTBC or STTC designs of each $\boldsymbol{X}_g$ GSTF code matrix, it becomes possible to achieve the full diversity order $N_t N_r(L+1)$ [162].

With regard to transmission rate, it is clear that $N_I = N_g N_{\text{sub}}$ symbols are transmitted over $N_g(L+1)N_x$ snapshots (channel uses); thus, the code rate in (8.14) can be rewritten as

$$\eta_{\text{STF}} = \frac{N_{\text{sub}}}{(L+1)N_x}. \tag{8.23}$$

Having reduced the STF code matrix construction to the design of N_g GSTF code matrices, it suffices to specify how one of the $\boldsymbol{X}_g$'s is constructed to achieve the maximum space-multipath diversity at the rate given by (8.23).

8.3.2 GSTF Block Codes

As we already mentioned, the GSTF block codes (abbreviated as GSTFBCs) developed in [162] combine LCFCs (cf. Section 3.5) with OSTBCs (cf. Section 3.3). The encoding process follows these steps:

S1) Given N_t, N_r, and L, choose N_g and $N_c = N_g(L+1)$.

S2) Select the OSTBC and determine its two parameters N_s and N_d according to [cf. Section 3.3]

$$(N_s, N_d) = \begin{cases} (2,2) & \text{if} \quad N_t = 2, \\ (3,4) & \text{if} \quad N_t = 3,4, \\ (N_t, 2N_t) & \text{if} \quad N_t > 4. \end{cases} \tag{8.24}$$

S3) Choose $N_{\text{sub}} = N_s(L+1)$, $N_x = N_d$, and demultiplex each $N_{\text{sub}} \times 1$ information symbol block $\boldsymbol{s}_g$ into $\{\boldsymbol{s}_{g,i} \in \mathbb{C}^{(L+1)\times 1}, i = 0, \ldots, N_s - 1\}$ subblocks so that

$$\boldsymbol{s}_g := [\boldsymbol{s}_{g,0}^T, \ldots, \boldsymbol{s}_{g,N_s-1}^T]^T. \tag{8.25}$$

[1] The reduced complexity effected by subcarrier grouping comes at the expense of reduced coding gains since ST codes of reduced length (not exceeding L) have to be used. Alternatively, coding across all subcarriers at reasonable complexity is possible by using low-density parity check (LDPC) or trellis codes with interleaving (see, e.g., [172, 173]).

S4) Precode using the $(L+1) \times (L+1)$ LCFC matrix $\boldsymbol{\Theta}_{\mathrm{LCF}}$ from Section 3.5 to obtain the precoded blocks

$$\tilde{\boldsymbol{s}}_{g,i} := \begin{bmatrix} \tilde{s}_{g,i,0} \\ \vdots \\ \tilde{s}_{g,i,L} \end{bmatrix} = \boldsymbol{\Theta}_{\mathrm{LCF}} \boldsymbol{s}_{g,i}. \tag{8.26}$$

S5) Use matrix pairs $\{\boldsymbol{A}_i, \boldsymbol{B}_i\}_{i=0}^{N_s-1}$ in (3.17) to construct per subcarrier l the OSTBC code matrix

$$\boldsymbol{X}_g(l) = \sum_{i=0}^{N_s-1} (\boldsymbol{A}_i \tilde{s}_{g,i,l} + \boldsymbol{B}_i \tilde{s}_{g,i,l}^*). \tag{8.27}$$

Putting steps S1 to S5 together, it is possible to express the STF matrix $\boldsymbol{X}_g$ as [162]

$$\boldsymbol{X}_g = \sum_{i=0}^{N_s-1} \left[(\boldsymbol{\Theta}_{\mathrm{LCF}} \boldsymbol{s}_{g,i})^T \otimes \boldsymbol{A}_i + (\boldsymbol{\Theta}_{\mathrm{LCF}} \boldsymbol{s}_{g,i})^{\mathcal{H}} \otimes \boldsymbol{B}_i \right], \tag{8.28}$$

where $\otimes$ stands for the Kronecker product.

With $N_x = N_d$, the rate for this GSTFB coded MIMO OFDM system is

$$\eta_{\mathrm{GSTFB}} = \frac{N_s}{N_d}. \tag{8.29}$$

According to (8.24), we have $\eta_{\mathrm{GSTFB}} < 1$ if $N_t > 2$. This implies that the GSTFB coding with more than two transmit-antennas incurs rate loss when compared to GSTFB coding with $N_t = 2$ transmit-antennas. This is inherited from the OSTBC, which is known to lose up to 50% in transmission rate when more than two transmit-antennas are deployed with generally complex constellations (cf. Chapter 3).

At this point it is useful to recapitulate the GSTFB encoding process with an example.

Example 8.2 Consider the GSTFB code design with $N_t = 2$, $L = 1$ and $(L+1) \times (L+1)$ LCFC matrix

$$\boldsymbol{\Theta}_{\mathrm{LCF}} = \frac{1}{\sqrt{2}} \begin{bmatrix} 1 & e^{j\pi/4} \\ 1 & e^{j5\pi/4} \end{bmatrix}. \tag{8.30}$$

We take $N_{\mathrm{sub}} = N_s(L+1) = 4$ symbols $[a, b, c, d]$, group them, and precode them to obtain

$$\tilde{\boldsymbol{s}}_{g,1} = \begin{bmatrix} a' \\ b' \end{bmatrix} = \boldsymbol{\Theta}_{\mathrm{LCF}} \begin{bmatrix} a \\ b \end{bmatrix} = \begin{bmatrix} \frac{1}{\sqrt{2}} a + \frac{1}{2}(1+j)b \\ \frac{1}{\sqrt{2}} a - \frac{1}{2}(1+j)b \end{bmatrix}, \tag{8.31}$$

$$\tilde{\boldsymbol{s}}_{g,2} = \begin{bmatrix} c' \\ d' \end{bmatrix} = \boldsymbol{\Theta}_{\mathrm{LCF}} \begin{bmatrix} c \\ d \end{bmatrix} = \begin{bmatrix} \frac{1}{\sqrt{2}} c + \frac{1}{2}(1+j)d \\ \frac{1}{\sqrt{2}} c - \frac{1}{2}(1+j)d \end{bmatrix}. \tag{8.32}$$

Since symbols $\{a', c'\}$ are ST coded on the first subcarrier and symbols $\{b', d'\}$ on the second subcarrier, the GSTFB code matrix is [cf. (8.27)]

$$\boldsymbol{X}_g = \Big[\underbrace{\begin{matrix} a' & -c'^* \\ c' & a'^* \end{matrix}}_{\boldsymbol{X}_g(0)} \quad \underbrace{\begin{matrix} b' & -d'^* \\ d' & b'^* \end{matrix}}_{\boldsymbol{X}_g(1)} \Big]. \tag{8.33}$$

Matrix $\boldsymbol{X}_g$ in this example implements Alamouti's ST coding per subcarrier.

Thanks to the orthogonality of OSTBCs, the GSTFB decoding at the receiver side has low complexity and proceeds in two steps:

S1) Linear ML-optimal OSTBC decoding per subcarrier to obtain $\{\hat{\tilde{\boldsymbol{s}}}_{g,i,l}\}_{i=1,l=0}^{N_s,L}$, as detailed in Section 3.3.

S2) Linear equalization or (near-)ML decoding based on the sphere decoding algorithm (SDA) to recover $\boldsymbol{s}_{g,i}$ from the decoded $\hat{\tilde{\boldsymbol{s}}}_{g,i}$, as in Section 3.5.

As we discussed in Chapter 5, SDA can have average complexity, which is polynomial in $L+1$ for a number of practical settings. Because L is small relative to N_c, complexity of GSTFB decoding is relatively low. While enjoying the two-stage (near-)optimal decoding at low complexity, the GSTFB codes in this section enable the maximum diversity of order $N_t N_r (L+1)$ at a rate not higher than one symbol per channel use (or considerably less with $N_t > 2$) [162].

8.3.3 GSTF Trellis Codes

In this section we present the STF trellis code (STFTC) design in [162]. The important step is to exploit the link between the STF representation and the ST model based on which we designed STTCs for flat fading MIMO channels in Chapter 3.

8.3.3.1 Space Virtual-Time Transmissions Upon recalling (8.20), which represents a GSTF codeword $\boldsymbol{X}_g$ as a set of code submatrices, $\boldsymbol{X}_g$ can be thought of as being transmitted from an equivalent (what we term) space-virtual-time (SVT) system with N_t transmit-antennas and N_r receive-antennas, where the virtual time dimension corresponds to the joint dimension of time and frequency as detailed next.

First, let $\boldsymbol{x}_t := [x_t^1, \ldots, x_t^{N_t}]^T \in \mathbb{C}^{N_t \times 1}$ be the symbol block transmitted from the N_t transmit-antennas during the tth virtual time interval, and let $\boldsymbol{y}_t := [y_t^1, \ldots, y_t^{N_r}]^T \in \mathbb{C}^{N_r \times 1}$ denote the corresponding block of received samples. The SVT system is modeled as

$$\boldsymbol{y}_t = \sqrt{\frac{\bar{\gamma}}{N_t}} \boldsymbol{H}_t \boldsymbol{x}_t + \boldsymbol{w}_t, \qquad t = 0, \ldots, N_x(L+1) - 1, \tag{8.34}$$

where $\boldsymbol{H}_t \in \mathbb{C}^{N_r \times N_t}$ is the MIMO channel matrix and $\boldsymbol{w}_t \in \mathbb{C}^{N_r \times 1}$ is the noise vector. To link the SVT system with the GSTF system, let us define [cf. (8.22)]

$$\boldsymbol{Y}_g := [\boldsymbol{Y}_g(0), \ldots, \boldsymbol{Y}_g(L)] \in \mathbb{C}^{N_r \times N_x(L+1)},$$
$$\boldsymbol{W}_g := [\boldsymbol{W}_g(0), \ldots, \boldsymbol{W}_g(L)] \in \mathbb{C}^{N_r \times N_x(L+1)},$$

and specify $\boldsymbol{y}_t$, $\boldsymbol{x}_t$, $\boldsymbol{w}_t$, and $\boldsymbol{H}_t$ as

$$\begin{aligned} \boldsymbol{y}_t &= [\boldsymbol{Y}_g]_{\zeta(t)}, & \boldsymbol{x}_t &= [\boldsymbol{X}_g]_{\zeta(t)}, \\ \boldsymbol{w}_t &= [\boldsymbol{W}_g]_{\zeta(t)}, & \boldsymbol{H}_t &:= \boldsymbol{H}_g(\tau(t)), \end{aligned} \tag{8.35}$$

where $[\boldsymbol{Y}_g]_{\zeta(t)}$ stands for the $\zeta(t)$th column of $\boldsymbol{Y}_g$. The index functions $\zeta(t)$ and $\tau(t)$ in (8.35) are given by (with $\lfloor \cdot \rfloor$ denoting integer floor)

$$\zeta(t) := N_x \left[t \mod (L+1)\right] + \left\lfloor \frac{t}{L+1} \right\rfloor + 1,$$
$$\tau(t) := t \mod (L+1),$$

respectively. It is not difficult to recognize that except for the difference in the ordering of transmissions, the model in (8.34) is mathematically equivalent to that in (8.22).

Although the underlying fading channels are time invariant, the equivalent channel $\boldsymbol{H}_t$ varies with the virtual time since GSTF codewords experience different subcarriers. Therefore, (8.34) can be thought of as an ST system transmitting over time-selective (but frequency-flat) fading channels. This provides an explicit link between each GSTF subsystem and the ST model for flat fading MIMO channels. Furthermore, (8.34) implies that: (i) since the channel varies with time (or more precisely with the virtual time), the SVT (or STF) system can potentially achieve diversity gain higher than $N_t N_r$; and (ii) it is possible to take advantage of existing ST coding techniques to design GSTFT code matrices.

The "smart-greedy" STTCs in [253] are purported to achieve acceptable performance for both flat fading time-invariant and time-varying MIMO channels. Because the time-varying channel $\boldsymbol{H}_t$ here is created artificially and its time variations are well structured, there is no need to design ST codes that are "smart" in the sense of [253]. Instead, they only have to be "greedy" to take advantage of the time variation in $\boldsymbol{H}_t$.

8.3.3.2 Code Construction Based on (8.34), the design of STFTCs reduces to building a trellis to generate $\boldsymbol{x}_t$'s continuously. Before outlining the design of such trellises, we should state the limitations of STFTCs in terms of transmission rate when the maximum diversity is achieved [162].

Proposition 8.1 *Suppose that each symbol x_t^μ transmitted belongs to the constellation set $\mathbb{X}_t$ of size $|\mathbb{X}_t|$. It is possible to design the grouped STFTC matrices enabling the maximum diversity order $G_{\mathrm{d,max}} = N_t N_r (L+1)$, but at a transmission rate no greater than $R_{\max} = \log_2 |\mathbb{X}_t|/(L+1)$ bits per channel use.*

The rate $R_{\max}$ is clearly related to the rate of the GSTF trellis code η_{GSTFT} via

$$\frac{\log_2 |\mathbb{X}_t|}{L+1} = R_{\max} = \eta_{\text{GSTFT}} \log_2 |\mathbb{S}|. \tag{8.36}$$

Proposition 8.1 implicitly suggests two possible design strategies:

Strategy 1 Design an STFTC with code rate $\eta_{\text{GSTFT}} = 1$, where the trellis outputs a single block $\boldsymbol{x}_t$ corresponding to each information symbol s_t with corresponding cardinalities $|\mathbb{S}| = 2^{R_{\max}}$ and $|\mathbb{X}_t| = 2^{R_{\max}(L+1)}$.

Strategy 2 Design a repetition-based STFTC with code rate $\eta_{\text{GSTFT}} = 1/(L+1)$, where the trellis outputs $L+1$ blocks $\boldsymbol{x}_t$ corresponding to each s_t and has cardinality $|\mathbb{S}| = |\mathbb{X}_t| = 2^{R_{\max}(L+1)}$.

These two design strategies are equivalent in the sense that the resulting codes achieve the same transmission rate $R_{\max}$. Moreover, both strategies expand either the constellation of the symbols transmitted or that of the information symbols. However, their implementations are different. In the first strategy, one looks for a trellis involving constellation expansion which is different from that of most existing STTCs. Simply taking advantage of existing STTC designs, we present next STFTCs using the second strategy.

Let $\mathcal{T}_{\text{ST}}(\cdot)$ denote a ST trellis encoder with code rate 1 and $|\mathbb{S}| = |\mathbb{X}_t| = 2^{R_{\max}(L+1)}$. According to strategy 2, the STF trellis encoder with $\eta_{\text{GSTFT}} = 1/(L+1)$ can be constructed from $\mathcal{T}_{\text{ST}}(\cdot)$ simply by repeating its output $L+1$ times. In other words, going back to each GSTF subsystem, we basically replicate transmissions over $L+1$ subcarriers. Since this approach to constructing STFTCs amounts to designing repetition-based STTCs for flat fading MIMO channels, the resulting designs are not expected to perform well. Similar comments apply to the related approach pursued when designing smart-greedy STTCs in [253].

Example 8.3 Let us consider a system with $N_t = 2$, $L = 1$, and $R = 2$ bits per channel use. In this case, we have $|\mathbb{S}| = 2^{R(L+1)} = 16$ and we can use the ST trellis depicted in Figure 3.7 (corresponding to [253, Figure 19]) to generate the STFTCs. Notice that the ST encoder output is repeated over two different OFDM subcarriers.

8.3.3.3 Decoding of GSTF Trellis Codes Because GSTF trellis codes are generated by a trellis, their decoding can be implemented efficiently by using Viterbi's algorithm. According to (8.34), the branch metric for $\boldsymbol{x}_t$ is given by

$$\sum_{\nu=1}^{N_r} \left| y_t^{\nu} - \sqrt{\frac{\overline{\gamma}}{N_t}} \sum_{\mu=1}^{N_t} [\boldsymbol{H}_t]_{\nu\mu} x_t^{\mu} \right|^2, \tag{8.37}$$

where $[\boldsymbol{H}_t]_{\nu\mu}$ denotes the (ν, μ)th element of $\boldsymbol{H}_t$. Similar to STTCs for flat fading MIMO channels, the decoding complexity of GSTF trellis codes is high since it grows exponentially in the number of trellis states as well as in the transmission rate.

8.3.4 Numerical Examples

We will test the performance of STF code designs with $N_t = 2$ and $N_r = 1$. The figure of merit is OFDM symbol error rate (OFDM-SER), averaged over 100,000 channel realizations. The transmission rate is fixed at $R_{\text{STF}} = 2$ bits per channel use in all test cases. While L is the channel order used for designing the codes, we distinguish between L and L_0, where L_0 is the true channel order for both channels. One could certainly choose to design a code targeting $L \neq L_0$ in order to trade off error performance with complexity. The random channels are generated according to two different channel models:

Multiray channel: This corresponds to channel taps that are i.i.d., zero mean, complex Gaussian with variance $1/(2L_0 + 2)$ per dimension.

HiperLAN/2 channel $(L_0 = 8)$: These random channels are based on the HiperLAN/2 channel model A corresponding to a typical office environment [63]. Each channel tap in the profile of channel model A is characterized by the Jakes' Doppler spectrum with mobile speed of 3 m/s.

Example 8.4 (Comparison with the SF codes in [24, 170]) In this example we compare the GSTF block and trellis coding schemes with the SF coding scheme in [24] (referred to as SF code 1) and that in [170] (referred to as SF code 2). The design of both GSTF block and trellis codes is based on $L = 1$, which could be different from L_0. We use QPSK modulation ($|\mathbb{S}| = 4$) for GSTF block coding and also for the two SF codes 1 and 2 and 16-QAM ($|\mathbb{S}| = 16$) for GSTF trellis coding, as described in Example 8.3. The transmission rate for all four schemes is therefore $R_{\text{STF}} = 2$ bits per channel use. The 16-state TCM code with effective length 3 and the 16-state ST trellis code in Figure 3.5 (corresponding to [253, Figure 5]) are used to generate the SF codes of [170] and [24], respectively. The number of subcarriers is set to $N_c = 64$ for all schemes. We first simulate all schemes in multiray channels with $L_0 = 1$. As expected, Figure 8.5 confirms that both GSTF trellis and block codes are able to achieve higher diversity gain than can SF codes. Whereas the GSTF block code outperforms SF codes consistently for all SNRs, the GSTF trellis code outperforms SF codes only for SNR values above 20 dB. The latter implies that the STF trellis code has poor coding gain, which is due to the fact that the chosen ST trellis code in Figure 3.7 (corresponding to [253, Figure 19]) has not been optimized with respect to the coding gain. Performance could be improved if both diversity and coding gains of GSTF trellis codes are maximized, which requires high-complexity computer search similar to the one in Section 3.2.3.

Using the same setup, the simulation can be repeated for the realistic HiperLAN/2 channels with $L_0 = 8$. As shown in Figure 8.6, both GSTF trellis and block codes enjoy higher diversity gain than those of the particular SF codes in [24, 170]. Error performance of the GSTF trellis code is close to that of SF codes, even for low SNR values. It is worth mentioning that the GSTF codes here are designed regardless of the

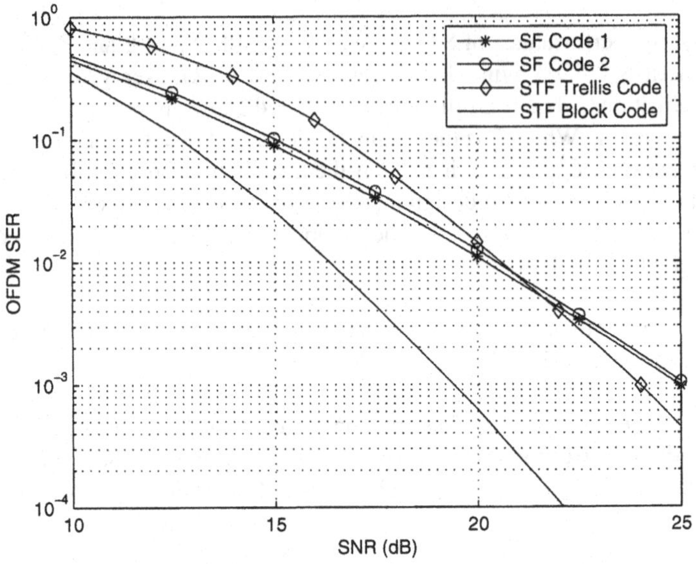

Figure 8.5 Comparison with SF codes (multiray channels)

real channel order $L_0 = 8$, channel correlation, and power profile of the HiperLAN/2 channels tested, which speaks for the robustness of the GSTF coding design.

Example 8.5 (Performance improvement with multipath diversity) To appreciate the importance of multipath diversity, we simulate the performance of GSTF block coding in multiray channels with different channel orders $L_0 = 1, 2, 3$. Assuming here perfect knowledge of L_0, we test GSTF block codes with $L = L_0$. The number of subcarriers is set to $N_c = 48$. We use QPSK constellation. Figure 8.7 shows that GSTF codes achieve higher diversity gain as the channel order increases, which confirms the importance of GSTF coding to enable the multipath diversity available.

The STF approach to designing OSBTCs and STTCs for MIMO OFDM transmissions over frequency-selective multi-antenna channels can afford low-complexity transceivers with maximum diversity (when OSBTCs are used) at the price of lowering the transmission rate below 1 symbol per channel use when more than two transmit-antennas are deployed. Since the available STTCs in [162] are basically repetition codes, this rate loss is not remedied even by the trellis designs, which are inevitably more complex than OSTBCs. In the ensuing section we present a class of low-complexity SF codes for MIMO OFDM, which are based on digital-phase sweeping and are capable of approaching a transmission rate of 1 symbol per channel use for any number of transmit-antennas while guaranteeing the full diversity of order $N_t N_r (L + 1)$.

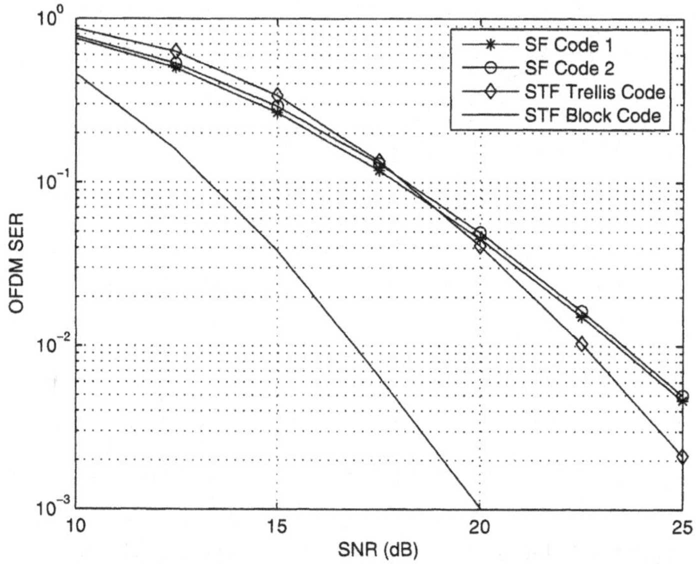

Figure 8.6 Comparison with SF codes (HiperLAN/2 channels)

8.4 DIGITAL-PHASE SWEEPING AND BLOCK CIRCULAR DELAY

The digital-phase sweeping (DPS) scheme we present in this section converts space diversity to multipath diversity and leads to an attractive MIMO OFDM system which has relatively low complexity and achieves the maximum possible diversity with minimal loss in transmission rate [184]. As we mentioned in Section 8.2, DPS-based MIMO OFDM belongs to the SF coded class for which $N_x = 1$; that is, the SF encoding involves symbols of only one block (one OFDM symbol), and the resulting SF code matrix $\boldsymbol{X}_{\text{DPS}}$ has dimensionality $N_t \times N_c$ [cf. (8.10)]. For this reason, and also because receiver processing will be performed on a block-by-block basis, we drop the block index n hereafter.

To outline the basic principles behind DPS, let us start as in the GSTF setup by parsing the stream of information symbols into blocks of size $N_c = N_g N_{\text{sub}}$. Each block $\boldsymbol{s}$ is divided into N_g subblocks $\{\boldsymbol{s}\}_{g=0}^{N_g-1}$ each of size N_{sub}, so that

$$\boldsymbol{s}_g = \left[[\boldsymbol{s}]_{N_{\text{sub}}g}, \ldots, [\boldsymbol{s}]_{N_{\text{sub}}(g+1)-1}\right]^T, \quad \forall g \in [0, N_g - 1], \qquad (8.38)$$

where $[\boldsymbol{s}]_i$ denotes the $(i+1)$st entry of $\boldsymbol{s}$. Using an LCFC matrix $\boldsymbol{\Theta}_{\text{LCF}}$ which is constructed as in Section 3.5, each subblock is precoded linearly to obtain

$$\boldsymbol{u}_g = \boldsymbol{\Theta}_{\text{LCF}} \boldsymbol{s}_g. \qquad (8.39)$$

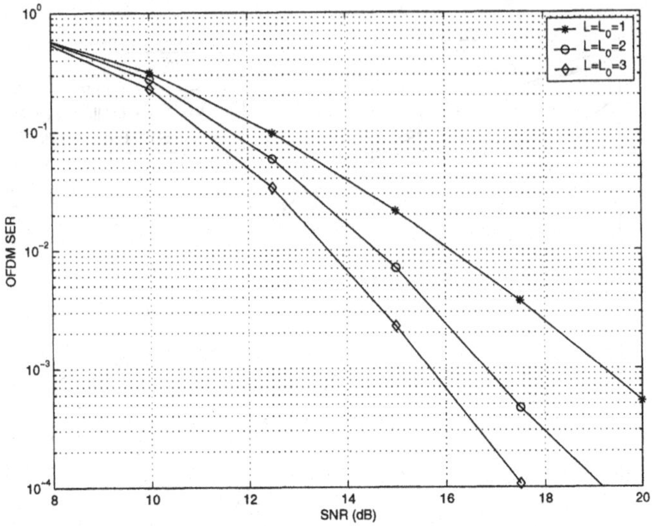

Figure 8.7 Importance of multipath diversity (multiray channels)

Subsequently, the subblocks $\{\boldsymbol{u}_g\}_{g=0}^{N_g-1}$ are multiplexed using a block interleaver with depth N_{sub} to yield the $N_c \times 1$ block

$$\boldsymbol{u} = \left[[\boldsymbol{u}_0]_0 \cdots [\boldsymbol{u}_{N_g-1}]_0, \ldots, [\boldsymbol{u}_0]_{N_{\text{sub}}-1} \cdots [\boldsymbol{u}_{N_g-1}]_{N_{\text{sub}}-1}\right]^T. \tag{8.40}$$

It is not difficult to verify that through linear precoding and interleaving each information block $\boldsymbol{s}$ and precoded block $\boldsymbol{u}$ are linked equivalently with the relationship

$$\boldsymbol{u} = \begin{bmatrix} \boldsymbol{I}_{N_g} \otimes \boldsymbol{\theta}_1^T \\ \vdots \\ \boldsymbol{I}_{N_g} \otimes \boldsymbol{\theta}_{N_{\text{sub}}}^T \end{bmatrix} \boldsymbol{s}, \tag{8.41}$$

where $\boldsymbol{\theta}_m^T$ is the mth row of $\boldsymbol{\Theta}_{\text{LCF}}$.

The next step is to apply the DPS operation on the $N_c \times 1$ precoded block $\boldsymbol{u} := [u(0), u(1), \ldots, u(N_c-1)]^T$. To this end we define for each transmit-antenna index $\mu = 1, \ldots, N_t$ the "phase constant" $\phi_\mu := -2\pi(\mu-1)(L+1)/N_c$ along with the corresponding diagonal matrix

$$\boldsymbol{\Phi}_\mu := \text{diag}[1, e^{j\phi_\mu}, \ldots, e^{j\phi_\mu(N_c-1)}]. \tag{8.42}$$

When left-multiplying the block $\boldsymbol{u}$, matrix $\boldsymbol{\Phi}_\mu$ implements digital-phase sweeping of its entries: hence, the acronym DPS. Based on the phase-swept blocks $\{\boldsymbol{\Phi}_\mu \boldsymbol{u}\}_{\mu=1}^{N_t}$, we form the DPS-based SF code matrix as

$$\boldsymbol{X}_{\text{DPS}} = \begin{bmatrix} (\boldsymbol{\Phi}_1 \boldsymbol{u})^T \\ \vdots \\ (\boldsymbol{\Phi}_{N_t} \boldsymbol{u})^T \end{bmatrix} \in \mathbb{C}^{N_t \times N_c}. \tag{8.43}$$

As in every MIMO OFDM system, each row of the $\boldsymbol{X}_{\text{DPS}}$ matrix is IFFT-processed and a CP is inserted before being transmitted from the corresponding antenna.

Let us now consider the symbols received after channel propagation at the output of the OFDM demodulator in the absence of noise. After FFT processing and CP removal, on the pth subcarrier of a receive-antenna ν we have [cf. (8.7)]

$$y^{\nu}(p) = \sqrt{\frac{\overline{\gamma}}{N_t}} \sum_{\mu=1}^{N_t} H_{\nu\mu}(p) x^{\mu}(p) = \sqrt{\frac{\overline{\gamma}}{N_t}} \sum_{\mu=1}^{N_t} H_{\nu\mu}(p) e^{j\phi_{\mu}p} u(p). \tag{8.44}$$

For future use, let us define the composite impulse response vector corresponding to the νth receive-antenna:

$$\boldsymbol{h}^{(\nu)} = \sqrt{\frac{1}{N_t}} \begin{bmatrix} \boldsymbol{h}_{\nu 1} \\ \vdots \\ \boldsymbol{h}_{\nu N_t} \end{bmatrix} \in \mathbb{C}^{N_t(L+1)\times 1}. \tag{8.45}$$

Substituting (8.8) into (8.44), we obtain

$$\begin{aligned} y^{\nu}(p) &= \sqrt{\frac{\overline{\gamma}}{N_t}} \left(\sum_{\mu=1}^{N_t} \sum_{l=0}^{L} h_{\nu\mu}(l) e^{-j2\pi[l+(\mu-1)(L+1)]p/N_c} \right) u(p) \\ &= \sqrt{\overline{\gamma}} \left(\sum_{l=0}^{N_t(L+1)-1} h^{(\nu)}(l) e^{-j2\pi lp/N_c} \right) u(p), \end{aligned} \tag{8.46}$$

where $h^{(\nu)}(l)$ is the $(l+1)$st entry of $\boldsymbol{h}^{(\nu)}$ in (8.45).

Comparing (8.46) with (8.6), we infer that each receive-antenna ν observes an equivalent OFDM system with a single transmit-antenna but over a concatenated channel $\boldsymbol{h}^{(\nu)}$ of length $N_t(L+1)$. Hence, the code design for MIMO OFDM reduces to that of a single-antenna OFDM system. This is a unique feature brought by the DPS-based MIMO OFDM system, which inherits all the benefits of OFDM. Indeed, $u(p)$ can be recovered by equalizing with a scalar division the noisy version of $y^{\nu}(p)$ in (8.46).

Having estimated $\boldsymbol{u}$, and thus its subblocks $\boldsymbol{u}_g$, we can obtain $\boldsymbol{s}_g$ from (8.39) using ML or suboptimal demodulation algorithms. The choice depends on the desirable error performance versus the complexity that can be afforded. The SDA presented in Chapter 5 offers in several practical cases a (near-)ML solution at reasonable complexity. Since complexity is polynomial in the size of the block we wish to demodulate, it should be clear how helpful the subcarrier grouping is. Had we sought demodulation of the entire block $\boldsymbol{s}$ from the $\boldsymbol{u}$ in (8.40), the complexity of SDA would have been polynomial in N_c instead of N_{sub} that demodulation from $\boldsymbol{u}_g$ in (8.39) can afford.

In addition to reduced complexity demodulation brought by subcarrier grouping, simplifications are also possible at the IFFT stage of the DPS-based MIMO OFDM transmitter. When considering the transmission of each row of the DPS-based SF

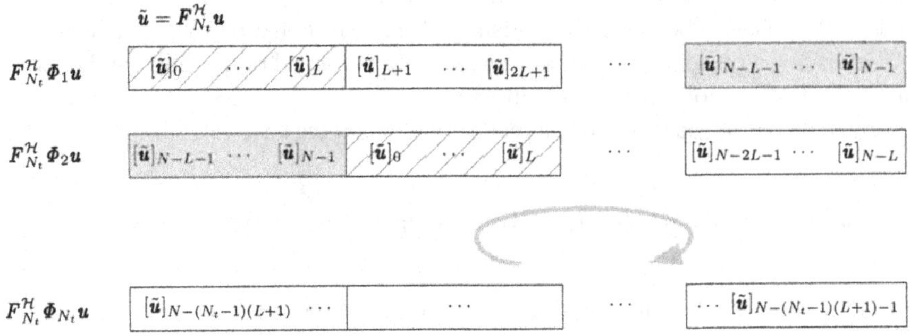

Figure 8.8 Block circular delay

code matrix in (8.43) from the corresponding transmit-antenna, it appears that one needs N_t IFFTs to transmit the N_t rows on the N_c subcarriers. Interestingly, we can show that only one IFFT suffices. To this end, consider the IFFT-processed block $\boldsymbol{F}_{N_c}^{\mathcal{H}} \boldsymbol{\Phi}_\mu \boldsymbol{u}$, corresponding to, say, the μth transmit-antenna, and define the permutation matrix

$$\boldsymbol{P}_\mu := \begin{bmatrix} \boldsymbol{0} & \boldsymbol{I}_{(\mu-1)(L+1)} \\ \boldsymbol{I}_{N-(\mu-1)(L+1)} & \boldsymbol{0} \end{bmatrix}. \tag{8.47}$$

Left-multiplying a column vector by $\boldsymbol{P}_\mu$ circularly delays the vector by $(\mu-1)(L+1)$ entries. Based on this, it is easy to verify the identity

$$\boldsymbol{F}_{N_c}^{\mathcal{H}} \boldsymbol{\Phi}_\mu \boldsymbol{u} = \boldsymbol{P}_\mu \boldsymbol{F}_{N_c}^{\mathcal{H}} \boldsymbol{u}. \tag{8.48}$$

The effect of (8.48) on a DPS-based MIMO OFDM transmitter is depicted in Figure 8.8. From this identity we can readily confirm that only one IFFT is needed at the MIMO transmitter to process $\boldsymbol{u}$ and forward to the N_t antennas circularly delayed versions of the IFFT-processed block $\boldsymbol{F}_{N_c}^{\mathcal{H}} \boldsymbol{u}$.

At this point we should also mention that only a single CP segment of size L suffices for the DPS-based MIMO OFDM system to cope with IBI. Clearly, the block circular delay diversity (BCDD) implemented afterward via the permutation matrix corresponding to each transmit-antenna does not incur additional rate loss. This is in sharp contrast with the generalized delay diversity (GDD) approach in Section 7.2.1 for single-carrier transmissions. Interestingly, both GDD and BCDD convert multiple FIR channels into a single, but longer FIR channel. In GDD, however, a total of $N_t(L+1)-1$ zeros are needed between adjacent blocks to cope with IBI. As a result, the transmission rate in the GDD scheme decreases as N_t increases [cf. (7.17)], while it remains invariant in BCDD regardless of the number of transmit-antennas, thanks to the circular delay operation.

By inspection of the $N_t \times N_c$ code matrix in (8.43), we deduce that the N_c information symbols in $\boldsymbol{s}$ which are used to form the $N_c \times 1$ precoded block $\boldsymbol{u}$ are transmitted over N_c time slots. Accounting also for the CP of size L, we find that the transmission rate in DPS-based MIMO OFDM is

$$\eta_{\text{DPS}} = \frac{N_c}{N_c + L} \qquad \text{symbols per channel use,} \tag{8.49}$$

which is close to one symbol per channel use, since practical systems typically have $N_c \gg L$.

To assess error performance at least for high SNR, let us examine the diversity order that DPS-based MIMO OFDM systems can achieve. The LCFC mapping from $\boldsymbol{s}$ to $\boldsymbol{u}$ in (8.40) is what endows the DPS-based scheme with full diversity. Indeed, if one selects $N_{\text{sub}} \geq N_t(L+1)$, then following the steps of the proof in Section 3.5, it can be shown that the maximum possible diversity of order $N_t N_r(L+1)$ is enabled by DPS [184]. If, on the other hand, the diversity is plenty and effecting it to its full order is not critical, selecting $N_{\text{sub}} < N_t(L+1)$ will trade diversity for reduced complexity.

Besides coherent MIMO OFDM, the DPS operation applied to LCF precoded blocks can enable the full diversity $N_t N_r(L+1)$ in differential MIMO OFDM transmissions as well. This can be established simply by combining two existing results. First, LCF precoding is known to effect the maximum possible multipath diversity in block differentially encoded single-antenna OFDM transmissions [159]. Second, we have seen that DPS converts space diversity to multipath diversity. Hence, applying DPS to LCF precoded and differentially encoded MIMO OFDM blocks enables the full space-multipath diversity of order $N_t N_r(L+1)$.

8.5 FULL-DIVERSITY FULL-RATE MIMO OFDM

So far in this chapter we have focused on error performance-oriented ST, SF or STF coded MIMO OFDM systems approaching rates of 1 symbol per channel use. We have, however, seen in Chapter 4 that upon combining LCF coding with proper ST multiplexing it is possible to design full-diversity full-rate (FDFR) ST coded transmissions over flat fading MIMO channels. Applying this FDFR design on a per subcarrier basis, we will see that rate-oriented MIMO OFDM based systems can be devised to achieve FDFR over frequency-selective MIMO channels. Since only one OFDM symbol will be used ($N_x = 1$), this construction, too, belongs to the class of SF coded MIMO OFDM systems.

8.5.1 Encoders and Decoders

Suppose for convenience that the number of OFDM subcarriers is

$$N_c = N_t(L+1). \tag{8.50}$$

If more subcarriers are needed, we choose N_c to be an integer (say, N_g) multiple of $N_t(L+1)$ and divide subcarriers into N_g groups so that each group contains $N_t(L+1)$ subcarriers. The ensuing steps specifying the encoder and the decoder can then be applied to each group of subcarriers.

Similar to the flat fading case, the FDFR encoder for frequency-selective MIMO channels includes two stages: LCF encoding, followed by ST multiplexing with circularly wrapped layers [cf. (4.10)]. To start, we parse the information symbol stream into blocks $\boldsymbol{s}$, each of size $N_t^2(L+1)$. We further divide every block $\boldsymbol{s}$ into N_t subblocks, $\{\boldsymbol{s}_k\}_{k=1}^{N_t}$, each of size $N_t(L+1)$. Every subblock is precoded with an LCF matrix $\boldsymbol{\Theta}_k$ designed as in (4.11) to obtain the LCF-encoded subblocks

$$\boldsymbol{u}_k = \boldsymbol{\Theta}_k \boldsymbol{s}_k, \qquad \forall k \in [1, N_t]. \tag{8.51}$$

Compared to the FDFR code construction for flat fading MIMO channels, where the precoder has size $N_t \times N_t$, matrix $\boldsymbol{\Theta}_k$ here has a larger size, $N_t(L+1) \times N_t(L+1)$.

The LCF-encoded symbols $u_k(n)$ from (8.51) are ST multiplexed to construct the $N_t \times N_c$ SF code matrix [182]

$$\begin{aligned}\boldsymbol{X} &:= [\boldsymbol{x}(0), \boldsymbol{x}(1), \ldots, \boldsymbol{x}(N_c-1)] \\ &= \begin{bmatrix} u_1(1) \cdots u_2(N_t) & u_1(N_t+1) & \cdots u_2(2N_t) & \cdots u_2(N_c) \\ u_2(1) \cdots u_3(N_t) & u_2(N_t+1) & \cdots u_3(2N_t) & \cdots u_3(N_c) \\ \vdots \quad \cdots \quad \vdots & \vdots & \cdots \quad \vdots & \\ u_{N_t}(1) \cdots u_1(N_t) & u_{N_t}(N_t+1) & \cdots u_1(2N_t) & \cdots u_1(N_c) \end{bmatrix}.\end{aligned} \tag{8.52}$$

Comparing X in (8.52) with the ST code matrix in (4.10), we recognize that the SF code matrix for FDFR MIMO OFDM consists of $L+1$ submatrices of size $N_t \times N_t$ (one per channel tap), each having the same structure as the FDFR code matrix in (4.10) for flat fading MIMO channels. Similar to all the MIMO OFDM designs that we studied in this chapter, every row of $\boldsymbol{X}$ in (8.52) is IFFT processed and a CP of size L is inserted before transmission from the corresponding antenna. Every layer in $\boldsymbol{X}$ is transmitted this way from all N_t transmit-antennas; and per antenna, each layer is spread across at least $L+1$ subcarriers. Intuitively, this form of SF multiplexing the LCF-coded symbols effects the maximum space-multipath diversity that is provided by the frequency-selective MIMO channel.

Corresponding to each column of the matrix $\boldsymbol{X}$, the symbols at the OFDM demodulator output across the N_r receive-antennas will obey the familiar flat fading model per subcarrier: namely, $\boldsymbol{y}(p) = \sqrt{\bar{\gamma}/N_t}\boldsymbol{H}(p)\boldsymbol{x}(p) + \boldsymbol{w}(p)$, where $\boldsymbol{H}(p)$ denotes the frequency response of the channel matrix at subcarrier p. Concatenating such vectors across all the N_c subcarriers, we arrive at the input-output relationship

$$\begin{aligned}\begin{bmatrix} \boldsymbol{y}(0) \\ \vdots \\ \boldsymbol{y}(N_c-1) \end{bmatrix} &= \sqrt{\frac{\bar{\gamma}}{N_t}} \begin{bmatrix} \boldsymbol{H}(0) & & \\ & \ddots & \\ & & \boldsymbol{H}(N_c-1) \end{bmatrix} \begin{bmatrix} \boldsymbol{x}(0) \\ \vdots \\ \boldsymbol{x}(N_c-1) \end{bmatrix} \\ &\quad + \begin{bmatrix} \boldsymbol{w}(0) \\ \vdots \\ \boldsymbol{w}(N_c-1) \end{bmatrix},\end{aligned} \tag{8.53}$$

which for $L = 0$ reduces to its flat fading counterpart in (4.12).

Given the vectors on the left-hand side of (8.53) and supposing that the channel matrices are available at the receiver, the decoder ultimately seeks optimal detection of the information block $\boldsymbol{s}$. Toward this objective, we collect the columns of $\boldsymbol{X}$ in a long vector which, similar to (4.16), can be related to $\boldsymbol{s}$ as

$$\begin{bmatrix} \boldsymbol{x}(0) \\ \vdots \\ \boldsymbol{x}(N_t-1) \\ \boldsymbol{x}(N_t) \\ \vdots \\ \boldsymbol{x}(N_c-1) \end{bmatrix} = \begin{bmatrix} (\boldsymbol{P}_1\boldsymbol{D}_\beta)\otimes\boldsymbol{\theta}_1^T \\ \vdots \\ (\boldsymbol{P}_{N_t}\boldsymbol{D}_\beta)\otimes\boldsymbol{\theta}_{N_t}^T \\ (\boldsymbol{P}_1\boldsymbol{D}_\beta)\otimes\boldsymbol{\theta}_{N_t+1}^T \\ \vdots \\ (\boldsymbol{P}_{N_t}\boldsymbol{D}_\beta)\otimes\boldsymbol{\theta}_{N_c}^T \end{bmatrix} \boldsymbol{s} := \boldsymbol{\Phi}\boldsymbol{s}, \tag{8.54}$$

where $\boldsymbol{P}_\mu$ and $\boldsymbol{D}_\beta$ are defined in (4.14).

Based on (8.54), since the channel and matrix $\boldsymbol{\Phi}$ are known, linear equalizers or nonlinear (near-)ML decoders can be used to recover $\boldsymbol{s}$ from $\{\boldsymbol{y}(p)\}_{p=0}^{N_c-1}$ in (8.53). The SDA of Chapter 5 offers the preferred decoder in a number of cases. As with flat fading MIMO channels, the complexity of SDA depends mainly on the size of $\boldsymbol{s}$, which for frequency-selective MIMO channels is $L+1$ times higher. [Recall that $N_I = N_t^2$ in the flat fading case, whereas $N_I = N_t^2(L+1)$ in the frequency-selective case.]

8.5.2 Diversity and Rate Analysis

Rearranging the per subcarrier received blocks $\{\boldsymbol{y}(p)\}_{p=0}^{N_c-1}$, we can rewrite (8.53) as

$$\begin{aligned}\overline{\boldsymbol{Y}} &:= [\boldsymbol{y}(0), \boldsymbol{y}(1), \ldots, \boldsymbol{y}(N_c-1)] \\ &= \sqrt{\frac{\overline{\gamma}}{N_t}}\overline{\boldsymbol{H}} \underbrace{\begin{bmatrix} \boldsymbol{F}_{1:L+1}\boldsymbol{D}_c^{(1)} \\ \vdots \\ \boldsymbol{F}_{1:L+1}\boldsymbol{D}_c^{(N_t)} \end{bmatrix}}_{:=\overline{\boldsymbol{X}}} + \overline{\boldsymbol{W}}\end{aligned} \tag{8.55}$$

where

$$\begin{aligned}\overline{\boldsymbol{H}} &:= \begin{bmatrix} \boldsymbol{h}_{11}^T & \cdots & \boldsymbol{h}_{1N_t}^T \\ \vdots & \cdots & \vdots \\ \boldsymbol{h}_{N_r1}^T & \cdots & \boldsymbol{h}_{N_rN_t}^T \end{bmatrix}, \\ \boldsymbol{h}_{\nu\mu} &:= [h_{\nu\mu}(0), \ldots, h_{\nu\mu}(L)]^T, \\ \boldsymbol{D}_c^{(\mu)} &:= \text{diag}[x^\mu(0), \cdots, x^\mu(N_c-1)],\end{aligned}$$

and the $(L+1)\times N_c$ matrix $\boldsymbol{F}_{1:L+1}$ is formed by the first $L+1$ rows of the scaled FFT matrix $\sqrt{N_c}\boldsymbol{F}_{N_c}$.

Equation (8.55) describes the input-output relationship of the OFDM-based FDFR multi-antenna system signaling over a frequency-selective MIMO channel. But it could also represent an ST coded system with $N_c = N_t(L+1)$ transmit-antennas signaling with the $N_c \times N_c$ ST code matrix $\overline{\boldsymbol{X}}$ over a flat fading MIMO channel. This input-output equivalence implies that based on our results for flat fading MIMO channels, the FDFR MIMO OFDM transmission will enable the full diversity $N_tN_r(L+1)$, provided that the matrix $\overline{\boldsymbol{X}}$ in (8.55) is designed such that $\det(\overline{\boldsymbol{X}} - \overline{\boldsymbol{X}}') \neq 0, \forall \boldsymbol{s} \neq \boldsymbol{s}'$.

As for the full rate aspect, recall that the code matrix in (8.52) carries $N_t^2(L+1)$ information symbols (size of $\boldsymbol{s}$) in $N_c = N_t(L+1)$ time slots (columns of $\boldsymbol{X}$). Accounting also for the CP of size L added at the OFDM modulator, we find that the FDFR MIMO OFDM system can attain transmission rate

$$\eta_{\text{FDFR}} = \frac{N_t N_c}{N_c + L} \qquad \text{symbols per channel use.} \tag{8.56}$$

Summarizing, and using the analogy with the flat fading MIMO case (cf. Proposition 4.1), we can readily establish the following result [182].

Proposition 8.2 *For any constellation of $\boldsymbol{s}$ carved from $\mathbb{Z}[j]$, with the ST encoder in (8.52), there exists at least one pair of $(\boldsymbol{\Theta}_{\text{LCF}}, \beta)$ in (4.11) which enables full diversity $N_tN_r(L+1)$ for the ST transmission in (8.55), with transmission rate $N_tN_c/(N_c+L)$ symbols per channel use.*

To illustrate the FDFR code matrix and demonstrate the LCFC construction for frequency-selective MIMO channels, it is useful to look at an example.

Example 8.6 (Two-antenna two-ray channels) Let us consider FDFR transmissions over two-ray channels with $(N_t, L) = (2, 1)$, and $N_c = 4$. Matrix $\overline{\boldsymbol{X}}$ in this case becomes

$$\overline{\boldsymbol{X}} = \begin{bmatrix} x^1(1) & x^1(2) & x^1(3) & x^1(4) \\ x^1(1) & jx^1(2) & -x^1(3) & -jx^2(4) \\ x^2(1) & x^2(2) & x^2(3) & x^2(4) \\ x^2(1) & jx^2(2) & -x^2(3) & -jx^2(4) \end{bmatrix}, \tag{8.57}$$

where $x^{\mu}(p)$ denotes the symbol transmitted from the μth antenna on the pth subcarrier. Considering the ST multiplexing matrix in (8.52), with LCF-encoded subblocks

$$\boldsymbol{u}_1 = \boldsymbol{\Theta}_{\text{LCF}} \boldsymbol{s}_1, \qquad \boldsymbol{u}_2 = \beta \boldsymbol{\Theta}_{\text{LCF}} \boldsymbol{s}_2, \tag{8.58}$$

we obtain the determinant of $\overline{\boldsymbol{X}}$ in terms of $\boldsymbol{s}_k$ and $(\boldsymbol{\Theta}_{\text{LCF}}, \beta)$ as

$$\begin{aligned} \det(\overline{\boldsymbol{X}}) = 2j \Bigg[& -2 \prod_{n=1}^{4} \boldsymbol{\theta}_n^T \boldsymbol{s}_1 + \beta^2 \Bigg(\prod_{n=1}^{2} \boldsymbol{\theta}_n^T \boldsymbol{s}_1 \prod_{n=3}^{4} \boldsymbol{\theta}_n^T \boldsymbol{s}_2 \\ & + (\boldsymbol{\theta}_1^T \boldsymbol{s}_1)(\boldsymbol{\theta}_4^T \boldsymbol{s}_1) \prod_{n=2}^{3} \boldsymbol{\theta}_n^T \boldsymbol{s}_2 + (\boldsymbol{\theta}_1^T \boldsymbol{s}_2)(\boldsymbol{\theta}_4^T \boldsymbol{s}_2) \prod_{n=2}^{3} \boldsymbol{\theta}_n^T \boldsymbol{s}_1 \\ & + \prod_{n=1}^{2} \boldsymbol{\theta}_n^T \boldsymbol{s}_2 \prod_{n=3}^{4} \boldsymbol{\theta}_n^T \boldsymbol{s}_1 \Bigg) - 2\beta^4 \prod_{n=1}^{4} \boldsymbol{\theta}_n^T \boldsymbol{s}_2 \Bigg]. \end{aligned} \tag{8.59}$$

If $s_k \in \mathbb{Z}[j]$, then (8.59) is a polynomial in β^{N_t} with coefficients in $\mathbb{Q}(j)(e^{j2\pi/N_t})(\alpha)$. To guarantee that all coefficients of β^{N_t} in $\det(\overline{\boldsymbol{X}})$ are nonzero $\forall \boldsymbol{s} \neq \boldsymbol{s}'$, we design $\boldsymbol{\Theta}_{\text{LCF}}$ according to the guidelines in Section 3.5. For this example we select $\boldsymbol{\Theta}_{\text{LCF}}$ as

$$\boldsymbol{\Theta}_{\text{LCF}} = \frac{1}{2}\begin{bmatrix} 1 & e^{j\pi/8} & e^{j2\pi/8} & e^{j3\pi/8} \\ 1 & e^{j5\pi/8} & e^{j10\pi/8} & e^{j15\pi/8} \\ 1 & e^{j9\pi/8} & e^{j18\pi/8} & e^{j27\pi/8} \\ 1 & e^{j13\pi/8} & e^{j26\pi/8} & e^{j39\pi/8} \end{bmatrix}. \tag{8.60}$$

With the LCFC matrix $\boldsymbol{\Theta}_{\text{LCF}}$ fixed, we can view $\det(\overline{\boldsymbol{X}})$ as a polynomial in β^{N_t} with coefficients in $\mathbb{Q}(\alpha)$. Using our design A in Chapter 4, we can select $\beta^{N_t} = e^{j\pi/32}$, such that $\det(\overline{\boldsymbol{X}}) \neq 0, \forall \boldsymbol{s} \neq \boldsymbol{s}'$.

Besides the transmission rate asserted by (8.56) and Proposition 8.2, we will complement the rate analysis with the mutual information for the OFDM-based FDFR system based on the input-output relationship (8.53).

Corollary 8.1 *If the information symbols $\boldsymbol{s} \sim \mathcal{CN}(\boldsymbol{0}, \boldsymbol{I}_{N_I})$, the mutual information of FDFR transmissions over frequency-selective MIMO OFDM channels is*

$$I_{\text{FDFR}}(\boldsymbol{X}; \boldsymbol{Y} | \{\boldsymbol{H}(p)\}_{p=0}^{N_c-1}) = \frac{1}{N_c + L} \sum_{p=0}^{N_c-1} \log_2 \det\left(\boldsymbol{I}_{N_r} + \frac{\overline{\gamma}}{N_t}\boldsymbol{H}(p)\boldsymbol{H}^{\mathcal{H}}(p)\right). \tag{8.61}$$

Comparing (8.61) with the mutual information in (7.12) or (7.15), we infer that a coded FDFR system can approach the capacity provided that the number of subcarriers N_c is sufficiently large. This makes the FDFR structure for MIMO OFDM very attractive for high-rate applications.

We wrap up this section with a word on rate-diversity-complexity trade-offs. Relative to flat fading MIMO channels, the diversity provided by frequency-selective MIMO channels is certainly higher but comes at the price of higher decoding complexity. We have seen in Chapter 4 how flexible the LCFC and multiplexing modules of the FDFR system can be to trade-off among rate, diversity, and complexity in the flat fading case. Similar designs for the frequency-selective MIMO case are equally important in practice and can be devised for the OFDM-based FDFR system by following the guidelines of Section 4.3.4 on a per subcarrier basis.

8.5.3 Numerical Examples

Let us consider a multi-antenna wireless system signaling over a frequency-selective MIMO channel with parameters $(N_t, N_r, L) = (2, 2, 1)$. The channel taps are uncorrelated, Rayleigh distributed, and for each channel the power across taps satisfies an exponentially decaying profile [182]. We compare the FDFR scheme for MIMO OFDM against two existing STF codes: VBLAST-OFDM [211] and the GSTFB scheme presented in Section 8.3.2. The block size is set to $N_c = N_t(L + 1) = 4$.

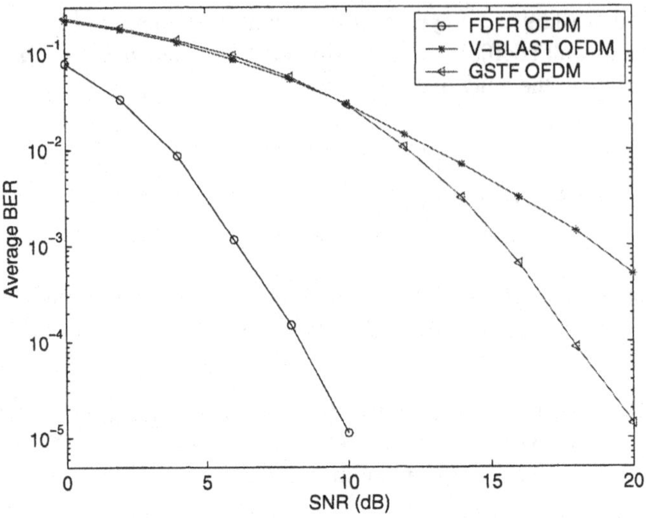

Figure 8.9 FDFR over frequency-selective MIMO channels

To fix the transmission rate at $R = 8/3$ bits per channel use, we use QPSK for the FDFR and VBLAST-OFDM schemes and 16-QAM for the GSTF scheme. At the receiver, we use the SDA for all three schemes. Figure 8.9 depicts the simulated error performance. Reading from the slopes of the BER curves, we infer that both FDFR-OFDM and GSTF achieve full diversity, whereas VBLAST-OFDM can attain only the receive-antenna diversity N_r. The big gap between FDFR and GSTF is due to the smaller constellation size that FDFR can afford. Notice that SDA's complexity of FDFR is $\mathcal{O}((N_t^2(L+1))^3)$, while for VBLAST-OFDM it is $\mathcal{O}(N_t^3)$ and for GSTF, $\mathcal{O}((L+1)^3)$.

8.6 CLOSING COMMENTS

In this chapter we outlined the basic operation and features of MIMO OFDM and presented a general space-time-frequency framework for coded MIMO OFDM systems. Depending on the dimension(s) across which coding takes place, we developed various coded transmission formats which fall under the categories of space-time, space-frequency or space-time-frequency, in general. From the STF block codes and the available trellis codes for frequency-selective MIMO channels, only the GSTFC scheme with two transmit-antennas exhibits desirable features in rate, diversity, and complexity. For any number of antennas, the DPS-based MIMO OFDM system is more attractive since it can effect full space-multipath diversity at reasonably low complexity and rates approaching 1 symbol per channel use. For high-rate applications, full-diversity $N_t N_r(L + 1)$ at full-rate N_t symbols per channel use was shown possible by generalizing the FDFR system of Chapter 4 to frequency-selective

MIMO channels. This chapter dealt with coherent MIMO-OFDM systems which rely on channel knowledge at the receiver side for data decoding. Differential and non-coherent ST coded MIMO-OFDM systems may be found, for example, in [26, 176] and references therein.

At this point it is of interest to comment on the advantages and limitations of the single-carrier systems in Chapter 7 in comparison with the multi-carrier OFDM-based systems of this chapter. Since OFDM in its discrete-time baseband equivalent model can be viewed either as a block multi-carrier transmission system or as a linearly precoded (with an IFFT precoder matrix) single-carrier block transmission system, there are no major differences in the model and the metrics used for their performance analysis. Both can attain the maximum space-multipath diversity and transmission rate; both can be cast in flat fading MIMO equivalent model forms and thus borrow some of the code design approaches from Chapter 3; and the rate outage probability can be used to benchmark the performance of both.

Differences originate from their practical implementation, and except for differences related to dimensionality, the pros and cons between these two classes of MIMO systems are inherited from their single-antenna counterparts. For a recent study comparing single-antenna OFDM with single-antenna single-carrier block transmission systems, we refer the reader to [291] and references therein. Briefly, their relative merits can be compared in the following four categories:

- *Complexity*: OFDM converts a frequency-selective channel to a set of flat fading subchannels, which can reduce receiver complexity considerably. In fact, with an IFFT at the transmitter and an FFT at the receiver, OFDM balances the complexity between the two ends: whereas a single-carrier block transmission has a simpler transmitter, where no IFFT is needed, but a more complex receiver, since if one wishes to have scalar equalization, two FFTs are needed at the receiver. Of course, differences in receiver complexity are more pronounced as the channel length increases. Besides equalization, OFDM also reduces complexity at the decoding module. Especially for joint equalization and channel decoding, single-carrier block systems with iterative (turbo) receivers tend to be more complex than their multi-carrier counterparts.

- *Peak to average power ratio*: As we explained in Section 8.1, IFFT processing at the OFDM modulator can increase the modulus of an information symbol by a factor as high as the number of subcarriers. To avoid nonlinear distortion effects at the power amplification stage, this necessitates backing-off the power, which renders amplification inefficient. For this reason, PAR-reduction techniques have to be incorporated in OFDM systems, which is not necessary for single-carrier systems.

- *Sensitivity to carrier-frequency offsets*: We also mentioned in Section 8.1 that OFDM is challenged by subcarrier drifts arising due to Doppler and mismatch between transmit-receive oscillators, both of which cause carrier-frequency offsets. Single-carrier systems are more robust to carrier-frequency offsets than are OFDM-based systems which by design exhibit performance sensitive to the resulting intercarrier interference [291]. By time-frequency duality, one would expect that single-carrier systems would be more sensitive to timing offsets than OFDM-based systems. The effect of timing offsets is relative to symbol duration as OFDM's sensitivity to frequency offsets is relative to subcarrier spacing.

- *Diversity*: Since uncoded OFDM does not enable the available multipath diversity in either single- or multi-antenna MIMO links, channel coding and/or linear precoding are absolutely necessary in wireless multi-carrier systems. Interestingly, single-carrier block transmissions can achieve multipath diversity without channel coding but with linear precoding as simple as padding zeros to insert a guard interval per block [291]. With channel coding, the comparative study in [291] suggests further that single-carrier systems may be preferable over multi-carrier systems when the code rate is high, and vice versa when the coding rate is low.

Both single- and multi-carrier approaches have been implemented in practical systems. Single-carrier modulations are commonly used over channels with a small number of taps. In the third-generation TDMA cellular standard EDGE (Enhanced Date Rates for GSM Evolution) [73], one can effectively model the symbol-rate discrete-time channel with four taps. On the other hand, multi-carrier modulation is typically invoked for transmission over channels with a large number of taps. For example, the HiperLAN/2 channel [63] has nine (16) taps in the symbol-rate discrete-time representation of its channel model A (B); see also [291, Table III]. Multi-carrier transmissions are also particularly useful when the channel length cannot be known beforehand, as in a cable TV broadcasting system, where a subscriber receives the same information from multiple sources but with different delays [223].

Whether single- or multi-carrier, the ST codes we designed so far for frequency-flat or frequency-selective MIMO channels rely on the time-invariance assumption; or realistically assume that the channel coherence interval is sufficiently long so that time variations are negligible within the block sizes considered. Clearly, this assumption becomes increasingly invalid as the mobility among wireless communicators increases. In the ensuing chapter we design ST codes for time-varying MIMO channels.

9

ST Codes for Time-Varying Channels

Whether frequency-flat or frequency-selective, the MIMO channels considered in previous chapters were assumed time invariant. On the other hand, mobility gives rise to time-varying (TV) channels which may vary too fast for the receiver to track and motivate designs that account explicitly for time variations. Besides mobility, variations in the MIMO channel arise also due to drifts in the phase and frequency mismatch between oscillators at the transmitter and receiver. Certainly, using the synchronization algorithms presented in Chapter 11, one can partially mitigate the mismatch between transmit-receive oscillators. But even after synchronization, residual phase and carrier frequency offsets are present and may degrade severely the performance of coherent as well as differentially ST coded systems.

These reasons justify the need to design ST codes for TV MIMO channels, which is the main theme of the present chapter. Critical to the design of such codes is a basis expansion model (BEM) that will be adopted to describe TV multi-antenna links. Assigning time variations to known Fourier bases, this model describes TV channels with (typically a few) parameters, the number of which depend on the sampling period as well as on the channel's delay and Doppler spread. In its parsimony, the BEM is more suitable than the popular Jakes model when it comes to designing ST codes and quantifying their diversity and coding gains. Furthermore, for single-carrier and multi-carrier block transmissions, the BEM leads to a neat duality between frequency- and time-selective MIMO models.

Based on this duality, it will become possible to transform ST block codes presented in Chapters 7 and 8 for frequency-selective channels, to obtain dual ST codes for use in time-selective MIMO channels. Specific designs will include zero padding (ZP), cyclic prefix (CP), and digital-phase sweeping (DPS) based ST codes, enabling the joint space-time Doppler (STDO) diversity provided by a time-selective MIMO channel. At rates not exceeding 1 symbol per channel use, these block STDO codes will be compared in terms of their complexity, diversity, and coding gains, and relative to the ST trellis codes designed for time-selective MIMO channels. A full-diversity full-rate (FDFR) STDO design will also be devised as the dual of the FDFR MIMO OFDM system presented in Section 8.5.

Beyond coherent transceivers, in this chapter we deal further with block differential STDO codes for time-selective MIMO channels. Based on the BEM, DPS-based ST codes will finally be presented for a very general class of TV channels: those exhibiting both time and frequency selectivity. This class of doubly selective channels is the most challenging but also the most rewarding in terms of degrees of freedom, and thus diversity, which is given by the product of the number of transmit-receive antennas times the number of BEM parameters.

9.1 TIME-VARYING CHANNELS

Transmissions over the wireless interface experience multipath propagation, due to which a number of reflected or scattered rays arrive at the receiving end coherently or noncoherently; see also Section 1.2 and [17, 214]. Each of the rays is characterized by an attenuation, a phase shift, and a time delay. The time-varying (TV) impulse response of the physical channel at baseband can be expressed generally as the superposition of clusters of rays arriving from different paths as

$$c^{(\mathrm{ch})}(t;\tau) = \sum_{p} \alpha_p(t) e^{j\theta_p(t)} \delta(\tau - \tau_p(t)), \tag{9.1}$$

where $\delta(\cdot)$ denotes Dirac's delta function and the index p denotes the pth cluster of rays (pth "path") arriving coherently with a common delay $\tau_p(t)$, amplitude $\alpha_p(t)$, and phase $\theta_p(t)$. The aggregation of rays in each cluster can be written as (see, e.g., [40, Chapter 3])

$$\alpha_p(t) e^{j\theta_p(t)} = \sum_{m} a_{m,p}(t) e^{j\phi_{m,p}(t)} e^{j2\pi f_{m,p}(t) t}, \tag{9.2}$$

where $f_{m,p}(t)$ denotes a possible frequency offset (capturing Doppler and/or transmit-receive oscillator drifts), $\phi_{m,p}(t) \in [-\pi, \pi)$ is the phase, and $a_{m,p}(t)$ is the amplitude of the mth ray in the pth cluster with delay $\tau_p(t)$.

The overall channel is defined as the convolution of the transmit filter $c^{(\mathrm{tr})}(t)$, the receive filter $c^{(\mathrm{rec})}(t)$, and the TV physical channel $c^{(\mathrm{ch})}(t;\tau)$. Since these three filters are linear, the continuous-time baseband equivalent channel can be written as

$$c(t;\tau) = \int_{-\infty}^{\infty} \int_{-\infty}^{\infty} c^{(\mathrm{tr})}(\tau - u - s) c^{(\mathrm{rec})}(s) c^{(\mathrm{ch})}(t - s; u)\, ds\, du. \tag{9.3}$$

Depending on how fast $c(t;\tau)$ changes in t and τ relative to the symbol period, the carrier frequency, and the channel coherence time, (9.3) can afford various simplifications which give rise to different channel models.

9.1.1 Channel Models

Suppose that the channel variation is relatively slow over the nonzero support of the receive filter which as the transmit filter is assumed to span 1 symbol period. Formally written, this means that:

A9.1) The TV impulse response $c^{(\mathrm{ch})}(t;u)$ remains invariant in t during the span of the receive-filter $c^{(\mathrm{rec})}(t)$.

Note that A9.1 is a practical assumption since the time span of the receive filter is typically short compared with the channel coherence time. If, on the other hand, the channel variations are extremely fast, the channel may exhibit both time and frequency selectivity, even if the delay spread is small.

Under A9.1, the overall impulse response in (9.3) can be written as

$$c(t;\tau) = \int_{-\infty}^{\infty} c^{(\mathrm{ch})}(t;u)\left(\int_{-\infty}^{\infty} c^{(\mathrm{tr})}(\tau - u - s)c^{(\mathrm{rec})}(s)\,ds\right) du.$$

With the definition $\psi(t) := (c^{(\mathrm{tr})} \star c^{(\mathrm{rec})})(t)$, the baseband equivalent channel can be represented by

$$c(t;\tau) = \int_{-\infty}^{\infty} c^{(\mathrm{ch})}(t;u)\psi(\tau - u)\,du. \tag{9.4}$$

Plugging (9.1) and (9.2) into (9.4), we arrive at the following continuous-time baseband equivalent channel [cf. (9.1)]

$$\begin{aligned} c(t;\tau) &= \sum_p \psi(\tau - \tau_p(t)) \sum_m a_{m,p}(t)e^{j\phi_{m,p}(t)}e^{j2\pi f_{m,p}(t)t} \\ &= \sum_p \psi(\tau - \tau_p(t))\alpha_p(t)e^{j\theta_p(t)}. \end{aligned} \tag{9.5}$$

Relative to the symbol duration, each ray's delay, amplitude, phase, and frequency offset in (9.2) change slowly as functions of time, which justifies the following assumption:

A9.2) With a sampling period T_s seconds chosen equal to the symbol period, the functional forms of $\tau_p(t)$, $a_{m,p}(t)$, $\phi_{m,p}(t)$, and $f_{m,p}(t)$ remain invariant over N symbol periods (i.e., over NT_s seconds).

Throughout this chapter, T_s will denote both the sampling period and the symbol period. Notice that A9.2 allows the channel parameters to vary across blocks of duration NT_s seconds. Without loss of generality, we henceforth consider transmissions in bursts (blocks) of duration $[0, NT_s)$ and drop the time index t (e.g., the parameters will be denoted as τ_p, $a_{m,p}$, $\phi_{m,p}$, and $f_{m,p}$).

Two important special cases of the channel model in (9.5) include:

Case 1: Frequency-Selective Channels Suppose that $\alpha_p(t)$ and $\theta_p(t)$ in (9.1) are time invariant over the kth block period of NT_s seconds. In this case, (9.5) reduces to the time-invariant form

$$c(t;\tau) = \sum_p \psi(\tau - \tau_p)\alpha_p e^{j\theta_p}, \quad \forall t \in [0, NT_s).$$

If the delays τ_p are comparable with or greater than the symbol period T_s, the sum in (9.1) contains multiple delayed deltas and the Fourier transform (from τ to f) of $c(t;\tau) \equiv c(\tau)$ is not constant in the frequency domain (i.e., the channel response exhibits frequency selectivity).

Case 2: Time-Selective Channels In this case, only one path (corresponding to $p = 0$) is discernible from the sum in (9.1) (i.e., all scattered rays arrive at the receiver almost simultaneously as a single cluster with common propagation delay τ_0). Selecting $\psi(\tau - \tau_0)$ to be a Nyquist pulse shaper (i.e., $\psi(nT_s - \tau_0) = 1$ if $n = 0$; and 0 otherwise), we can simplify (9.5) to

$$c(t;\tau_0) = \sum_m a_{m,0} e^{j\phi_{m,0}} e^{j2\pi f_{m,0} t}, \quad \forall t \in [0, NT_s). \tag{9.6}$$

If $\omega_m := 2\pi f_{m,0}$ and m takes on a finite number of values, (9.6) coincides with the deterministic basis expansion model (BEM) in [92].

It is of interest to relate (9.6) with the Jakes model which has been widely used to characterize rapidly fading wireless channels. Toward this objective, we define the maximum Doppler shift (spread) as $f_{\max} := v_{\max} f_c / v_{\text{light}}$, where $v_{\max}$ is the maximum relative velocity between the transmitter and the receiver, v_{light} is the speed of light, and f_c stands for the carrier frequency. With β_m denoting uniformly distributed random variables over $[-\pi, \pi)$, we let $f_m := f_{\max}\cos(\beta_m)$ model omnidirectional arrivals of the rays at the receiver end. Then (9.6) reduces to the Jakes model [127, Page 65]:

$$h_J(t) = \sum_m a_m e^{j\phi_m} e^{j2\pi f_{\max}\cos(\beta_m)t}, \quad \forall t \in [0, NT_s). \tag{9.7}$$

If the scattering is rich, the number of paths in (9.7) can be large. Albeit useful for performance analysis studies, the Jakes model presents formidable challenges to channel estimation because it entails a prohibitively large (theoretically infinite) number of parameters. These challenges are avoided by a truncated basis expansion model; this is what we abbreviate as BEM and derive in the remainder of this section.

Consider block transmissions with each block containing N symbols, transmitted over channels with maximum Doppler spread $f_{\max}$. Since $f_{\max}$ does not depend on the cluster index p or ray index m, $f_{\max}$ is the maximum of all $f_{m,p}$'s. The Fourier transform (from t to f) of $c(t;\tau)$ in (9.5) over the time interval $[0, NT_s)$ is

$$C(f;\tau) = \sum_p \psi(\tau - \tau_p) \sum_m a_{m,p} NT_s e^{j\phi_{m,p}} e^{-j\pi NT_s(f - f_{m,p})}$$
$$\times \operatorname{sinc}(NT_s(f - f_{m,p})),$$

where $\text{sinc}(x) := (\sin \pi x)/\pi x$. For the block of N symbols which is time limited (NT_s seconds), we sample $C(f;\tau)$ in the frequency domain with period $1/NT_s$, to obtain

$$C\left(\frac{q}{NT_s};\tau\right) = \sum_p \psi(\tau - \tau_p) \sum_m a_{m,p} NT_s e^{j\phi_{m,p}} e^{j\pi(NT_s f_{m,p} - q)} \times \text{sinc}\left(q - NT_s f_{m,p}\right), \quad q \in (-\infty, \infty).$$

Notice that $C(f;\tau)$ is a superposition of sinc functions each with a different frequency shift. Although the bandwidth of $c(t;\tau)$ is theoretically infinite, we practically have that $C(f;\tau) \approx 0$, for $f \notin [-f_{\max}, f_{\max}]$ (see [127]); hence, $C(q/(NT_s);\tau) = 0$ for $q \notin [-Q/2, Q/2]$, where

$$Q := 2\lceil f_{\max} T_s N \rceil, \quad f_{\max} \geq \max_{p,m} f_{m,p}. \tag{9.8}$$

To proceed with the BEM derivation, we impose the following design condition, which will turn out to affect the Doppler-induced diversity order:

C9.1) With $f_{\max}$ and T_s given, we choose N such that $2 f_{\max} N T_s \geq 1$.

Thanks to C9.1, there are at least $Q \geq 1$ sinc samples in $\{C(q/(NT_s);\tau)\}_{q=-Q/2}^{Q/2}$. We will see soon that Q also dictates the number of bases in the BEM.

Restricting the frequency-domain samples over $[-Q/2, Q/2]$ and taking the inverse Fourier transform, we obtain the TV impulse response model:

$$c_B(t;\tau) = \sum_{q=-Q/2}^{Q/2} \left[\sum_p \psi(\tau - \tau_p) \sum_m a_{m,p} NT_s e^{j\phi_{m,p}} e^{j\pi(NT_s f_{m,p} - q)} \times \text{sinc}\left(q - NT_s f_{m,p}\right) \right] e^{j2\pi q t/NT_s}.$$

As in every model, errors emerge when fitting $c(t;\tau)$ with $c_B(t;\tau)$. Sampling $c_B(t;\tau)$ in t and τ with sampling period T_s, we are led naturally to a discrete-time BEM with $Q+1$ complex exponential bases $\{e^{j2\pi q/N}\}_{q=-Q/2}^{Q/2}$; that is,

$$h_B(n;l) := c_B(nT_s; lT_s),$$

where n is the sample index, $l \in [0, L]$, and L is the discrete-time equivalent channel order, which is defined as

$$L = \lceil \tau_{\max}/T_s \rceil, \quad \tau_{\max} := \max_p \{\tau_p\}. \tag{9.9}$$

In compact form, the BEM for *doubly selective* fading channels is given by

$$h(n;l) = \sum_{q=0}^{Q} h_q(l) e^{j\omega_q n}, \quad l \in [0, L], \tag{9.10}$$

where $\omega_q := 2\pi(q - Q/2)/N$, and

$$h_q(l) := \sum_p \psi(lT_s - \tau_p) \sum_m - a_{m,p} NT_s e^{j\phi_{m,p}} e^{j\pi(NT_s f_{m,p} - (q-Q/2))} \times \mathrm{sinc}\left(q - Q/2 - NT_s f_{m,p}\right).$$

The parsimonious representation of general TV channels offered by the finite number of BEM parameters is instrumental in at least three directions: (i) estimation of TV channels for equalization and demodulation at the receiver, (ii) quantification of the diversity order that TV channels can provide, and (iii) design of ST codes for TV channels. Notwithstanding, none of these tasks can be accomplished with the infinitely parameterized Jakes model.

If the paths arrive at the receiver almost simultaneously (i.e., $\tau_{\max} \ll T_s$), the doubly selective fading channel becomes a time-selective frequency-flat channel ($L = 0$) and the BEM reduces to

$$h(n) := \sum_{q=0}^{Q} h_q e^{j\omega_q n}, \tag{9.11}$$

where $h_q := h_q(0)$.

Next, we show that the BEM is as important for time-selective channels as the tapped-delay line is for frequency-selective channels.

9.1.2 Time-Frequency Duality

Let $x(n)$ denote the nth transmitted symbol and $y(n)$ the corresponding received symbol. For a time-selective frequency-flat channel, the input-output relationship is

$$y(n) = \sqrt{\bar{\gamma}}\, h(n)x(n) + w(n), \quad \forall n \in [0, N-1], \tag{9.12}$$

where $w(n)$, as usual denotes AWGN and $\bar{\gamma}$ is the average SNR. Collecting N symbols, we can write the matrix-vector counterpart of (9.12) as

$$\boldsymbol{y} = \sqrt{\bar{\gamma}}\, \boldsymbol{D}_h \boldsymbol{x} + \boldsymbol{w}, \tag{9.13}$$

where $\boldsymbol{y} := [y(0) \cdots y(N-1)]^T$, $\boldsymbol{x} := [x(0) \cdots x(N-1)]^T$, $\boldsymbol{w} := [w(0) \cdots w(N-1)]^T$, and $\boldsymbol{D}_h := \mathrm{diag}[h(0) \cdots h(N-1)]$.

It is well known that circulant matrices can be diagonalized by (I)FFT matrices [99, Page 202]. Using this property and recalling that the BEM in (9.11) has its basis on the FFT grid, we can rewrite $\boldsymbol{D}_h$ as

$$\boldsymbol{D}_h = \sum_{q=0}^{Q} h_q \boldsymbol{D}_q = \boldsymbol{F}_N \boldsymbol{H} \boldsymbol{F}_N^{\mathcal{H}}, \tag{9.14}$$

where $\boldsymbol{D}_q := \mathrm{diag}[1, e^{j\omega_q}, \ldots, e^{j\omega_q(N-1)}]$, $\boldsymbol{H}$ is an $N \times N$ circulant matrix with first column $[h_{Q/2} \ \cdots \ h_0 \ 0 \ \cdots \ 0 \ h_Q \ \cdots \ h_{Q/2+1}]^T$, and $\boldsymbol{F}_N$ denotes the N-point normalized FFT matrix. With FFT processing at the transmitter and IFFT processing at the receiver, (9.13) can be rewritten as

$$\bar{\boldsymbol{y}} = \sqrt{\bar{\gamma}}\boldsymbol{F}^{\mathcal{H}}\boldsymbol{D}_h\boldsymbol{F}\bar{\boldsymbol{x}} + \bar{\boldsymbol{w}} = \sqrt{\bar{\gamma}}\boldsymbol{H}\bar{\boldsymbol{x}} + \bar{\boldsymbol{w}}, \tag{9.15}$$

where $\bar{\boldsymbol{y}} := \boldsymbol{F}^{\mathcal{H}}\boldsymbol{y}$, $\bar{\boldsymbol{x}} := \boldsymbol{F}^{\mathcal{H}}\boldsymbol{x}$, and $\bar{\boldsymbol{w}} := \boldsymbol{F}^{\mathcal{H}}\boldsymbol{w}$. We have already seen in Chapters 7 and 8 that for block transmissions over frequency-selective channels, one can insert (at the transmitter) and remove (at the receiver) a cyclic prefix (CP) to render the channel equivalent to a circulant matrix. Then the circulant matrix can be diagonalized by FFT and IFFT operations [cf. (8.4)]. Equations (9.14) and (9.15) suggest the converse direction: thanks to the BEM, it is possible to convert the diagonal time-selective channel $\boldsymbol{D}_h$ to a circulant matrix after IFFT and FFT operations. The $Q+1$ BEM coefficients are dual to the $L+1$ channel taps of a frequency-selective channel. We summarize this time-frequency duality between channel models in the following:

Property 9.1 *Based on the BEM in* (9.11), *a block transmission over time-selective channels can be viewed equivalently as a transmission over frequency-selective channels after FFT processing at the transmitter and IFFT processing at the receiver. An equivalent input-output relationship involving a diagonal (or circulant) channel matrix can come from either a time-selective or a frequency-selective channel with appropriate (I)FFT operations.*

The usefulness of the duality asserted by Property 9.1 is that available results regarding block transmission systems derived for frequency-selective channels can be directly mapped to obtain corresponding results for time-selective channels, and vice versa. In particular, Property 9.1 will allow us to design ST codes for time- and doubly selective MIMO channels starting from codes that we designed in Chapters 7 and 8 for frequency-selective MIMO channels.

Generalizing Property 9.1 to doubly selective channels with TV impulse response modeled as in (9.10), we have that:

Property 9.2 *A time- and frequency-selective channel with $Q+1$ bases and $L+1$ taps can be viewed as a dual time- and frequency-selective channel with $L+1$ bases and $Q+1$ taps.*

9.1.3 Doppler Diversity

Based on the duality offered by the BEM, one can quantify the diversity order provided by time- and doubly selective channels by mapping results from frequency-selective channels. Indeed, recalling (9.15) and Proposition 7.1 established for the model in (7.8), it follows readily from the duality that:

Proposition 9.1 *If the channel coefficients $\{h_q\}_{q=0}^{Q}$ in* (9.11) *are complex Gaussian distributed and their correlation matrix $\boldsymbol{R}_h := E[\boldsymbol{h}\boldsymbol{h}^{\mathcal{H}}]$ with $\boldsymbol{h} := [h_0, \ldots, h_Q]^T$ has rank r_h, the Doppler diversity order provided by the time-selective channel is*

$G_d = r_h$. When $\boldsymbol{R}_h$ has full rank equal to its dimension (i.e., $r_h = Q+1$), the maximum diversity order is $G_{d,\max} = Q+1$ (i.e., the number of bases in the BEM determines the maximum possible diversity order).

When the delay spread of a TV channel is greater than or comparable with the sampling period, the channel also shows frequency selectivity. In this case, based on the channel model in (9.10), the input-output relationship is given by

$$y(n) = \sqrt{\bar{\gamma}} \sum_{l=0}^{L} h(n;l)x(n-l) + w(n). \tag{9.16}$$

If L zeros are padded after every block of N symbols, interblock interference can be avoided and the last input-output relationship can be written in matrix-vector form as

$$\boldsymbol{y} = \sqrt{\bar{\gamma}}\boldsymbol{H}\boldsymbol{x} + \boldsymbol{w}, \tag{9.17}$$

where

$$\boldsymbol{H} = \sum_{q=0}^{Q} \boldsymbol{D}_q \boldsymbol{H}_q,$$

with $\boldsymbol{D}_q := \text{diag}[1, \exp(j\omega_q), \ldots, \exp(j\omega_q(N-1))]$ and $\boldsymbol{H}_q$ a lower-triangular Toeplitz matrix with first column $[h_q(0), \ldots, h_q(L), 0, \ldots, 0]^T$. Upon defining the impulse response vector for doubly selective channels as

$$\boldsymbol{h} := [h_0(0), \ldots, h_Q(0), \ldots, h_Q(L)]^T \in \mathbb{C}^{(L+1)(Q+1)}, \tag{9.18}$$

we can argue as in Proposition 9.1 to establish that:

Proposition 9.2 *If the channel coefficients $\{h_q(l)\}$ in* (9.10) *are Gaussian distributed and the correlation matrix $\boldsymbol{R}_h := E[\boldsymbol{h}\boldsymbol{h}^{\mathcal{H}}]$ of the channel vector in* (9.18) *has rank r_h, the maximum diversity order (a.k.a. multipath-Doppler diversity order) that the doubly selective channel in* (9.10) *can provide is $G_d = r_h$. When $\boldsymbol{R}_h$ has full rank $r_h = (L+1)(Q+1)$, the maximum diversity gain is $G_{d,\max} = (L+1)(Q+1)$.*

It is worth stressing that the multipath-Doppler diversity provided by a doubly selective channel is the product of the Doppler diversity order times the multipath diversity order. In what follows we describe multi-antenna systems capable of collecting the diversity provided by TV channels as asserted by Propositions 9.1 and 9.2.

9.2 SPACE-TIME-DOPPLER BLOCK CODES

In this section we consider the design of ST codes for multi-antenna transmissions over time-selective but frequency-flat channels. Suppose that the channel from the

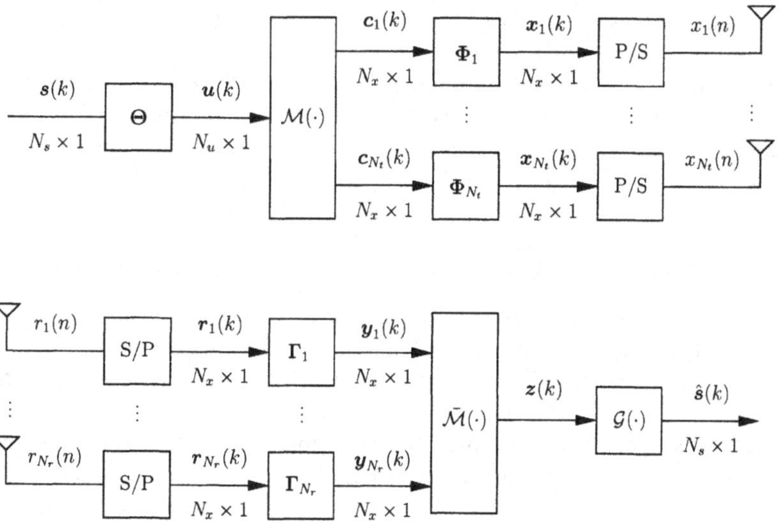

Figure 9.1 Unified discrete-time model for STDO designs

μth transmit-antenna to the νth receive-antenna is modeled using the BEM as [cf. (9.11)]

$$h_{\nu\mu}(n) = \sum_{q=0}^{Q} h_q^{(\nu,\mu)} e^{j\omega_q n}, \tag{9.19}$$

where the channel coefficients $h_q^{(\nu,\mu)}$ are known at the receiver but not at the transmitter, and remain invariant over blocks containing N symbols.

The BEM-based duality will enable us to map the ST codes from Chapters 7 and 8 to time-selective MIMO channels. Among the various ST codes in Chapters 7 and 8, we pursue those based on LCFC. The reason is that they are the only ones capable of effecting the maximum diversity available at reasonable complexity without major penalty in transmission rate. A unified discrete-time transmitter and receiver structure that we will use to construct these LCFC-based designs for time-selective MIMO channels is shown in Figure 9.1. The information-bearing symbols $\{s(n)\}$ are drawn from a finite alphabet $\mathbb{S}$ and grouped in blocks of size $N_s \times 1$: $\boldsymbol{s}(k) := [s(kN_s), \ldots, s((k+1)N_s - 1)]^T$. Each block $\boldsymbol{s}(k)$ is linearly precoded by the $N_u \times N_s$ matrix $\boldsymbol{\Theta}$, resulting in $\boldsymbol{u}(k) := \boldsymbol{\Theta}\boldsymbol{s}(k)$. This operation will be termed the outer space-time-Doppler (STDO) encoder. Each block $\boldsymbol{u}(k)$ is further transformed into N_t blocks $\{\mathbf{c}_\mu(k)\}_{\mu=1}^{N_t}$, each of size $N_x \times 1$ by a mapper $\mathcal{M}(\cdot)$: $\{\mathbf{c}_\mu(k)\}_{\mu=1}^{N_t} := \mathcal{M}(\boldsymbol{u}(k))$. This operation will be termed the middle STDO encoder. Each block $\boldsymbol{c}_\mu(k)$ is finally linearly processed by the $N_x \times N_x$ matrix $\boldsymbol{\Phi}_\mu$, resulting in $\boldsymbol{x}_\mu(k) := \boldsymbol{\Phi}_\mu \boldsymbol{c}_\mu(k)$, which is transmitted from the μth antenna. This operation will be termed the inner STDO encoder. Not all STDO designs rely on all three (outer, middle, inner) stages of the unified structure in Figure 9.1. If one stage (e.g., the inner stage) is inactive, we simply set $\boldsymbol{\Phi}_\mu = \boldsymbol{I}_\mu$.

The sequence $x_\mu(n)$ obtained after parallel-to-serial (P/S) conversion of the blocks $\{\boldsymbol{x}_\mu(k)\}$ is then pulse-shaped, carrier modulated, and transmitted from the μth transmit-antenna. The nth sample at the νth antenna's receive-filter output is

$$r_\nu(n) = \sqrt{\frac{\bar{\gamma}}{N_t}} \sum_{\mu=1}^{N_t} h_{\nu\mu}(n) x_\mu(n) + w_\nu(n), \quad \forall \nu \in [1, N_r], \tag{9.20}$$

where $h_{\nu\mu}(n)$ is the TV impulse response of the time-selective channel from the μth transmit-antenna to the νth receive-antenna (notice the channel dependence on n); and $w_\nu(n)$ is complex AWGN at the νth receive-antenna with mean zero and variance N_0.

At each receive-antenna, the symbol-rate sampled sequence $r_\nu(n)$ at the receive filter output is serial-to-parallel-converted to form the $N_x \times 1$ blocks $\boldsymbol{r}_\nu(k) := [r_\nu(kN_x), r_\nu(kN_x + 1), \ldots, r_\nu(kN_x + N_x - 1)]^T$. The matrix-vector counterpart of (9.20) can then be expressed as

$$\boldsymbol{r}_\nu(k) = \sqrt{\frac{\bar{\gamma}}{N_t}} \sum_{\mu=1}^{N_t} \boldsymbol{D}_H^{(\nu,\mu)}(k) \boldsymbol{x}_\mu(k) + \boldsymbol{w}_\nu(k), \quad \forall \nu \in [1, N_r], \tag{9.21}$$

where $\boldsymbol{D}_H^{(\nu,\mu)}(k)$ is an $N_x \times N_x$ diagonal channel matrix that obeys the BEM in (9.19):

$$\boldsymbol{D}_H^{(\nu,\mu)}(k) := \sum_{q=0}^{Q} h_q^{(\nu,\mu)} \boldsymbol{D}_q(k), \tag{9.22}$$

with $\boldsymbol{D}_q(k) := \text{diag}[\exp(j\omega_q kN_x), \exp(j\omega_q(kN_x + 1)), \ldots, \exp(j\omega_q(kN_x + N_x - 1))]$ and $\boldsymbol{w}_\nu(k)$'s are corresponding i.i.d. AWGN noise vectors. Each block $\boldsymbol{r}_\nu(k)$ is linearly processed by the $N_x \times N_x$ matrix $\boldsymbol{\Gamma}_\nu$ to yield $\boldsymbol{y}_\nu(k) := \boldsymbol{\Gamma}_\nu \boldsymbol{r}_\nu(k)$. This operation constitutes the inner STDO decoder. The blocks $\boldsymbol{y}_\nu(k)$ are further "demapped" to a block $\boldsymbol{z}(k)$ by $\bar{\mathcal{M}}(\cdot)$: $\boldsymbol{z}(k) := \bar{\mathcal{M}}(\{\boldsymbol{y}_\nu(k)\}_{\nu=1}^{N_r})$. This operation is termed the middle STDO decoder. The block $\boldsymbol{z}(k)$ is finally decoded by $\mathcal{G}(\cdot)$ to obtain an estimate of $\boldsymbol{s}(k)$ as: $\hat{\boldsymbol{s}}(k) := \mathcal{G}(\boldsymbol{z}(k))$. This operation implements the outer STDO decoder.

In Chapters 7 and 8 we have seen that for frequency-selective MIMO channels, the maximum space-multipath diversity of order $N_t N_r (L+1)$ can be enabled with LCFC. For the time-selective MIMO channel, after invoking the time-frequency duality, we can view the $Q+1$ bases as $L+1$ taps and using Propositions 7.1 and 3.2, verify the following result [188]:

Proposition 9.3 *Consider* (N_t, N_r) *multi-antenna transmissions through time-selective channels adhering to a BEM as in* (9.11) *with* $Q + 1$ *bases. If the correlation matrix of the channel coefficients has rank* r_h, *the maximum diversity order of transmissions in* (9.21) *is* $G_{d,\max} = r_h \leq N_t N_r (Q + 1)$. *For LCF-coded* *systems, if $\boldsymbol{R}_h$ has full rank $r_h = N_r N_t(Q+1)$, the maximum coding gain is $G_{c,\max} = (\det(\boldsymbol{R}_h))^{1/r_h}\Delta_{\min}^2/N_t$, where $\Delta_{\min}^2$ denotes the minimum Euclidean distance in the constellation $\mathbb{S}$.*

Next, we show how to design the inner, middle, and outer STDO coders and decoders in order to collect the maximum order of the joint space-Doppler diversity. Since in the following we work on a block-by-block basis, we drop the block index k.

9.2.1 Duality-Based STDO Codes

By appealing to the time-frequency duality offered by the BEM, in this subsection, we demonstrate how to "transform" space-time-frequency (STF) codes developed for frequency-selective channels to obtain STDO codes for time-varying MIMO channels.

Suppose that we select the inner STDO (de)coder in the unifying transceiver of Figure 9.1 as

$$\boldsymbol{\Phi}_\mu = \boldsymbol{F}_N, \quad \forall \mu \in [1, N_t] \quad \text{and} \quad \boldsymbol{\Gamma}_\nu = \boldsymbol{F}_N^{\mathcal{H}}, \quad \forall \nu \in [1, N_r]. \tag{9.23}$$

Then, based on (9.14), (9.21), and (9.23), we obtain

$$\begin{aligned} \boldsymbol{y}_\nu = \boldsymbol{F}_N^{\mathcal{H}} \boldsymbol{r}_\nu &= \sqrt{\frac{\bar{\gamma}}{N_t}} \sum_{\mu=1}^{N_t} \boldsymbol{F}_N^{\mathcal{H}} \boldsymbol{D}_H^{(\nu,\mu)} \boldsymbol{F}_N \boldsymbol{c}_\mu + \boldsymbol{F}_N^{\mathcal{H}} \boldsymbol{w}_\nu \\ &= \sqrt{\frac{\bar{\gamma}}{N_t}} \sum_{\mu=1}^{N_t} \boldsymbol{H}_{\nu\mu} \boldsymbol{c}_\mu + \boldsymbol{\eta}_\nu, \quad \forall \nu \in [1, N_r], \end{aligned} \tag{9.24}$$

where $\boldsymbol{H}_{\nu\mu}$ is a circulant matrix formed by the coefficients $\{h_q^{(\nu,\mu)}\}_{q=0}^{Q}$.

Thanks to the (I)FFT inner (de)coder, the time-selective model has been converted in (9.24) to an equivalent model of a frequency-selective channel. Intuitively, the ST codes introduced in Chapters 7 and 8 for frequency-selective channels can be used to achieve the space-Doppler diversity. Next, we present two such paradigms of STDO codes.

9.2.1.1 CP-Based Design The encoding and decoding modules of this CP-based approach are depicted in Figure 9.2. The design of the outer STDO encoder starts by dividing each information-bearing block $\boldsymbol{s}$ into N_o subblocks where each subblock has length N_I. Each subblock can then be viewed as an OFDM-processed STF code block. For each subblock, grouped LCFC is used by further partitioning the subblock into N_g groups with $N_I = N_g N_{\text{sub}}$, and left-multiplying each group by $\boldsymbol{\Theta}_{\text{sub}}$. The outer STDO encoder is thus implementing an LCFC operation with aggregate LCF matrix $\boldsymbol{\Theta}$ given by

$$\boldsymbol{\Theta} = \boldsymbol{I}_{N_o} \otimes \bar{\boldsymbol{\Theta}},$$

where the $N_I \times N_I$ matrix $\bar{\boldsymbol{\Theta}}$ is formed using $\boldsymbol{\Theta}_{\text{sub}}$ similar to the way that the precoder was formed from $\boldsymbol{\Theta}_{\text{LCF}}$ in (8.41); see also Section 8.3 for a detailed description of the grouping approach.

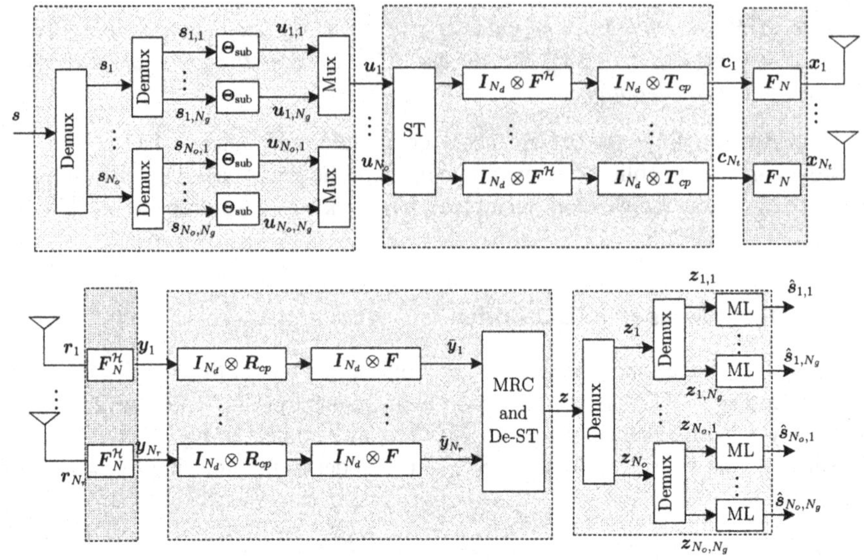

Figure 9.2 CP-based STDO transceiver design

The design of the middle STDO encoder proceeds in two stages. The first stage implements an OSTBC which, as we saw in Chapters 3 and 8, operates on each subblock (or OFDM block) to enable the spatial diversity. The OSTBC yields N_d subblocks from the N_o subblocks and has rate $\eta_{\text{OSTBC}} = N_o/N_d$. The second stage is the same as the MIMO OFDM design, which includes IFFT and CP insertion. Considering N_d subblocks together, the encoder in the second stage can be written as $(\boldsymbol{I}_{N_d} \otimes \boldsymbol{F}^{\mathcal{H}})(\boldsymbol{I}_{N_d} \otimes \boldsymbol{T}_{\text{cp}})$, where $\boldsymbol{F}^{\mathcal{H}}$ has size $N_I \times N_I$ and $\boldsymbol{T}_{\text{cp}} := [\boldsymbol{T}_1, \boldsymbol{I}_{N_I}, \boldsymbol{T}_2]^T$ is the CP insertion matrix with

$$\boldsymbol{T}_1 := [\boldsymbol{0}_{(Q/2)\times(N_I-Q/2)}, \boldsymbol{I}_{Q/2}]^T, \quad \text{and} \quad \boldsymbol{T}_2 := [\boldsymbol{I}_{Q/2}, \boldsymbol{0}_{(Q/2)\times(N_I-Q/2)}]^T.$$

Correspondingly at the receiver, we design the middle STDO decoder $\bar{\mathcal{M}}(\cdot)$ following the reverse order of the two encoding stages. Specifically, we remove the CP and perform FFT by premultiplying the received block on each antenna with $(\boldsymbol{I}_{N_d} \otimes \boldsymbol{F})(\boldsymbol{I}_{N_d} \otimes \boldsymbol{R}_{\text{cp}})$, where $\boldsymbol{R}_{\text{cp}} := [\boldsymbol{0}_{N_I\times(Q/2)}, \boldsymbol{I}_{N_I}, \boldsymbol{0}_{N_I\times(Q/2)}]$ is a matrix description of the CP removal operation. The maximum ratio combining (MRC) operation corresponds to the linear processing in (3.31), which decouples the groups to reduce decoding complexity. This and the remaining decoding steps follow those of the GSTFB system in Chapter 8.

Since the equivalent channel matrix between $\boldsymbol{z}_i$ and $\boldsymbol{s}_i$ is diagonal, we can write the gth group of $\boldsymbol{z}_i$, denoted by $\boldsymbol{z}_{i,g}$, as

$$\boldsymbol{z}_{i,g} = \sqrt{\frac{\bar{\gamma}}{N_t}} \left(\sum_{\nu=1}^{N_r} \sum_{\mu=1}^{N_t} \boldsymbol{D}_{H,g}^{(\nu,\mu)} (\boldsymbol{D}_{H,g}^{(\nu,\mu)})^* \right)^{1/2} \boldsymbol{\Theta}_{\text{sub}} \boldsymbol{s}_{i,g} + \boldsymbol{\xi}_{i,g},$$

where $\boldsymbol{\Theta}_{\text{sub}}$ is an $N_{\text{sub}} \times N_{\text{sub}}$ matrix designed according to Section 3.5 and $\boldsymbol{D}_{H,g}^{(\nu,\mu)}$ is a diagonal matrix generated by $\{h_g^{(\nu,\mu)}\}_{g=0}^{Q}$ as detailed in Chapter 8. Again, ML decoding of $\boldsymbol{s}_{i,g}$ can be performed via sphere decoding with block size N_{sub}.

To assess the achievable diversity, let us collect the channel coefficients into one $N_t N_r(Q+1) \times 1$ vector:

$$\boldsymbol{h} := [h_0^{(1,1)}, h_1^{(1,1)}, \ldots, h_Q^{(N_r,N_t)}]^T,$$

and define $\boldsymbol{R}_h := E[\boldsymbol{h}\boldsymbol{h}^{\mathcal{H}}]$. When $N_{\text{sub}} \geq Q+1$, the maximum diversity order $r_h := \text{rank}(\boldsymbol{R}_h)$ is guaranteed. Furthermore, when $r_h = N_r N_t(Q+1)$ and we select $N_{\text{sub}} = Q+1$, the coding gain for this CP-based scheme satisfies [cf. Proposition 3.2]

$$(\ln 2)(\det(\boldsymbol{R}_h))^{1/r_h} \frac{N_I \Delta_{\min}^2}{(N_I+Q)N_t} \leq G_c \leq (\det(\boldsymbol{R}_h))^{1/r_h} \frac{N_I \Delta_{\min}^2}{(N_I+Q)N_t}, \quad (9.25)$$

where the upper bound is achieved when $N_{\text{sub}} = Q+1$ satisfies a certain algebraic property [313]. As N_I increases, the coding gain of the CP-based design in (9.25) approaches the maximum coding gain.

In a nutshell, the following can be asserted about the performance of the CP-based design [188]:

Proposition 9.4 *CP-based STDO block codes guarantee the maximum space-Doppler diversity $G_d = r_h$ at low decoding complexity when the group size satisfies $N_{\text{sub}} \geq Q+1$. When the channel correlation matrix $\boldsymbol{R}_h$ has full rank $N_t N_r(Q+1)$, the CP-based design achieves the maximum coding gain $G_{c,\max}$ of LCF-coded systems, asymptotically in N_I. The transmission rate of the CP-based design is $\eta_{OSTBC} = N_I/(N_I+Q)$, where η_{OSTBC} is the rate of OSTBC specified in Section 3.3.1.*

9.2.1.2 ZP-Based Design In this approach, ZP replaces the CP guard. Instead of OFDM, this design relies on the single-carrier block transmission (ZP-only) system described in Section 7.3.1.2. Similar to the CP-based design, there are two stages of the middle STDO codec. The first stage implements the OSTBC, which is similar to the CP-based approach but relies on the time-reversal operation of Section 7.3.1 to guarantee Doppler diversity; the second stage eliminates interblock interference by padding zeros after each subblock.

The encoding and decoding modules of the ZP-based scheme are depicted in Figure 9.3; and the major results on its error and rate performance are summarized in the following:

Proposition 9.5 *ZP-based STDO block codes enable the maximum space-Doppler diversity $G_d = r_h$. When the channel correlation matrix $\boldsymbol{R}_h$ has full rank $N_t N_r(Q+1)$, the ZP-based design achieves the maximum coding gain $G_{c,\max}$ of LCF-coded systems at transmission rate $\eta_{OSTBC} = N_I/(N_I+Q)$, where η_{OSTBC} is the rate of the corresponding OSTBC specified in Section 3.3.1.*

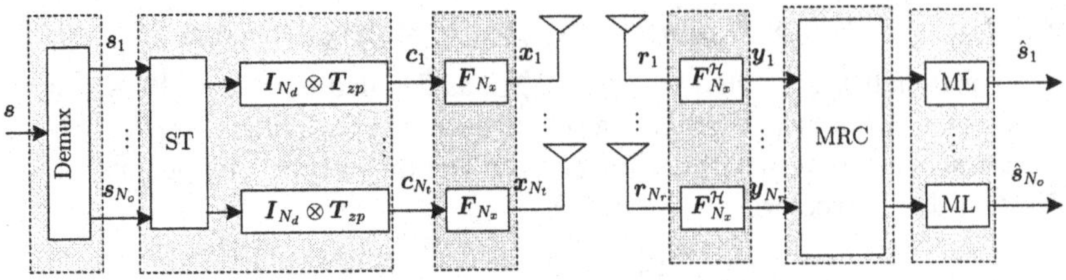

Figure 9.3 ZP-based STDO transceiver design

Recapitulating, the ST codes designed for frequency-selective MIMO channels can be used for time-selective MIMO channels with minor modifications needed to abide by the duality property. The relative merits of CP- and ZP-based designs are analogous to those outlined for single- and multi-carrier ST codes in the closing comments of Chapter 8.

9.2.2 Phase Sweeping Design

As asserted by Proposition 9.3, diversity increases when the channel varies faster (which corresponds to a larger Q). This observation motivates the present section's phase sweeping approach to designing STDO codes for time-selective MIMO channels. It will turn out that a phase sweeping operation at the transmitter can effectively increase the time variation of the intended channel. Interestingly, this operation can be viewed as *dual* to the delay diversity operation dealt with in Sections 3.1 and 7.2, which was originally developed for converting multiple frequency-flat channels to a single frequency-selective channel. In this section we show that phase sweeping converts multiple time-varying channels to a single but faster time-varying channel.

9.2.2.1 Analog Phase Sweeping Design Historically, the analog phase sweeping (a.k.a. intentional frequency offset) operation was introduced in [111]; and later combined with channel coding in [139] to further improve error performance over time-varying channels. To illustrate the basic idea, consider the two-transmit-antenna analog configuration depicted in Figure 9.4. The waveform $s(t)$ directed to the lower antenna element is modulated by a sweeping frequency f_s in addition to the carrier frequency f_c that is present in both antenna elements [111, 139].

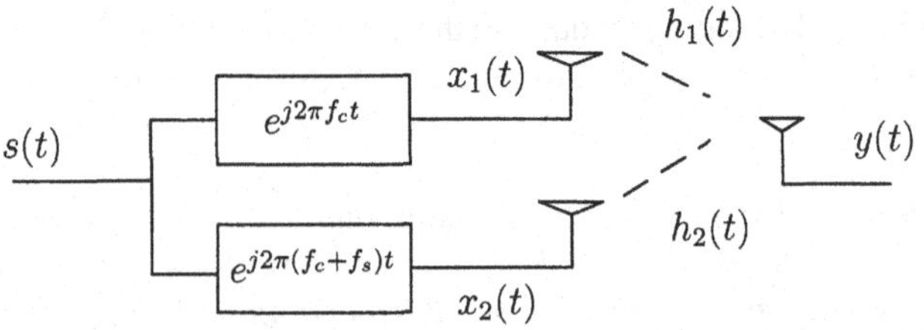

Figure 9.4 Analog phase sweeping design

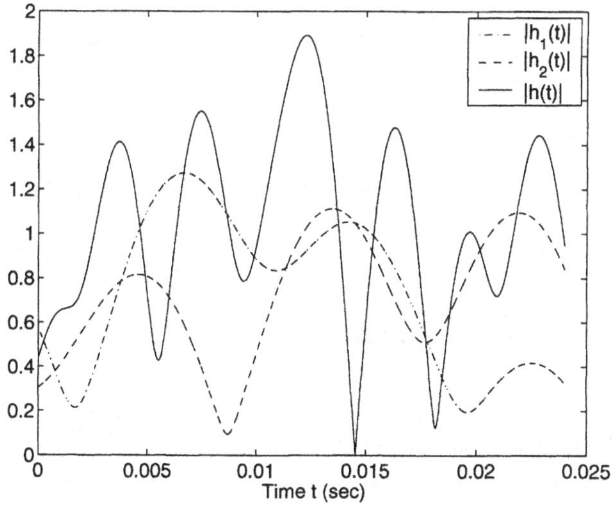

Figure 9.5 Phase sweeping effect on channel variability

The resulting input-output relationship in continuous time is

$$y(t) = h_1(t)x_1(t) + h_2(t)x_2(t) + w(t) = h(t)s(t)e^{j2\pi f_c t} + w(t),$$

where $w(t)$ is AWGN and the equivalent channel is defined as

$$h(t) := h_1(t) + h_2(t)e^{j2\pi f_s t}. \tag{9.26}$$

In Figure 9.5 we depict realizations of $h(t)$, $h_1(t)$, and $h_2(t)$ using the Jakes model in (9.7) with carrier frequency $f_c = 2$ GHz, sampling period $T_s = 4$ μs, and mobile velocity 10 m/s. When $f_s = 200$ Hz, we observe that the equivalent channel $h(t)$ in (9.26) exhibits faster time variation than $h_1(t)$ and $h_2(t)$. As we have seen in Section 9.1.3, when the channel changes faster, it can provide higher Doppler diversity order. Notice that dual to delay diversity, the phase sweeping operation provides a means of converting antenna diversity to Doppler diversity.

The price paid for the simplicity of the analog phase sweeping operation is bandwidth expansion. In Figure 9.6 we use an example to illustrate this bandwidth loss. By inspecting the spectrum of the original signal $s(t)$ along with the spectra of the waveforms transmitted from the two antennas, it becomes clear that modulation with the intentional frequency offset introduces excess bandwidth. Furthermore, without an explicit channel model, the analog phase sweeping approaches in [111, 139] are unable to collect the maximum diversity and coding gains. These limitations motivate the introduction of digital-phase sweeping (DPS) .

9.2.2.2 Digital Phase Sweeping Design In this section we rely on DPS to design codes for time-selective MIMO channels. The system model for the DPS-based design is described by the unified structure of Figure 9.1, where the middle

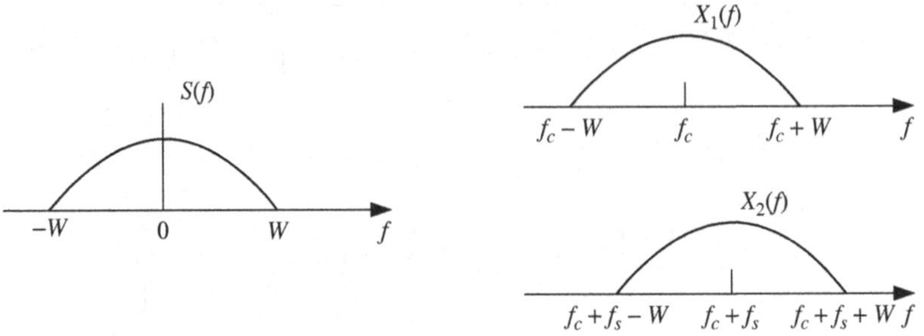

Figure 9.6 Analog phase sweeping costs extra bandwidth

STDO encoder $\mathcal{M}(\cdot)$ is now just a power splitter. By allocating the power of each transmitted block equally across antennas, we obtain $\boldsymbol{c}_\mu = \boldsymbol{u}/\sqrt{N_t}$, $\forall \mu \in [1, N_t]$. This means that for the DPS-based transceiver, we have $N_u = N_x$. Using (9.21), $\boldsymbol{r}_\nu$ and $\boldsymbol{s}$ can then be related via

$$\boldsymbol{r}_\nu = \frac{1}{\sqrt{N_t}} \sum_{\mu=1}^{N_t} \boldsymbol{D}_H^{(\nu,\mu)} \boldsymbol{\Phi}_\mu \boldsymbol{\Theta} \boldsymbol{s} + \boldsymbol{w}_\nu, \quad \nu \in [1, N_r]. \tag{9.27}$$

We wish to quantify how the DPS-based system described by (9.27) effects a faster time variation and thereby enables a higher space-Doppler diversity order. To this end, we start with (9.22), from where we notice that different channels share the same exponential bases but have different channel coefficients. Suppose next that we shift the $Q+1$ bases of each channel corresponding to one of the N_t transmit antennas so that all the bases become consecutive on the FFT grid of complex exponentials, as shown in Figure 9.7 for $Q+1 = N_t = 3$. Then we can view the N_t channels corresponding to each receive-antenna as one equivalent time-selective channel with $N_t(Q+1)$ bases. To realize this intuition, we select the matrices $\{\boldsymbol{\Phi}_\mu\}_{\mu=1}^{N_t}$, which determine the inner STDO encoder as

$$\boldsymbol{\Phi}_\mu = \text{diag}[1, e^{j\phi_\mu}, \ldots, e^{j\phi_\mu(N_x-1)}], \quad \forall \mu \in [1, N_t],$$

where $\phi_\mu := 2\pi(\mu-1)(Q+1)/N_x$. As $\boldsymbol{\Phi}_1 = \boldsymbol{I}$, the exponentials of the channel corresponding to the first ($\mu = 1$) transmit-antenna remain unchanged since $\phi_1 = 0$. But those corresponding to the second channel ($\mu = 2$) are shifted from their original location at $\{\boldsymbol{D}_q\}_{q=0}^{Q}$ to $\{\boldsymbol{D}_q\}_{q=Q+1}^{2Q+1}$, after multiplication with the DPS matrix $\boldsymbol{\Phi}_2$ that takes place at the second transmit-antenna (i.e., $\{\boldsymbol{D}_q\boldsymbol{\Phi}_2\}_{q=0}^{Q} = \{\boldsymbol{D}_q\}_{q=Q+1}^{2Q+1}$). Proceeding similarly with all N_t DPS matrices, it follows that (9.27) can be rewritten as

$$\boldsymbol{r}_\nu = \frac{1}{\sqrt{N_t}} \sum_{q=0}^{N_t(Q+1)-1} h_q^{(\nu)} \boldsymbol{D}_q \boldsymbol{\Theta} \boldsymbol{s} + \boldsymbol{w}_\nu, \quad \forall \nu \in [1, N_r], \tag{9.28}$$

where $h_q^{(\nu)} := h_{q \bmod (Q+1)}^{(\nu, \lfloor q/(Q+1) \rfloor + 1)}$. Comparing (9.27) with (9.28), we arrive at:

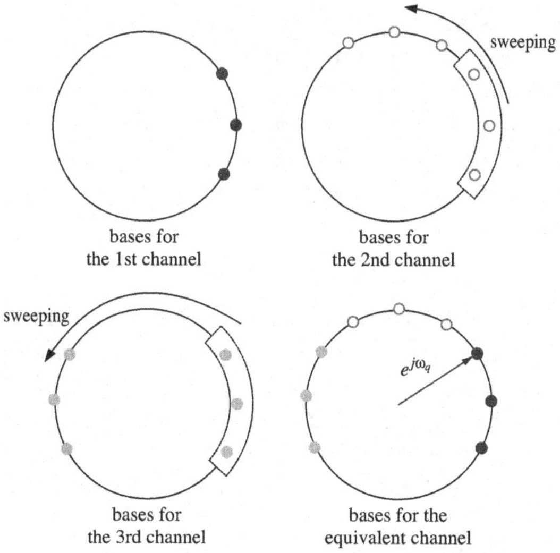

Figure 9.7 DPS illustration: black, hollow, and gray circles are shifted FT bases from three channels

Property 9.3 *DPS converts an N_t transmit-antenna system, where each channel can be expressed via $Q+1$ exponential bases, to a single transmit-antenna system, where the equivalent but faster channel is expressed by $N_t(Q+1)$ exponential bases.*

Notice that since $\boldsymbol{\Phi}_\mu$ operates in the digital domain, the sweeping wraps the phases around $[-\pi, \pi)$, which explains why DPS does not incur bandwidth expansion.

Remark 9.1 To avoid overlapping the shifted bases, we should make sure that $N_x > N_t(Q+1)$. From the definition of $Q := 2\lceil f_{\max} N_x T_s \rceil$ in (9.8) and for fixed $f_{\max}$ and N_x, we can adjust the sampling period T_s to satisfy this condition. As for each receive-antenna, recall that we have $N_t(Q+1)$ unknown BEM coefficients corresponding to N_t channels every N_x symbols. Hence, selecting N_x to exceed $N_t(Q+1)$ also guarantees that the number of unknowns is less than the number of equations. Therefore, even from a channel estimation point of view, this condition is reasonable.

With the equivalence established by Property 9.3, the outer STDO codec, which is determined by $\boldsymbol{\Theta}$ and $\mathcal{G}(\cdot)$, can be any single-input codec that achieves the maximum diversity gain for the single-transmit-antenna time-selective channels corresponding to each receive-antenna. If ML decoding is used in the demodulator $\mathcal{G}(\cdot)$, the maximum diversity gain r_h is achieved, provided that the LCF precoder matrix $\boldsymbol{\Theta}$ is designed in such a way that $\boldsymbol{\Theta}\boldsymbol{e}$ has at least $N_t(Q+1)$ nonzero entries, for all possible error vectors $\boldsymbol{e} := \boldsymbol{s} - \boldsymbol{s}' \neq \mathbf{0}$ [183, Proposition 2]. However, as N_t increases, the block length N increases as well. ML decoding for the entire $N \times 1$ block entails high computational complexity. To reduce the decoding complexity, we can partition the design of the outer STDO encoder $\boldsymbol{\Theta}$ in groups of smaller size and use grouped LCFC-OFDM as in Chapter 8; see [163] for details.

If we adopt the grouped LCFC-OFDM approach, the error performance of the DPS-based design depends on the selection of the group size N_{sub}. When $N_{\text{sub}} \geq N_t(Q+1)$, the maximum diversity order is achieved.

We can now summarize the diversity and coding claims pertaining to the DPS-based design as follows:

Proposition 9.6 *The maximum achievable STDO diversity order $G_d = r_h$ is guaranteed by the DPS-based design when the group size is selected as $N_{\text{sub}} \geq N_t(Q+1)$. When the channel correlation matrix $\boldsymbol{R}_h$ has full rank $r_h = N_r N_t(Q+1)$, the DPS-based design also enables the maximum possible coding gain among all LCF-coded transmissions that is given in closed form by $G_{c,\max} = (\det(\boldsymbol{R}_h))^{1/r_h}\Delta^2_{\min}/N_t$. As far as transmission rate, the DPS-based design attains 1 symbol per channel use.*

In fact, the group size N_{sub} controls the trade-off between error performance and decoding complexity. When $N_{\text{sub}} \leq N_t(Q+1)$, as N_{sub} decreases, the decoding complexity decreases while the diversity order is decreasing. By adjusting N_{sub}, we can balance the affordable complexity with the error performance required.

Similar to analog phase sweeping, matrices $\{\boldsymbol{\Phi}_\mu\}_{\mu=1}^{N_t}$ in (9.27) effect digital-phase sweeping in block transmissions, which increases the variation (and thus the available diversity) of time-selective MIMO channels. Summarizing, the differences between DPS and analog phase sweeping based designs are as follows:

1. Combined with grouped LCFC, DPS collects not only space-diversity as in [111], but also Doppler diversity.

2. DPS design does not consume extra bandwidth.

3. Combined with grouped LCFC, the DPS can afford low decoding complexity.

Remark 9.2 Comparing the three block STDO designs (CP-, ZP-, and DPS-based), we note that (i) all schemes guarantee the maximum diversity gain; (ii) DPS- and ZP-based schemes also achieve the maximum coding gain, while the CP-based scheme achieves the maximum coding gain asymptotically (as N_I increases); (iii) to guarantee the maximum diversity gain, the CP-based scheme can afford the lowest decoding complexity; and (iv) to deal with interblock interference, CP- and ZP-based approaches rely on CP or ZP guards, which consume extra bandwidth compared with the DPS scheme, which does not require any guard. Furthermore, along with the OSTBC design benefits (cf. Section 3.3), the CP- and ZP-based STDO codecs also inherit its limitation in suffering up to 50% rate loss, when $N_t > 2$ antennas are signaling with complex constellations. Notwithstanding, the DPS attains a transmission rate of 1 symbol per channel use for any N_t.

9.3 SPACE-TIME-DOPPLER FDFR CODES

In Chapter 4 we considered FDFR designs for quasi-static flat fading MIMO channels. A more challenging problem is the design of LCF encoders capable of ensuring FDFR, even when the underlying channels are time varying. Let us first suppose that the MIMO channel is time selective but frequency flat. In this case, the input-output relationship can be written as [cf. (2.9)]

$$\boldsymbol{y}(n) = \sqrt{\frac{\bar{\gamma}}{N_t}}\boldsymbol{H}(n)\boldsymbol{x}(n) + \boldsymbol{w}(n), \tag{9.29}$$

where we notice that the $N_r \times N_t$ channel matrix $\boldsymbol{H}(n)$ is allowed to vary along the time index n. Interestingly, the input-output relationship (9.29) coincides with (8.53) derived in Section 8.5 for frequency-selective channels. Therefore, at least in principle, the designs for frequency-selective channels can be used for time-selective channels as well.

Thanks to the BEM in (9.11) and the time-frequency duality in (9.14), one can view a time-selective channel with $Q+1$ bases as a frequency-selective channel with $Q+1$ taps. Relying on this time-frequency duality, the FDFR design for time-selective MIMO channels can be obtained from the one developed for frequency-selective MIMO channels. We summarize this assertion as follows [182]:

Proposition 9.7 *For any constellation of* $\boldsymbol{s}$ *carved from the Gaussian integer ring* $\mathbb{Z}[j]$ *(if* p *and* q *are integers, then* $p + qj$ *belongs to* $\mathbb{Z}[j]$*), with the ST encoder-multiplexer formed by LCFC and circularly wrapped layers, there exists at least one pair of* $(\boldsymbol{\Theta}, \beta)$ *in* (4.11) *which enables full diversity* $[N_t N_r(Q+1)$ *if each channel provides Doppler diversity* $Q+1]$ *for the ST transmission in* (9.29) *at full rate* N_t *symbols per channel use.*

Remark 9.3 Note that as the block size N_x (and thus Q) increases, the Doppler diversity increases. However, the price paid here is twofold: decoding delay and complexity. Later we illustrate via simulations that when the Doppler diversity order is high enough (say, $Q > 3$), error performance with multiple antennas does not improve much by further increasing diversity.

9.4 SPACE-TIME-DOPPLER TRELLIS CODES

We have seen so far how OSTBC- and DPS-based designs enable the space-Doppler diversity available with time-selective MIMO channels. Space-time trellis codes (STTCs) have also been designed to enable space-Doppler diversity. Using a distance (as opposed to a rank) criterion, space-Doppler diversity has been assessed in [253] for fast-fading MIMO channels whose impulse response is invariant over each symbol period but is allowed to vary independently from symbol to symbol. A generalized design criterion encompassing both distance and rank criteria can be found in [75] for block-fading (a.k.a. quasi-static) MIMO channels. We present this criterion for the model in (9.11), where block fading means that $h(n)$ is time invariant over a block period but is allowed to vary from block to block.

9.4.1 Design Criterion

Suppose that the STTC codeword length is N_x while the channel remains invariant during N_{coh} symbols. Starting from the scalar system model in Chapter 2, we can rewrite the input-output relationship of a block-fading MIMO channel as

$$y_\nu(n) = \sum_{\mu=1}^{N_t} \sqrt{\frac{\bar{\gamma}}{N_t}} h_{\nu\mu}(\lfloor n/N_{\text{coh}} \rfloor) x_\mu(n) + w_\nu(n). \tag{9.30}$$

Upon defining

$$\begin{aligned}
\boldsymbol{X}(m) &:= \begin{bmatrix} x_1(mN_{\text{coh}}) & \cdots & x_{N_t}(mN_{\text{coh}}) \\ \vdots & & \vdots \\ x_1((m+1)N_{\text{coh}}-1) & \cdots & x_{N_t}((m+1)N_{\text{coh}}-1) \end{bmatrix}, \\
\boldsymbol{h}(m) &:= [h_{11}(m), \ldots, h_{1N_t}(m), \ldots, h_{N_r N_t}(m)]^T, \\
\boldsymbol{y}(m) &:= [y_1(mN_{\text{coh}}), \ldots, y_1((m+1)N_{\text{coh}}-1), \ldots, y_{N_r}((m+1)N_{\text{coh}}-1)]^T, \\
\boldsymbol{w}(m) &:= [w_1(mN_{\text{coh}}), \ldots, w_1((m+1)N_{\text{coh}}-1), \ldots, w_{N_r}((m+1)N_{\text{coh}}-1)]^T,
\end{aligned}$$

we can re-express (9.30) in vector-matrix form as

$$\boldsymbol{y}(m) = \sqrt{\frac{\bar{\gamma}}{N_t}} (\boldsymbol{X}(m) \otimes \boldsymbol{I}_{N_r}) \boldsymbol{h}(m) + \boldsymbol{w}(m), \quad \forall m \in [0, N_x/N_{\text{coh}} - 1].$$

Because $\boldsymbol{h}(m)$ is independent across m, following the pairwise error probability (PEP) approach in Chapter 2, we find that the diversity depends on

$$d = \sum_{m=0}^{N_x/N_{\text{coh}}-1} \text{rank}(\boldsymbol{X}(m) - \boldsymbol{X}'(m)),$$

where the transmitted code matrix $\boldsymbol{X}(m)$ is decoded as $\boldsymbol{X}'(m)$ and there exists an m such that $\boldsymbol{X}(m) \neq \boldsymbol{X}'(m)$. Quantity d denotes what is referred to as the sum-of-ranks criterion. The diversity gain effected is

$$G_d = \min_{\forall \boldsymbol{X}(m) \neq \boldsymbol{X}'(m)} dN_r.$$

Notice that for $N_{\text{coh}} = 1$, matrix $\boldsymbol{X}(m)$ reduces to a vector, and the sum of ranks criterion is then equivalent to the distance criterion in [253] for fast-fading channels. When $N_{\text{coh}} = N_x$, we have $N_x/N_{\text{coh}} = 1$ and the channel remains time invariant over the entire codeword. In this case, the sum of ranks criterion is equivalent to the rank criterion in Chapter 2.

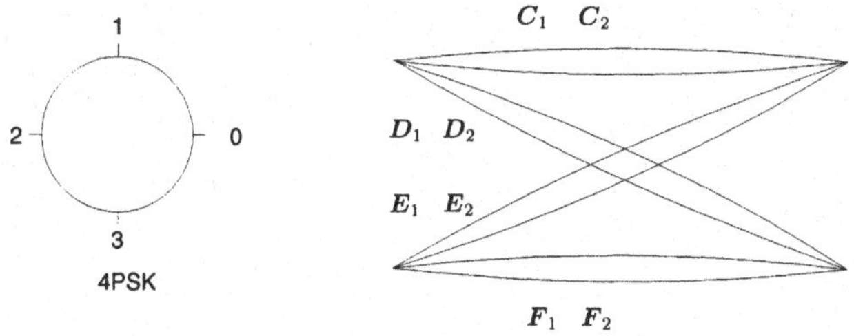

Figure 9.8 2-state 4-PSK smart-greedy code

9.4.2 Smart-Greedy Codes

One paradigm of the space-time-Doppler trellis code (STDOTC) design is provided by the smart-greedy code in [253]. Smart-greedy codes are designed for both quasi-static block fading and time-varying MIMO channels and rely on a hybrid rank-distance criterion. We will consider a smart-greedy code at transmission rate 1 bit per channel use which is based on multiple trellis-coded modulation (MTCM). The latter is well motivated for fading channels even in the single-antenna case [58]. To differentiate MTCM from TCM, we use a double line to depict multiple paths in Figure 9.8. Furthermore, we will use a matrix to represent the codewords on each branch. Subscripts 1 and 2 stand for the first and second transmit-antennas, respectively, and the four rows denote the four parallel paths per branch.

$$C_1 = \begin{bmatrix} 0 & 0 & 0 \\ 0 & 1 & 1 \\ 0 & 2 & 2 \\ 0 & 3 & 3 \end{bmatrix}, \quad C_2 = \begin{bmatrix} 0 & 0 & 0 \\ 1 & 1 & 0 \\ 2 & 2 & 0 \\ 3 & 3 & 0 \end{bmatrix}, \quad D_1 = \begin{bmatrix} 2 & 0 & 0 \\ 2 & 1 & 1 \\ 2 & 2 & 2 \\ 2 & 3 & 3 \end{bmatrix}, \quad D_2 = \begin{bmatrix} 0 & 2 & 2 \\ 1 & 3 & 2 \\ 2 & 0 & 2 \\ 3 & 1 & 2 \end{bmatrix},$$
$$E_1 = \begin{bmatrix} 1 & 0 & 0 \\ 1 & 1 & 1 \\ 1 & 2 & 2 \\ 1 & 3 & 3 \end{bmatrix}, \quad E_2 = \begin{bmatrix} 0 & 3 & 3 \\ 1 & 0 & 3 \\ 2 & 1 & 3 \\ 3 & 2 & 3 \end{bmatrix}, \quad F_1 = \begin{bmatrix} 3 & 0 & 0 \\ 3 & 1 & 1 \\ 3 & 2 & 2 \\ 3 & 3 & 3 \end{bmatrix}, \quad F_2 = \begin{bmatrix} 0 & 1 & 1 \\ 1 & 2 & 1 \\ 2 & 3 & 1 \\ 3 & 0 & 1 \end{bmatrix}.$$

Notice that we have two states here, and from one state to the next there are four paths. If the channel is static, then based on the rank criterion, this smart-greedy code enables second-order diversity. But if the channel is fast-fading independently from symbol to symbol, then using the distance criterion, it turns out that the same code enables third-order diversity.

9.5 NUMERICAL EXAMPLES

Example 9.1 (Comparisons among the three STDO block codecs) We compare DPS-, CP-, and ZP-based schemes with $N_t = 2$ transmit antennas, $Q + 1 = 3$ bases per channel, and BEM coefficients generated to be i.i.d. Gaussian with mean zero and variance $1/(Q+1)$. We choose QPSK modulation and set the number of

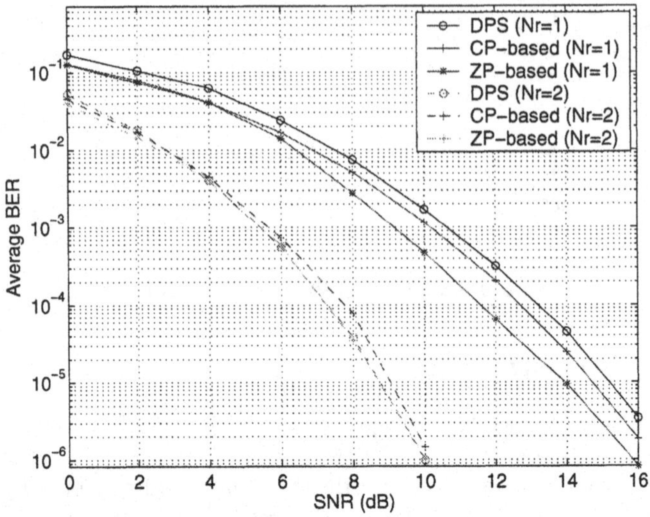

Figure 9.9 Comparisons among the three block STDO codecs when $N_t = 2$

information symbols per block to $N_s = 24$. For DPS, the block length is $N_x = 24$, while for CP- and ZP-based schemes the block length is $N_x = 28$, to account for the CP and ZP guards, respectively. The linear precoder with grouping is employed for DPS- and CP-based schemes with group sizes $N_{\text{sub}} = 6$ and $N_{\text{sub}} = 3$, respectively. Figure 9.9 depicts the BER performance of these three codecs. The sphere decoding algorithm (SDA) is adopted for demodulation. We observe that (i) from the slope of the BER curves for $N_r = 1$, all three schemes guarantee the maximum diversity order $G_{d,\max} = N_t(Q+1) = 6$; (ii) with either $N_r = 1$ or 2, the ZP-based scheme exhibits the best performance among the three; (iii) compared with CP, the performance of DPS incurs about 0.5 dB loss at high SNR for $N_r = 1$; and (iv) as N_r increases, the performance difference among the three schemes diminishes at high SNR.

When $N_t > 2$, the CP- and ZP-based schemes inherit OSTBC's rate loss up to 50% [162, 253, 332]. In contrast, the DPS scheme attains the full rate for any N_t. Figure 9.10 depicts the performance of the three STDO code designs with $N_t = 4$. For the CP- and ZP-based schemes, we select the block ST code as in [249, Equation (38)], which loses 50% of the rate. To maintain fairness in rates, we select QPSK for CP- and ZP-based schemes and BPSK for the DPS-based scheme with the same symbol power. The information block length is $N_s = 36$. From Figure 9.10 we verify that DPS outperforms both CP- and ZP-based designs. Notice that even in this case, CP- and ZP-based systems have a lower rate ($9/11$ bit per channel use) than the DPS-based system (1 bit per channel use).

Example 9.2 (Comparisons with the ST codes in [253]) In this example we compare the DPS scheme with the smart-greedy (SG) code proposed in [253] for $(N_t, N_r) = (2, 1)$. To maintain the same rate, we select BPSK for the DPS scheme,

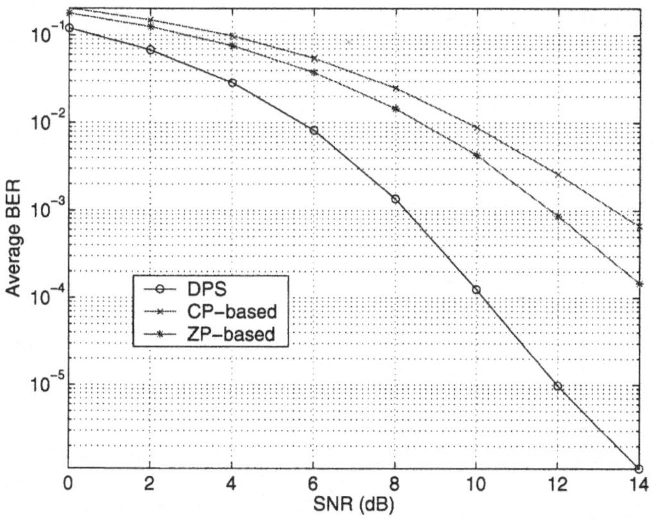

Figure 9.10 Comparisons among the three block STDO codecs when $N_t = 4$

and from the SG class we use the STDOTC in [253, Example 3.9.2]. Each channel has $Q + 1 = 3$ bases, and the channel coefficients are generated i.i.d. with mean zero and variance $1/(Q + 1)$. First, we consider the uncoded setup. The information block length is $N_s = N_x = 30$. The number of groups for DPS is set to $N_g = 5$, so that these two schemes have comparable decoding complexity. Figure 9.11 depicts the BER versus SNR comparison for the two schemes. It is evident that DPS outperforms the smart-greedy STDOTC because the former guarantees the full space-Doppler diversity.

Furthermore, we consider the coded case when both schemes rely on a $(7, 3)$ Reed-Solomon coder with block interleaving. The number of information bits is 90, which implies that the length of the coded block of bits is 210. We select the depth of the block interleaver to be 42. For the DPS design, we partition the coded bits into five blocks and divide each block into seven groups. The simulated error performance depicted in Figure 9.11 (dashed lines) confirms that the DPS scheme still outperforms the smart-greedy codes considerably.

9.6 SPACE-TIME-DOPPLER DIFFERENTIAL CODES

So far in this chapter we have seen how *coherent* block, layered, or trellis STDO codes can take advantage of the available space-Doppler diversity. Recall also that to bypass the need for MIMO channel estimation, in Chapter 6 we introduced *differential* ST codes capable of effecting the space diversity provided by flat fading MIMO channels. Here we design differential codes that can also benefit from the Doppler diversity.

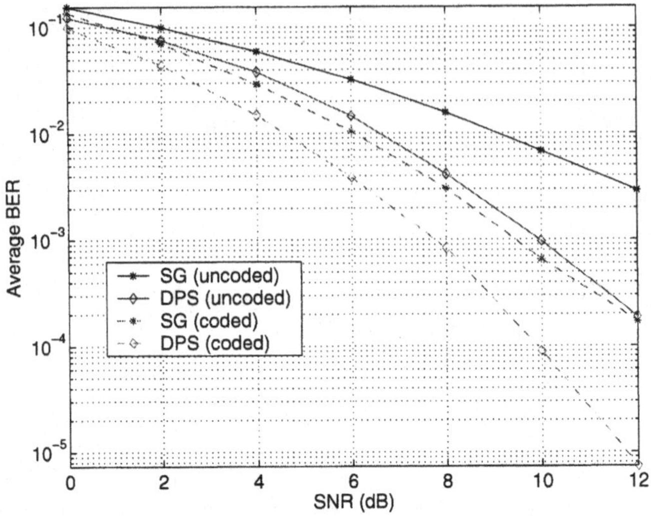

Figure 9.11 Comparison of DPS with the smart-greedy (SG) codes in [253]

For time-selective channels, differential designs are particularly challenging because (i) the channels may change from symbol to symbol, and (ii) the differential encoders must be designed properly; otherwise, the power per transmitted block may diverge or diminish with time. Existing approaches designed to deal with time-selective channels employ either multiple symbol detection (MSD) [112, 229] or decision-feedback differential detection (DF-DD) [47, 227–229, 277].

In this section we focus on single-antenna differentially encoded transmissions over time-selective channels. Using the DPS technique, however, extensions to multi-antenna differentially encoded transmissions over time-selective MIMO channels become readily available. The block diagram for the single-antenna system model is depicted in Figure 9.12.

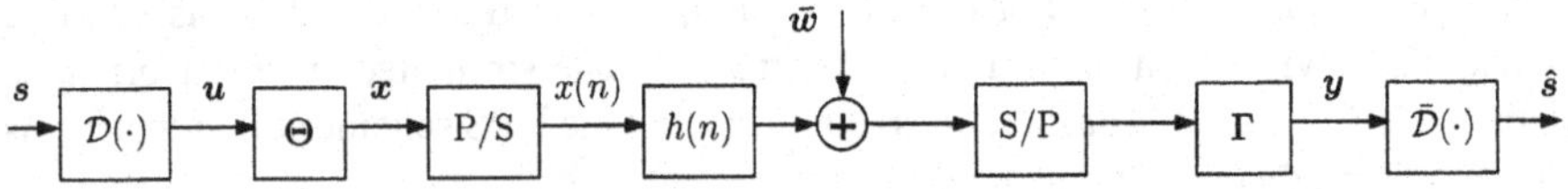

Figure 9.12 Discrete-time baseband equivalent block model of a differential system

Because of the time-frequency duality in (9.14), we already mentioned that (9.13) could have resulted from an OFDM-based transmission over a frequency-selective channel with $Q+1$ taps, whose channel matrix $\tilde{\boldsymbol{H}}$ is circulant due to the insertion and removal of the CP. We will find this time-frequency duality useful in the ensuing design steps. By inspection of Figure 9.12, we can write the matrix-vector input-output relationship as

$$\boldsymbol{y} = \sqrt{\bar{\gamma}}\,\boldsymbol{\Gamma D}_h \boldsymbol{\Theta u} + \boldsymbol{\Gamma}\bar{\boldsymbol{w}}, \tag{9.31}$$

where $\boldsymbol{y}$ is an $N_x \times 1$ vector, $\boldsymbol{\Theta}$ denotes an $N_x \times N_u$ encoder, $\boldsymbol{\Gamma}$ is an $N_u \times N_x$ linear decoder, $\boldsymbol{D}_h$ is the diagonal matrix in (9.13), and $\boldsymbol{u}$ is the $N_u \times 1$ vector output of the differential encoder $\mathcal{D}(\boldsymbol{s})$. To understand how the differential system in Figure 9.12 works, we need to specify the differential encoder and decoder.

9.6.1 Inner Codec

The role of the inner encoder $\boldsymbol{\Theta}$ in Figure 9.12 is to map the time-selective channel $(\boldsymbol{D}_h)$ to a frequency-selective equivalent channel $(\tilde{\boldsymbol{H}})$ and then partition the large channel matrix $\tilde{\boldsymbol{H}}$ into N_o identical submatrices each of size $N_I \times N_I$, with Q guard symbols between them. To this effect, we select the information block size N_u to be an integer multiple of N_o (i.e., $N_u = N_o N_I$) and subsequently choose the inner encoder $\boldsymbol{\Theta}$ as

$$\boldsymbol{\Theta} = \boldsymbol{F}_{N_x}(\boldsymbol{I}_{N_o} \otimes \boldsymbol{T}_{\text{cp}})(\boldsymbol{I}_{N_o} \otimes \boldsymbol{F}_{N_I}^{\mathcal{H}}), \tag{9.32}$$

where $\boldsymbol{T}_{\text{cp}} := [\boldsymbol{T}_1, \boldsymbol{I}_{N_I}, \boldsymbol{T}_2]^T$ is the matrix inserting a CP of length Q per subblock of size N_I using the $N_I \times N_I$ identity operator $\boldsymbol{I}_{N_I}$ and the matrices

$$\begin{aligned} \boldsymbol{T}_1 &:= [\boldsymbol{0}_{(Q/2)\times(N_I - Q/2)}, \boldsymbol{I}_{Q/2}]^T \\ \boldsymbol{T}_2 &:= [\boldsymbol{I}_{Q/2}, \boldsymbol{0}_{(Q/2)\times(N_I - Q/2)}]^T. \end{aligned}$$

Checking dimensionalities of the matrices forming $\boldsymbol{\Theta}$ in (9.32), we can verify that $N_x = N_o(N_I + Q) = N_u + N_o Q$.

At the receiver we design the inner decoder as

$$\boldsymbol{\Gamma} = (\boldsymbol{I}_{N_o} \otimes \boldsymbol{F}_{N_I})(\boldsymbol{I}_{N_o} \otimes \boldsymbol{R}_{\text{cp}})\boldsymbol{F}_{N_x}^{\mathcal{H}}, \tag{9.33}$$

where $\boldsymbol{R}_{\text{cp}} := [\boldsymbol{0}_{N_I \times (Q/2)}, \boldsymbol{I}_{N_I}, \boldsymbol{0}_{N_I \times (Q/2)}]$ is the matrix implementation of the CP removal operation per subblock. Notice that if $\bar{\boldsymbol{w}}$ is AWGN, $\boldsymbol{\Gamma}\bar{\boldsymbol{w}}$ remains white since $\boldsymbol{\Gamma}$ is unitary.

Using (9.15) and (9.33), it follows by direct substitution that the equivalent aggregate channel becomes

$$\boldsymbol{\Gamma}\boldsymbol{D}_h\boldsymbol{\Theta} = \boldsymbol{I}_{N_o} \otimes \tilde{\boldsymbol{D}}_h, \tag{9.34}$$

where

$$\begin{aligned} \tilde{\boldsymbol{D}}_h &:= \text{diag}[H(0), H(1), \ldots, H(N_I - 1)], \\ H(n) &:= \sum_{q=0}^{Q} h_q e^{-j2\pi(q-Q/2)n/N_I}. \end{aligned}$$

The effect of the inner codec is threefold: (i) to convert the time-selective channel into an equivalent frequency-selective channel (via $\boldsymbol{F}_{N_x}$ and $\boldsymbol{F}_{N_x}^{\mathcal{H}}$); (ii) to partition each block containing N_u information symbols into N_o subblocks; and (iii) to apply OFDM per subblock. Note that over the N_o subblocks, the equivalent channel matrix is the same, namely $\tilde{\boldsymbol{D}}_h$, since the BEM coefficients remain invariant per block of N_u

information symbols. This enables application of block differential encoding *across* the N_o subblocks of the same block, in a fashion that can be thought of as the time-frequency dual of that applied in Chapter 8 for frequency-selective channels; see also [159]. What makes this possible is the BEM, which entails quasi-static coefficients per block while allowing for rapid variations within the block through the bases.

9.6.2 Outer Differential Codec

Let us now turn our attention to the differential encoder $\mathcal{D}(\cdot)$ and decoder $\bar{\mathcal{D}}(\cdot)$ in Figure 9.12. Substituting $\boldsymbol{x} := \boldsymbol{\Theta u}$ and inserting (9.34) into (9.31), we can rewrite the input-output model as

$$\boldsymbol{y} = \sqrt{\bar{\gamma}}\,(\boldsymbol{I}_{N_o} \otimes \tilde{\boldsymbol{D}}_h)\boldsymbol{u} + \boldsymbol{w}, \tag{9.35}$$

where $\boldsymbol{w} := \boldsymbol{\Gamma}\bar{\boldsymbol{w}}$. In accordance with the inner codec, let us partition the vectors $\boldsymbol{s}$, $\boldsymbol{u}$, and $\boldsymbol{y}$ into N_o equally long subblocks (i.e., $\boldsymbol{s} = [\boldsymbol{s}_0^T, \ldots, \boldsymbol{s}_{N_o-1}^T]^T$, and similarly for $\boldsymbol{u}$, $\boldsymbol{y}$, and $\boldsymbol{w}$). Now we can write (9.35) in a per subblock basis as

$$\boldsymbol{y}_b = \sqrt{\bar{\gamma}}\,\tilde{\boldsymbol{D}}_h\,\boldsymbol{u}_b + \boldsymbol{w}_b, \;\; \forall b \in [0, N_o - 1]. \tag{9.36}$$

Before entering the inner encoder, we differentially encode each subblock using the matrix-vector recursion

$$\boldsymbol{u}_b = \begin{cases} \boldsymbol{F}_b \boldsymbol{u}_{b-1} & \text{if } 1 \le b \le N_o - 1, \\ \boldsymbol{1}_{N_I} & \text{if } b = 0, \end{cases} \tag{9.37}$$

where the $N_I \times N_I$ diagonal matrix $\boldsymbol{F}_b \in \mathbb{X}$ conveys the information and is chosen to correspond one-to-one with the bth subblock $\boldsymbol{s}_b$. It is not difficult to verify that the length of $\boldsymbol{s}$ is the same as that of $\boldsymbol{u}$ (i.e., $N_s = N_u$). Suppose that each entry of $\boldsymbol{s}_b$ is chosen from a finite alphabet with cardinality 2^R (each symbol contains R information bits). The set $\mathbb{X}$ must therefore have cardinality $|\mathbb{X}| = 2^{RN_I}$. A simple design comprises a commutative group $\mathbb{X}$ of diagonal matrices with 2^{RN_I} elements so as to make it cyclic (cf. Section 6.2.2). With this design of $\mathbb{X}$, if we select $N_I \ge Q+1$, the maximum Doppler diversity can be guaranteed. However, when N_I and/or R is large, the drawback of the differential scheme in (9.37) is twofold: (i) its design entails a computationally complex search; and (ii) its decoding complexity is high. In the following we modify the differential design by partitioning the bases. This partitioning can be thought of as the dual of the subcarrier grouping approach we encountered with the MIMO OFDM system in Section 8.3.1. The resulting scheme will turn out to enjoy full diversity and at the same time flexibility to trade off decoding complexity for performance.

Suppose that we further divide $\boldsymbol{u}_b$ into N_g groups each with length N_{sub} (i.e., $N_I = N_g N_{\text{sub}}$). The gth group is given as $[\boldsymbol{u}_b(g)]_k := u_b(g + N_g k), k \in [0, N_{\text{sub}} - 1]$. The idea is to apply the differential encoder (9.37) per group across all the subblocks; that is, to have

$$\boldsymbol{u}_b(g) = \begin{cases} \boldsymbol{F}_{b_g} \boldsymbol{u}_{b-1}(g) & \text{if } 1 \le b \le N_I - 1, \\ \boldsymbol{1}_{N_{\text{sub}}} & \text{if } b = 0, \end{cases} \tag{9.38}$$

where $\forall g \in [0, N_g - 1]$, the diagonal matrix $\boldsymbol{F}_{b_g} \in \mathbb{X}$ conveys the information from $\boldsymbol{s}_b(g)$. Now to support rate R, we only need to design a set $\mathbb{X}$ with cardinality $2^{RN_{\text{sub}}} < 2^{RN_I}$. The design of $\mathbb{X}$ is given in Chapter 6. Clearly, when $Q = 0$ (i.e., when the channel is time invariant over one block), vectors and matrices in (9.38) become scalars and the differential design here reduces to the scalar DPSK outlined in Section 6.2.1.

To derive the outer block differential decoder, we use (9.36) to write the input-output relationship for the gth group of the bth subblock as

$$\boldsymbol{y}_b(g) = \sqrt{\bar{\gamma}}\, \tilde{\boldsymbol{D}}_h(g)\, \boldsymbol{u}_b(g) + \boldsymbol{w}_b(g), \tag{9.39}$$

where $\boldsymbol{y}_b(g)$ and $\boldsymbol{w}_b(g)$ are defined similar to $\boldsymbol{u}_b(g)$, and $\tilde{\boldsymbol{D}}_h(g)$ is the channel submatrix from $\tilde{\boldsymbol{D}}_h$ corresponding to the group g. By interchanging $\tilde{\boldsymbol{D}}_h(g)$ with $\boldsymbol{u}_b(g)$, we have $\tilde{\boldsymbol{D}}_h(g)\boldsymbol{u}_b(g) = \boldsymbol{D}_{u_b}(g)\tilde{\boldsymbol{h}}(g)$, where $\boldsymbol{D}_{u_b}(g) := \text{diag}[\boldsymbol{u}_b(g)]$, and $\tilde{\boldsymbol{h}}(g)$ contains the diagonal elements of $\tilde{\boldsymbol{D}}_h(g)$. Since $\boldsymbol{F}_{b_g}$ is unitary, $\boldsymbol{D}_{u_b}(g)$ is also unitary. Relating $\boldsymbol{y}_b(g)$ with $\boldsymbol{y}_{b-1}(g)$, we find that [cf. (9.39)]

$$\boldsymbol{y}_b(g) = \boldsymbol{D}_{u_b}(g)\boldsymbol{D}^{\mathcal{H}}_{u_{b-1}}(g)\left[\boldsymbol{y}_{b-1}(g) - \boldsymbol{w}_{b-1}(g)\right] + \boldsymbol{w}_b(g). \tag{9.40}$$

Substituting $\boldsymbol{D}_{u_b}(g) = \boldsymbol{F}_{b_g}\boldsymbol{D}_{u_{b-1}}(g)$ into (9.40), we obtain

$$\boldsymbol{y}_b(g) = \boldsymbol{F}_{b_g}\boldsymbol{y}_{b-1}(g) + \boldsymbol{w}'_b(g), \tag{9.41}$$

where $\boldsymbol{w}'_b(g) := \boldsymbol{w}_b(g) - \boldsymbol{F}^{\mathcal{H}}_{b_g}(g)\boldsymbol{w}_{b-1}(g)$ is AWGN because $\boldsymbol{D}_{u_{b-1}}(g)$ is a unitary matrix.

Based on (9.41), the ML detector for $\boldsymbol{F}_{b_g}$ is

$$\hat{\boldsymbol{F}}_{b_g} = \arg\min_{\boldsymbol{F}\in\mathbb{X}} \left\| \boldsymbol{y}_b(g) - \boldsymbol{F}\boldsymbol{y}_{b-1}(g) \right\|_{\text{F}}, \quad \forall g \in [0, N_g - 1], \tag{9.42}$$

where $\|\cdot\|_{\text{F}}$ denotes the Frobenius norm. Using the one-to-one demapping $\boldsymbol{F}_{b_g} \rightarrow \boldsymbol{s}_b(g)$, we can obtain the information-conveying group $\hat{\boldsymbol{s}}_b(g)$ from $\hat{\boldsymbol{F}}_{b_g}$ with the maximum possible Doppler diversity provided by the time-selective channel.

On top of an initial training symbol (here block symbol) that is needed by all differential designs, the CP insertion per subblock causes extra bandwidth loss in the method just presented. To improve the bandwidth efficiency, [186] introduced a block differential scheme that avoids the CP guards and thus operates on blocks of size $N_x = N_u$. Block differential encoding is applied directly in [186, Scheme BD-II] along with decision feedback or Viterbi decoding.

9.7 ST CODES FOR DOUBLY SELECTIVE CHANNELS

As we established in Section 9.1.3, doubly selective channels provide multipath-Doppler diversity. In this section we show how to enable the maximum possible joint space-multipath-Doppler diversity when multiple antennas are employed.

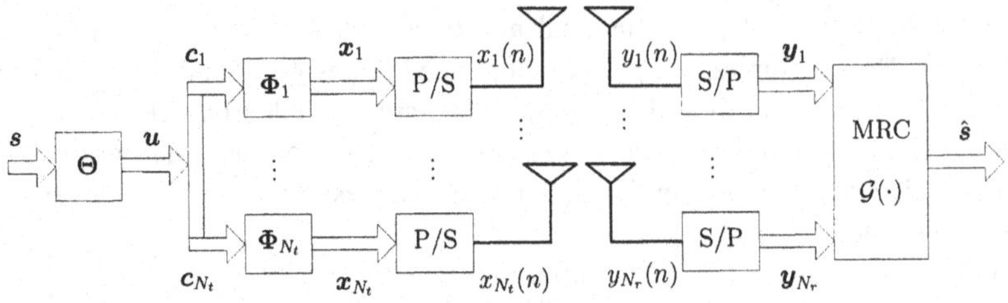

Figure 9.13 Discrete-time transceiver model for doubly selective MIMO channels

At each receive-antenna, the symbol-rate sampled sequence $y_\nu(n)$ at the receive filter output is serial-to-parallel-converted to form $N \times 1$ blocks $[\boldsymbol{y}_\nu]_n := y_\nu(n)$. Suppose that guard intervals are inserted per transmitted block to remove the interblock interference as usual. The pertinent input-output relationship can then be expressed in matrix-vector form as

$$\boldsymbol{y}_\nu = \sqrt{\bar{\gamma}} \sum_{\mu=1}^{N_t} \boldsymbol{H}_{\nu\mu} \boldsymbol{x}_\mu + \boldsymbol{w}_\nu, \quad \forall \nu \in [1, N_r], \tag{9.43}$$

where the channel matrix is now

$$\boldsymbol{H}_{\nu\mu} := \sum_{q=0}^{Q} \boldsymbol{D}_q \boldsymbol{H}_q^{(\nu,\mu)} \tag{9.44}$$

with $\boldsymbol{D}_q := \text{diag}[1, \exp(j\omega_q), \ldots, \exp(j\omega_q(N-1))]$ and $\boldsymbol{H}_q^{(\nu,\mu)}$ a lower-triangular Toeplitz matrix with first column $[h_q^{(\nu,\mu)}(0), \ldots, h_q^{(\nu,\mu)}(L), 0, \ldots, 0]^T$. Each block $\boldsymbol{y}_\nu$ is decoded by $\mathcal{G}(\cdot)$ to obtain an estimate of $\boldsymbol{s}$ as: $\hat{\boldsymbol{s}} := \mathcal{G}(\{\boldsymbol{y}_\nu\}_{\nu=1}^{N_r})$.

For the doubly selective MIMO channel described by the BEM equation (9.43), the following result has been established in [181]:

Proposition 9.8 *Consider (N_t, N_r) multi-antenna transmissions over doubly selective channels adhering to a BEM with $Q+1$ bases and $L+1$ taps. If the channel taps are Gaussian distributed and the correlation matrix of the $N_t N_r(L+1)(Q+1)$ channel coefficients has rank r_h, the maximum diversity order of transmissions adhering to* (9.43) *is $G_d = r_h \leq N_t N_r (L+1)(Q+1)$.*

The transceiver structure in Figure 9.13 can achieve this high-order space-Doppler-multipath diversity [181]. The transmit-power is allocated equally, by setting $\boldsymbol{c}_\mu = \boldsymbol{u}/\sqrt{N_t}$, $\forall \mu \in [1, N_t]$. Using (9.43), $\boldsymbol{y}_\nu$ and $\boldsymbol{s}$ can then be related via

$$\boldsymbol{y}_\nu = \sqrt{\frac{\bar{\gamma}}{N_t}} \sum_{\mu=1}^{N_t} \boldsymbol{H}_{\nu\mu} \boldsymbol{\Phi}_\mu \boldsymbol{\Theta} \boldsymbol{s} + \boldsymbol{w}_\nu, \quad \forall \nu \in [1, N_r]. \tag{9.45}$$

Different channels in (9.44) share the same exponential bases but have different channel coefficients. Similar to the time-selective case, the idea is to shift the $Q+1$ bases of each channel corresponding to one of the N_t transmit antennas so that all the bases become consecutive on the FFT grid of complex exponentials. This converts the N_t channels corresponding to each receive-antenna to an equivalent single doubly selective channel with $N_t(Q+1)$ bases. Analytically, this conversion is effected using DPS matrices $\{\boldsymbol{\Phi}_\mu\}_{\mu=1}^{N_t}$, which are designed as

$$\boldsymbol{\Phi}_\mu := \operatorname{diag}[1, e^{j\phi_\mu}, \ldots, e^{j\phi_\mu(N-1)}], \quad \forall \mu \in [1, N_t], \tag{9.46}$$

with $\phi_\mu := 2\pi(\mu-1)(Q+1)/N$. Recalling the structure of $\boldsymbol{H}_q^{(\nu,\mu)}$ in (9.44), we have

$$\boldsymbol{H}_q^{(\nu,\mu)}\boldsymbol{\Phi}_\mu = \boldsymbol{\Phi}_\mu \bar{\boldsymbol{H}}_q^{(\nu,\mu)}, \quad \forall q, \nu, \mu,$$

where $\bar{\boldsymbol{H}}_q^{(\nu,\mu)}$ has the same structure as $\boldsymbol{H}_q^{(\nu,\mu)}$, but its first column is $[h_q^{(\nu,\mu)}(0), h_q^{(\nu,\mu)}(1)e^{-j\phi_\mu}, \ldots, h_q^{(\nu,\mu)}(L)e^{-j\phi_\mu L}, 0, \ldots, 0]^T$.

Considering the DPS matrices in (9.46), equation (9.45) can be rewritten as

$$\boldsymbol{y}_\nu = \sqrt{\frac{\bar{\gamma}}{N_t}} \sum_{q=0}^{N_t(Q+1)-1} \boldsymbol{D}_q \boldsymbol{H}_q^{(\nu)} \boldsymbol{\Theta} \boldsymbol{s} + \boldsymbol{w}_\nu, \quad \forall \nu \in [1, N_r], \tag{9.47}$$

where $\boldsymbol{H}_q^{(\nu)} := \bar{\boldsymbol{H}}_{\operatorname{mod}(q,Q+1)}^{(\nu,\lfloor q/(Q+1)\rfloor+1)}$.

Comparing (9.47) with the model (9.15) for time-selective frequency-flat channels, we deduce that this DPS design transforms the N_t transmit-antenna setup, where each channel can be expressed via $Q+1$ exponential bases into a single transmit-antenna setup where the equivalent channel can be expressed by $N_t(Q+1)$ Fourier bases. Hence, relying on the precoder $\boldsymbol{\Theta}$ and $\mathcal{G}(\cdot)$, we can now apply any single-input multi (single)-output method for doubly selective channels to ensure the maximum diversity gain. Here, we adopt the following time-frequency precoder with time-frequency guards [183]

$$\boldsymbol{\Theta} = (\boldsymbol{F}_{P+N_t(Q+1)-1}^{\mathcal{H}} \boldsymbol{T}_f) \otimes \boldsymbol{T}_t, \tag{9.48}$$

where $\boldsymbol{T}_f := [\boldsymbol{I}_P, \boldsymbol{0}_{P\times(Nt(Q+1)-1)}]^T$ places the frequency guard, $\boldsymbol{T}_t := [\boldsymbol{I}_K, \boldsymbol{0}_{K\times L}]^T$ inserts the time guard, and P and K are defined so that $N = [P + N_t(Q+1) - 1](K+L)$. It turns out that the LCF precoder in (9.48) enables the full multipath-Doppler diversity; see [183] for a detailed proof.

9.7.1 Numerical Examples

In this section we verify numerically that the diversity provided by doubly selective channels is indeed multiplicative in the number of degrees of freedom involved per dimension (space, multipath, Doppler).

Example 9.3 We consider first a single-antenna system with carrier frequency $f_o = 900$ MHz. With maximum mobile velocity $v_{\max} = 96$ km/h, the maximum Doppler shift is $f_{\max} = 80$ Hz. Using these parameters, the Jakes model

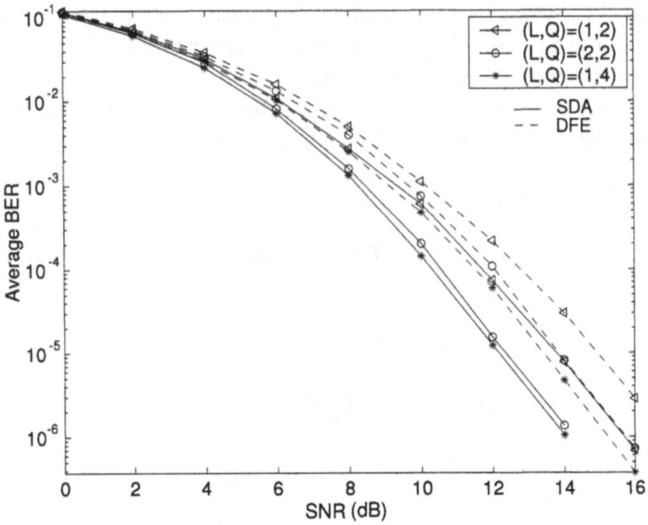

Figure 9.14 Multipath-Doppler diversity illustration

is generated according to [127, Page 68] for each tap. The $L+1$ channel taps are generated as independent Gaussian random variables with multipath power intensity profile $\phi_c(\tau) = \exp(-0.1\tau/T_s)$. The parameters of the time-frequency LCF precoder in (9.48) are set to $(P, K) = (4, 4)$. For a fixed (L, Q) pair, we select $N = (P+Q)(K+L)$ and then we choose T_s. SDA or block MMSE-decision-feedback equalization (DFE) is used at the receiver for demodulation. Figure 9.14 shows that the diversity increases as L and/or Q increase. There is less than a 2 dB difference in SNR between the DFE and the SDA. Observing the slope of the error curve when $(L, Q) = (1, 2)$, we deduce that the diversity order is $(L+1)(Q+1) = 6$. Diversity orders 9 and 10 show up for the other two cases $[(L, Q) = (2, 2)$ and $(L, Q) = (1, 4)]$ for SNR values higher than 16 dB, which is out of the SNR range of this figure. Recall that when quantifying the diversity order, the channel was assumed to obey the BEM throughout this chapter. In this example, however, the channel was generated according to the Jakes model. Figure 9.14 validates the BEM and corroborates the diversity claims that are based on it.

Example 9.4 Here we test a multi-antenna system with $(N_t, N_r) = (2, 1)$ and $\boldsymbol{\Theta}$ selected as in (9.48). The BEM coefficients are generated as i.i.d. zero mean Gaussian random variables with variance $1/[(L+1)(Q+1)]$. Figure 9.15 depicts the simulated error performance at the demodulator output when either the SDA or ZF equalization is employed. We verify that as Q or L increases, the diversity order increases. SDA outperforms ZF equalization only by about 1 dB when BER $= 10^{-3}$ and $(Q, L) = (2, 2)$. This suggests that when the channels' potential diversity order is high, even linear equalizers can afford acceptable BER performance while offering reduced-complexity demodulation.

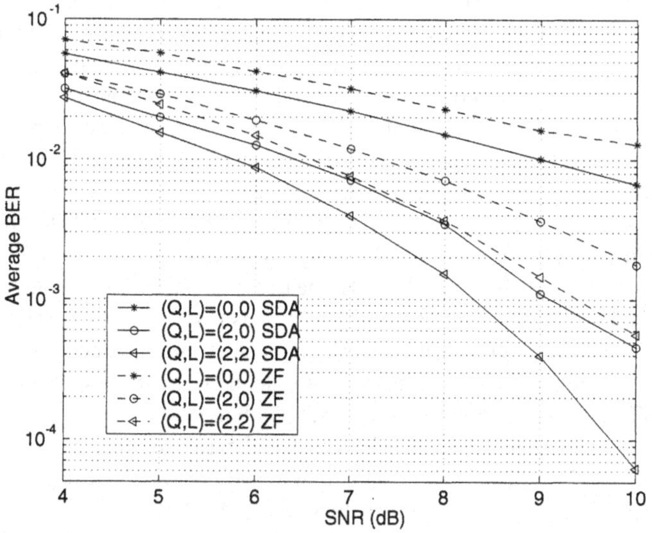

Figure 9.15 Space-multipath-Doppler diversity

9.8 CLOSING COMMENTS

Mismatch between transmit-receive oscillators as well as mobility-induced Doppler effects give rise to time-varying wireless channels. Depending on the extent of the delay-Doppler spread relative to the data rate and the operational carrier frequency, TV channels can be time- or doubly selective (i.e., time and frequency-selective). In the multi-antenna context of this book, the extra (Doppler) dimension provided by TV effects presents challenges when it comes to acquiring channel state information but also provides additional degrees of freedom. In conjunction with the space-time (and possibly multipath) dimensions, these degrees of freedom enhance the potential for improving error and rate performance when ST codes are tailored for operation over TV MIMO channels.

High-rate operation ensuring reliable error performance at high mobility is the main scope of the IEEE802.20 mobile broadband wireless access (MBWA) standard. With peak rates in excess of 1 Mbps, MBWA supports various vehicular mobility classes up to 250 km/h in a metropolitan area network to meet the needs of business and residential end-user markets. The ST codes presented in this chapter are in response to these needs. Their design relies on a basis expansion model which offers a parsimonious description of time- and doubly selective channels over a finite time horizon using a TV Fourier basis and a finite number of time-invariant BEM parameters. For an (N_t, N_r) multi-antenna system, the number of BEM parameters ($Q+1$ for the Doppler dimension and $L+1$ for the multipath dimension) quantify the diversity order provided by time-selective MIMO channels as $N_tN_r(Q+1)$ and the diversity order of doubly selective MIMO channels as $N_tN_r(Q+1)(L+1)$. Furthermore, the BEM manifests a neat duality between time- and frequency-selective channel models based on which ST codes designed for frequency-selective MIMO channels can be mapped to ST codes for time- and doubly selective MIMO channels.

Block ST codes for coherent and differentially encoded transmissions were designed based on this duality to enable the maximum space-time-Doppler diversity provided by time-selective MIMO channels. These block codes are dual to codes designed for frequency-selective channels based on LCFC and offer flexibility to achieve error rate-complexity trade-offs which compare favorably with STDO trellis codes. The latter can also be derived for block fading TV MIMO channels based on modified design criteria. Among the block STDO code designs, the one based on digital-phase sweeping is particularly attractive because it avoids losses in transmission rate while achieving maximum diversity and coding gains at affordable decoding complexity. The DPS operation plays also an instrumental role in the design of maximum diversity achieving ST codes for doubly selective MIMO channels.

Besides 1 symbol per channel use, the BEM-based duality guides further the design of FDFR ST codes for time-selective MIMO channels. Coherent as well as differential FDFR designs for doubly selective MIMO channels are topics of current investigation, and preliminary works in these directions can be found in [37,38,76,78] and references therein.

10

Joint Galois- and Linear Complex-Field ST Codes

The ST coded systems introduced and analyzed so far did not account explicitly for channel coding, assuming tacitly that the underlying channel code which is always present in practical systems is to be designed separately from the ST code chosen. To appreciate the limitations of this separate design, let us recall that in a MIMO fading channel the maximum diversity enabled by ST coding is proportional to the number of transmit-antennas times the number of receive-antennas. Consequently, the diversity achieved by ST coding is limited by the number of antennas and the associated deployment cost and complexity that a system can afford. But even if a sufficiently large number of antennas can be deployed, ST coding can at best render multiple flat fading channels equivalent to an additive white Gaussian noise (AWGN) channel. As a result, a multi-antenna system with ST coding alone can at best perform as an uncoded single-antenna system in AWGN.

On the other hand, channel coding allows for error control (EC) through which a system can meet a prescribed error performance. For AWGN channels, EC coding is the standard means of approaching the performance dictated by capacity without overexpanding bandwidth. For fading channels, EC (i.e., channel) codes usually offer larger coding gains than ST codes. They also enable diversity gains even for single-antenna transmissions. However, the diversity gain enabled by EC coding is determined by the minimum Hamming distance (or free distance) of the code used. The higher the diversity gains, the longer the codewords or the larger the memory that has to be used, which comes at the price of increased decoding complexity and longer decoding delays. As a result, EC coding alone is not as effective as ST coding in enabling the diversity provided by fading channels.

The individual limitations of ST and EC coding motivate their joint consideration over fading channels with the goal of combining their strengths: namely, the advantages of ST coding in enabling the diversity with the large coding gains that EC codes can offer when the channel is (close to) AWGN. Although in principle any ST code can be combined with any EC code, in this chapter we concentrate on the linear complex field (LCF) ST codes introduced in Sections 3.5 and 4.3 because they are available for any number of transmit-receive antennas in closed form and they do not sacrifice bandwidth. Since EC codes are defined over the Galois field (GF), their combination with LCF will be referred to as *joint GF-LCF coding*.

Following the classification of individual ST codes in Chapters 3 and 4, we also develop error performance- and rate-oriented joint GF-LCF ST codes by combining EC codes with STLCF and layered ST (BLAST or FDFR) codes, respectively. It will turn out that joint GF-LCF ST coding enables a diversity gain that is multiplicative in what can be achieved by ST and EC coding alone. It will further be shown that GF-LCF ST codes can be decoded via low-complexity iterative (a.k.a. turbo) decoding, which will allow the GF-coded FDFR and V-BLAST classes to approach MIMO capacity. Simulations will also confirm that joint GF-LCF ST coding leads to multi-antenna systems most promising for practical deployment since they offer desirable trade-offs in rate, error performance, and complexity.

10.1 GF-LCF ST CODES

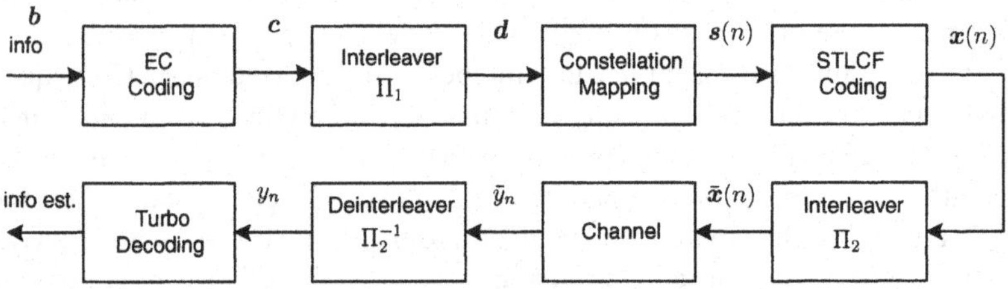

Figure 10.1 Block diagram of a GF-LCF ST coded system

Figure 10.1 depicts a GF-LCF ST coded wireless system with N_t transmit-antennas and one receive-antenna. The flat fading channels from each transmit-antenna to the receive-antenna are assumed to satisfy assumptions A2.1 and A2.2 in Chapter 2. At the transmitter side, a stream of information bits $\boldsymbol{b} := (\ldots, b_{-1}, b_0, b_1, \ldots)$ is first encoded using an EC encoder. The EC-coded bit sequence $\boldsymbol{c} := (\ldots, c_{-1}, c_0, c_1, \ldots)$ passes through interleaver Π_1, which outputs the permuted bit sequence $\boldsymbol{d} := (\ldots, d_{-1}, d_0, d_1, \ldots)$. The role of Π_1 is to ensure that consecutive symbols will experience diversified channel gains as described in Section 1.4.2; see also Figure 1.5. Permuted bits $\{d_n\}$ are mapped to constellation symbols $s_n \in \mathbb{S}$ drawn from the signal constellation $\mathbb{S}$. Possible constellations include BPSK, QPSK, or QAM. Groups of $\log_2 |\mathbb{S}|$ consecutive bits d_n are mapped to one constellation symbol. When the signal constellation is multilevel (i.e., $|\mathbb{S}| > 2$), the bit mapper adopted will affect the coding gain, depending on the decoding scheme. For conventional noniterative decoding, Gray mapping is optimum in the sense of minimizing the bit error probability; but if iterative bit-demapping is used, then anti-Gray mapping outperforms Gray mapping [258]. For the GF-LCF ST coded system under consideration, we assume that the bit mapper has been optimized according to [258].

After mapping bits to symbols, successive blocks of N_t symbols $\boldsymbol{s}(n) := [s_{nN_t}, s_{nN_t+1}, \ldots, s_{nN_t+N_t-1}]^T$ are further encoded by an $N_t \times N_t$ matrix $\boldsymbol{\Theta}_{\text{LCF}}$, designed as in Section 3.5, to obtain an LCF-coded block $\boldsymbol{x}(n) = \boldsymbol{\Theta}_{\text{LCF}}\boldsymbol{s}(n)$. After LCF coding, each block $\boldsymbol{x}(n)$ is serialized to $x_{nN_t}, x_{nN_t+1}, \ldots, x_{nN_t+N_t-1}$. The precoded symbol sequence $\boldsymbol{x} := (\ldots, x_{-1}, x_0, x_1, \ldots)$ is interleaved by a second interleaver Π_2. The role of Π_2 is to decorrelate the channels in the time domain, although this can be achieved equally well by increasing the size of Π_1. The output of Π_2, denoted by $\bar{\boldsymbol{x}} := (\ldots, \bar{x}_{-1}, \bar{x}_0, \bar{x}_1, \ldots)$, is then parsed into blocks $\bar{\boldsymbol{x}}(n) := [\bar{x}_{nN_t}, \bar{x}_{nN_t+1}, \ldots, \bar{x}_{nN_t+N_t-1}]^T$, each of which is treated as $\boldsymbol{x}_{\text{LCF}}$ in (3.54) and transmitted symbol-by-symbol after pulse shaping and carrier modulation as usual.

According to A2.1 and A2.2, the channel between the μth transmit-antenna and the receive-antenna is modeled as a complex Gaussian random variable h_μ whose value remains constant over blocks containing N_t transmitted symbols. With $\bar{x}_n$ transmitted symbol-by-symbol [cf. (3.54)], at the receiver end the $(nN_t + m)$th received sample is given by

$$\bar{y}_{nN_t+\mu-1} = \sqrt{\bar{\gamma}}\, h_\mu \bar{x}_{nN_t+\mu-1} + \bar{w}_{nN_t+\mu-1}, \; n \in [-\infty, \infty], \; \mu \in [1, N_t], \quad (10.1)$$

where $\bar{w}_n$ is the i.i.d. complex AWGN with zero mean and unit variance. The received samples $\bar{y}_n$ are deinterleaved by Π_2^{-1} to yield

$$y_n = \sqrt{\bar{\gamma}}\alpha_n x_n + w_n \quad (10.2)$$

where $y_n := \Pi_2^{-1}[\bar{y}_n]$, $\alpha_n := \Pi_2^{-1}[h_n]$ and $w_n := \Pi_2^{-1}[\bar{w}_n]$. With a well designed Π_2 of sufficiently large depth, the deinterleaved channel coefficients α_n can safely be assumed to be uncorrelated (thus independent) Gaussian random variables with zero mean and unit variance.

10.1.1 Separate Versus Joint GF-LCF ST Coding

To appreciate the value of combining GF with LCF coding, it is instructive to reflect on their individual merits. Supposing that the channel fading is sufficiently fast to provide ample diversity, a properly designed STLCF code matrix $\boldsymbol{\Theta}_{\text{LCF}}$ of size $N_t \times N_t$ can enable diversity of order N_t. In the limit $(N_t \to \infty)$, STLCF coding converts a flat Rayleigh-fading MISO channel to an AWGN channel [29]. The price paid is twofold. First, to guarantee collection of the full diversity, the receiver must rely on ML decoding, which incurs exponential complexity in the precoder size N_t. Although in certain cases the sphere decoding algorithm can approach ML performance in polynomial time (cf. Chapter 5), STLCF coding comes at the expense of (at times prohibitively) high decoding complexity. Second, since the ideal STLCF encoder with $N_t \to \infty$ renders the fading channel equivalent to an AWGN channel, its error performance cannot exceed that of an uncoded single-antenna transmission over an AWGN channel. This implies that STLCF coding alone cannot approach the performance dictated by the capacity of a MIMO channel.

To reduce the ML decoding complexity and since diversity orders beyond 4 show up for impractically high SNR values, it is reasonable to limit ourselves to small STLCF codes of size $N_t \leq 4$. We will see later that there is practically little to be gained by using codes of size $N_t > 4$. For $N_t \leq 4$ and small-size constellations such as BPSK or QPSK, one can even afford an exhaustive ML search at the decoding stage.

For $N_t = 2$ or $N_t = 4$, the STLCF codes in (3.69) were shown in Section 3.5.5 to enable the maximum diversity (2 or 4) and maximum coding gain (1 or $1/2$), for QAM constellations. Besides avoiding bandwidth loss, another nice property that these codes possess is that all their entries have norm $1/\sqrt{N_t}$. This property will be used when deriving the performance of GF-LCF ST coded systems in Section 10.1.2.

Relative to LCF codes, GF-based EC codes achieve larger coding gains not only over fading channels but also over AWGN channels, for which LCF-based codes bring no coding gains at all [178]. When used over fading channels, EC coding also enables diversity gains. For example, a linear block code with minimum Hamming distance $d_{\min}$ is capable of effecting diversity of order $d_{\min}$. For convolutional codes, the diversity gain equals d_{free}, the free distance of the code. But the free distance of an (n, k) convolutional code with memory m is upper bounded by [296, Page 575]

$$d_{\text{free}} \leq (m+1)n. \tag{10.3}$$

Equation (10.3) suggests that to increase d_{free} while keeping the code size n (and thus decoding delay) fixed, one can increase the memory m. The resulting diversity will then increase at best linearly with the memory m but at the price of an exponential increase in decoding complexity, because the number of trellis states grows as 2^m for binary codes.

In a nutshell, LCF codes of relatively small size are preferable for collecting high diversity at reasonable complexity, and GF codes have an edge because their large coding gain becomes increasingly critical as a fading channel comes close to an AWGN channel. Wedding GF with LCF coding will merge their strengths and minimize their individual limitations. Specifically, it will become possible to combine the large coding gains of GF-based EC codes with the diversity gains enabled by LCF codes. Both diversity and coding gains will be collected at the receiver using iterative decoding of concatenated GF-LCF codes, which will turn out to be only a few times more complex than the sum of their individual complexities. The performance of the resulting GF-LCF ST coded system will go beyond that of AWGN, allowing it to approach the performance dictated by the MIMO channel capacity.

10.1.2 Performance Analysis

To analyze the performance of GF-LCF ST coded systems, we rely on the union bound. For simplicity in exposition, only convolutional EC coding will be treated here for the GF based part. The analysis carries over, however, to any linear code with slight modifications.

We start with the bit-error enumerating function $B(Z)$ of the EC code in Figure 10.1, defined as $B(Z) := \sum_{w=d_{\min}}^{\infty} B_w Z^w$, where B_w is the total number of nonzero information bits on the weight-w error events. Using $B(Z)$ and pairwise error probability (PEP) analysis, the BER will be upper bounded using union bound techniques [13].

Let $\Delta_{\min}$ denote the minimum Euclidean distance between constellation points (i.e., $\Delta_{\min} := \min\{|s_1 - s_2| : s_1, s_2 \in \mathbb{S}, s_1 \neq s_2\}$). Clearly, $\Delta_{\min}$ depends on the constellation used. To make the PEP analysis tractable, we suppose that the GF-LCF ST coded system operates under the following simplifying conditions:

C10.1) The EC coded bits are mapped to different constellation symbols.

C10.2) The constellation symbols of an error event are coded in different STLCF blocks.

Conditions C10.1 and C10.2 are approximately true, thanks to the interleaver Π_1, especially for small weight error events or when the depth of Π_1 is large enough. For this reason, we refer to C10.1 and C10.2 as conditions of perfect interleaving.

10.1.2.1 ***Pairwise Error Probability*** Consider the EC-coded sequence $\tilde{\boldsymbol{c}} := (\ldots, \tilde{c}_{-1}, \tilde{c}_0, \tilde{c}_1, \ldots)$, which is to be interleaved, mapped to constellation symbols, and STLCF coded before transmission. Let $\boldsymbol{c} := (\ldots, c_{-1}, c_0, c_1, \ldots)$ denote another EC-coded sequence of Hamming distance w away from $\tilde{\boldsymbol{c}}$. When coded convolutionally, $\boldsymbol{c}$ will entail a trellis path that diverges from the path of $\tilde{\boldsymbol{c}}$ and re-merges back to it (we assume that the EC code is not catastrophic). Thanks to the interleaver Π_1, the symbols in $\boldsymbol{c}$ that are different from $\tilde{\boldsymbol{c}}$, say $c_{j_1}, c_{j_2}, \ldots, c_{j_w}$, $1 \leq j_1, j_2, \ldots, j_w < \infty$, will be scattered after interleaving and mapped to different constellation symbols, say $s_{n_1}, s_{n_2}, \ldots, s_{n_w}$. Proper design of the interleaver Π_1 ensures that for small w's the symbol indices $n_1, n_2, \ldots, n_w$ will be separated by at least N_t (i.e., $|n_a - n_b| > N_t, \forall a, b \in [1, w], a \neq b$); hence, the constellation symbols $s_{n_1}, s_{n_2}, \ldots, s_{n_w}$ will enter different blocks $\boldsymbol{s}(i_1), \boldsymbol{s}(i_2), \ldots, \boldsymbol{s}(i_w)$. Let $m_a := n_a \bmod N_t$ denote the column index of $\boldsymbol{\Theta}_{\text{LCF}}$ that s_{n_a} will be multiplied with, and define $l_a := n_a - m_a$, where $a \in [1, w]$.

All symbols in $\{s_n\}$ except $s_{n_a}, a \in [1, w]$, will be the same as their corresponding symbols $\{\tilde{s}_n\}$ in the sequence $\tilde{\boldsymbol{c}}$. If we subtract the received sample $\tilde{y}_n$ which corresponds to $\tilde{\boldsymbol{c}}$, from y_n, which corresponds to $\boldsymbol{c}$, we will obtain [cf. (10.2)]

$$\Delta y_{l_a+k} = \sqrt{\bar{\gamma}}\, \alpha_{l_a+k} \theta_{k,m_a} \Delta s_{n_a}, \; a = 1, 2, \ldots, w, \;\; k = 0, 1, \ldots, N_t - 1,$$

where $\Delta y_n := y_n - \tilde{y}_n$ and $\Delta s_n := s_n - \tilde{s}_n$. For $n \neq l_a + k$, we have $\Delta y_n = 0$.

The PEP $P_2(\tilde{\mathbf{c}} \to \mathbf{c}|h)$ conditioned on channel h (h is a generic symbol denoting the entire channel realization) is determined by the squared Euclidean distance,

$$d^2(y, \tilde{y}) = \bar{\gamma} \sum_{a=1}^{w} \sum_{k=0}^{N_t-1} |\alpha_{l_a+k} \theta_{k,m_a} \Delta s_{n_a}|^2. \tag{10.4}$$

By definition of $\Delta_{\min}$, it is true that $|\Delta s_{n_a}| \geq \Delta_{\min}$. Since all entries of $\boldsymbol{\Theta}_{\text{LCF}}$ have identical norm $1/\sqrt{N_t}$, it follows that $d^2(y, \tilde{y}) \geq \bar{\gamma}(\Delta_{\min}^2/N_t) \sum_{a=1}^{w} \sum_{k=0}^{N_t-1} |\alpha_{l_a+k}|^2$. Therefore, the PEP can be upper bounded by

$$P_2(\tilde{\mathbf{c}} \to \mathbf{c}|h) \leq Q\left(\sqrt{\frac{\bar{\gamma}}{2}\left(\frac{\Delta_{\min}^2}{N_t}\right) \sum_{a=1}^{w} \sum_{k=0}^{N_t-1} |\alpha_{l_a+k}|^2}\right). \tag{10.5}$$

Using the Chernoff bound $Q(x) \leq (1/2)e^{-x^2/2}$ and averaging with respect to the complex Gaussian distribution of the uncorrelated α's, we obtain

$$P_2(\tilde{\mathbf{c}} \to \mathbf{c}) \leq \frac{1}{2}\left(1 + \frac{\bar{\gamma}}{4}\frac{\Delta_{\min}^2}{N_t}\right)^{-wN_t}. \tag{10.6}$$

Equation (10.6) shows that diversity of order wN_t emerges in the PEP.

10.1.2.2 Bit Error Probability Applying the union bound to all error events, we can bound the BER as [13, Page 718]

$$\begin{aligned} P_b &\leq \sum_{w=d_{\min}}^{\infty} \frac{1}{2} B_w \left(1 + \frac{\bar{\gamma}\Delta_{\min}^2}{2N_t}\right)^{-wN_t} \\ &= \frac{1}{2} B\left(\left(1 + \frac{\bar{\gamma}\Delta_{\min}^2}{2N_t}\right)^{-N_t}\right). \end{aligned} \tag{10.7}$$

In deriving (10.7), we have implicitly used the uniformity of bit errors, that is, that the error properties of the coded sequence depend only on the weights rather than the distances (or locations) of the nonzero symbols in the error events. Under C10.1 and C10.2, this property holds true for the PEP upper bound in (10.7).

The upper bound in (10.7) manifests a *multiplicative* diversity effect since the slope of the BER curve at high SNR is $N_t d_{\min}$. A convolutionally coded (or any general EC-coded) transmission without STLCF coding will have a performance upper bound given by (10.7) with $N_t = 1$, simply because it can be viewed as a GF-LCF ST coded transmission with $N_t = 1$.

For convenience, we denote an EC-coded transmission with (respectively, without) STLCF coding by GF-LCF (EC-only). For $N_t \geq 2$, the gain in SNR of a GF-LCF system as compared to an EC-only system can be evaluated from (10.7) by equating the argument of the $B(\cdot)$ function in (10.7) with $(1 + \Delta_0^2\bar{\gamma}/4)^{-1}$, where Δ_0 is the minimum constellation distance needed for the EC-only system to achieve the same performance as that of the GF-LCF system. The ratio $G_{\text{LCF}} := \Delta_0^2/\Delta_{\min}^2$ measures the gain in SNR introduced by STLCF coding and can readily be found to be

$$G_{\text{LCF}} = \frac{\bar{\gamma}\Delta_0^2}{2N_t\left[(1+\bar{\gamma}\Delta_0^2/2)^{1/N_t} - 1\right]} \overset{N_t\to\infty}{\longrightarrow} \frac{\bar{\gamma}\Delta_0/2}{\ln(1+\bar{\gamma}\Delta_0/2)}. \tag{10.8}$$

Notice that G_{LCF} is monotonically increasing with N_t. Using BPSK and a rate 1/2 convolutional code with $\Delta_{\min} = 2$, Figure 10.2 depicts the gain G_{LCF} at $\bar{\gamma} = 8$ dB of the EC-LCF system, or $\Delta_0^2 = 12.6$. Notice that the gain for $N_t = 4$ is already quite large (2.7 dB). Further increasing N_t results in no more than a 0.8 dB gain in SNR, because the ultimate gain with $N_t = \infty$ is about 3.5 dB.

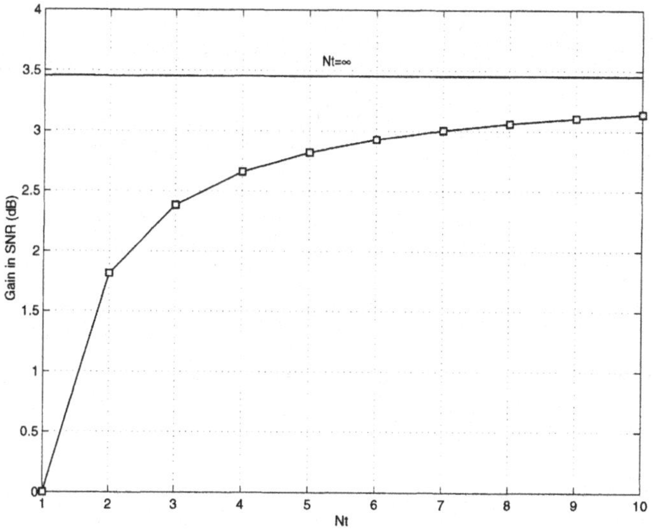

Figure 10.2 Gain of LCF coding

Example 10.1 Consider a rate-1/2 convolutional code with generators $(133, 171)$, BPSK constellation, and $N_t = 4$. For BPSK, we have $\Delta_{\min} = 2$. The code has $d_{\text{free}} = 10$ and $B(Z) = 36Z^{10} + 211Z^{12} + 1404Z^{14} + 11,633Z^{16} + \cdots$ which are obtained using the transfer function method described in [13, Page 544]. Under the perfect interleaving conditions C10.1 and C10.2, the BER can then be upper bounded by $P_b \leq (1/2)B\left([1+\bar{\gamma}/4]^{-4}\right)$.

Using the puncturing matrix

$$P_{3/4} = \begin{bmatrix} 1 & 1 & 0 \\ 1 & 0 & 1 \end{bmatrix},$$

we can obtain a punctured code of rate 3/4 from the rate-1/2 code, having $d_{\text{free}} = 5$ and $B_{3/4}(Z) = (42Z^5 + 201Z^6 + 1492Z^7 + 10469Z^8 + \cdots)/3$. We depict the performance upper bound (10.7) for these rate-1/2 and rate-3/4 codes with $N_t = 1, 2, 4$ in Figure 10.3. The $N_t = 1$ curves correspond to convolutionally coded systems without LCF coding. We can deduce from the slope of the curves that there is a multiplicative diversity gain introduced by STLCF coding. Also, the gains at SNR = 8 dB of the rate-1/2 convolutional code corroborate those of Figure 10.2.

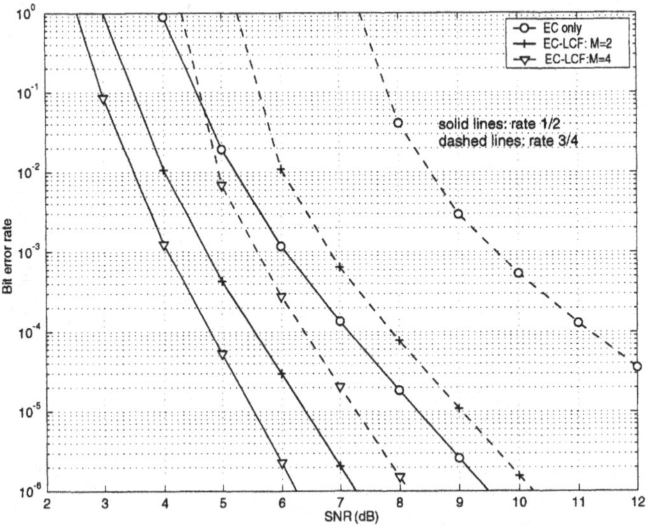

Figure 10.3 Comparison of EC-coded systems with and without STLCF coding

10.1.3 Turbo Decoding

Exhaustive search over all possible input sequences certainly offers exact ML detection of joint GF-LCF coded transmissions. The presence of the interleaver Π_1 in between the EC and STLCF coders prevents application of most efficient ML decoders to this joint detection problem. However, it has been widely demonstrated that iterative (turbo) decoding is quite effective in dealing with such joint ML detection problems. In what follows we show how the turbo principle applies to GF-LCF ST decoding.

10.1.3.1 Turbo Principle The turbo principle [15] amounts to passing soft information between two maximum a posteriori (MAP) soft-input soft-output (siso) modules iteratively [14]. Other approximate MAP modules, such as the Max-Log MAP and the modified soft-output Viterbi algorithm (SOVA), can also be used [279]. We consider only bit-by-bit MAP decoding. When the constellation is multilevel, iterative bit demapping can also be incorporated [258]. The soft information can be in the form of the likelihood function, log-likelihood ratios (LLRs), or just the probability distribution of an encoder's input and output symbols.

Each siso module is a four-port device that accepts as input the probability distribution of both the input and output of a coder, and outputs an updated probability distribution of both the input and the output, based on the module's input distribution and the coder structure. To decode GF-LCF ST coded transmissions, one needs two siso modules, denoted as siso-EC and siso-LCF in Figure 10.4.

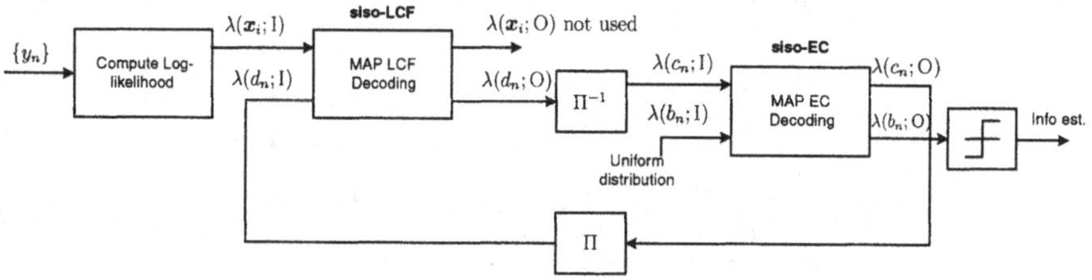

Figure 10.4 Turbo decoding with siso modules

Inputs and outputs of the siso modules are in the form of LLRs, denoted by $\lambda(\cdot\,; \mathrm{I})$ for inputs and $\lambda(\cdot\,; \mathrm{O})$ for outputs. For example,

$$\lambda(d_n; \mathrm{I}) = \log \frac{\mathrm{P}[d_n = +1; \mathrm{I}]}{\mathrm{P}[d_n = -1; \mathrm{I}]}, \tag{10.9}$$

$$\lambda(\boldsymbol{x}_i; \mathrm{I})|_{\boldsymbol{x}_i = \boldsymbol{x}} = \log \frac{\mathrm{P}[\boldsymbol{x}_i = \boldsymbol{x}; \mathrm{I}]}{\mathrm{P}[\boldsymbol{x}_i = \boldsymbol{v}; \mathrm{I}]}, \quad \boldsymbol{x} \in \mathcal{X}, \tag{10.10}$$

where $\boldsymbol{v}$ is any fixed vector belonging to the set $\mathcal{X} := \{\boldsymbol{x} | \boldsymbol{x} = \boldsymbol{\Theta}_{\mathrm{LCF}} \boldsymbol{s},\ \boldsymbol{s} \in \mathbb{S}^{N_t \times 1}\}$.

The output of each siso module will be interleaved (or deinterleaved) to be used as input to the other module. The decoding algorithm can be described as follows:

S1) Obtain $\{y_n\}$ from deinterleaver Π_2^{-1}.

S2) Compute the LLR of $\{\lambda(\boldsymbol{x}_i; \mathrm{I})\}$.

S3) Set both $\lambda(d_n; \mathrm{I})$ and $\lambda(b_n; \mathrm{I})$ to zero, $\forall n$.

Begin iterative decoding.

S4) Execute the siso-LCF module to produce $\{\lambda(d_n; \mathrm{O})\}$.

S5) Deinterleave $\{\lambda(d_n; \mathrm{O})\}$ to obtain $\lambda(c_n; \mathrm{I})$.

S6) Execute the siso-EC module to produce $\{\lambda(c_n; \mathrm{O})\}$ and $\{\lambda(b_n; \mathrm{O})\}$.

S7) Interleave $\{\lambda(c_n; \mathrm{O})\}$ to obtain $\{\lambda(\boldsymbol{x}_i; \mathrm{I})\}$. If the number of iterations is less than a maximum number allowed, then go to S4. Otherwise, go to the next step.

End iterative decoding.

S8) Compare $\lambda(b_n; \mathrm{O})$ with zero and output the decisions.

10.1.3.2 siso-LCF Module The siso-LCF module implements MAP decoding of the constellation-mapped and STLCF-coded symbols. With reference to Figure 10.4, the siso-LCF module outputs the *extrinsic information* [14, 15] $\{\lambda(\boldsymbol{x}_i; \mathrm{O})\}$ and $\{\lambda(d_n; \mathrm{O})\}$, using as input the a priori information $\{\lambda(\boldsymbol{x}_n; \mathrm{I})\}$ and $\{\lambda(d_n; \mathrm{I})\}$. The extrinsic information of a symbol (or symbol vector) is its LLR given other symbols' (or symbol vectors') prior information, along with the structure of the STLCF encoder. In our case, the outputs $\{\lambda(\boldsymbol{x}_i; \mathrm{O})\}$ are not used; hence, we only need to evaluate $\lambda(d_n; \mathrm{O})$, which is the LLR of d_n given $\{d_k : k \neq n\}$'s, the priors $\{\lambda(d_k; \mathrm{I}) : k \neq n\}$, and the priors $\{\lambda(\boldsymbol{x}_i; \mathrm{I})\}$. The LLR $\lambda(d_n; \mathrm{O})$ can be computed as

$$\begin{aligned}\lambda(d_n; \mathrm{O}) &= \log \frac{P[d_n = +1 | \{\lambda(d_k; \mathrm{I}) : k \neq n\}, \{\lambda(\boldsymbol{x}_i; \mathrm{I})\}]}{P[d_n = -1 | \{\lambda(d_k; \mathrm{I}) : k \neq n\}, \{\lambda(\boldsymbol{x}_i; \mathrm{I})\}]} \\ &= \underbrace{\log \frac{P[d_n = +1 | \{\lambda(d_k; \mathrm{I})\}, \{\lambda(\boldsymbol{x}_i; \mathrm{I})\}]}{P[d_n = -1 | \{\lambda(d_k; \mathrm{I})\}, \{\lambda(\boldsymbol{x}_i; \mathrm{I})\}]}}_{:=\Lambda(d_n; \mathrm{O})} - \lambda(d_n; \mathrm{I}), \end{aligned} \tag{10.11}$$

where the equation follows from the definition of conditional probability and $\Lambda(d_n; \mathrm{O})$ is the complete a posteriori probability (APP) of d_n, which can be calculated by using [292, Equation (21)].

10.1.3.3 siso-EC Module The siso-EC module obviously depends on the EC code used. For convolutional codes, the soft information $\lambda(c_n; \mathrm{O})$ and $\lambda(b_n; \mathrm{O})$ can be obtained by optimum or suboptimum siso decoding algorithms, including the BCJR algorithm, the SOVA, Log-MAP, or max-Log-MAP alternatives [279]. Among them, BCJR and Log-MAP are optimum decoders. The complexity of Log-MAP is lower than that of BCJR and is approximately four times higher that of Viterbi's algorithm [275].

10.1.3.4 Computational Complexity In this subsection we analyze the arithmetic complexity of the decoding algorithm in terms of the approximate number of flops. We do not consider the finite word-length effects and the associated complexity issues. We also ignore the complexity introduced by the interleavers.

The number of flops per bit can be found as

$$\begin{aligned} N_{\text{flop}} &= \frac{4|\mathbb{S}|^{N_t}}{\log_2 |\mathbb{S}|} + N_{\text{iter}}(3N_t|\mathbb{S}|^{N_t} \log_2 |\mathbb{S}| + F_{\text{EC}}) \\ &\approx N_{\text{iter}}(3N_t|\mathbb{S}|^{N_t} \log_2 |\mathbb{S}| + F_{\text{EC}}), \end{aligned} \tag{10.12}$$

where F_{EC} is the number of flops per bit needed in siso-EC and N_{iter} denotes the number of iterations. A typical value for N_{iter} is 3.

Example 10.2 Consider again a rate-1/2 convolutional code with generators $(133, 171)$, BPSK constellation ($|\mathbb{S}| = 2$), and $N_t = 4$. The convolutional code has 64 states. This implies that the complexity per bit will be a small multiple of 64

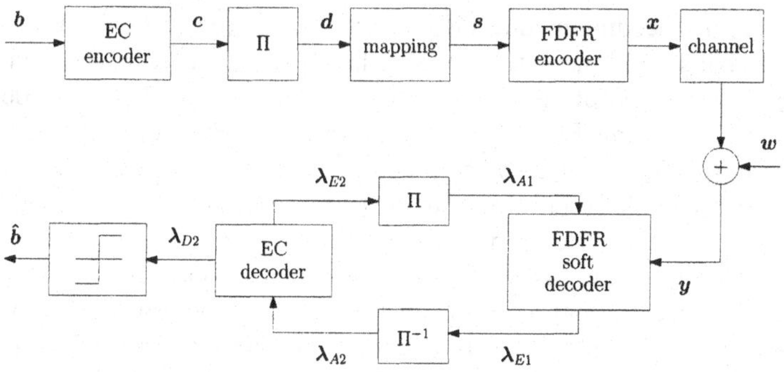

Figure 10.5 Coded FDFR with QPSK signaling

flops; "small" here could be 8 — a factor of 2 for the add-compare-select operation and a factor of 4 for Log-MAP decoding using the generalized Viterbi algorithm [275]. In this case, we have $F_{\text{EC}} \approx 64 \times 8 = 512$. According to (10.12), if $N_{\text{iter}} = 3$, then $N_{\text{flop}} \approx 2112$, which is about four times that of a system relying on convolutional coding only. To achieve the same amount of diversity increase (i.e., from $d_{\text{free}} = 9$ to $N_t d_{\text{free}} = 36$) by convolutional coding alone, the code memory would need to increase roughly $N_t = 4$ times from 6 to 24 [cf. (10.3)] and the complexity increase would be more than four times that required for GF-LCF decoding.

10.2 GF-LCF LAYERED ST CODES

In Section 10.1.3 we established that joint GF-LCF ST coding and decoding can improve the error performance of ST coded systems in terms of diversity and coding gains. However, we have seen in Chapter 4 that layered ST codes can also improve the rate of multi-antenna systems and have the potential to approach the capacity limits of MIMO channels. Similar to error performance-oriented ST codes, layered ST codes alone cannot even outperform single-antenna uncoded systems in AWGN. These considerations motivate combining GF coding with layered ST coding. Within the class of layered ST codes, our emphasis will be on combining EC coding with the full-diversity full-rate (FDFR) architecture of Chapter 4, which provides flexible trade-offs in complexity, rate, and error performance. The resulting joint GF-LCF ST coded FDFR system will also be compared with the canonical GF-coded V-BLAST architecture, which has been tested frequently in practical settings [70, 299].

10.2.1 GF-LCF ST FDFR Codes: QPSK Signaling

The GF-coded FDFR system for QPSK-modulated transmissions is depicted in Figure 10.5. At the transmitter side, an ECC module is serially concatenated with an FDFR module, and a random interleaver lies in between. Soft turbo decoding between the

ECC and FDFR decoding modules is performed at the receiver end. The effects of the MIMO channel and the FDFR code are handled simultaneously at the decoding stage. In the remainder of this section, we use the terms FDFR block to denote the unit in FDFR processing, ECC stream for the unit in the ECC encoder, and frame for the set of information bits that will be processed by the ECC module, the interleaver, and the FDFR module serially without dependence on other frames. We can also think of a frame as the system's processing unit.

A frame of information bits $\boldsymbol{b}$ with length K_c is first encoded by the ECC module to yield $\boldsymbol{c}$ and then goes through a random interleaver Π. Interleaved bits $\boldsymbol{d}$ are mapped to a frame $\boldsymbol{s}$ of length N_c containing symbols drawn from a certain constellation. Frame $\boldsymbol{s}$ is first partitioned into FDFR blocks $\{\boldsymbol{s}(k)\}_{k=1}^{K}$ each containing N_t^2 symbols, where k is used to index the FDFR block and K denotes the number of blocks. Blocks $\{\boldsymbol{s}(k)\}_{k=1}^{K}$ are fed to the FDFR module, which outputs FDFR-coded symbols that are transmitted over the N_t antennas.

At the receiver end, turbo decoding is carried out to achieve an overall near-ML performance. Two modules, indexed by subscripts 1 and 2, perform soft decoding of the FDFR-MIMO and GF coding parts, respectively. Extrinsic information about $\boldsymbol{c}$, denoted as $\boldsymbol{\lambda}_E$, from one decoding module is (de-)interleavered to yield a priori information about $\boldsymbol{c}$, denoted as $\boldsymbol{\lambda}_A$, for the other module. After a certain number of iterations or after a prescribed BER is achieved, a hard decision $\hat{\boldsymbol{b}}$ is obtained based on the a posteriori information about $\boldsymbol{b}$, denoted as $\boldsymbol{\lambda}_{D2}$, from the ECC decoding module.

Inside each module the optimal MAP decoder is implemented over the GF and the real/complex field (R/CF), respectively. Over either field, MAP decoding incurs complexity that in general grows exponentially with the problem size [i.e., the memory length of a convolutional code (CC) or the block size and the constellation size of the R/CF code]. Several near-optimum algorithms with polynomial complexity have been developed for decoding GF codes [e.g., Viterbi's algorithm for CCs and message-passing algorithms for turbo codes (TCs)] [220]. Algorithms for decoding R/CF codes at polynomial complexity include the sphere decoding algorithm (SDA) and various alternatives mentioned in Chapter 5. In this section we adopt the log-MAP algorithm to decode CCs and TCs as in [220], along with hard and soft (SoD-HSD) SDAs that will be used to decode R/CF-coded transmissions. Soft decoders will allow GF-coded FDFR ST systems to approach MIMO capacity.

To this end, we explain briefly how the SoD-HSD algorithm of Section 5.5.2 can be used for soft decoding of QPSK-modulated GF-coded FDFR transmissions. Starting from the equivalent model in (4.16), defining [cf. (4.16)]

$$\boldsymbol{H}_{\text{eq}} = \boldsymbol{\mathcal{H}}\boldsymbol{\Phi} \tag{10.13}$$

and separating the real and imaginary parts of the matrices and vectors in (4.16), we obtain the real equivalent model:

$$\tilde{\boldsymbol{y}} = \begin{bmatrix} \Re(\boldsymbol{y}) \\ \Im(\boldsymbol{y}) \end{bmatrix} = \sqrt{\frac{\bar{\gamma}}{N_t}} \begin{bmatrix} \Re(\boldsymbol{H}_{\text{eq}}) & -\Im(\boldsymbol{H}_{\text{eq}}) \\ \Im(\boldsymbol{H}_{\text{eq}}) & \Re(\boldsymbol{H}_{\text{eq}}) \end{bmatrix} \begin{bmatrix} \Re(\boldsymbol{s}) \\ \Im(\boldsymbol{s}) \end{bmatrix} + \begin{bmatrix} \Re(\boldsymbol{w}) \\ \Im(\boldsymbol{w}) \end{bmatrix} = \sqrt{\frac{\bar{\gamma}}{N_t}} \tilde{\boldsymbol{H}} \tilde{\boldsymbol{s}} + \tilde{\boldsymbol{w}}, \tag{10.14}$$

where we dropped the block index k for notational simplicity. Each entry of $\tilde{\boldsymbol{s}}$, $\tilde{s}_n$ ($n = 1, \ldots, 2N_t^2$) is equal to either $+1$ or -1. Let us also define the a priori information, the a posteriori information given $\tilde{\boldsymbol{y}}$, and the extrinsic information of $\tilde{s}_n$, respectively, as

$$\lambda_A(\tilde{s}_n) := \ln \frac{P(\tilde{s}_n = +1)}{P(\tilde{s}_n = -1)},$$
$$\lambda_D(\tilde{s}_n|\tilde{\boldsymbol{y}}) := \ln \frac{P(\tilde{s}_n = +1|\tilde{\boldsymbol{y}})}{P(\tilde{s}_n = -1|\tilde{\boldsymbol{y}})},$$
$$\lambda_E(\tilde{s}_n|\tilde{\boldsymbol{y}}) := \lambda_D(\tilde{s}_n|\tilde{\boldsymbol{y}}) - \lambda_A(\tilde{s}_n).$$

If $\boldsymbol{\lambda}_A := [\lambda_A(\tilde{s}_1), \ldots, \lambda_A(\tilde{s}_{2N_t^2})]^T$ denotes the a priori vector of $\tilde{\boldsymbol{s}}$, then based on (5.16)–(5.18) we can apply the SoD-HSD scheme to decode GF-coded FDFR transmissions.

10.2.2 GF-LCF ST FDFR Codes: QAM Signaling

The SoD-HSD scheme in Section 10.2.1 was derived when the entries of $\tilde{\boldsymbol{s}}$ were binary (i.e., ± 1). In fact, the SoD-HSD algorithm cannot be applied directly to QAM signaling because after the complex-to-real conversion in (10.14), $\tilde{\boldsymbol{s}}$ is a $2N_t^2 \times 1$ vector containing PAM symbols with $M_b = (\log_2 |\mathbb{S}|)/2$ bits per symbol if we suppose natural bit mapping.

Since different bits in $\tilde{s}_k$ are received with generally different SNRs, we adopt a bit-level multi-stream coded layered ST transmission for QAM signaling, as depicted in Figure 10.6. At the transmitter, the stream of information bits $\boldsymbol{b}$ is first divided into M_b substreams $\{\boldsymbol{b}_m\}_{m=1}^{M_b}$. Each substream is coded with an EC code and scrambled through a random interleaver to generate $\boldsymbol{u}_i$. From each substream, we take one interleaved bit to form a PAM symbol consisting of M_b bits ($\tilde{\boldsymbol{s}} = \sum_{i=1}^{M_b} 2^{i-1}\boldsymbol{u}_i$). Two PAM symbols are combined further to form a QAM symbol. The QAM symbol vectors are then transmitted using FDFR. At the receiver end, iterative decoding is performed in a layered fashion per bit level. In the FDFR channel decoding module, the complex FDFR block model is first converted to the real block model as in (10.14). When decoding one bit level, interference from other bit levels is treated as Gaussian noise, and based on the a priori information, the mean and covariance matrix of the equivalent noise are estimated as in [175]. This reduces decoding of each bit level to an equivalent QPSK decoding problem in the presence of colored Gaussian noise. After prewhitening the noise, the SoD-HSD scheme can readily be applied with the extrinsic information exchanged through the interleaver/deinterleaver between the ECC decoding module and the FDFR decoding module, as in Section 10.2.1. In what follows, we elaborate further on the decoding process.

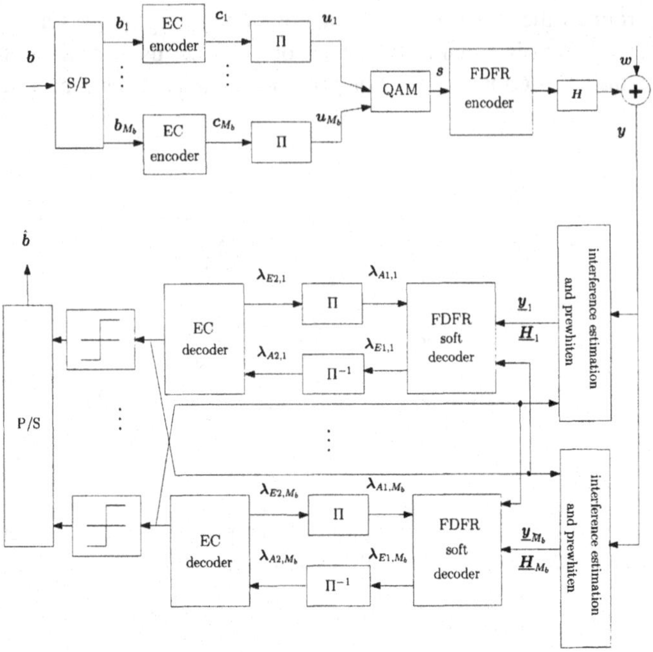

Figure 10.6 Bit-level multi-stream system model for QAM signaling

Probabilistic interference estimation: When decoding the mth bit level, we rewrite (10.14) as

$$\tilde{\boldsymbol{y}} = 2^{m-1}\sqrt{\frac{\bar{\gamma}}{N_t}}\tilde{\boldsymbol{H}}\boldsymbol{u}_m + \sum_{i=1,i\neq m}^{M_b} 2^{i-1}\sqrt{\frac{\bar{\gamma}}{N_t}}\tilde{\boldsymbol{H}}\boldsymbol{u}_i + \tilde{\boldsymbol{w}}$$

$$= \sqrt{\frac{\bar{\gamma}}{N_t}}\boldsymbol{H}_m\boldsymbol{u}_m + \boldsymbol{w}_m, \tag{10.15}$$

where $\boldsymbol{H}_m := 2^{m-1}\tilde{\boldsymbol{H}}$ is the equivalent block code matrix for the mth bit level, and $\boldsymbol{w}_m := \sqrt{\bar{\gamma}/N_t}\sum_{i=1,i\neq m}^{M_b}\boldsymbol{H}_i\boldsymbol{u}_i + \tilde{\boldsymbol{w}}$ is the equivalent noise encompassing both interference and the actual noise.

Let $\boldsymbol{p}_i$ denote the probability vector of the binary vector $\boldsymbol{u}_i$ with kth entry $p_{i,k} := \mathrm{P}(u_{i,k} = +1)$. Given $\boldsymbol{\lambda}_{A,i}$, the a priori information vector for $\boldsymbol{u}_i$, we obtain $\boldsymbol{p}_i = (1/2)[1 + \tanh(\boldsymbol{\lambda}_{A,i}/2)]$. Since the entries of $\{\boldsymbol{u}_i\}_{i=1}^{M_b}$ are assumed independent, we can estimate the mean and the covariance matrix of $\boldsymbol{w}_m$ using $\{\boldsymbol{p}_i\}_{i=1,i\neq m}^{M_b}$. However, if soft information from lower-power bit levels is used to decode a certain bit level, extensive simulations indicate that the decoding error probability becomes worse. In fact, as the number of iterations increases at high SNR, the number of errors actually blows up. In a 64-QAM test using turbo ECC with frame length 9,000 in a 4×4 transmit-receive antenna setup, at SNR = 13 dB the number of errors per iteration counted in one frame is (0, 2554, 2310, 3398) for four iterations. The possible reasons for the "error-blowup" problem are: (i) The power difference between adjacent bit levels is 6 dB which renders the soft information from bit levels with lower power unreliable to be used by higher power levels; and (ii) decoding takes place from the highest power level to the lowest. When decoding the $(m-1)$st bit level, a priori information $\boldsymbol{p}_m$ is utilized from the mth level. Therefore, the output $\boldsymbol{p}_{m-1}$ from the $(m-1)$st level already contains a priori information about the mth level. Going back to decode the mth level again and using $\boldsymbol{p}_{m-1}$, the a priori information about the mth level is reused, which causes error propagation. Although there are no rigorous proofs for these observations, both speculations match the fact that the problem disappears after the estimation method is modified as follows: When decoding the mth bit level with transmit power $2^{2(m-1)}$, we use only the soft information $\boldsymbol{p}_i$ from the bit levels with higher power ($i = m+1, \ldots, M_b$) and we assume that $\boldsymbol{p}_i = 0.5 \cdot \mathbf{1}$ with $\mathbf{1}$ denoting the vector with all entries equal to 1 for the bit levels with lower power ($i = 1, \ldots, m-1$), as if we had no a priori information about these levels. Specifically, we use

$$\bar{\boldsymbol{w}}_m := E(\boldsymbol{w}_m) = \sum_{i=1, i\neq m}^{M_b} \sqrt{\frac{\bar{\gamma}}{N_t}} \boldsymbol{H}_i E(\boldsymbol{u}_i) = \sqrt{\frac{\bar{\gamma}}{N_t}} \sum_{i=m+1}^{M_b} \boldsymbol{H}_i (2\boldsymbol{p}_i - 1), \quad (10.16)$$

$$\begin{aligned} \boldsymbol{C}_m &:= \mathrm{Cov}(\boldsymbol{w}_m) = \mathrm{Cov}(\tilde{\boldsymbol{w}}) + \sum_{i=1, i\neq m}^{M_b} \boldsymbol{H}_i \,\mathrm{Cov}(\boldsymbol{u}_i) \boldsymbol{H}_i^T \\ &= \frac{1}{2}\boldsymbol{I} + \frac{\bar{\gamma}}{N_t} \sum_{i=m+1}^{M_b} \boldsymbol{H}_i \,\mathrm{diag}[4\boldsymbol{p}_i \odot (1 - \boldsymbol{p}_i)] \boldsymbol{H}_i^T + \frac{\bar{\gamma}}{N_t} \sum_{i=1}^{m-1} \boldsymbol{H}_i \boldsymbol{H}_i^T, \quad (10.17) \end{aligned}$$

where diag[$\cdot$] denotes a diagonal matrix with a vector in parentheses as its main diagonal, $\odot$ stands for elementwise multiplication, and $\mathrm{Cov}(\boldsymbol{x})$ gives the covariance matrix of $\boldsymbol{x}$. Equations (10.16) and (10.17) show that the noise in (10.15) is not white. Therefore, one needs to whiten the noise so that the SoD-HSD scheme can be applied.

Prewhitening the equivalent noise: Before applying the SoD-HSD scheme to the mth bit level, by subtracting $\bar{\boldsymbol{w}}_m$ from both sides of (10.15) and left-multiplying them with $\boldsymbol{C}_m^{-1/2}$, we obtain the equivalent zero mean white Gaussian noise model

$$\underline{\boldsymbol{y}}_m = \sqrt{\frac{\bar{\gamma}}{N_t}} \underline{\boldsymbol{H}}_m \boldsymbol{u}_m + \underline{\boldsymbol{w}}_m, \quad (10.18)$$

where $\underline{\boldsymbol{y}}_m := \boldsymbol{C}_m^{-1/2}(\tilde{\boldsymbol{y}}_m - \bar{\boldsymbol{w}}_m)$, $\underline{\boldsymbol{H}}_m := \boldsymbol{C}_m^{-1/2}\boldsymbol{H}_m$, and $\underline{\boldsymbol{w}}_m := \boldsymbol{C}_m^{-1/2}(\boldsymbol{w}_m - \bar{\boldsymbol{w}}_m)$. With the noise $\underline{\boldsymbol{w}}_m$ being zero mean white Gaussian with identity covariance matrix, the SoD-HSD scheme can be applied to compute the extrinsic information of $\boldsymbol{u}_m$. The iterative multi-stream decoding is given as follows.

Let the superscript (t) index time, T denote the number of received vectors in a coded frame, and subscripts 1 and 2 denote the index of the MIMO and ECC decoding modules, respectively.

Iterative decoding steps for the multi-stream system:

S1) Initialization: $\boldsymbol{p}_m^{(t)} = 0.5 \cdot \mathbf{1}$ and $\boldsymbol{\lambda}_{A1,m}^{(t)} = \mathbf{0}$, $m = 1, \ldots, M_b, t = 1, \ldots, T$.

S2) One iteration:

(a) $m = M_b$.

(b) In the MIMO channel decoding module:

(i) $t = 1$.

(ii) Convert the QAM model (10.14) to the QPSK model as (10.15), and estimate $\bar{\boldsymbol{w}}_m^{(t)}$ and $\boldsymbol{C}_m^{(t)}$ according to (10.16) and (10.17).

(iii) Prewhiten the noise $\boldsymbol{w}_m^{(t)}$ as in (10.18) and decode $\boldsymbol{u}_m^{(t)}$ with SoD-HSD scheme and output extrinsic information vector $\boldsymbol{\lambda}_{E1,m}^{(t)}$.

(iv) $t = t + 1$; if $t \leq T$, return to (ii).

(c) Deinterleave $\boldsymbol{\lambda}_{E1,m}^{(t)}$, $t = 1, \ldots, T$ to obtain the a priori information $\boldsymbol{\lambda}_{A2,m}$ for the ECC decoding module.

(d) In the ECC decoding module: Use a soft decoding scheme depending on the ECC used. Output the extrinsic information $\boldsymbol{\lambda}_{E2,m}$.

(e) Interleave $\boldsymbol{\lambda}_{E2,m}$ to obtain the a priori information $\boldsymbol{\lambda}_{A1,m}^{(t)}$ and $\boldsymbol{p}_m^{(t)}$, $t = 1, \ldots, T$ for the MIMO channel decoding module.

(f) $m = m - 1$; if $m \geq 1$, return to (b).

S3) Return to S2 until a desired performance is achieved or the number of iterations reaches a certain number.

10.2.3 Performance Analysis

Similar to Section 10.1.2, we resort to a PEP approach to analyze the performance of the GF-coded FDFR system. Let us suppose for now that QPSK-modulated symbols are transmitted over a MIMO channel which remains constant over each FDFR-coded block but is allowed to vary independently from block to block. Consider two different information bit frames $\boldsymbol{b}$ and $\tilde{\boldsymbol{b}}$, each with length K_c. At the output of the ECC module they yield two codewords $\boldsymbol{c}$ and $\tilde{\boldsymbol{c}}$ with length N_c. If these two codewords differ in w positions, so do the interleaved codewords. Although these w positions could be in close proximity for a certain interleaver and a certain pair of codewords, considering the fact that the interleaver Π is random with a different realization per frame, these w positions will probably be sufficiently far apart, provided that the interleaver depth is sufficiently long. Under this assumption, we can henceforth consider that after constellation mapping, the two symbol frames $\boldsymbol{s}$ and $\tilde{\boldsymbol{s}}$ (see Figure 10.5) still have d different symbols, and in any FDFR block the vectors $\boldsymbol{s}(k)$ and $\tilde{\boldsymbol{s}}(k)$ differ in at most one symbol, where $k \in [1, K]$ is used to index the FDFR block and K is the number of FDFR blocks.

After LCF coding, layered ST mapping and propagation through the MIMO channel $\boldsymbol{H}(k)$, the equivalent channel matrix seen by the kth FDFR block is $\boldsymbol{H}_{\text{eq}}(k)$ [cf. (10.13)]. The resulting symbol vectors are $\{\boldsymbol{z}(k) = \sqrt{\bar{\gamma}/N_t}\boldsymbol{H}_{\text{eq}}(k)\boldsymbol{s}(k)\}_{k=1}^{K}$ and $\{\tilde{\boldsymbol{z}}(k) = \sqrt{\bar{\gamma}/N_t}\boldsymbol{H}_{\text{eq}}(k)\tilde{\boldsymbol{s}}(k)\}_{k=1}^{K}$. Out of K blocks, only d of them are different. Without causing confusion, we will use $\{\boldsymbol{z}(k_a)\}_{a=1}^{w}$ and $\{\tilde{\boldsymbol{z}}(k_a)\}_{a=1}^{w}$ to denote them. When $\boldsymbol{s}(k_a)$ and $\tilde{\boldsymbol{s}}(k_a)$ are different in the nth symbol, the Euclidean distance between $\boldsymbol{z}(k_a)$ and $\tilde{\boldsymbol{z}}(k_a)$ is

$$||\boldsymbol{z}(k_a) - \tilde{\boldsymbol{z}}(k_a)||^2 = \frac{\bar{\gamma}}{N_t}||\boldsymbol{h}_{\text{eq},n}(k_a)||^2|\Delta s_n(k_a)|^2, \tag{10.19}$$

where $\boldsymbol{h}_{\text{eq},n}(k_a)$ is the nth column of the equivalent channel matrix $\boldsymbol{H}_{\text{eq}}(k_a)$ and $|\Delta s_n(k_a)|^2$ is the Euclidean distance between the two different symbols $s_n(k_a)$ and $\tilde{s}_n(k_a)$. With $\Delta_{\min}^2$ standing for the minimum Euclidean distance between two symbols, we have that $|\Delta s_n(k_a)|^2 \geq \Delta_{\min}^2$.

Since in the FDFR system $\boldsymbol{H}_{\text{eq}}(k_a) = \boldsymbol{\mathcal{H}}(k_a)\boldsymbol{\Phi}$, it follows that (cf. (4.14) and (4.15))

$$||\boldsymbol{h}_{\text{eq},n}(k_a)||^2 = ||\boldsymbol{\mathcal{H}}(k_a)\boldsymbol{\phi}_n||^2 = \frac{1}{N_t}\sum_{\nu=1}^{N_r}\sum_{\mu=1}^{N_t}|h_{\nu\mu}(k_a)|^2, \tag{10.20}$$

where the $\boldsymbol{\phi}_n$ is the nth column of $\boldsymbol{\Phi}$, $h_{\nu\mu}(k_a)$ is the (ν,μ)th entry of the $N_r \times N_t$ channel matrix $\boldsymbol{H}(k_a)$, and $\boldsymbol{\mathcal{H}}(k_a) = \boldsymbol{I}_{N_t} \otimes \boldsymbol{H}(k_a)$. Notice that (10.20) is true because all entries of $\boldsymbol{\Theta}_g$ have equal norm $1/\sqrt{N_t}$.

The overall Euclidean distance of these two sequences can be written as

$$\sum_{a=1}^{w}||\boldsymbol{z}(k_a) - \tilde{\boldsymbol{z}}(k_a)||^2 \geq \frac{\bar{\gamma}}{N_t}\frac{\Delta_{\min}^2}{N_t}\sum_{a=1}^{w}\sum_{\nu=1}^{N_r}\sum_{\mu=1}^{N_t}|h_{\nu\mu}(k_a)|^2. \tag{10.21}$$

The PEP for a given channel realization, $P_2(\boldsymbol{c} \to \tilde{\boldsymbol{c}}|\boldsymbol{H}(k))$, can therefore be upper bounded as

$$P_2(\boldsymbol{c} \to \tilde{\boldsymbol{c}}|\boldsymbol{H}(k)) \leq Q\left(\sqrt{\frac{\bar{\gamma}}{N_t}\frac{\Delta_{\min}^2\sum_{a=1}^{w}\sum_{\nu=1}^{N_r}\sum_{\mu=1}^{N_t}|h_{\nu\mu}(k_a)|^2}{2N_t}}\right),$$

where $Q(x) := (1/\sqrt{2\pi})\int_x^\infty \exp(-t^2/2)\,dt$. Using the Chernoff bound $Q(x) \leq (1/2)\exp(-x^2/2)$ and averaging over all h's, we obtain the average PEP as

$$P_2(\boldsymbol{c} \to \tilde{\boldsymbol{c}}) \leq \frac{1}{2}\left[\int \exp\left(-\frac{\bar{\gamma}}{N_t}\frac{\Delta_{\min}^2}{4N_t}\alpha^2/2\right)f(\alpha)\,d\alpha\right]^{wN_tN_r},$$

where $f(\alpha)$ is the pdf of the channel amplitude $\alpha := |h_{\nu\mu}|$ between each transmit/receive antenna pair (ν,μ), which is assumed to be spatially i.i.d. In the case of Rayleigh fading, the average PEP is

$$P_2(\boldsymbol{c} \to \tilde{\boldsymbol{c}}) \leq \frac{1}{2}\left(1 + \frac{\bar{\gamma}}{N_t}\frac{\Delta_{\min}^2}{4N_t}\right)^{-wN_tN_r}. \tag{10.22}$$

Equation (10.22) shows that the PEP exhibits diversity order wN_tN_r. After applying the union bound to all error events, we can upper-bound the average BER of GF-coded FDFR as [13]

$$P_b^F \leq \frac{1}{2} \sum_{w=d_{\min}}^{N_c} B_w \left(1 + \frac{\bar{\gamma}}{N_t} \frac{\Delta_{\min}^2}{4N_t}\right)^{-wN_tN_r}, \tag{10.23}$$

where $d_{\min}$ is the minimum Hamming distance of the block ECC, or the free distance of the CC, and B_w is the average number of bit errors associated with error events of distance d. By definition, we have

$$B_w = \sum_{\ell=1}^{K_c} \frac{\ell}{K_c} A_{\ell,w}, \tag{10.24}$$

where $A_{\ell,w}$ denotes the number of error events with input sequence weight ℓ and output sequence weight w and K_c is the number of information bits per input frame. Equation (10.23) reveals that a maximum diversity order $d_{\min}N_tN_r$ is possible with a GF-coded FDFR system.

Although we assumed that the MIMO channel remains constant over an FDFR block, this is not always the case. When MIMO channels are both spatially and temporally independent, with slight modifications it can easily be shown that the diversity order $d_{\min}N_tN_r$ is still enabled by GF-coded FDFR transmissions. In the case of temporally correlated time-varying MIMO channels, a second interleaver Π_2 must be appended after the FDFR module and before transmission to decorrelate the MIMO channels (see Figure 10.1). If the channels change fast enough or the frame size is large enough, the equivalent channels can still be treated as being independent. However, if the channels change very slowly, the full diversity they provide is less than $d_{\min}N_tN_r$. In the extreme case when the MIMO channels remain invariant over the entire frame but vary independently from frame to frame, the overall diversity order is that enabled by the FDFR ST codes only: namely, N_tN_r for uncoded FDFR (i.e., no extra diversity is gained with GF codes). This is because (10.21) now becomes

$$\sum_{a=1}^{w} ||\boldsymbol{z}(k_a) - \tilde{\boldsymbol{z}}(k_a)||^2 \geq \frac{\bar{\gamma}}{N_t} \frac{w\Delta_{\min}^2}{N_t} \sum_{\nu=1}^{N_r} \sum_{\mu=1}^{N_t} |h_{\nu\mu}(k_a)|^2. \tag{10.25}$$

The average BER of GF-coded FDFR over Rayleigh frame-fading channels is thus

$$\begin{aligned} P_b^F &\leq \frac{1}{2} \sum_{w=d_{\min}}^{N_c} B_w \left(1 + \frac{\bar{\gamma}}{N_t} \frac{w\Delta_{\min}^2}{4N_t}\right)^{-N_tN_r} \\ &\leq \frac{1}{2} \left(1 + \frac{\bar{\gamma}}{N_t} \frac{d_{\min}\Delta_{\min}^2}{4N_t}\right)^{-N_tN_r} \sum_{w=d_{\min}}^{N_c} B_w. \end{aligned} \tag{10.26}$$

Clearly, N_tN_r is the diversity enabled by uncoded FDFR transmissions (see Section 4.3). Different choices of ECC will affect B_w in (10.23), and therefore ECC will improve error performance only through the coding gain. When the underlying MIMO channel is block fading exhibiting L independent realizations per frame, the diversity order of GF-coded FDFR is at most $\min(L, d_{\min})N_tN_r$. Note that the term block in block fading refers to the channel coherence time. For QAM transmissions, the same argument applies for each bit level and allows one to prove that the GF-coded FDFR system still enjoys full diversity gains.

10.2.4 GF-LCF FDFR Versus GF-Coded V-BLAST

Since V-BLAST is the simplest layered ST code achieving full rate and GF-coded V-BLAST has been shown to be capacity approaching (see Figure 5.9 and [118]), which means that it is asymptotically optimal, we use coded V-BLAST as a benchmark. The latter can be obtained by replacing the FDFR module in Figure 10.5 with a serial-to-parallel converter and adopting the SoD-HSD algorithm as in (5.16)–(5.18). GF-coded FDFR and V-BLAST have the same transfer rate, $\eta_{\text{GF}} N_t \log_2|\mathbb{S}|$ bits per channel use, for a given ECC rate η_{GF} and constellation size $|\mathbb{S}|$. When fading channels are both spatially and temporally i.i.d., the average BER of GF-coded V-BLAST can be upper-bounded similarly to that for GF-coded FDFR by

$$P_b^V \leq \sum_{w=d_{\min}}^{N_c} \frac{B_w}{2}\left(1 + \frac{\bar{\gamma}}{N_t}\frac{\Delta_{\min}^2}{4}\right)^{-wN_r}. \tag{10.27}$$

The diversity order of GF-coded V-BLAST is thus $d_{\min}N_r$ (i.e., a fraction $1/N_t$ of the diversity achieved by the GF-coded FDFR system). To achieve identical BER performance, the $\Delta_{\min}^2$ distance needed is Δ_v^2 for coded V-BLAST and Δ_f^2 for GF-coded FDFR. The gain of FDFR over V-BLAST is defined as $G_{\text{LCF}} := \Delta_v^2/\Delta_f^2$. Comparing (10.23) with (10.27), it follows that the LCF gain of GF-coded FDFR over GF-coded V-BLAST is approximately

$$G_{\text{LCF}} := \frac{\Delta_v^2}{\Delta_f^2} = \frac{\frac{\bar{\gamma}\Delta_v^2}{4N_t}}{N_t\left[\left(1+\frac{\bar{\gamma}\Delta_v^2}{4N_t}\right)^{1/N_t} - 1\right]}. \tag{10.28}$$

The gain in (10.28) is similar to the gain that the joint GF-LCF coded system offers over the system with GF coding alone (EC-only) in (10.8). Here $\bar{\gamma}/N_t$ plays the role of $\bar{\gamma}$ in (10.8). This similarity is quite reasonable, since from (10.20) we deduce that the diversity factor N_t is due to the LCF code matrix with size N_t. From (10.28) we also infer that increasing N_t will increase the LCF gain. However, this gain increment decreases as N_t increases. Considering the complexity increment as N_t increases, FDFR is more useful when N_t is small (e.g., $N_t = 2$).

When the MIMO channel is fading on a frame-by-frame basis, neither GF-coded V-BLAST nor FDFR can collect diversity w. It can be derived from (10.26) that to achieve the same BER performance, the LCF gain that GF-coded FDFR enjoys over

GF-coded V-BLAST is

$$G_{\mathrm{LCF}} := \frac{\Delta_v^2}{\Delta_f^2} = \frac{\frac{\bar{\gamma} d_{\min} \Delta_v^2}{4N_t}}{N_t \left[\left(1 + \frac{\bar{\gamma} d_{\min} \Delta_v^2}{4N_t} \right)^{1/N_t} - 1 \right]}. \tag{10.29}$$

Comparing (10.28) with (10.29), we deduce that for a given Δ_v^2, GF-coded FDFR exhibits larger gain over GF-coded V-BLAST in frame-by-frame fading channels relative to the gain over fast fading channels.

10.2.5 Numerical Examples

In this section we present simulations with different ECCs for QPSK-modulated MIMO transmissions to compare BER of GF-coded V-BLAST and FDFR systems. Except for Example 10.4, both systems use the single-stream ECC structure. We use the SoD-HSD algorithm of Section 5.5.2 [284]. The interleaver size is chosen to be 2^{10}. The spatially i.i.d. channels remain invariant on a per FDFR block basis but change independently from FDFR block to block. In each figure, curves with the same marker correspond to BER for the same system (either GF-coded V-BLAST or FDFR) at different iterations. SNR is defined as the energy of an information bit over noise power.

Notice that the equivalent channel matrix $\boldsymbol{H}_{\text{eq}}$ in (10.13) is an $N_r N_t \times N_t^2$ complex matrix, which implies a considerable increase in decoding complexity relative to V-BLAST with block size N_t for the same antenna setup.

Example 10.3 Figures 10.7 and 10.8 depict BER performance comparisons between GF-coded V-BLAST and FDFR when relatively weak codes (CC and high-rate TC) are used in the ECC module in a 2×2 antenna setup. BER curves after three iterations are shown in Figure 10.7. We observe that with rate-$1/2$ CC, gain after the second iteration is negligible. The same behavior is observed with rate-$3/4$ TC, although the BER curve of the third iteration is not shown in Figure 10.8. Rate-$1/2$ CC with memory 2 is generated using the feedback polynomial $G_r(D) = 1 + D + D^2$ and the feedforward polynomial $G(D) = 1 + D^2$. About 1.5 dB gain at BER = 10^{-4} is offered by the FDFR design when rate-$1/2$ CC is used. We also test a rate-$3/4$ parallel concatenated convolutional code (PCCC) in Figure 10.8 which consists of two CC modules parameterized as before. The puncturing pattern used retains all systematic bits and takes 1 bit every 6 bits from each coded stream. Five iterations are performed inside the decoding module for the rate-$3/4$ TC. Two outer iterations are performed between the two decoding modules. Figure 10.8 shows the BER comparison for rate-3/4 TC in a 2×2 setup. In this case, GF-coded FDFR outperforms GF-coded V-BLAST by about 2 dB.

Example 10.4 In this simulation, we use a rate-$1/2$ PCCC constructed as a rate-$3/4$ PCCC except for the puncturing pattern. Here we keep all the systematic bits and

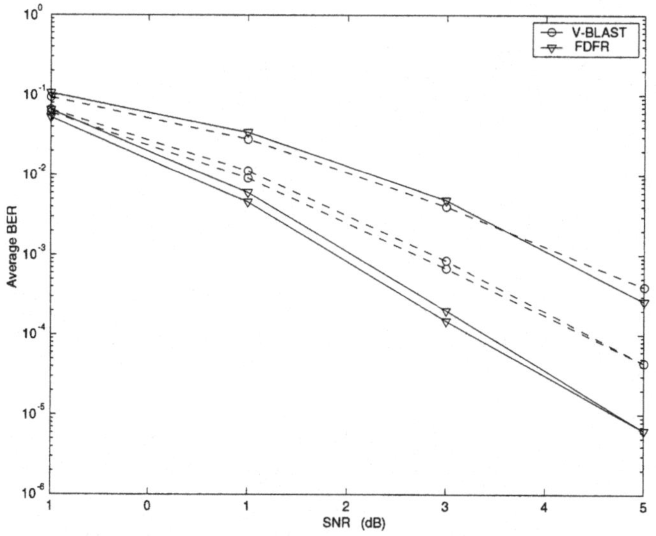

Figure 10.7 Rate-1/2 CC with FDFR vs. rate-1/2 CC with V-BLAST in a 2×2 setup

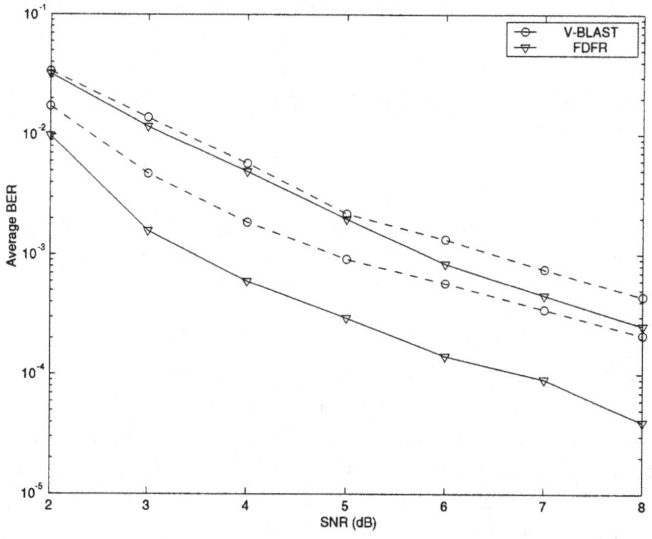

Figure 10.8 Rate-3/4 turbo codes with FDFR vs. rate-3/4 turbo codes with V-BLAST in a 2×2 setup

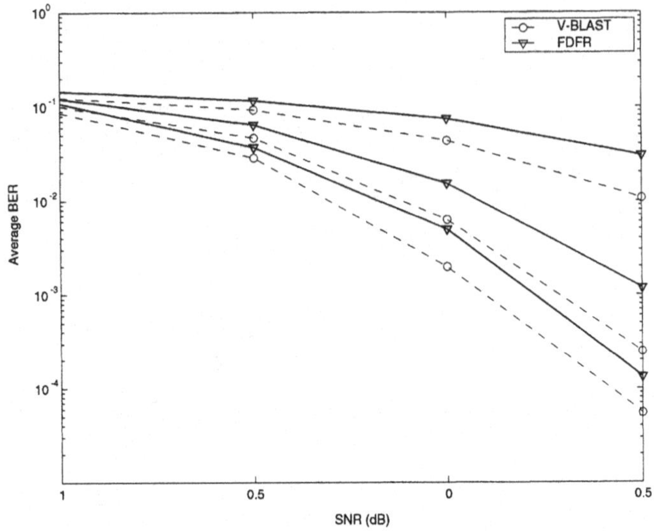

Figure 10.9 Rate-$1/2$ TCs with FDFR vs. rate-$1/2$ TCs with V-BLAST in a 2×2 setup

every other bit from each coded stream. Figure 10.9 depicts the comparison in a 2×2 setup with three outer iterations. Five iterations are performed inside the decoding module for the rate-1/2 TC. We see no performance improvement with FDFR, which is reasonable when such a strong code is used with $N_r = N_t$. Thanks to the larger diversity order, uncoded FDFR noticeably outperforms uncoded V-BLAST at high SNR. However, when strong codes are used (as is the case with rate-$1/2$ TC), the diversity gains are already high at low SNR. In this case, the larger diversity order enabled by FDFR brings no advantage.

Figure 10.10 shows the same comparison in a 2×5 setup. By increasing N_r, GF-coded FDFR and V-BLAST both benefit from the extra energy collected and the extra diversity provided by the extra receive-antennas, but the turbo gain between iterations becomes smaller. We observe further that in the 2×5 case, even the coded ST system with rate-$1/2$ TC can benefit from the FDFR design by about 0.5 dB over the coded V-BLAST system.

In these numerical examples, CCs or TCs were considered in the ECC module. A serially concatenated scheme consisting of an outer low-density parity-check (LDPC) code and an inner LCF code for reliable communications over wireless fading channels has been studied in [318]. The mutual information between the input and the output of the uncorrelated Rayleigh flat fading channel with and without LCF coding were compared to corroborate that joint LDPC-LCF coding can bring considerable SNR gains over LDPC-only coded systems even with LCF coders of size as small as 2 or 4 [318]. One extra benefit when using LDPC codes in the ECC module is that interleaver Π_1 in Figure 10.1 can be avoided, since interleaving is a built-in component in an LDPC code.

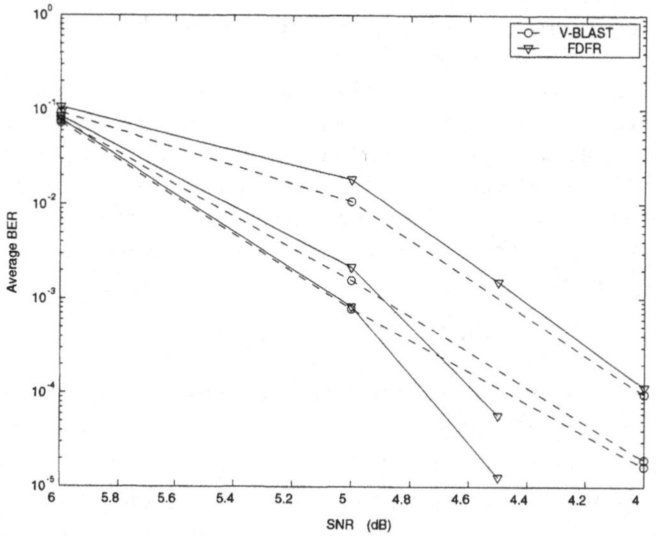

Figure 10.10 Rate-$1/2$ TCs with FDFR vs. rate-$1/2$ TCs with V-BLAST in a 2×5 setup

10.3 GF-LCF CODED MIMO OFDM

In Sections 10.1 and 10.2 we saw that joint GF-LCF coding can be exploited to improve the error performance of multi-antenna systems signaling over flat fading MIMO channels. Joint GF-LCF coding has also been applied to single-antenna OFDM transmissions over frequency-selective channels in [292]. In this section we generalize the major results of [292] to the MIMO OFDM setup.

10.3.1 Joint GF-LCF Coding and Decoding

When the channel is frequency selective, MIMO OFDM modulation and demodulation converts it to a set of parallel (correlated) flat fading channels (see Chapter 8). This allows per subcarrier application of joint GF-LCF encoding and decoding to MIMO OFDM systems.

Similar to the flat fading case, the information bit stream $\boldsymbol{b}$ at the transmitter is first encoded using a GF-based EC code, which can be a block, convolutional, turbo, or LDPC code (see Figure 10.1). The output stream $\boldsymbol{c}$ is then interleaved using Π_1 and mapped to constellation symbols. If trellis-coded modulation (TCM) is used in the ECC module, the constellation mapping is not needed since the mapping is already incorporated in the TCM. Furthermore, if an LDPC code is used the interleaver is not necessary, as we commented earlier.

After constellation mapping, the information-bearing symbol stream is encoded by an LCF code matrix $\boldsymbol{\Theta}_{\text{LCF}}$. As before, to keep the decoding complexity low, we use only LCF matrices of size $M = 2$ or 4.

After the LCF block encoding by $\boldsymbol{\Theta}_{\text{LCF}}$, the symbol stream $\bar{\boldsymbol{x}}$ is interleaved by Π_2 to decorrelate symbols in the subcarrier domain, and the interleaver output $\boldsymbol{x}$ is transmitted over the MIMO channel; see also Figure 8.3. For simplicity, we assume that the symbols are transmitted over the multiple antennas alternately, using one transmit-antenna per time slot. This mode corresponds to an antenna switching transmission per subcarrier. From (8.13), we know that the input-output relationship on each subcarrier is given by

$$\boldsymbol{Y} = \sqrt{\bar{\gamma}}\boldsymbol{H}\boldsymbol{X} + \boldsymbol{W} \Rightarrow \bar{\boldsymbol{y}} = \sqrt{\bar{\gamma}}\bar{\boldsymbol{H}}\boldsymbol{x} + \boldsymbol{w}, \tag{10.30}$$

where after dropping the subcarrier index, $\boldsymbol{X}$ denotes a diagonal matrix with $\boldsymbol{x}$ on its main diagonal and $\bar{\boldsymbol{H}}$ is given in (3.76). It can easily be verified that $\bar{\boldsymbol{H}}^{\mathcal{H}}\bar{\boldsymbol{H}}$ is an $N_t \times N_t$ diagonal matrix with (μ, μ)th entry $\sum_{\nu=1}^{N_r} |h_{\nu\mu}|^2$. The MRC step in Figure 10.11 is to perform $\bar{\boldsymbol{H}}^{\mathcal{H}}$ on both sides of the second equation in (10.30). Note that the noise is colored in this case. A simple whitening step can be used to make all the noise variables have the unit variance. This converts the MIMO frequency-selective channel to a diagonal one with input-output relationship per entry equivalent to (10.2) except for the pdf of the equivalent channel coefficient. Assuming that the entries of $\boldsymbol{H}$ are i.i.d., the resulting N_t SISO channels are also independent. The SNR at the μth subchannel output is given by $\sum_{\nu=1}^{N_r} |h_{\nu\mu}|^2$. Since all N_t subchannels have identical statistics, they can be viewed as N_t uses of the same SISO flat fading channel whose SNR equals the squared sum of N_r Rayleigh-distributed random variables. The pdf of the equivalent SISO channel is therefore chi-square with $2N_r$ degrees of freedom, in which case the channel obeys a Nakagami-m model with parameter $m = N_r$ [236]. The equivalent model is depicted in Figure 10.11. Thanks to OFDM and antenna switching, the equivalent system becomes a serial SISO system equipped with joint GF-LCF coding.

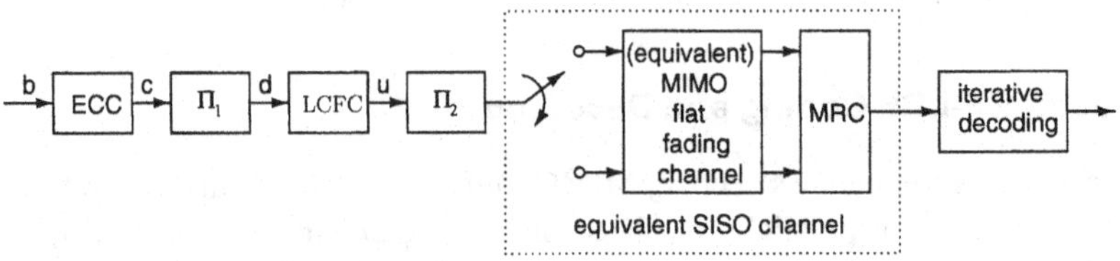

Figure 10.11 Joint GF-LCF coded MIMO OFDM model

Selecting the size of the interleaver Π_2 depends on how fast the channel varies. For slowly varying channels, Π_2 must have depth large enough to decorrelate the channel sufficiently across subcarriers. Under this condition, the performance analysis is similar to the one provided in Section 10.1.2. If the minimum Hamming (or free) distance of the ECC is $d_{\min}$, the diversity order enabled by this joint GF-LCF coded system is $d_{\min} M N_r$. Notice the *multiplicative* effect that joint GF-LCF coding has on the enabled diversity order: The overall diversity is the product of $d_{\min}$, the diversity achieved by the convolutional code alone, times M, the diversity achieved by the LCFC alone, times the receive diversity N_r. The number of transmit antennas N_t does not play an apparent role here, because we assumed that antenna switching is used and the equivalent SISO channel has been decorrelated sufficiently. However, N_t will show up if we consider the performance of other MIMO OFDM schemes, such as the FDFR-OFDM system described in Section 8.5.

At the receiver we use iterative decoding, whereby information is exchanged between two siso modules that we abbreviate as siso-LCF and siso-GF; see Figure 10.4. The siso-LCF module performs maximum a posteriori (MAP) decoding of the LCF block code $\boldsymbol{\Theta}_{\text{LCF}}$ and produces soft information about the GF-coded symbols. The siso-GF module implements MAP decoding of the EC code. The detailed decoding method and complexity analysis are the same as those in Section 10.1.3.

10.3.2 Numerical Examples

Example 10.5 Relying on two transmit-antennas, here we compare the GF-LCF coded antenna switching scheme with the Alamouti code [6], which also has rate 1 symbol per channel use. Both schemes are encoded using a rate-$1/2$ convolutional code with generators $(7, 5)$. Figure 10.12 depicts the error performance of the GF-LCF scheme after two iterations along with that of the GF-coded Alamouti system. It can be seen that there is virtually no difference in the BER.

It is also possible to use LCF coding and Alamouti coding jointly, in which case the antenna switching strategy is replaced by the Alamouti code. The result is also depicted in Figure 10.12. Although the additional diversity gain shows up at high SNR, combining the two offers a coding gain of only about 0.5 dB at low SNR.

Example 10.6 We simulated in this test the GF-LCF coded MIMO OFDM system in a HiperLAN/2 setup using the HiperLAN/2 channel model A [63, 64] with carrier frequency 5.2 GHz and speed 3 m/s. We used the equivalent correlated flat fading MIMO model on each subcarrier. Π_1 is chosen to be a random interleaver with a delay corresponding to 16,384 information bits, while Π_2 is chosen to be a 4×16 block interleaver. Transmit-antenna switching is used. Figure 10.13 depicts the results with LCF code sizes $M = 4$ and $M = 1$ (i.e., without LCF coding). As GF code, a rate-$3/4$ convolutional code is invoked, punctured from the rate-$1/2$ $(171, 133)$ code used in the HiperLAN/2 standard [64]. We can see that with LCF coding, we obtain at BER $= 10^{-3}$ an SNR gain of 2 dB with $M = 1$ and 3 dB with $M = 4$, as compared to the system using EC coding only.

10.4 CLOSING COMMENTS

In this chapter we introduced a joint GF-LCF coding scheme for multi-antenna ST communications. The general transmitter model comprised an outer error control

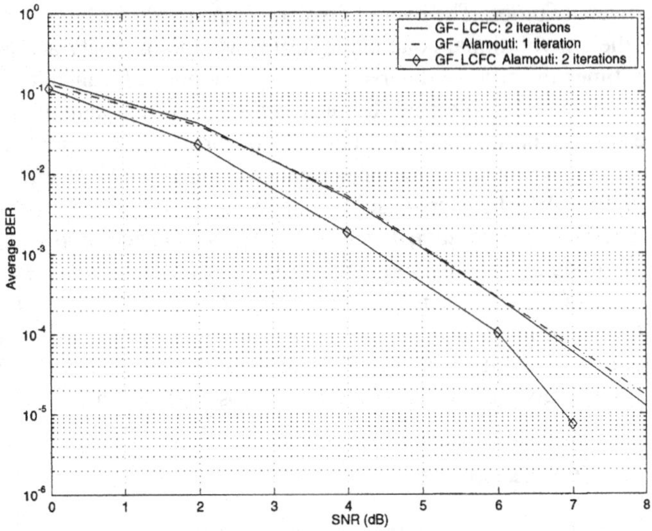

Figure 10.12 Comparison with the Alamouti code

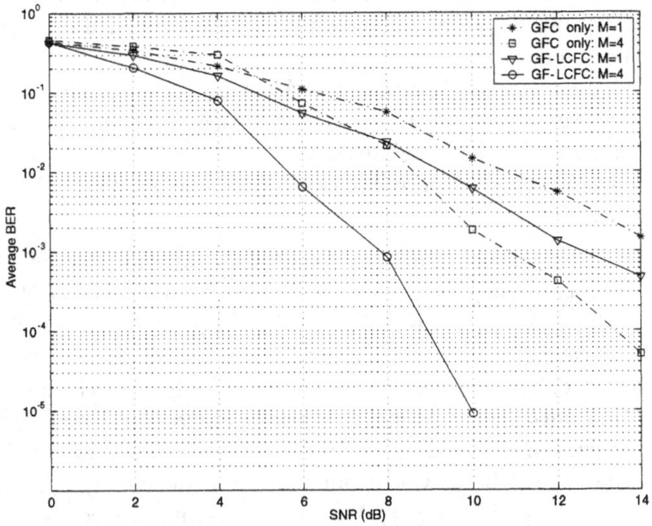

Figure 10.13 HiperLAN/2 simulations

code serially concatenated with an inner ST code and an interleaver in between. Although various options are available for the inner and outer codes, STLCF, FDFR, or V-BLAST schemes were considered for the inner code, and convolutional or turbo codes were tested as candidate outer codes. At the receiver side, turbo decoding was introduced and shown capable of approaching the performance benchmarked by the MIMO capacity. For GF-coded FDFR, with N_t transmit-antennas, N_r receive-antennas, rate $\eta_{\rm GF}$ error control code, and constellation size $|\mathbb{S}|$, it was shown possible to achieve high rate ($\eta_{\rm GF} N_t \log_2|\mathbb{S}|$ bits per channel use) and high diversity order (up to $d_{\min} N_t N_r$). Compared with GF-coded V-BLAST, GF-coded FDFR improves performance without sacrificing rate when weak outer codes are used. The price paid is a moderate increase in complexity.

Relying on OFDM to convert a frequency-selective MIMO channel to a set of parallel flat fading (but generally correlated) MIMO channels, the GF-LCF coding scheme was applied on a per subcarrier basis to frequency-selective MIMO channels. The GF-LCF coded symbols were transmitted from multiple antennas using antenna switching. Alternative means of transmitting the jointly coded symbols include ST orthogonal designs, delay diversity, antenna phase sweeping, and FDFR layered schemes. It was established that the diversity achieved by the jointly coded system with transmit antenna switching is the product of the free distance of the code, times the LCF encoder size, times the number of receive antennas. When the interleaver is ideal, the number of transmit antennas does not affect the diversity gain of the antenna switching scheme.

In summary, joint GF-LCF ST coders and turbo decoders are practically very appealing in wireless multi-antenna communications over MIMO fading channels because they collect the multidimensional diversity and are capable of bringing the performance of ST coded systems close to the MIMO capacity limit at affordable complexity.

11

MIMO Channel Estimation and Synchronization

Apart from the differential and noncoherent ST codes of Chapter 6, which do not require channel state information (CSI), all ST codes we dealt with so far require knowledge of the MIMO channel for coherent decoding. Besides channel estimation, coherent decoding of ST coded transmissions also requires timing and frequency synchronization at the receiver. Because a timing offset introduces a pure delay convolutional channel, timing synchronization can be lumped into the channel estimation task. Carrier-frequency offsets (CFOs), on the other hand, manifest themselves into multiplicative channels, and mitigating them requires separate treatment. As we alluded to in the closing comments of Chapter 8, CFO estimation is needed for frequency synchronization and constitutes a performance-critical factor, especially for multi-carrier transmissions.

Similar to single-antenna systems, depending on how much the receiver knows regarding transmitter parameters, MIMO channel and CFO estimators can be classified into data-aided (a.k.a. training or pilot-based), non-data aided (a.k.a. blind) and hybrid schemes, including semi-blind and decision-directed schemes. The latter aim at joint channel estimation and demodulation. Training-based schemes require fewer received samples than do blind schemes, and are more popular for their rapid acquisition and reliable performance, which is measured by the estimators' minimum mean-square error (MMSE). They are particularly appealing for bursty links such as those encountered with packet-switched networks. On the other hand, blind alternatives are a better fit for noncooperative tactical point-to-point links and also when traffic is continuous (as in HDTV broadcasting), because they do not interrupt transmission and do not sacrifice bandwidth.

The subjects of the present chapter are MIMO channel estimation and frequency synchronization of multi-antenna transmissions. Our primary concern is with frequency-selective MIMO channels, but we also indicate how the algorithms developed can also be specialized to frequency-flat and time-selective MIMO channels. We present MIMO channel and CFO estimators for both single- and multi-carrier multi-antenna systems. Besides preamble-based and optimal training schemes, we touch upon (semi-)blind and decision-directed algorithms. Whether designed for training-based or (semi-)blind operation, our exposition focuses on those estimators that find universal applicability to all coherent ST codes. Relative to single-antenna systems, the MIMO setting is more challenging because more parameters need to be estimated and the transmit-power must be shared among multiple antennas. Certainly, potential degradation in estimation performance due to these factors can be compensated by the spatial diversity available.

11.1 PREAMBLE-BASED CHANNEL ESTIMATION

Prior to information transmission, wireless as well as wireline communication systems typically rely on symbols that are known to both transmitter and receiver in order to estimate the underlying channel. Because these training or pilot symbols are placed at the beginning of each information-bearing frame (or session), they are referred to as the preamble.

Let us suppose first that CFO is absent and consider a wireless multi-antenna link over a frequency-flat MIMO channel obeying (2.16). Supposing further that the preamble occupies N_x time slots (columns of $\boldsymbol{X}$), the input-output relationship is a familiar one from Chapter 2:

$$\boldsymbol{Y} = \sqrt{\frac{\bar{\gamma}}{N_t}} \boldsymbol{H}\boldsymbol{X} + \boldsymbol{W}, \tag{11.1}$$

the only difference here being that the $N_t \times N_x$ matrix $\boldsymbol{X}$ is known to both the transmitter and the receiver. Recall also from Chapters 7, 8, and 9 that the input-output relationships of single- and multi-carrier systems over frequency- or time-selective MIMO channels can be brought to this flat fading matrix form. Hence, preamble-based estimators apply to all these MIMO fading channels.

Concatenating vectors $\{\boldsymbol{y}_\nu\}_{\nu=1}^{N_r}$ from all receive-antennas and using Kronecker product ($\otimes$) notation, we can rewrite (11.1) as

$$\boldsymbol{y} = \sqrt{\frac{\bar{\gamma}}{N_t}} (\boldsymbol{I}_{N_r} \otimes \boldsymbol{X}^T)\boldsymbol{h} + \boldsymbol{w}, \tag{11.2}$$

where $\boldsymbol{y} := [\boldsymbol{y}_1^T, \ldots, \boldsymbol{y}_{N_r}]^T$, $\boldsymbol{y}_\nu^T$ is the νth row of $\boldsymbol{Y}$, $\boldsymbol{h} := [\boldsymbol{h}_1^T, \ldots, \boldsymbol{h}_{N_r}]^T$, $\boldsymbol{h}_\nu^T$ is the νth row of $\boldsymbol{H}$, and similarly for the AWGN vector $\boldsymbol{w}$. Given $\boldsymbol{X}$ and $\boldsymbol{y}$, the goal is to estimate the channel vector $\boldsymbol{h}$ in (11.2).

Treating the realization $\boldsymbol{h}$ as a deterministic vector, the least-squares (LS) channel estimator is given by

$$\hat{\boldsymbol{h}}_{\text{LS}} = \sqrt{\frac{N_t}{\bar{\gamma}}}(\boldsymbol{I}_{N_r} \otimes \boldsymbol{X}^T)^\dagger \boldsymbol{y}, \tag{11.3}$$

where † denotes the matrix pseudo-inverse. Alternatively, if $\boldsymbol{h}$ is viewed as a zero-mean random vector with correlation matrix $\boldsymbol{R}_h := E[\boldsymbol{h}\boldsymbol{h}^{\mathcal{H}}]$ known at the receiver, one can use the linear minimum mean-square error (LMMSE) estimator

$$\hat{\boldsymbol{h}}_{\text{LMMSE}} = \sqrt{\frac{\bar{\gamma}}{N_t}}\left(\boldsymbol{R}_h^{-1} + \frac{\bar{\gamma}}{N_t}\boldsymbol{I}_{N_r} \otimes (\boldsymbol{X}^*\boldsymbol{X}^T)\right)^{-1}(\boldsymbol{I}_{N_r} \otimes \boldsymbol{X}^*)\boldsymbol{y}. \tag{11.4}$$

Being linear functions of the received vector $\boldsymbol{y}$, the estimators in (11.3) and (11.4) are simple to compute and their performance is assessed by the corresponding MMSEs $E[\|\boldsymbol{h} - \hat{\boldsymbol{h}}\|^2]$, which can be expressed as

$$\begin{aligned} \sigma_{\text{LS}}^2 &= \text{tr}\left[\frac{N_t}{\bar{\gamma}}(\boldsymbol{I}_{N_r} \otimes \boldsymbol{X}^*\boldsymbol{X}^T)^{-1}\right], \\ \sigma_{\text{LMMSE}}^2 &= \text{tr}\left[\left(\boldsymbol{R}_h^{-1} + \frac{\bar{\gamma}}{N_t}(\boldsymbol{I}_{N_r} \otimes \boldsymbol{X}^*\boldsymbol{X}^T)\right)^{-1}\right]. \end{aligned} \tag{11.5}$$

If one fixes the transmit-power, the MMSEs increase with the number of transmit-antennas N_t, which is intuitively reasonable since more parameters need to be estimated with the same number of training symbols. For a given N_t and preamble size, one can certainly lower the MMSE in channel estimation by increasing the transmit-power to effect a larger $\bar{\gamma}$. These features of LS and LMMSE estimators and their MMSE performance are common to all MIMO system identification schemes that are based on input-output samples. What is unique in the context of estimating MIMO communication channels is that transmit-power per burst is limited and the number of input samples (preamble size) must be as small as possible to minimize bandwidth loss. Indeed, the more power or bandwidth is spent for training, the less will remain for transmitting information-bearing symbols. This will degrade performance with respect to the ultimate metrics of a communication system: error probability and data rate. These considerations motivate criteria and design of training patterns which jointly optimize channel estimation performance along with the utilization of bandwidth and power resources at the transmitter.

11.2 OPTIMAL TRAINING-BASED CHANNEL ESTIMATION

The goal in this section is to design optimal training sequences for reliable estimation of MIMO channels, which unlike the simple preamble-based designs in (11.3) or (11.4), also optimize allocation of transmit-power and rate resources between training and information-bearing symbols. Optimization is carried out for block transmissions with a fixed number of symbols and fixed power per block. Training symbols

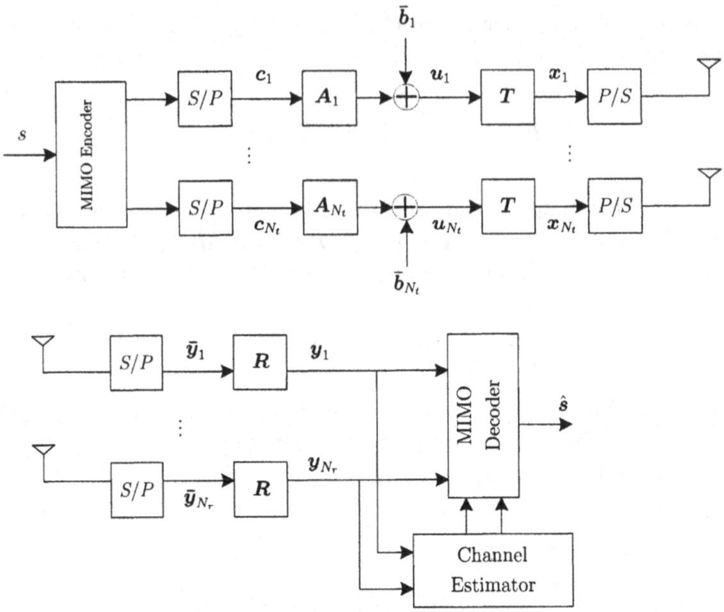

Figure 11.1 Discrete-time baseband equivalent MIMO model for channel estimation

are not placed at the beginning of the block a fortiori. The focus is on estimation of frequency-selective MIMO channels, bearing in mind that frequency-flat MIMO channels constitute a special case, while time-selective ones can be handled in a dual fashion as described in Chapter 9. Only the LMMSE estimator will be considered, since the LS estimator in (11.3) can be obtained from the LMMSE estimator as a special case after setting $\boldsymbol{R}_h^{-1} = \boldsymbol{0}$.

Consider the multi-antenna system depicted in Figure 11.1 with N_t transmit-antennas and N_r receive-antennas. Before transmission at rate $1/T_s$, the information-bearing symbols are grouped in blocks of size $N_s \times 1$, with the kth block denoted as $\boldsymbol{s}(k) := [s(kN_s), s(kN_s+1), \ldots, s(kN_s+N_s-1)]^T$.

Each block $\boldsymbol{s}(k)$ is first encoded and/or multiplexed in space and time. The resulting N_t blocks $\{\boldsymbol{c}_\mu(k)\}_{\mu=1}^{N_t}$ have size $N_c \times 1$ and each is directed to one transmit-antenna. At the μth transmit-antenna, $\boldsymbol{c}_\mu(k)$ is processed by a matrix $\boldsymbol{A}_\mu$ of size $N_u \times N_c$, where $N_u \geq N_c$. Training blocks $\bar{\boldsymbol{b}}_\mu(k)$ of size $N_u \times 1$, which are known to both transmitter and receiver, are then superimposed to $\boldsymbol{A}_\mu \boldsymbol{c}_\mu(k)$ after the ST coder to form $\boldsymbol{u}_\mu(k) := \boldsymbol{A}_\mu \boldsymbol{c}_\mu(k) + \bar{\boldsymbol{b}}_\mu(k)$. This affine model is fairly general: It encompasses linear (pre)coding via $\boldsymbol{A}_\mu$ as well as inserted training symbols by having nonzero entries in $\bar{\boldsymbol{b}}_\mu(k)$ where $\boldsymbol{A}_\mu \boldsymbol{c}_\mu(k)$ has zero entries.

To allow for block-by-block processing free of interblock interference (IBI) at the receiver, redundant (guard) symbols are added to $\boldsymbol{u}_\mu(k)$ via an $N \times N_u$ matrix $\boldsymbol{T}$ to generate blocks $\boldsymbol{x}_\mu(k) := \boldsymbol{T}\boldsymbol{u}_\mu(k)$ of length $N > N_u$. The type and size of the added redundancy will be specified soon. The blocks $\boldsymbol{x}_\mu(k)$ are then parallel-to-serial (P/S)-converted, pulse-shaped, carrier-modulated, and transmitted from the μth transmit-antenna.

Let $h_{\nu\mu}(l)$, $l \in [0, L]$ denote the discrete-time baseband equivalent channel that includes transmit-receive filters as well as the frequency-selective propagation effects between the μth transmit-antenna and the νth receive-antenna. With $\tau_{\max}^{(\nu,\mu)}$ denoting the delay spread of the channel between the μth transmit-antenna and the νth receive-antenna, the MIMO channel's maximum delay spread is defined as $\tau_{\max} := \max\{\tau_{\max}^{(\nu,\mu)}\}$. With T_s denoting the sampling period, the maximum channel order is then $L := \lceil \tau_{\max}/T_s \rceil$.

The samples at the νth receive-antenna filter output can be expressed as

$$\bar{y}_\nu(n) = \sum_{\mu=1}^{N_t} \sum_{l=0}^{L} h_{\nu\mu}(l) x_\mu(n-l) + w_\nu(n), \tag{11.6}$$

where $w_\nu(n)$ is AWGN with zero mean and variance σ^2. Samples $\bar{y}_\nu(n)$ are serial-to-parallel-converted to form $N \times 1$ blocks $\bar{\boldsymbol{y}}_\nu(k) := [\bar{y}_\nu(kN), \bar{y}_\nu(kN+1), \ldots, \bar{y}_\nu(kN+N-1)]^T$. Selecting $N \geq L$, one can write the matrix-vector counterpart of (11.6) as (see, e.g., [286])

$$\bar{\boldsymbol{y}}_\nu(k) = \sum_{\mu=1}^{N_t} \left[\boldsymbol{H}_{\nu\mu} \boldsymbol{x}_\mu(k) + \boldsymbol{H}_{\nu\mu}^{\text{ibi}} \boldsymbol{x}_\mu(k-1) \right] + \boldsymbol{w}_\nu(k), \tag{11.7}$$

where $\boldsymbol{w}_\nu(k) := [w_\nu(kN), w_\nu(kN+1), \ldots, w_\nu(kN+N-1)]^T$, $\boldsymbol{H}_{\nu\mu}$ is an $N \times N$ lower-triangular Toeplitz matrix with first column $[h_{\nu\mu}(0), \ldots, h_{\nu\mu}(L), 0, \ldots, 0]^T$ and $\boldsymbol{H}_{\nu\mu}^{\text{ibi}}$ is an $N \times N$ upper triangular Toeplitz matrix with first row $[0, \ldots, 0, h_{\nu\mu}(L), \ldots, h_{\nu\mu}(1)]$. The $\boldsymbol{x}_\mu(k-1)$-dependent term in (11.7) captures the IBI due to the channel delay spread. Compared with (11.1), the reason we do not explicitly include the average SNR term $\bar{\gamma}/N_t$ in (11.7) is because we want to allow training and information symbols per transmitted block to possibly have different power.

To enable block-by-block channel estimation and symbol decoding, it is necessary to remove the IBI. As depicted in Figure 11.1, this can be accomplished by left-multiplying $\bar{\boldsymbol{y}}_\nu(k)$ with the IBI-removing matrix $\boldsymbol{R}$ to obtain

$$\boldsymbol{y}_\nu(k) = \sum_{\mu=1}^{N_t} \left[\boldsymbol{R}\boldsymbol{H}_{\nu\mu} \boldsymbol{T} \boldsymbol{u}_\mu(k) + \boldsymbol{R}\boldsymbol{H}_{\nu\mu}^{\text{ibi}} \boldsymbol{T} \boldsymbol{u}_\mu(k-1) \right] + \boldsymbol{R}\boldsymbol{w}_\nu(k). \tag{11.8}$$

We have seen in Chapters 7 and 8 that the two prevalent and easy-to-implement guard options for IBI elimination are zero padding (ZP) and cyclic prefix (CP). Either one of the two requires a specific pair of matrices $(\boldsymbol{T}, \boldsymbol{R})$ to insert the corresponding guard at the transmitter and remove it at the receiver. In this section, we wish to design optimal training strategies which are *universally* applicable to all coherent ST coded systems. For this reason, optimal pairs $(\boldsymbol{A}_\mu, \bar{\boldsymbol{b}}_\mu)$ must be sought for training patterns placed *after* the ST coded symbols $\forall \mu \in [1, N_t]$.

For both ZP- and CP-based block transmissions, we specifically seek to optimize selection of the number of training symbols per block, placement of training symbols, and power allocation between training and information symbols. We also look for training patterns that are optimal *for any* block size.

Before delving into derivations, it is useful to outline the major steps of the approach taken to this end:

S1) Design the matrices $\{\boldsymbol{A}_\mu\}_{\mu=1}^{N_t}$ and vectors $\{\bar{\boldsymbol{b}}_\mu\}_{\mu=1}^{N_t}$ to decouple channel estimation from symbol detection; this step is motivated by identifiability and low-complexity considerations.

S2) Perform LMMSE channel estimation based on the training symbols [cf. (11.4)].

S3) Derive upper and lower bounds on the average (ergodic) capacity to benchmark the rate when training-based channel estimation is accounted for.

S4) Link the lower bound on capacity with the channel MMSE to obtain a jointly optimal criterion for training.

S5) Derive the optimal placement of information-bearing and training symbols and the optimal power allocation between the two per transmit-antenna.

Since the ensuing channel estimators and symbol detectors operate on a block-by-block basis, henceforth we drop the block index k.

11.2.1 ZP-Based Block Transmissions

Capitalizing on the fact that the nonzero entries of $\boldsymbol{H}_{\nu\mu}^{\text{ibi}}$ in (11.7) are confined to its last L columns, the ZP-based transmission eliminates IBI by appending L trailing zeros to each block $\boldsymbol{x}_\mu$ transmitted. Analytically, we select the matrices $\boldsymbol{T}$ and $\boldsymbol{R}$ as

$$\boldsymbol{T} = \boldsymbol{T}_{\text{zp}} := [\boldsymbol{I}_{N_u} \ \boldsymbol{0}_{N_u \times L}]^T, \quad \boldsymbol{R} = \boldsymbol{R}_{\text{zp}} := \boldsymbol{I}_{N_u+L},$$

where $\boldsymbol{T}_{\text{zp}}$ is the zero-padding matrix, which appends L zeros to the $N_u \times 1$ vector $\boldsymbol{u}_\mu$ when multiplying it from the left. It is clear that the length of $\boldsymbol{x}_\mu = \boldsymbol{T}_{\text{zp}}\boldsymbol{u}_\mu$ is now $N := N_u + L$. Upon defining the $N \times N_u$ matrix $\bar{\boldsymbol{H}}_{\nu\mu} := \boldsymbol{H}_{\nu\mu}\boldsymbol{T}_{\text{zp}}$, the IBI-free counterpart of the input-output relationship in (11.7) is

$$\boldsymbol{y}_\nu = \sum_{\mu=1}^{N_t} (\bar{\boldsymbol{H}}_{\nu\mu}\boldsymbol{A}_\mu\boldsymbol{c}_\mu + \bar{\boldsymbol{H}}_{\nu\mu}\bar{\boldsymbol{b}}_\mu) + \boldsymbol{w}_\nu, \tag{11.9}$$

where $\bar{\boldsymbol{H}}_{\nu\mu}$ is an $N \times N_u$ Toeplitz matrix with first column $[h_{\nu\mu}(0), \ldots, h_{\nu\mu}(L), 0, \ldots, 0]^T$. Recall now that convolution between two vectors can be represented as the product of a Toeplitz matrix with a vector. Because convolution is a commutative operation, we deduce that $\bar{\boldsymbol{H}}_{\nu\mu}\bar{\boldsymbol{b}}_\mu = \bar{\boldsymbol{B}}_\mu \boldsymbol{h}_{\nu\mu}$, where $\bar{\boldsymbol{B}}_\mu$ is an $N \times (L+1)$ Toeplitz matrix having first column $[\bar{b}_\mu(1), \ldots, \bar{b}_\mu(N_u), 0, \ldots, 0]^T$, with $\bar{b}_\mu(m)$ being the mth entry of $\bar{\boldsymbol{b}}_\mu$ and $\boldsymbol{h}_{\nu\mu} := [h_{\nu\mu}(0), h_{\nu\mu}(1), \ldots, h_{\nu\mu}(L)]^T$. As a result, (11.9) can be expressed as

$$\begin{aligned} \boldsymbol{y}_\nu &= \sum_{\mu=1}^{N_t} (\bar{\boldsymbol{H}}_{\nu\mu}\boldsymbol{A}_\mu \boldsymbol{c}_\mu + \bar{\boldsymbol{B}}_\mu \boldsymbol{h}_{\nu\mu}) + \boldsymbol{w}_\nu \\ &= \left[\bar{\boldsymbol{H}}_{\nu,1}\boldsymbol{A}_1 \cdots \bar{\boldsymbol{H}}_{\nu,N_t}\boldsymbol{A}_{N_t}\right]\boldsymbol{c} + \bar{\boldsymbol{B}}\boldsymbol{h}_\nu + \boldsymbol{w}_\nu, \end{aligned} \tag{11.10}$$

where $\boldsymbol{h}_\nu := [\boldsymbol{h}_{\nu,1}^T, \ldots, \boldsymbol{h}_{\nu,N_t}^T]^T$, $\boldsymbol{c} := [\boldsymbol{c}_1^T, \ldots, \boldsymbol{c}_{N_t}^T]^T$, and $\bar{\boldsymbol{B}} := [\bar{\boldsymbol{B}}_1, \ldots, \bar{\boldsymbol{B}}_{N_t}]$. Concatenating the N_r received vectors into a single block $\boldsymbol{y} := [\boldsymbol{y}_1^T, \ldots, \boldsymbol{y}_{N_r}^T]^T$, we have

$$\boldsymbol{y} = \boldsymbol{\Phi}\boldsymbol{c} + (\boldsymbol{I}_{N_r} \otimes \bar{\boldsymbol{B}})\boldsymbol{h} + \boldsymbol{w}, \tag{11.11}$$

where $\boldsymbol{h} := [\boldsymbol{h}_1^T, \ldots, \boldsymbol{h}_{N_r}^T]^T$ and the $N_rN \times N_tN_c$ matrix $\boldsymbol{\Phi}$ is defined as

$$\boldsymbol{\Phi} := \begin{bmatrix} \bar{\boldsymbol{H}}_{1,1}\boldsymbol{A}_1 & \cdots & \bar{\boldsymbol{H}}_{1,N_t}\boldsymbol{A}_{N_t} \\ \vdots & \ddots & \vdots \\ \bar{\boldsymbol{H}}_{N_r,1}\boldsymbol{A}_1 & \cdots & \bar{\boldsymbol{H}}_{N_r,N_t}\boldsymbol{A}_{N_t} \end{bmatrix}. \tag{11.12}$$

Since $\boldsymbol{\Phi}$ depends on $\boldsymbol{h}$, we infer from (11.11) that estimating $\boldsymbol{h}$ and recovering $\boldsymbol{c}$ from $\boldsymbol{y}$ is a nonlinear estimation problem. Its joint ML-optimal solution is computationally complex because it requires iterative search, and convergence to the global optimum cannot be ensured, and in certain cases even identifiability may be impossible to establish (see, e.g., [204]). This suggests designing the pair $(\boldsymbol{A}_\mu, \bar{\boldsymbol{b}}_\mu)$ so that channel estimation can be decoupled from symbol decoding. Besides ensuring identifiability and consistent estimation at a relatively low degree of complexity, this decoupling allows the designed training patterns for channel estimation to have universal applicability to all coherent ST decoders.

11.2.1.1 Decoupling Channel Estimation From Symbol Decoding If the unknown term $\boldsymbol{\Phi}\boldsymbol{c}$ were absent from (11.11), we could have estimated $\boldsymbol{h}$ from the training and received samples using the LMMSE estimator. The decoupling needed to accomplish this amounts to requiring the condition

$$\begin{aligned} &\left(\sigma^2 \boldsymbol{R}_h^{-1} + \boldsymbol{I}_{N_r} \otimes (\bar{\boldsymbol{B}}^{\mathcal{H}}\bar{\boldsymbol{B}})\right)^{-1} (\boldsymbol{I}_{N_r} \otimes \bar{\boldsymbol{B}}^{\mathcal{H}})\boldsymbol{\Phi}\boldsymbol{c} = \boldsymbol{0} \\ &\Leftrightarrow (\boldsymbol{I}_{N_r} \otimes \bar{\boldsymbol{B}}^{\mathcal{H}})\boldsymbol{\Phi}\boldsymbol{c} = \boldsymbol{0}. \end{aligned} \tag{11.13}$$

Noting that (11.13) must hold true for any transmitted data $\boldsymbol{c}$, we infer after recalling the definitions of $\bar{\boldsymbol{B}}$ and $\boldsymbol{\Phi}$ that the selection of $\boldsymbol{B}$ and $\boldsymbol{A}_\mu$'s should satisfy

$$\bar{\boldsymbol{B}}_{\mu_1}^{\mathcal{H}} \bar{\boldsymbol{H}}_{\nu\mu_2}\boldsymbol{A}_{\mu_2} = \boldsymbol{0}_{(L+1)\times N_c}, \quad \forall \mu_1, \mu_2 \in [1, N_t], \; \nu \in [1, N_r]. \tag{11.14}$$

Expressing $\boldsymbol{A}_{\mu_2}$ in terms of its N_c columns, we have

$$\begin{aligned} \bar{\boldsymbol{H}}_{\nu\mu_2}\boldsymbol{A}_{\mu_2} &= \left[\bar{\boldsymbol{H}}_{\nu\mu_2}\boldsymbol{a}_{\mu_2 1} \cdots \bar{\boldsymbol{H}}_{\nu\mu_2}\boldsymbol{a}_{\mu_2 N_c}\right] \\ &= \left[\boldsymbol{A}_{\mu_2 1}\boldsymbol{h}_{\nu\mu_2} \cdots \boldsymbol{A}_{\mu_2 N_c}\boldsymbol{h}_{\nu\mu_2}\right], \end{aligned} \tag{11.15}$$

where each $N \times (L+1)$ matrix $\boldsymbol{A}_{\mu_2 n}$ is Toeplitz generated by the vector $\boldsymbol{a}_{\mu_2,n}$. Because (11.14) must hold true for *all* channel realizations, the decoupling goal mandates $(\boldsymbol{A}_\mu, \bar{\boldsymbol{b}}_\mu)$ pairs satisfying

$$\bar{\boldsymbol{B}}_{\mu_1}^{\mathcal{H}} \boldsymbol{A}_{\mu_2,n} = \boldsymbol{0}_{(L+1)\times(L+1)}, \quad \forall \mu_1, \mu_2 \in [1, N_t], \text{ and } \forall n \in [1, N_c]. \tag{11.16}$$

Equation (11.16) implies that the matrix of training symbols must be orthogonal to (possibly linearly precoded) information-bearing symbols. Orthogonality between the corresponding training and information subblocks must hold true for each transmit-antenna but also across transmit-antennas. The implication of this condition is summarized in the next lemma, which is proved in [191].

Lemma 11.1 *Pairs $(\boldsymbol{A}_\mu, \bar{\boldsymbol{b}}_\mu)$ that guarantee decoupling of channel estimation from symbol detection, for any block length N must have the form*

$$\boldsymbol{A}_\mu = \begin{bmatrix} \boldsymbol{\Theta} \\ \boldsymbol{0}_{(N_u - N_c')\times N_c} \end{bmatrix}, \quad \bar{\boldsymbol{b}}_\mu = \begin{bmatrix} \boldsymbol{0}_{N_c'+L} \\ \boldsymbol{b}_\mu \end{bmatrix}, \tag{11.17}$$

where $\boldsymbol{\Theta}$ is an $N_c' \times N_c$ matrix that optionally precodes (if $\boldsymbol{\Theta} \neq \boldsymbol{I}_{N_c}$) the information-bearing block $\boldsymbol{c}_\mu$ linearly, N_c' is the number of symbols that $\mathbf{c}$ is mapped to, and the $N_b \times 1$ vector $\boldsymbol{b}_\mu$ contains all the nonzero entries of $\bar{\boldsymbol{b}}_\mu$ with $N_b = N_u - N_c' - L$.

It is worth emphasizing that we neither assumed insertion of training symbols a fortiori, nor imposed time-division multiplexing (TDM) of training with information subblocks at the outset. Lemma 11.1 reveals, however, that to separate channel estimation from symbol decoding, we need to insert at least L zeros per antenna between the information-bearing subblock $\boldsymbol{c}_\mu$ and the training subblock $\bar{\boldsymbol{b}}_\mu$ in the transmitted block $\boldsymbol{x}_\mu = \boldsymbol{T}_{\text{zp}}(\boldsymbol{A}_\mu \boldsymbol{c}_\mu + \bar{\boldsymbol{b}}_\mu)$. Note that the L guard zeros could in principle be replaced by L known symbols. However, besides requiring more power allocated to guard symbols, this replacement causes interference from training symbols to information symbols.

Since the information blocks $\boldsymbol{c}_\mu$ have been encoded, in order to retain the structure[1] of $\boldsymbol{c}_\mu$ and reduce decoding complexity, we simply choose $N_c' = N_c$, and $\boldsymbol{\Theta} = \boldsymbol{I}_{N_c}$. For this design, we then have

$$\bar{\boldsymbol{B}}_\mu := \begin{bmatrix} \boldsymbol{0}_{(N_c+L)\times(L+1)} \\ \boldsymbol{B}_\mu \end{bmatrix} \text{ and } \bar{\boldsymbol{B}} := \begin{bmatrix} \boldsymbol{0}_{(N_c+L)\times N_t(L+1)} \\ \boldsymbol{B} \end{bmatrix}, \tag{11.18}$$

where $\boldsymbol{B} := [\boldsymbol{B}_1, \ldots, \boldsymbol{B}_{N_t}]$ is an $(N_b+L)\times N_t(L+1)$ matrix and $N_b + L = N_u - N_c'$.

Using (11.17) and (11.18), we can partition $\boldsymbol{y}_\nu$ in (11.10) as

$$\boldsymbol{y}_\nu = \begin{bmatrix} \boldsymbol{y}_{\nu,c} \\ \boldsymbol{y}_{\nu,b} \end{bmatrix} = \begin{bmatrix} \sum_{\mu=1}^{N_t} \boldsymbol{H}_c^{(\nu,\mu)} \boldsymbol{c}_\mu + \boldsymbol{w}_{\nu,c} \\ \sum_{\mu=1}^{N_t} \boldsymbol{B}_\mu \boldsymbol{h}_{\nu\mu} + \boldsymbol{w}_{\nu,b} \end{bmatrix} = \begin{bmatrix} \boldsymbol{H}_c^{(\nu)} \boldsymbol{c} + \boldsymbol{w}_{\nu,c} \\ \boldsymbol{B}\boldsymbol{h}_\nu + \boldsymbol{w}_{\nu,b} \end{bmatrix}, \tag{11.19}$$

[1] Different from [71, 72], for example, where the training sequence is designed *before* ST mapping and is tailored to ST trellis codes, the approach of designing training *after* ST mapping offers flexibility to accommodate various ST coded multi-antenna systems. It also retains the performance and capacity features of any ST code chosen, while facilitating transmitter and receiver design.

where $\boldsymbol{H}_c^{(\nu,\mu)} := [\boldsymbol{I}_{N_c+L}, \boldsymbol{0}_{(N_c+L)\times(N_b+L)}]\bar{\boldsymbol{H}}_{\nu\mu}\boldsymbol{A}_\mu$ extracts the first $(N_c + L)$ rows and N_c columns of the channel matrix $\bar{\boldsymbol{H}}_{\nu\mu}$, and $\boldsymbol{H}_c^{(\nu)} := [\boldsymbol{H}_c^{(\nu,1)}, \ldots, \boldsymbol{H}_c^{(\nu,N_t)}]$. Notice that the $(N_c + L) \times 1$ vector $\boldsymbol{y}_{\nu,c}$ contains the information-bearing symbols while the $(N_b + L) \times 1$ vector $\boldsymbol{y}_{\nu,b}$ contains the training symbols. Stacking vectors $\boldsymbol{y}_{\nu,b}$ from all receive-antennas into $\boldsymbol{y}_b := [\boldsymbol{y}_{1,b}^T, \ldots, \boldsymbol{y}_{N_r,b}^T]^T$, the LMMSE channel estimator can be written as

$$\hat{\boldsymbol{h}} = \left(\sigma^2 \boldsymbol{R}_h^{-1} + \boldsymbol{I}_{N_r} \otimes (\boldsymbol{B}^{\mathcal{H}}\boldsymbol{B})\right)^{-1} (\boldsymbol{I}_{N_r} \otimes \boldsymbol{B}^{\mathcal{H}})\boldsymbol{y}_b. \tag{11.20}$$

Comparing (11.20) with the LMMSE estimator in (11.4), we notice that the $\boldsymbol{B}$ matrix in (11.20) has a different structure from the preamble-based matrix $\boldsymbol{X}$. Furthermore, since $\hat{\boldsymbol{h}}$ in (11.20) depends only on the training blocks $\{\boldsymbol{b}_\mu\}_{\mu=1}^{N_t}$, we have indeed accomplished the goal of decoupling channel estimation from symbol decoding. This offers flexibility in designing optimal training for MIMO channels without modifying ST code matrices, which can be designed separately to satisfy application-specific trade-offs among complexity, rate, and error performance.

To assert optimality in selecting the training matrix $\boldsymbol{B}$, we further assume that:

A11.1) The channel coefficients $h_{\nu\mu}(l)$ are uncorrelated complex Gaussian distributed and the covariance matrix $\boldsymbol{R}_{h_\nu} := E[\boldsymbol{h}_\nu \boldsymbol{h}_\nu^{\mathcal{H}}]$ is identical $\forall \nu \in [1, N_r]$ (i.e., $\boldsymbol{R}_h$ is block diagonal and can be written using Kronecker product notation as $\boldsymbol{R}_h = \boldsymbol{I}_{N_r} \otimes \boldsymbol{R}_{h_\nu}$ with $\mathrm{tr}[\boldsymbol{R}_h] = N_t N_r$).

Because neither matrix $\boldsymbol{R}_h$ nor any CSI is assumed available at the transmitter, A11.1 does not affect the optimal design of the training sequence. Upon defining the channel estimation error as $\check{\boldsymbol{h}} := \boldsymbol{h} - \hat{\boldsymbol{h}}$, we can express its correlation as (cf. A11.1)

$$\boldsymbol{R}_{\check{h}} = \boldsymbol{I}_{N_r} \otimes \left(\boldsymbol{R}_{h_1}^{-1} + \frac{1}{\sigma^2}\boldsymbol{B}^{\mathcal{H}}\boldsymbol{B}\right)^{-1}. \tag{11.21}$$

Based on A11.1, it can be shown further that the MSE $\sigma_{\check{h}}^2$ is lower bounded by

$$\begin{aligned} \sigma_{\check{h}}^2 &= N_r \,\mathrm{tr}\left[\left(\boldsymbol{R}_{h_1}^{-1} + \frac{1}{\sigma^2}\boldsymbol{B}^{\mathcal{H}}\boldsymbol{B}\right)^{-1}\right] \\ &\geq N_r \sum_{m=1}^{N_t(L+1)} \left([\boldsymbol{R}_{h_1}^{-1}]_{m,m} + \frac{1}{\sigma^2}[\boldsymbol{B}^{\mathcal{H}}\boldsymbol{B}]_{m,m}\right)^{-1}, \end{aligned} \tag{11.22}$$

where the equality holds if and only if $\boldsymbol{B}^{\mathcal{H}}\boldsymbol{B}$ is a diagonal matrix. The bound in (11.22) implies that the following condition is required to minimize the channel MSE $\sigma_{\check{h}}^2$ over all training patterns:

C11.1) For fixed N_b and N_c, the training symbols inserted must enforce the matrix $\boldsymbol{B}^{\mathcal{H}}\boldsymbol{B}$ to be diagonal.

For single-input single-output channels, C11.1 coincides with those in [59, 65] and it is also assumed a fortiori by [1]. In addition to the TDM-based orthogonality between information and training subblocks dictated by Lemma 11.1, condition C11.1 asserts MMSE optimality in channel estimation when the training subblocks are mutually orthogonal across transmit-antennas.

So far, we have seen that orthogonality between information and training subblocks ensures low-complexity channel estimation decoupled from symbol detection; while under C11.1, orthogonality among training subblocks across antennas guarantees that the LMMSE channel estimator in (11.20) will have the smallest MSE. What is left to specify is the number and placement of pilot symbols within each training subblock as well as the power allocation in each training and information subblock. These parameters affect the performance of MIMO channel estimation, the effective transmission rate, the mutual information and the bit error rate (BER). In what follows we select these training parameters by optimizing an ergodic (average) capacity bound.

11.2.1.2 Ergodic Capacity Bounds Because no exact expression for the average capacity is available when channel estimation is taken into account, optimality criteria for designing training sequences have to rely on lower bounds of the average capacity. Besides using these lower bounds to select optimal training parameters, we also invoke an upper bound of the average capacity to serve as a benchmark for the maximum rate achievable by single-carrier transmissions over frequency-selective MIMO channels.

Collecting the N_r received symbol blocks corresponding to all receive-antennas, the information-bearing part of (11.19) becomes

$$\boldsymbol{y}_c = \boldsymbol{H}_c \boldsymbol{c} + \boldsymbol{w}_c, \tag{11.23}$$

where $\boldsymbol{w}_c := [\boldsymbol{w}_{1,c}^T, \ldots, \boldsymbol{w}_{N_r,c}^T]^T$, $\boldsymbol{y}_c := [\boldsymbol{y}_{1,c}^T, \ldots, \boldsymbol{y}_{N_r,c}^T]^T$, and

$$\boldsymbol{H}_c := \begin{bmatrix} \boldsymbol{H}_c^{(1,1)} & \cdots & \boldsymbol{H}_c^{(1,N_t)} \\ \vdots & \ddots & \vdots \\ \boldsymbol{H}_c^{(N_r,1)} & \cdots & \boldsymbol{H}_c^{(N_r,N_t)} \end{bmatrix} \in \mathbb{C}^{N_r(N_c+L)\times N_t N_c}.$$

Let $\hat{\boldsymbol{h}}$ stand for any estimator of $\boldsymbol{h}$, $\mathcal{P}$ the total transmit-power per block, $\mathcal{P}_c$ the power allocated to the information part, and $\mathcal{P}_b$ the power allocated to the training part. Since training symbols $\boldsymbol{b}$ do not convey information, for a fixed power $\mathcal{P}_c := E[\|\boldsymbol{c}\|^2]$, the mutual information between transmitted and received symbols in (11.23) is given by $\mathcal{I}(\boldsymbol{y}_c; \boldsymbol{c}|\hat{\boldsymbol{h}})$. The channel capacity averaged over the random channel $\boldsymbol{h}$ is by definition [256]

$$C := \frac{1}{N} E\left[\max_{p_c(\cdot),\, \mathcal{P}_c} \mathcal{I}(\boldsymbol{y}_c; \boldsymbol{c}|\hat{\boldsymbol{h}})\right] \quad \text{bits per channel use (pcu)}, \tag{11.24}$$

where $p_c(\cdot)$ denotes the probability density function of $\boldsymbol{c}$.

When the channel estimate is perfect (i.e., $\hat{\boldsymbol{h}} \equiv \boldsymbol{h}$), the upper bound on the capacity can be obtained for a Gaussian $\boldsymbol{c}$ with $\boldsymbol{R}_c := E[\boldsymbol{c}\boldsymbol{c}^{\mathcal{H}}]$ as (see, e.g., [1, 187, 205])

$$\bar{C} := \frac{1}{N} E\left[\max_{\boldsymbol{R}_c} \log_2 \det\left(\boldsymbol{I}_{N_r(N_c+L)} + \frac{1}{\sigma^2} \boldsymbol{H}_c \boldsymbol{R}_c \boldsymbol{H}_c^{\mathcal{H}}\right)\right] \quad \text{bits pcu.} \tag{11.25}$$

Recall that $\boldsymbol{c}$ is obtained by encoding each information symbol block $\boldsymbol{s}$ in space and time. When $\boldsymbol{s}$ is Gaussian, $\boldsymbol{c}$ will also be approximately Gaussian for many ST coders with block or layered structure, as in Chapters 3 and 4. Even if $\boldsymbol{s}$ is drawn from a finite alphabet, an LCF precoded $\boldsymbol{c}$ is also approximately Gaussian distributed provided that the block size is large enough. In these cases, we can safely assume that:

A11.2) The information-bearing symbol block $\boldsymbol{c}$ is zero-mean Gaussian with covariance $\boldsymbol{R}_c = \bar{\mathcal{P}}_c \boldsymbol{I}_{N_t N_c}$, where $\bar{\mathcal{P}}_c := \mathcal{P}_c/(N_t N_c)$ denotes the normalized power.

For certain ST codes (e.g., OSTBCs), A11.2 does not hold true. Nonetheless, the ensuing analysis carries over to these cases but with a different $\boldsymbol{R}_c$. Under A11.2, the capacity upper bound (11.25) becomes

$$\bar{C} := \frac{1}{N} E\left[\log_2 \det\left(\boldsymbol{I}_{N_r(N_c+L)} + \frac{\bar{\mathcal{P}}_c}{\sigma^2}\boldsymbol{H}_c\boldsymbol{H}_c^{\mathcal{H}}\right)\right] \quad \text{bits per channel use.} \tag{11.26}$$

Having established the upper bound in (11.26) when the channel is perfectly known, let us now turn our attention to the case where only channel estimates are available. We will see that in this case, an approximate lower bound can be derived. When the estimate $\hat{\boldsymbol{h}}$ is imperfect, we have $\boldsymbol{y}_c = \hat{\boldsymbol{H}}_c\boldsymbol{c} + \boldsymbol{v}$ [cf. (11.23)], where $\boldsymbol{v} := \breve{\boldsymbol{H}}_c\boldsymbol{c} + \boldsymbol{w}_c$ and $\breve{\boldsymbol{H}}_c := \boldsymbol{H}_c - \hat{\boldsymbol{H}}_c$. The correlation matrix of $\boldsymbol{v}$ is given by

$$\begin{aligned}\boldsymbol{R}_v &:= E[\boldsymbol{v}\boldsymbol{v}^{\mathcal{H}}] = E[\breve{\boldsymbol{H}}_c\boldsymbol{c}\boldsymbol{c}^{\mathcal{H}}\breve{\boldsymbol{H}}_c^{\mathcal{H}}] + \sigma^2\boldsymbol{I}_{N_r(N_c+L)} \\ &= \bar{\mathcal{P}}_c E[\breve{\boldsymbol{H}}_c\breve{\boldsymbol{H}}_c^{\mathcal{H}}] + \sigma^2\boldsymbol{I}_{N_r(N_c+L)}\end{aligned} \tag{11.27}$$

and can be approximated as [191]

$$\boldsymbol{R}_v \approx \left(\frac{\bar{\mathcal{P}}_c\sigma_{\breve{h}}^2}{N_r} + \sigma^2\right)\boldsymbol{I}_{N_r(N_c+L)}. \tag{11.28}$$

From (11.28), we can readily verify that as $\sigma_{\breve{h}}^2$ decreases, $\boldsymbol{R}_v$ decreases[2].

By employing the LMMSE channel estimator and using A11.1 and A11.2, it has been shown in [187, Lemma 2] that the capacity in (11.24) is lower bounded by

$$C \geq \underline{C} := \frac{1}{N} E\left[\log_2 \det\left(\boldsymbol{I}_{N_r(N_c+L)} + \bar{\mathcal{P}}_c\boldsymbol{R}_v^{-1}\hat{\boldsymbol{H}}_c\hat{\boldsymbol{H}}_c^{\mathcal{H}}\right)\right] \quad \text{bits pcu.} \tag{11.29}$$

Using $\boldsymbol{R}_v$ from (11.28), we will select training parameters so that $\underline{C}$ is maximized. Intuitively speaking, we look for optimal training parameters to improve the lower bound on capacity and also the quality of the LMMSE channel estimator through the

[2]For two positive semi-definite matrices $\boldsymbol{R}_{v1}$ and $\boldsymbol{R}_{v2}$, we say that $\boldsymbol{R}_{v1} \leq \boldsymbol{R}_{v2}$ if and only if $\boldsymbol{R}_{v2} - \boldsymbol{R}_{v1}$ is a positive semi-definite matrix.

associated MSE. To solidify this intuition, it is useful to recall the following result from [187, Appendix C]:

Lemma 11.2 *Suppose that C11.1, A11.1 and A11.2 hold true while the information symbol power $\bar{\mathcal{P}}_c$ and the training (information) block lengths $N_b(N_c)$ are fixed. Then at high SNR, maximizing $\underline{C}$ in* (11.29) *is equivalent to minimizing $\boldsymbol{R}_v$ in* (11.28).

Lemma 11.2 implies that $\underline{C}$ increases as $\boldsymbol{R}_v$ decreases, which occurs when $\sigma_{\tilde{h}}^2$ decreases, according to (11.28). In other words, improved channel estimation (lower channel MSE) induces a higher average capacity bound.

11.2.1.3 Optimal Training Parameters Under C11.1, $\sigma_{\tilde{h}}^2$ is given by

$$\sigma_{\tilde{h}}^2 = N_r \sum_{m=1}^{N_t(L+1)} \left([\boldsymbol{R}_{h_1}^{-1}]_{m,m} + \frac{1}{\sigma^2} [\boldsymbol{B}^{\mathcal{H}} \boldsymbol{B}]_{m,m} \right)^{-1}.$$

This channel MSE expression can be further lower bounded as

$$\sigma_{\tilde{h}}^2 \geq N_r \sum_{m=1}^{N_t(L+1)} \left([\boldsymbol{R}_{h_1}^{-1}]_{m,m} + \frac{1}{\sigma^2} \frac{\mathcal{P}_b}{N_t} \right)^{-1},$$

where the equality holds if and only if $\boldsymbol{B}^{\mathcal{H}}\boldsymbol{B} = (\mathcal{P}_b/N_t)\boldsymbol{I}_{N_t(L+1)}$. Condition C11.1 can now be modified as follows:

C11.1′) For fixed N_b and N_c, the training symbols should be inserted so that $\boldsymbol{B}^{\mathcal{H}}\boldsymbol{B} = (\mathcal{P}_b/N_t)\,\boldsymbol{I}_{N_t(L+1)}$.

By the definition of $\boldsymbol{B}$, this modified condition implies that the Toeplitz training matrices should be mutually orthogonal; that is,

$$\boldsymbol{B}_{\mu_1}^{\mathcal{H}} \boldsymbol{B}_{\mu_2} = \frac{\mathcal{P}_b}{N_t} \boldsymbol{I}_{L+1} \delta(\mu_1 - \mu_2), \quad \forall \mu_1, \mu_2 \in [1, N_t], \tag{11.30}$$

where $\delta(\cdot)$ denotes the Kronecker delta function. Since according to C11.1′ the $(N_b + L) \times N_t(L+1)$ matrix $\boldsymbol{B}$ has full column rank, the minimum possible number of training symbols is $N_b = N_t(L+1) - L$, which suggests only one nonzero entry in each training subblock per transmit-antenna. In fact, for fixed $\mathcal{P}_c$ and $\mathcal{P}_b$, as N_b increases beyond $N_t(L+1) - L$, the average capacity bound $\underline{C}$ decreases. Intuitively, this is because the channel MSE is fixed when the transmit-power is fixed, while increasing N_b decreases the information rate under the orthogonal pilot structure.

To summarize, we have shown that for fixed $\mathcal{P}_c$ and $\mathcal{P}_b$ the optimal placement of information and training subblocks per transmit-antenna, say the μth one, is $[\boldsymbol{c}_\mu^T\ \boldsymbol{0}_L^T\ \boldsymbol{b}_\mu^T]^T$, where $\boldsymbol{b}_\mu$ is selected to satisfy (11.30) with length $N_b = N_t(L+1) - L$. Clearly, one selection of the training subblock satisfying C11.1′ is $\boldsymbol{b}_\mu = [\boldsymbol{0}_{(\mu-1)(L+1)}^T\ \ b\ \ \boldsymbol{0}_{(N_t-\mu)(L+1)}^T]^T$. For this optimal choice, the corresponding training matrix is

$$\boldsymbol{B}_\mu = \sqrt{\frac{\mathcal{P}_b}{N_t}} [\boldsymbol{0}_{(L+1)\times(\mu-1)(L+1)}\ \ \boldsymbol{I}_{L+1}\ \ \boldsymbol{0}_{(L+1)\times(N_t-\mu)(L+1)}]^T.$$

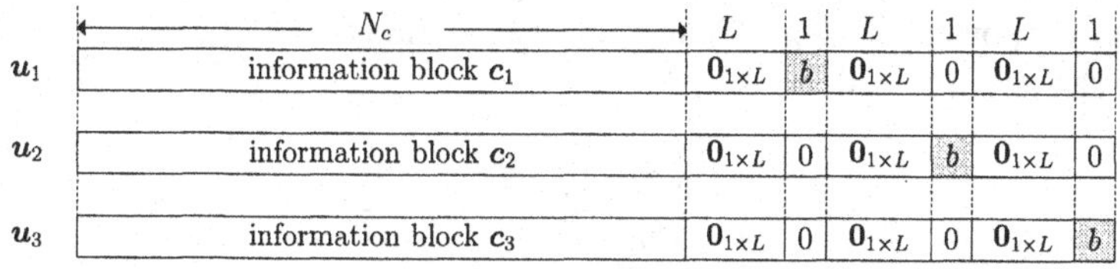
N_c
L
1
L
1
L
1
u_1
information block c_1
$0_{1\times L}$
b
$0_{1\times L}$
0
$0_{1\times L}$
0
u_2
information block c_2
$0_{1\times L}$
0
$0_{1\times L}$
b
$0_{1\times L}$
0
u_3
information block c_3
$0_{1\times L}$
0
$0_{1\times L}$
0
$0_{1\times L}$
b

Figure 11.2 Optimal pattern for ZP-based transmissions with $N_t = 3$

Since this placement of information and training subblocks satisfies C11.1′ , the channel MSE is achieved and the average capacity lower bound $\underline{C}$ is maximized.

An example of optimal placement with $N_t = 3$ transmit-antennas and a frequency-selective MIMO channel of order L is shown in Figure 11.2. We observe that L zeros are inserted between the information subblock and the training subblock in each antenna to ensure decoupling of symbol detection from channel estimation according to Lemma 11.1. At the same time, the training subblock per antenna has size $N_b = 3(L+1) - L = 2L + 3$ and all training subblocks across antennas are orthogonal to minimize the channel MSE in accordance with C11.1′. Since the training subblocks can be permuted across antennas without altering the optimality of the training pattern, we should remark that the design of $\boldsymbol{b}_\mu$ is not unique for a given (N_t, L) pair.

Having optimized the placement of information and training symbols, we pursue next optimal allocation of the power available per block. To this end, let us normalize the estimated channel using $\sigma^2_{\hat{H}_c} := \text{tr}\big(E[\hat{\boldsymbol{H}}_c \hat{\boldsymbol{H}}_c^{\mathcal{H}}]\big)$, to obtain

$$\hat{\bar{\mathbf{H}}}_c := \frac{1}{\sigma_{\hat{H}_c}} \hat{\boldsymbol{H}}_c. \tag{11.31}$$

Substituting (11.28) and (11.31) into (11.29) yields the capacity lower bound when optimal placement is employed:

$$\underline{C} = \frac{1}{N} E\left[\log_2 \det\big(\boldsymbol{I}_{N_r(N_c+L)} + \rho_{\text{eff}} \hat{\bar{\mathbf{H}}}_c \hat{\bar{\mathbf{H}}}_c^{\mathcal{H}}\big)\right] \quad \text{bits pcu,} \tag{11.32}$$

where the effective SNR ρ_{eff} accounting for channel estimation errors is defined as

$$\rho_{\text{eff}} := \frac{N_r \bar{\mathcal{P}}_c \sigma^2_{\hat{H}_c}}{\bar{\mathcal{P}}_c \sigma^2_{\tilde{h}} + N_r \sigma^2}, \tag{11.33}$$

where $\sigma^2_{\hat{H}_c} = \text{tr}\left(E[\hat{\boldsymbol{H}}_c \hat{\boldsymbol{H}}_c^{\mathcal{H}}]\right) = N_c(N_t N_r - \sigma^2_{\tilde{h}})$, [188]. Moreover, with $\psi_l^{(1,\mu)} := E\left[h_{1,\mu}(l) h^*_{1,\mu}(l)\right]$, we have

$$\sigma^2_{\tilde{h}} = N_r \sum_{\mu=1}^{N_t} \sum_{l=0}^{L} \frac{1}{\mathcal{P}_b \psi_l^{(1,\mu)}/(N_t \sigma^2) + 1} \psi_l^{(1,\mu)}. \tag{11.34}$$

With total power $\mathcal{P} = \mathcal{P}_c + \mathcal{P}_b$ and power allocation factor $\alpha_{\text{zp}} := \mathcal{P}_c/\mathcal{P} \in (0,1)$, (11.33) can be expressed as

$$\rho_{\text{eff}} = \frac{\alpha_{\text{zp}} N_r \mathcal{P}(N_t N_r - \sigma_{\tilde{h}}^2)}{\alpha_{\text{zp}} \mathcal{P} \sigma_{\tilde{h}}^2 / N_c + N_r N_t \sigma^2}. \tag{11.35}$$

Because $\sigma_{\tilde{h}}^2$ is a function of $\psi_l^{(1,\mu)}$, it is difficult to optimize ρ_{eff} in (11.35) with respect to α_{zp}. For this reason, we optimize separately three simplified cases for which ρ_{eff} does not depend on CSI parameters.

Case 1: Low SNR **($\mathcal{P}/N_t \ll \sigma^2$)** In this case, (11.34) becomes $\sigma_{\tilde{h}}^2 \approx N_t N_r - N_r(1-\alpha_{\text{zp}})\mathcal{P}/\sigma^2$ after using Taylor's expansion and the approximation

$$N_t - \sum_{\mu=1}^{N_t} \sum_{l=0}^{L} [\varphi_l^{(1,\mu)}]^2 \frac{\mathcal{P}_b}{N_t \sigma^2} \approx N_t - \frac{\mathcal{P}_b}{\sigma^2}.$$

Upon substituting into (11.35), we obtain

$$\rho_{\text{eff}} \approx \frac{\alpha_{\text{zp}}(1-\alpha_{\text{zp}}) N_c N_r \mathcal{P}^2}{\alpha_{\text{zp}} \mathcal{P}[N_t \sigma^2 - (1-\alpha_{\text{zp}})\mathcal{P}] + N_c N_t \sigma^4}. \tag{11.36}$$

Differentiating ρ_{eff} with respect to α_{zp}, one can find that at low SNR, the optimal power allocation factor is [188]

$$\alpha_{\text{zp}} \approx \frac{1}{2}, \tag{11.37}$$

which implies that the power in the optimal training sequence must be allocated equally between the information subblock and the training subblock.

Case 2: High SNR **($\mathcal{P}/N_t \gg \sigma^2$)** In this case, (11.34) becomes $\sigma_{\tilde{h}}^2 \approx N_r N_t^2 (L+1)\sigma^2/\mathcal{P}_b$ and upon substituting into (11.35), we obtain

$$\rho_{\text{eff}} \approx \frac{\alpha_{\text{zp}} N_r \mathcal{P}\left[1 - N_t(L+1)\sigma^2/\mathcal{P}_b\right]}{N_t(L+1)\alpha_{\text{zp}}\sigma^2/[N_c(1-\alpha_{\text{zp}})] + \sigma^2}. \tag{11.38}$$

Differentiating ρ_{eff} with respect to α_{zp}, one can find that at high SNR, the optimal power allocation factor is

$$\alpha_{\text{zp}} = \frac{1 - \sqrt{1 - (1-\lambda)(1 - N_c\sigma^2\lambda/\mathcal{P}}}{1-\lambda} \approx \frac{1}{1+\sqrt{\lambda}}, \tag{11.39}$$

where $\lambda := N_t(L+1)/N_c$.

Case 3: Identically distributed channel coefficients $[\psi_l^{(1,\mu)} = 1/(L+1)]$
In this case, (11.34) becomes

$$\sigma_{\tilde{h}}^2 = \frac{N_t N_r}{(1-\alpha_{\text{zp}})\mathcal{P}/[N_t(L+1)\sigma^2]+1},$$

and correspondingly, (11.35) yields

$$\rho_{\text{eff}} = \frac{\alpha_{\text{zp}} N_r \mathcal{P}\left(1 - \frac{1}{(1-\alpha_{\text{zp}})\mathcal{P}/[N_t(L+1)\sigma^2]+1}\right)}{\frac{\alpha_{\text{zp}}\mathcal{P}}{N_c}\frac{1}{(1-\alpha_{\text{zp}}\mathcal{P})/[N_t(L+1)\sigma^2]+1} + \sigma^2}. \tag{11.40}$$

Differentiating ρ_{eff} with respect to α_{zp}, we find that with identically distributed channel coefficients, the optimal power allocation factor is

$$\alpha_{\text{zp}} = \frac{\beta - \sqrt{\beta}\sqrt{\beta-(1-\lambda)}}{1-\lambda}, \tag{11.41}$$

where $\beta := 1 + N_t(L+1)\sigma^2/\mathcal{P}$.

In addition to guiding the selection of optimal training parameters for ZP-based single-carrier MIMO transmissions, (11.26) and (11.32) are interesting on their own, since they offer simple closed forms for bounding the average capacity of frequency-selective MIMO fading channels. Note also that the optimal placement and design of training symbols in Figure 11.2 are not unique. The structure of training blocks can be shuffled among the N_t transmit-antennas without affecting the channel MSE or the capacity lower bound. The optimal training parameters for ZP-based block transmissions are summarized in Table 11.1.

Table 11.1 SUMMARY OF DESIGN PARAMETERS FOR ZP-BASED SCHEMES

Parameters	Optimal training
Placement of training subblock	TDM with info. subblock (zero guard of size L)
Structure of training subblock	$\boldsymbol{b}_\mu = [\mathbf{0}^T_{(\mu-1)(L+1)}\ b\ \mathbf{0}^T_{(N_t-\mu)(L+1)}]^T, \forall\mu$
Number of training symbols	$N_t(L+1)$ per block
Power allocation ratio	$\alpha_{\text{zp}} = \sqrt{N_c}/[\sqrt{N_c} + \sqrt{N_t(L+1)}]$

11.2.2 CP-Based Block Transmissions

Instead of having L zeros at the end of each block $\boldsymbol{x}_\mu$, an alternative way to eliminate IBI is by adding a CP of length L at the transmitter and removing it at the receiver. This is how OFDM blocks are transmitted, and the MIMO channel estimator in this subsection is clearly applicable to the MIMO OFDM systems dealt with in Chapter 8. The CP-insertion and CP-removal matrices are defined, respectively, as

$$\boldsymbol{T} = \boldsymbol{T}_{\text{cp}} := \begin{bmatrix}\boldsymbol{I}_{\text{cp}} & \boldsymbol{I}_{N_u}\end{bmatrix}^T \text{ and } \boldsymbol{R} = \boldsymbol{R}_{\text{cp}} := \begin{bmatrix}\mathbf{0}_{N_u\times L} & \boldsymbol{I}_{N_u}\end{bmatrix},$$

where I_{cp} contains the last L columns of I_{N_u}. In this case, the channel matrix becomes

$$\tilde{H}_{\nu\mu} = R_{\text{cp}} H_{\nu\mu} T_{\text{cp}},$$

which is an $N_u \times N_u$ columnwise circulant matrix with first column $[h_{\nu\mu}(0), \ldots, h_{\nu\mu}(L), 0, \ldots, 0]^T$. The input-output relationship in matrix-vector form is [cf. (11.9)]

$$y_\nu = \sum_{\mu=1}^{N_t} (\tilde{H}_{\nu\mu} A_\mu c_\mu + \tilde{H}_{\nu\mu} \bar{b}_\mu) + \zeta_\nu, \tag{11.42}$$

where $\zeta_\nu := R_{\text{cp}} w_\nu$. As in the ZP case, the roles of $\tilde{H}_{\nu\mu}$ and $\bar{b}_\mu$ can be interchanged, that is, $\tilde{H}_{\nu\mu}\bar{b}_\mu = \tilde{B}_\mu h_{\nu\mu}$, where $\tilde{B}_\mu$ is an $N_u \times (L+1)$ columnwise circulant matrix with first column $[\bar{b}(1), \ldots, \bar{b}(N_u)]^T$. Concatenating the N_r channel vectors and defining $\tilde{B} := [\tilde{B}_1, \ldots, \tilde{B}_{N_t}]$, we have [cf. (11.11)]

$$y = \tilde{\Phi} c + (I_{N_r} \otimes \tilde{B}) h + \zeta,$$

where the $N_r N_u \times N_t N_c$ matrix $\tilde{\Phi}$ has the same structure as the $N_r N \times N_t N_c$ matrix in (11.12) but with $\tilde{H}_{\nu\mu}$'s in place of $\bar{H}_{\nu\mu}$'s. Furthermore, for the purpose of decoupling symbol detection from channel estimation, we now need

$$\tilde{B}_{\mu_1}^{\mathcal{H}} \tilde{H}_{\nu\mu_2} A_{\mu_2} = \mathbf{0}, \quad \forall \mu_1, \mu_2 \in [1, N_t] \text{ and } \nu \in [1, N_r]. \tag{11.43}$$

As in (11.15), we express A_{μ_2} in terms of its N_c columns as

$$\begin{aligned} \tilde{H}_{\nu\mu_2} A_{\mu_2} &= \left[\tilde{H}_{\nu\mu_2} a_{\mu_2,1}, \ldots, \tilde{H}_{\nu\mu_2} a_{\mu_2,N_c}\right] \\ &= \left[\tilde{A}_{\mu_2,1} h_{\nu\mu_2}, \ldots, \tilde{A}_{\mu_2,N_c} h_{\nu\mu_2}\right], \end{aligned} \tag{11.44}$$

where $\tilde{A}_{\mu_2,n}$ is an $N_u \times (L+1)$ columnwise circulant matrix with first column the $N_u \times 1$ vector $a_{\mu_2,n}$. For (11.43) to hold true over *all* channel realizations, the $(A_\mu, \bar{b}_\mu)$ pairs must satisfy

$$\tilde{B}_{\mu_1}^{\mathcal{H}} \tilde{A}_{\mu_2,n} = \mathbf{0}, \quad \forall \mu_1, \mu_2 \in [1, N_t] \text{ and } \forall n \in [1, N_c], \tag{11.45}$$

which leads to the counterpart of Lemma 11.1 for CP-based transmissions:

Lemma 11.3 *The $(A_\mu, \bar{b}_\mu)$ pairs that guarantee decoupling of channel estimation from symbol detection must satisfy* (11.45) *and are given by*

$$A_\mu = F_{N_u}^{\mathcal{H}} P_A \Theta, \qquad \bar{b}_\mu = F_{N_u}^{\mathcal{H}} P_b b_\mu, \tag{11.46}$$

where Θ is an $N_c' \times N_c$ matrix that optionally precodes (if $\Theta \neq I_{N_c}$) the information-bearing block c_μ linearly, b_μ consists of the $N_b := N_u - N_c' - L$ possibly nonzero *training symbols from the μth transmit-antenna, and P_A, P_b are permutation matrices satisfying $P_A^{\mathcal{H}} P_b = 0_{N_c' \times N_b}$.*

There are different choices for (P_A, P_b). One simple choice we will adopt is to form P_A using the first N_c' columns of I_{N_u} and P_b using the next N_b columns of I_{N_u}. Notice that the permutation matrices play the role of assigning OFDM carriers to (possibly precoded) information and training symbols (pilot tones). With this interpretation, Lemma 11.3 shows that the optimal transmission pattern assigns different frequency tones to information-bearing and training symbols in order to separate channel estimation from symbol decoding in the subcarrier domain. As a result, the optimal affine model $Ac_\mu + b_\mu$ entails frequency-division multiplexing (FDM) the ST codewords c_μ with the training blocks b_μ. As with the TDM structure of the ZP-based transmissions of Section 11.2.1, the FDM-based separability was not assumed at the outset. It follows naturally from the CP-based IBI removal and the objective of decoupling symbol detection from MIMO channel estimation.

Since the objective here is to design an optimal training sequence per block, we will not worry about the precoder but for simplicity choose it as $\Theta = I_{N_c}$ with $N_c' = N_c$. Since Lemma 11.3 dictates IFFT processing at the transmitter, it is natural to employ FFT processing at the receiver and end up with a MIMO OFDM system. The FFT processed block at the νth receiver is

$$z_\nu := F_M y_\nu = \sum_{\mu=1}^{N_t} (D_H^{(\nu,\mu)} P_A c_\mu + D_H^{(\nu,\mu)} P_b b_\mu) + \xi_\nu, \tag{11.47}$$

where the diagonal matrix $D_H^{(\nu,\mu)} = F_M H_{\nu,\mu} F_M^{\mathcal{H}}$. With the mutually orthogonal permutation matrices P_A and P_b, we can now easily isolate the training and information-bearing parts:

$$\begin{aligned} z_{\nu,b} &:= P_b^T z_\nu = \sum_{\mu=1}^{N_t} P_b^T D_H^{(\nu,\mu)} P_b b_\mu + \xi_{\nu,b}, \\ z_{\nu,c} &:= P_A^T z_\nu = \sum_{\mu=1}^{N_t} P_A^T D_H^{(\nu,\mu)} P_A c_\mu + \xi_{\nu,c}, \end{aligned} \tag{11.48}$$

where the noise terms are defined as $\xi_{\nu,b} := P_b^T \xi_\nu$ and $\xi_{\nu,c} := P_A^T \xi_\nu$. Based on the specific permutation matrices chosen and using the definition of $D_H^{(\nu,\mu)}$, we can further verify that

$$P_b^T D_H^{(\nu,\mu)} P_b b_\mu = \text{diag}\{P_b^T F_{0:L} h_{\nu\mu}\} b_\mu = B_\mu P_b^T F_{0:L} h_{\nu\mu},$$

with $B_\mu := \text{diag}\{b_\mu\}$ being an $N_b \times N_b$ diagonal matrix and $F_{0:L}$ denoting the first $L+1$ columns of F_M. The input-output relationship corresponding to the training part becomes

$$z_{\nu,b} = B h_\nu + \xi_{\nu,b}, \tag{11.49}$$

where $\boldsymbol{B} := \sqrt{N_u}\,[\boldsymbol{B}_1\boldsymbol{P}_b^T\boldsymbol{F}_{0:L},\ldots,\boldsymbol{B}_{N_t}\boldsymbol{P}_b^T\boldsymbol{F}_{0:L}]$. Notice that similar to $\boldsymbol{\Phi}$ in the ZP case, $\boldsymbol{B}$ is the same $\forall\nu \in [1, N_r]$.

Concatenating vectors $\boldsymbol{z}_{\nu,b}$ from all receive-antennas into $\boldsymbol{z}_b := [\boldsymbol{z}_{1,b}^T,\ldots,\boldsymbol{z}_{N_r,b}^T]^T$, the LMMSE channel estimator is now given by

$$\hat{\boldsymbol{h}} = \left(\sigma^2\boldsymbol{R}_h^{-1} + \boldsymbol{I}_{N_r}\otimes(\boldsymbol{B}^{\mathcal{H}}\boldsymbol{B})\right)^{-1}(\boldsymbol{I}_{N_r}\otimes\boldsymbol{B}^{\mathcal{H}})\boldsymbol{z}_b. \tag{11.50}$$

Upon defining the $N_c\times N_c$ diagonal matrix $\boldsymbol{D}_{H_c}^{(\nu,\mu)} := \boldsymbol{P}_A^T\boldsymbol{D}_H^{(\nu,\mu)}\boldsymbol{P}_A$ and collecting N_rN_t such matrices into a block matrix

$$\boldsymbol{D}_{H_c} := \begin{bmatrix} \boldsymbol{D}_{H_c}^{(1,1)} & \cdots & \boldsymbol{D}_{H_c}^{(1,N_t)} \\ \vdots & \ddots & \vdots \\ \boldsymbol{D}_{H_c}^{(N_r,1)} & \cdots & \boldsymbol{D}_{H_c}^{(N_r,N_t)} \end{bmatrix}_{N_rN_c\times N_tN_c},$$

we arrive at the overall input-output relationship corresponding to the information part,

$$\boldsymbol{z}_c = \boldsymbol{D}_{H_c}\boldsymbol{c} + \boldsymbol{\xi}_c. \tag{11.51}$$

Comparing (11.50) and (11.51) with (11.20) and (11.23) derived for ZP-based transmissions, we find that the LMMSE estimators and the overall input-output relationships share the same form in both cases. Therefore, the analysis of ZP carries over to CP.

Not surprisingly, we find that C11.1$'$ also applies to the CP case and the minimum channel MSE

$$\sigma_{\tilde{h}}^2 = N_r\sum_{m=1}^{N_t(L+1)}\left([\boldsymbol{R}_h^{-1}]_{m,m} + \frac{1}{\sigma^2}\frac{\mathcal{P}_b}{N_t}\right)^{-1}$$

is identical to that in the ZP case and can be achieved if and only if $\boldsymbol{B}^{\mathcal{H}}\boldsymbol{B} = (\mathcal{P}_b/N_t)\boldsymbol{I}_{N_t(L+1)}$; that is,

$$N_u\boldsymbol{F}_{0:L}^{\mathcal{H}}\boldsymbol{P}_b\boldsymbol{B}_{\mu_1}^{\mathcal{H}}\boldsymbol{B}_{\mu_2}\boldsymbol{P}_b^T\boldsymbol{F}_{0:L} = \frac{\mathcal{P}_b}{N_t}\boldsymbol{I}_{L+1}\delta(\mu_1-\mu_2), \quad \forall\mu_1,\mu_2\in[1,N_t]. \tag{11.52}$$

Using the fact that $\boldsymbol{P}_b^T\boldsymbol{P}_b = \boldsymbol{I}_{N_b}$ and the definition $\tilde{\boldsymbol{D}}_{B_\mu} := \sqrt{N_u}\,\boldsymbol{P}_b\boldsymbol{B}_\mu^{\mathcal{H}}\boldsymbol{P}_b^T$, we can rewrite (11.52) as

$$\boldsymbol{F}_{0:L}^{\mathcal{H}}\tilde{\boldsymbol{D}}_{B_{\mu_1}}\tilde{\boldsymbol{D}}_{B_{\mu_2}}^{\mathcal{H}}\boldsymbol{F}_{0:L} = \frac{\mathcal{P}_b}{N_t}\boldsymbol{I}_{L+1}\delta(\mu_1-\mu_2), \quad \forall\mu_1,\mu_2\in[1,N_t]. \tag{11.53}$$

It can be verified that when $\mu_1 = \mu_2$, (11.53) is satisfied if $K > L$ of the N_b pilot tones are equispaced and equipowered with $\mathcal{P}_b/(KN_t)$ and the remaining $N_b - K$ pilots are zero. When $\mu_1 \neq \mu_2$, setting $\tilde{\boldsymbol{D}}_{B_{\mu_1}}^{\mathcal{H}}\tilde{\boldsymbol{D}}_{B_{\mu_2}} = \boldsymbol{0}$ will satisfy (11.53). Based on these arguments, we deduce that one pilot tone cannot be shared by more than one

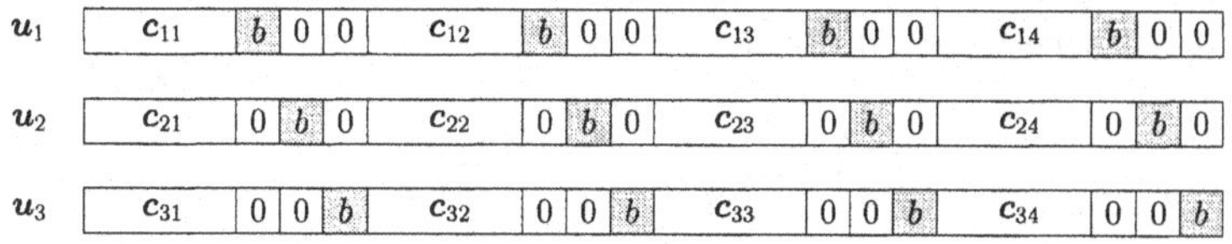

Figure 11.3 Optimal pattern for CP-based transmissions with $L = 3, N_t = 3$

nonzero training symbol on different transmit-antennas. In other words, the nonzero pilot tones of the N_t transmit-antennas must be mutually orthogonal: $\boldsymbol{B}_{\mu_1}^{\mathcal{H}} \boldsymbol{B}_{\mu_2} = \mathbf{0}$, $\forall \mu_1 \neq \mu_2$.

Recall now that the number of nonzero pilot tones in each transmit-antenna is K, and is lower bounded by $L+1$. For the same channel MSE and fixed $\mathcal{P}_c$ and $\mathcal{P}_b$, as the number of nonzero pilot tones increases, the capacity lower bound decreases. Therefore, the number of nonzero pilot tones should be chosen to be the minimum possible (i.e., $K = L+1$). Summarizing, we have shown that for fixed $\mathcal{P}_c$ and $\mathcal{P}_b$, the following placement is optimal: The $L+1$ subblocks of pilot tones for the μth transmit-antenna must be inserted equispaced in the output of the ST encoder; each of the training subblocks has structure $[\mathbf{0}_{\mu-1}^T \; b \; \mathbf{0}_{N_t-\mu}^T]^T$ with length N_t; and all pilot symbols must be equipowered with $\bar{\mathcal{P}}_b := \mathcal{P}_b / N_t(L+1)$.

An example of the optimal placement of pilot tones with $N_t = 3$ transmit-antennas and frequency-selective MIMO channels of order $L = 3$ is shown in Figure 11.3, where $\boldsymbol{c}_{\mu n}$ denotes the nth information-containing part for the μth transmit-antenna. Since $L+1 = 4$ here, there are four nonzero training symbols per block, and they are guarded by zeros to ensure orthogonality of the training segments across transmit-antennas and decoupling of symbol detection from channel estimation. As in the ZP case, this is one optimal placement but is not unique. For instance, shuffling the rows of the transmitted matrix among the N_t transmit-antennas yields other placements that are also optimal.

With the optimal training placement, the capacity lower bound for CP-based block transmissions is given by [cf. (11.29)]

$$\underline{C} := \frac{1}{N} E\left[\log_2 \det\left(\boldsymbol{I}_{N_r N_c} + \bar{\mathcal{P}}_c \boldsymbol{R}_v^{-1} \hat{\boldsymbol{D}}_{H_c} \hat{\boldsymbol{D}}_{H_c}^{\mathcal{H}}\right)\right] \quad \text{bits pcu,} \qquad (11.54)$$

where $\bar{\mathcal{P}}_c = \mathcal{P}_c / N_t N_c$. With $\mathcal{P}$ denoting the total power per block, notice that the sum $\mathcal{P}_b + \mathcal{P}_c$ is the "total" power, excluding the CP (i.e., $\mathcal{P}_b + \mathcal{P}_c = \mathcal{P} N_u / N$), and $\hat{\boldsymbol{D}}_{H_c}$ is the LMMSE estimate of $\boldsymbol{D}_{H_c}$ in (11.51). Defining the normalization factor $\sigma_{\hat{D}_{H_c}}^2 := \operatorname{tr}\left(E[\hat{\boldsymbol{D}}_{H_c} \hat{\boldsymbol{D}}_{H_c}^{\mathcal{H}}]\right)$, we have the normalized channel matrix

$$\hat{\bar{\boldsymbol{D}}}_{H_c} := \frac{1}{\sigma_{\hat{D}_{H_c}}} \hat{\boldsymbol{D}}_{H_c}.$$

Equation (11.54) can then be expressed as

$$\underline{C} = \frac{1}{N} E\left[\log_2 \det\left(\boldsymbol{I}_{N_r N_c} + \rho_{\text{eff}}^{\text{cp}} \hat{\bar{\boldsymbol{D}}}_{H_c} \hat{\bar{\boldsymbol{D}}}_{H_c}^{\mathcal{H}}\right)\right] \quad \text{bits pcu,} \qquad (11.55)$$

where the effective SNR is given by

$$\rho_{\text{eff}}^{\text{cp}} := \frac{\mathcal{P}_b \bar{\mathcal{P}}_c \sigma_{\hat{D}_{H_c}}^2}{[N_t^2(L+1)\bar{\mathcal{P}}_c + \mathcal{P}_b]\sigma^2}. \tag{11.56}$$

Differentiating $\rho_{\text{eff}}^{\text{cp}}$ with respect to $\alpha_{\text{cp}} := \mathcal{P}_c/\mathcal{P} \in (0,1)$, we obtain the optimal power allocation factor as

$$\alpha_{\text{cp}} = \frac{\sqrt{N_c}}{\sqrt{N_c} + \sqrt{N_t(L+1)}} \frac{N_u}{N}. \tag{11.57}$$

As in the ZP case, the placement of optimal pilot tones is not unique. The structure of each of the $L+1$ subblocks can be interchanged among transmit-antennas as long as the pilot tones from different transmit-antennas are distinct (and thus orthogonal), and those from the same transmit-antenna are equispaced. We summarize the optimal design parameters for CP-based schemes in Table 11.2.

Table 11.2 SUMMARY OF DESIGN PARAMETERS FOR CP-BASED SCHEMES

Parameters	Optimal training
Placement of training symbols	FDM, equispaced subblocks (each of size N_t)
Structure of training subblocks	$\boldsymbol{b}_\mu = [\mathbf{0}_{(\mu-1)}^T \; b \; \mathbf{0}_{(N_t-\mu)}^T]^T, \mu \in [1, N_t]$
Number of subblocks per block	$L+1$ training and $L+1$ info. subblocks
Number of training symbols	$N_t(L+1)$ per block
Power allocation ratio	$\alpha_{\text{cp}} = \frac{\sqrt{N_c}}{\sqrt{N_c} + \sqrt{N_t(L+1)}} \frac{N_u}{N}$

11.2.3 Special Cases

In the preceding sections we designed two optimal training schemes for frequency-selective MIMO channels, which are tailored for ZP- and CP-based block transmissions. The ZP-based scheme leads to single-carrier MIMO transmissions with TDM between training and information symbols, while the CP-based scheme leads to multi-carrier MIMO transmissions with FDM between training and information symbols. Both designs are optimal in terms of minimizing channel MSE (without assuming any CSI available at the transmitter besides the channel order L) and maximizing a lower bound on average capacity. In both cases, the optimal designs ended up possessing (time- or frequency-domain) orthogonality among training blocks *across* different transmit-antennas.

As alluded to in the closing comments of Chapter 8, there are trade-offs in selecting ZP versus CP in practice. When the ST mapper output features constant modulus, the ZP-based approach results in a two-level constant-modulus transmission, whereas CP-based OFDM generally results in transmitted blocks with a wide dynamic range in the order of the block length. The training symbols in ZP are simply attached to the head or tail of the blocks transmitted, whereas the pilot tones in CP are inserted equispaced into the blocks transmitted. Furthermore, ZP-based transmissions are more tolerant to frequency offsets, in contrast to CP-OFDM transmissions [291]. On the other hand, CP-OFDM transmissions provide easier implementation and lower-complexity equalization and symbol decoding than ZP-based transmissions, which involve relatively more complicated processing at the receiver side.

The L guard symbols needed to enable block-by-block decoding in either ZP- or CP-based systems can be also beneficial in the ST mapper design. For instance, the ZP guard adopted to guarantee decoupling of channel estimation from symbol decoding also plays an instrumental role in ST code design for the single-carrier MIMO systems in Chapter 7. Similarly, the CP guard can be shared with the STF codes designed in Chapter 8. The optimal training patterns presented here merge well with any ST mapper without sacrificing their achievable diversity gains.

So far, we dealt with frequency-selective MIMO channels. We now consider three special cases: single-input single-output frequency-selective, MIMO flat fading, and time-varying MIMO channels.

Special Case 1: Single-antenna frequency-selective channels In this setup, the ZP transmitted block $\boldsymbol{x} = \boldsymbol{T}_{\text{zp}}\boldsymbol{A}\boldsymbol{c} + \bar{\boldsymbol{b}}$ becomes

$$\boldsymbol{x} = [\boldsymbol{c}^T,\ \mathbf{0}_L^T,\ b,\ \mathbf{0}_L^T]^T, \tag{11.58}$$

where the subscript μ has been dropped since we deal with a single transmit-antenna and a single receive-antenna. This design has the same structure as the one in [1, Theorem 4]. It was stated in [270, Theorem 1] that the optimal number of training pilots is equal to the channel length $L+1$. The apparent discrepancy with (11.58), which requires $2L+1$ training slots comes from the fact that the redundancy of length L, which is needed to achieve IBI-free block transmission, is included in the training sequence here but is not accounted for in [270]. At high SNR, the power allocation factor $\alpha_{\text{zp}} = \sqrt{N_c}/(\sqrt{N_c} + \sqrt{L+1})$ coincides with what has been reported in [1, Theorem 5] and [270, Theorem 1].

For CP-based single-antenna transmissions, the MIMO results corroborate that OFDM with equispaced and equipowered pilot tones per block are optimal; and the optimal power allocation factor is $\alpha_{\text{cp}} = (N_u/N)\sqrt{N_c}/(\sqrt{N_c}+\sqrt{L+1})$. The same conclusions were reached in both [1, Theorems 2 and 3] and [205] for single-input single-output channels.

Special Case 2: Frequency-flat MIMO channels When the MIMO channel is frequency flat, we have $L = 0$, which implies that for both ZP- and CP-based transmissions, one needs only a single pilot per block per antenna. For such channels the placement of training blocks has no effect on the MMSE or the average capacity as long as the orthogonality among the N_t antennas is preserved. For this special case, a similar conclusion was reached in [108]. At high SNR, the optimal power allocation factor $\alpha_{\text{cp}} = \alpha_{\text{zp}} = \sqrt{N_c}/(\sqrt{N_c} + \sqrt{N_t})$ coincides with that in [108, Corollary 1].

Special Case 3: Time-varying MIMO channels A duality was established in Chapter 9 between time- and frequency-selective MIMO channels. Based on this duality, the CP-based setup can be viewed as an equivalent system model with time-selective MIMO channels. If we view the $Q+1$ bases of a time-selective channel as the corresponding $L+1$ taps of a frequency-selective channel, the optimal training results here still hold true for the time-selective channel. The extra benefit with time-selective MIMO channels is that there is no need for ZP or CP. The design of optimal training sequences for time-selective channels whose impulse response can be modeled as an autoregressive (AR) process can be found in [60]. Channel capacity analysis with imperfect channel knowledge is also considered in [195]. Additional applications of pilot-assisted transmissions are given in [261].

11.2.4 Numerical Examples

In this subsection we illustrate the optimal training-based channel estimators with numerical examples.

Example 11.1 (Average capacity bound with channel estimation) In this example, we adopt the training schemes summarized in Sections 11.2.1 and 11.2.2. We depict the relationship between the bounds in (11.55) and (11.32) on average capacity, and test the effect that several training parameters have on average capacity.

(a) Average capacity versus SNR: We select $(N_t, N_r) = (2, 2)$, $L = 6$, and the transmitted block length $N = 62$. The average capacity bounds versus SNR for both ZP and CP cases are plotted in Figure 11.4. As expected, the capacity bounds increase monotonically as SNR increases. The bounds for ZP exceed those for CP, which is due mainly to the power loss incurred by the CP.

(b) Average capacity versus N_t: Here we fix $N_r = 3, L = 6, N = 69$ and the total power per block $\mathcal{P}$. Figure 11.5 depicts the average capacity bounds versus the number of transmit-antennas. At high SNR, when $1 \leq N_t \leq 3$, the average capacity increases with the number of transmit-antennas and starts decreasing when $N_t > 3$. This is because training symbols take over a larger and larger portion of the transmitted block, leaving less and less of the block to be used for transmitting information symbols.

(c) Average capacity versus L: The relationship between average capacity bounds and channel order L is depicted in Figure 11.6. We fix $N_t = N_r = 2, N = 59$ and the total power per block $\mathcal{P}$. Notice that the capacity bounds for both ZP and CP decrease monotonically as L increases, since channel estimation takes increasing amounts of both time and power. As L increases, the gap between ZP and CP capacity increases, simply because of the power loss due to the CP.

(d) Average capacity versus N_c: Here we select $N_t = N_r = 2$ and $L = 6$, and test the effect of the information block length N_c. As N_c increases, the total block length N increases since $N = N_c + N_t(L+1) + L$. From Figure 11.7 we observe that average capacity bounds increase monotonically with increasing N_c. For large N_c, the information transmission dominates the entire block period.

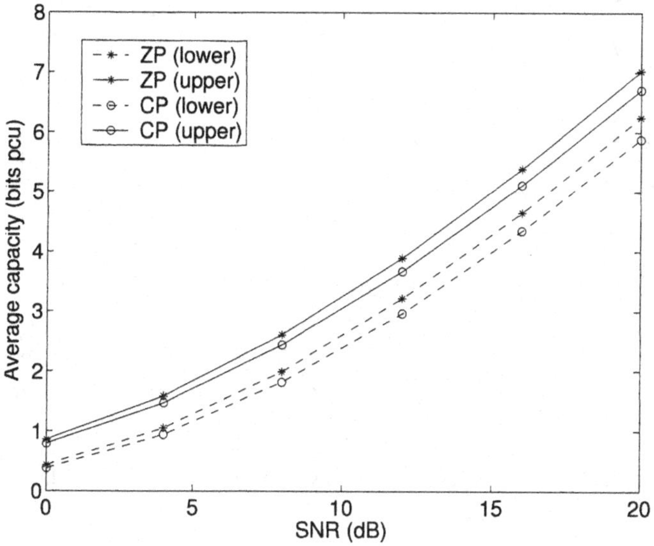

Figure 11.4 Average capacity bounds vs. SNR ($N_t = 2$, $N_r = 2$, $L = 6$)

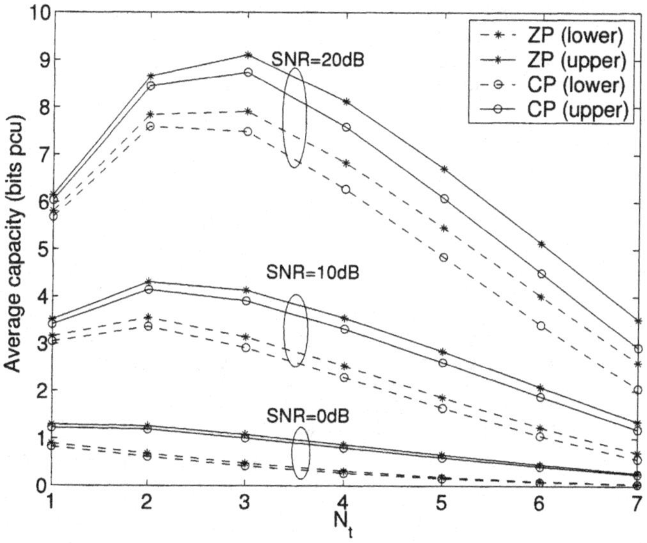

Figure 11.5 Average capacity bounds vs. number of transmit-antennas ($N_r = 3$, $L = 6$)

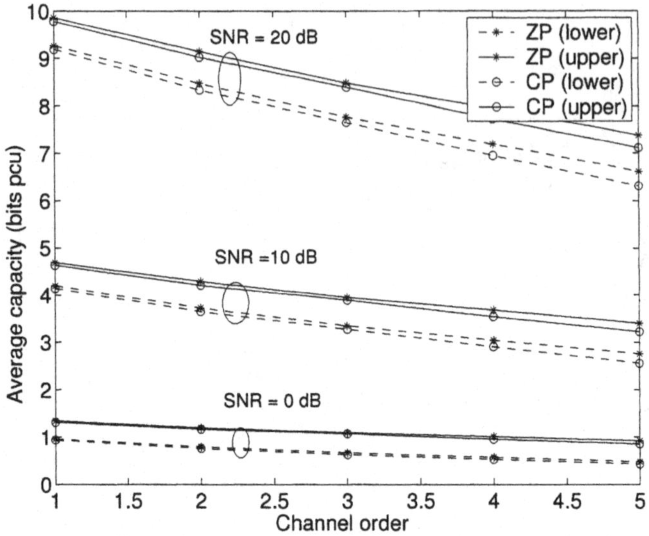

Figure 11.6 Average capacity bounds vs. channel order L ($N_t = 2$, $N_r = 2$)

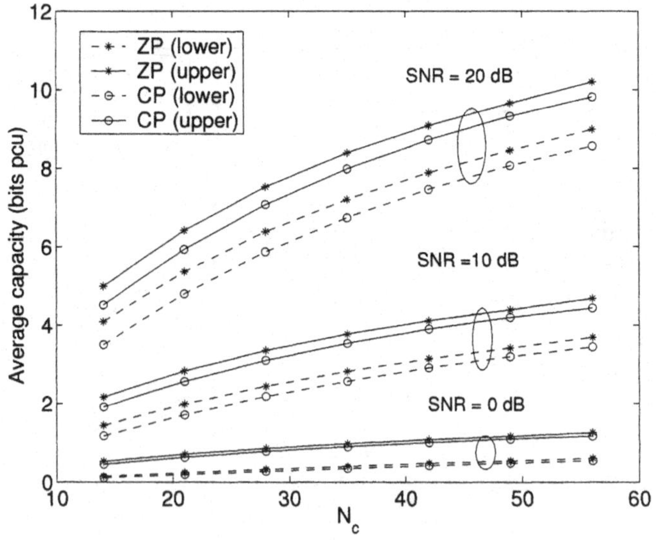

Figure 11.7 Average capacity bounds vs. symbol block length ($N_t = 2$, $N_r = 2$, $L = 6$)

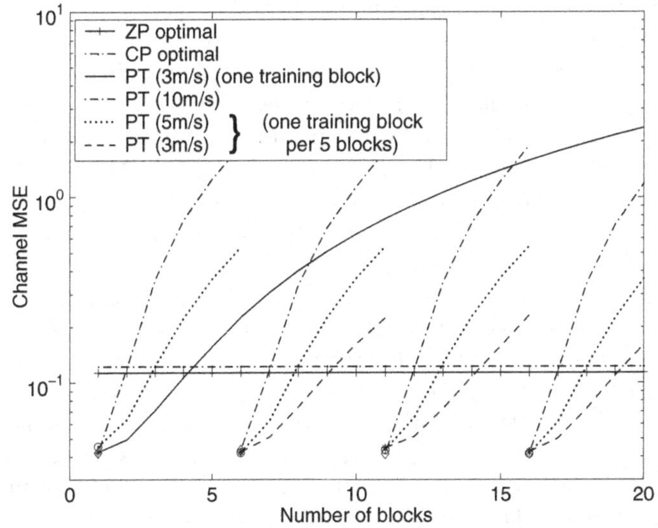

Figure 11.8 Channel MSE comparison between ZP- and CP-based optimal training and the preamble-based training scheme (denoted as "PT") in [150] ($N_t = 2$, $N_r = 1$, $L = 6$)

Example 11.2 (Channel estimation comparisons) To test the optimal training-based channel estimator, we use as a figure of merit the channel MSE $E[\|\hat{\boldsymbol{h}} - \boldsymbol{h}\|^2]$. We compare the optimal training approach with the preamble-based training scheme in [150] when the channel is slowly time varying but still changes from symbol to symbol. The training blocks for the preamble scheme are selected according to [150, Equation (12)]. Each channel tap is generated using the Jakes model with a terminal speed of 3, 5, or 10 m/s and a carrier frequency of 5.2 GHz. The variances of channel taps satisfy an exponential power profile. To maintain the same transmission rate, every training block is followed by four information blocks for the preamble scheme of [150]. Figure 11.8 depicts the channel MMSE versus the number of blocks transmitted at SNR $=$ 10 dB. Although the channel is slowly time varying, the MSE of optimal training schemes remains invariant from block to block while that of [150] increases very fast. Notice that [150] yields smaller MSEs at the beginning of each retraining burst because the entire transmit-power is used for training.

11.3 (SEMI-)BLIND CHANNEL ESTIMATION

Channel estimation based on training symbols is a data-aided approach requiring fewer samples relative to non-data-aided (a.k.a. blind) alternatives, which rely on received samples only. Blind channel estimators, on the other hand, do not sacrifice transmission bandwidth and are more appropriate for "cold-startup" CSI acquisition and uninterrupted information transmission. Hybrid schemes are referred to as semi-blind and trade-off rapid acquisition for bandwidth loss.

A number of blind algorithms have been developed for estimating MIMO channels. Relying on nonredundant (and nonconstant modulus) precoding, [25] introduced a blind channel identification and equalization algorithm for MIMO OFDM multi-antenna systems using cyclostationary statistics. For MIMO OFDM systems, a deterministic constant-modulus (CM) blind channel estimator proposed in [160] can identify MIMO channels under certain conditions on the channel zeros.

A subspace-based semi-blind algorithm was developed in [339] to identify MIMO OFDM channels. This algorithm possesses three attractive features: (i) it can be applied to arbitrary signal constellations; (ii) by proper system design, it guarantees channel identifiability regardless of channel zero locations; and (iii) it can identify multiple channels up to a scalar ambiguity only, which can be resolved with minimal training. Similar to most blind methods, the subspace algorithm requires a sufficient number of blocks to obtain reliable channel estimates. To facilitate convergence and enable tracking of slow channel variations, a semi-blind implementation of the subspace-based method was also devised in [339] by capitalizing on training sequences, which are usually provided for synchronization or quick channel acquisition in practical systems.

11.4 JOINT SYMBOL DETECTION AND CHANNEL ESTIMATION

We have discussed training-based and (semi-)blind approaches to acquiring channel state information. Besides received samples, training-based methods use the training symbols to estimate the channel. On the other hand, blind or semi-blind methods rely on statistical or structural properties of the information symbols. A hybrid approach is offered by the decision-directed class of algorithms, which are also capable of joint channel estimation and symbol detection.

11.4.1 Decision-Directed Methods

During the initialization phase, decision-directed algorithms estimate the channel based on preamble training. Using the estimated channel, symbols are subsequently decoded within a block of size not exceeding the coherence interval of the channel. Relying on the decoded (as opposed to training) symbols, a refined channel estimate is obtained. The process proceeds by alternating between channel estimation and symbol decoding. Decision-directed iterations of this type can clearly cope with slowly varying channels, at least for high SNR. Certainly, if the channel varies from symbol to symbol and/or the SNR is below a certain threshold, erroneous decisions lead to unreliable channel estimates, which in turn propagate errors in decoding.

A popular alternative for solving the nonlinear problem of simultaneous ML symbol detection and channel estimation involves the expectation-maximization (EM) algorithm, which has guaranteed convergence to at least a local optimum of the ML objective function. The EM algorithm was used in [171] to decode MIMO OFDM transmissions with OSTBC. Assuming that the fading process of the MIMO channel remains constant over the duration of one OSTBC transmission and exploiting the orthogonal channels effected by OSTBCs in conjunction with OFDM, the EM-based receiver proposed by [171] enjoys low complexity. But when the channel varies within one ST codeword, channel tracking and ST decoding become more challenging.

In the following subsection we use an example system with two transmit-antennas and one receive-antenna to illustrate a decision-directed approach for decoding OSTBC transmissions where tracking of (even rapidly) time-varying channels can be accomplished through the use of Kalman filtering.

11.4.2 Kalman Filtering-Based Methods

Allowing the MIMO impulse response to vary from symbol to symbol, let $h_i(n)$, $i = 1, 2$, denote the time-selective channel from the ith transmit-antenna to the receive-antenna. Two consecutive received samples can be expressed as

$$y(2n) = \sqrt{\frac{\bar{\gamma}}{2}}\,[h_1(2n)s(2n) + h_2(2n)s(2n+1)] + w(2n), \tag{11.59}$$

$$y(2n+1) = \sqrt{\frac{\bar{\gamma}}{2}}\,[-h_1(2n+1)s^*(2n+1) + h_2(2n+1)s^*(2n)] + w(2n+1),$$

where the noise $w(n)$ is complex AWGN with zero mean and unit variance.

We wish to recover $s(n)$ from $y(n)$ and make sure that the space diversity provided by the two transmit-antennas is collected at the receiver. Without imposing any structure on $h_\mu(n)$ however, this goal is ill-posed simply because for every two incoming received samples, two extra unknowns, $h_1(n)$ and $h_2(n)$, appear in addition to the two unknown symbols, $s(2n)$ and $s(2n+1)$. Fortunately, many wireless channels exhibit structured variations that can be fit parsimoniously with finitely parameterized, yet time-varying models such as the basis expansion model we saw in Chapter 9. An alternative approach we adopt here uses a first-order AR recursion which provides a sufficiently accurate description of the rapidly fading channel [283]. Specifically, we suppose that $h_\mu(n)$ is a zero-mean unit-variance complex Gaussian process obeying the AR recursion

$$h_\mu(n) = \alpha h_\mu(n-1) + v_\mu(n), \quad \mu = 1, 2, \tag{11.60}$$

where $v_\mu(n)$ denotes AWGN with zero mean and covariance $\sigma_v^2/2$ per dimension, uncorrelated with $h_\mu(n-1)$. The channel coefficient α can be estimated as detailed in [262]. Assuming that $h_\mu(0)$ is zero-mean, unit-variance complex Gaussian and using (11.60), we can easily verify that

$$\sigma_v^2 = 1 - |\alpha|^2 \quad \text{and} \quad \alpha = E[h_\mu(n)h_\mu^*(n-1)], \tag{11.61}$$

where $E(\cdot)$ stands for expectation.

As discussed in Chapter 9, wireless channel variations are caused mainly by two sources: one due to Doppler effects arising from relative motion between the transmitter and the receiver, and the other due to the CFO originating from the mismatch between transmitter-receiver oscillators. With f_o denoting the CFO and T_s the information symbol duration, we can express $h_\mu(n)$ as

$$h_\mu(n) = \bar{h}_\mu(n)\exp(j2\pi f_o T_s n), \tag{11.62}$$

where $\bar{h}_\mu(n)$ accounts for the Doppler effects. According to the Jakes model [127], we have $E[\bar{h}_\mu(n)\bar{h}_\mu^*(n-1)] = J_0(2\pi f_{\max} T_s)$, where $J_0(\cdot)$ is the zeroth-order Bessel function and $f_{\max}$ denotes the maximum Doppler shift. Recalling (11.61), the AR coefficient α is related to $f_{\max}$ and f_o via

$$\alpha = J_0(2\pi f_{\max} T_s)\exp(j2\pi f_o T_s). \tag{11.63}$$

Having justified the channel model described by (11.60), we proceed to specify the ST decoder used in the joint channel and symbol estimation.

11.4.2.1 Alamouti-Based ST Decoding Let $\boldsymbol{y}(n) := [y(2n), y^*(2n+1)]^T$ and rewrite (11.59) in a matrix-vector form

$$\boldsymbol{y}(n) = \sqrt{\frac{\bar{\gamma}}{2}}\,\boldsymbol{H}(n)\boldsymbol{s}(n) + \boldsymbol{w}(n), \tag{11.64}$$

where $\boldsymbol{w}(n) := [w(2n),\ w^*(2n+1)]^T$, $\boldsymbol{s}(n) := [s(2n),\ s(2n+1)]^T$, and the channel matrix

$$\boldsymbol{H}(n) := \begin{bmatrix} h_1(2n) & h_2(2n) \\ h_2^*(2n+1) & -h_1^*(2n+1) \end{bmatrix}. \tag{11.65}$$

Similar to the decoder in (3.31), we will recover $\boldsymbol{s}(n)$ from $\boldsymbol{y}(n)$ using the decision vector $\boldsymbol{z}(n) := [z(2n), z(2n+1)]^T$, given by

$$\boldsymbol{z}(n) = \boldsymbol{H}^{\mathcal{H}}(n)\boldsymbol{y}(n). \tag{11.66}$$

Notice that when the $h_\mu(n)$'s are time invariant, $\boldsymbol{H}(n)$ reduces to a (scaled) unitary matrix and the detection rule (11.66) is ML, as discussed in Section 3.3.2. After recognizing that $\boldsymbol{H}(n)$ is near unitary in the mean sense, we will show that (11.66) indeed leads to near-ML performance even with time-selective channels. Based on the definition (11.65), it follows by direct substitution that

$$\boldsymbol{H}^{\mathcal{H}}(n)\boldsymbol{H}(n) = \begin{bmatrix} \rho_1(n) & \epsilon(n) \\ \epsilon^*(n) & \rho_2(n) \end{bmatrix}, \tag{11.67}$$

where $\rho_1(n) := |h_1(2n)|^2 + |h_2(2n+1)|^2$, $\rho_2(n) := |h_1(2n+1)|^2 + |h_2(2n)|^2$, and $\epsilon(n) := h_1^*(2n)h_2(2n) - h_1^*(2n+1)h_2(2n+1)$. In most wireless applications, the product $f_{\max}T_s$ is typically small (e.g., $f_{\max}T_s < 0.004$ in [283]). Thus, we deduce from (11.61) and (11.63) that

$$|\alpha|^2 \approx 1 \quad \text{and} \quad \sigma_v^2 \approx 0. \tag{11.68}$$

Using (11.68), it follows readily that in the mean sense we have $E\{\epsilon(n)\} \approx 0$ and $E\{\rho_1(n)\} \approx E\{\rho_2(n)\}$. Using the latter along with (11.67), we infer that $\boldsymbol{H}(n)$ is (within a scale) a near-unitary matrix in the mean sense. Notice that if $f_{\max} = 0$, $\boldsymbol{H}(n)$ is exactly orthogonal, regardless of the value of f_o. This proves a not so widely known result: that Alamouti's ST block code is insensitive to CFO. In fact, we can go one step further and use Alamouti's ST coding for transmissions through more general time-selective channels, and similar to [6], we can decode the information symbols using (11.66).

The ST decoder in (11.66) assumes that the channel coefficients $h_\mu(n)$ are known at the receiver. Notice that estimating the latter is challenging since the MIMO channel we deal with here is time selective. Fortunately, the AR model in (11.60) lends itself to a state-space representation, which enables application of the Kalman filter (KF) for tracking the time-varying channel.

11.4.2.2 KF-Based Channel Tracking The KF-based receiver starts with a training phase to acquire initial $h_\mu(n)$ estimates and then reverts to a decision-directed mode during which symbol detection and channel tracking steps are implemented in an alternating fashion. In the training mode, the receiver knows the symbols transmitted, whereas in the decision-directed mode, the decoded symbols replace the information symbols. Supposing that initial channel estimates are available by using any of the approaches discussed earlier, we will demonstrate the KF-based decision-directed operation.

Viewing the channel vector $\boldsymbol{h}(n) := [h_1(n),\ h_2(n)]^T$ as a state, we obtain from (11.60) the state equation

$$\boldsymbol{h}(n) = \boldsymbol{A}\boldsymbol{h}(n-1) + \boldsymbol{v}(n), \tag{11.69}$$

where $\boldsymbol{A} := \mathrm{diag}(\alpha,\ \alpha)$ and $\boldsymbol{v}(n) := [v_1(n),\ v_2(n)]^T$. Using (11.59), received samples obey the measurement equation

$$y(n) = \sqrt{\frac{\bar{\gamma}}{2}}\,\bar{\boldsymbol{s}}^T(n)\boldsymbol{h}(n) + w(n), \tag{11.70}$$

where $\bar{\boldsymbol{s}}(n) := [s(n),\ s(n+1)]^T$ when n is even; and $\bar{\boldsymbol{s}}(n) := [-s^*(n),\ s^*(n-1)]^T$ when n is odd. Note that in the joint detection-estimation problem described by (11.69) and (11.70), both $\boldsymbol{h}(n)$ and $\bar{\boldsymbol{s}}(n)$ are unknown. Clearly, knowing the decoded symbols $\bar{\boldsymbol{s}}(n)$ and the observations $y(n)$, the channel $\boldsymbol{h}(n)$ can be obtained using a standard KF predictor [135, Page 448]. However, detecting $\bar{\boldsymbol{s}}(n)$ relies on estimates of $\boldsymbol{h}(n)$, which in turn require knowledge of $\bar{\boldsymbol{s}}(n)$. This implies that an iterative method should be sought to obtain alternately either $\bar{\boldsymbol{s}}(n)$ or $\boldsymbol{h}(n)$. By appealing to (11.68), a coarse prediction of $\boldsymbol{h}(n)$ can be obtained directly from (11.60). Let $\boldsymbol{h}(n|m)$ denote the predicted channel at time n based on the state and/or the observation at time m. The coarse channel prediction is given by the recursions

$$\begin{aligned} \boldsymbol{h}(2n|2n-1) &= \alpha\, \boldsymbol{h}(2n-1|2n-1), \\ \boldsymbol{h}(2n+1|2n-1) &= \alpha^2\, \boldsymbol{h}(2n-1|2n-1), \end{aligned} \tag{11.71}$$

that are initialized by $\boldsymbol{h}(1|1)$, which is obtained during the training mode. Next, we use the coarse channel estimates and (11.66) to obtain coarse symbol estimates for $\bar{\boldsymbol{s}}(2n)$ and $\bar{\boldsymbol{s}}(2n+1)$, denoted by $\bar{\boldsymbol{s}}^{(c)}(2n)$ and $\bar{\boldsymbol{s}}^{(c)}(2n+1)$, respectively.

Replacing $\bar{\boldsymbol{s}}(n)$ by $\bar{\boldsymbol{s}}^{(c)}(n)$, we rely on the KF to obtain refined channel estimates $\boldsymbol{h}(2n|2n)$. Based on $\boldsymbol{h}(2n|2n)$, we perform KF once more to obtain $\boldsymbol{h}(2n+1|2n+1)$. With refined $\boldsymbol{h}(2n|2n)$ and $\boldsymbol{h}(2n+1|2n+1)$, refined estimates $\bar{\boldsymbol{s}}^{(r)}(2n)$ and $\bar{\boldsymbol{s}}^{(r)}(2n+1)$ are obtained with diversity gain from (11.66). We summarize the algorithm for channel tracking and symbol decoding, in the following steps:

Initialization: Find $\boldsymbol{h}(1|1)$ using a training based algorithm;

S1) Obtain $\boldsymbol{h}(2n|2n-1)$ and $\boldsymbol{h}(2n+1|2n-1)$ using (11.71).

S2) Use (11.64) and (11.66) to decode $\bar{\boldsymbol{s}}^{(c)}(2n)$ and $\bar{\boldsymbol{s}}^{(c)}(2n+1)$.

S3) Perform KF to retrieve $\boldsymbol{h}(2n|2n)$ and $\boldsymbol{h}(2n+1|2n+1)$ using $\bar{\boldsymbol{s}}^{(c)}(2n)$ and $\bar{\boldsymbol{s}}^{(c)}(2n+1)$.

S4) Decode $\bar{\boldsymbol{s}}^{(r)}(2n)$ and $\bar{\boldsymbol{s}}^{(r)}(2n+1)$ based on $\boldsymbol{h}(2n|2n)$ and $\boldsymbol{h}(2n+1|2n+1)$.

S5) If necessary, iterate over S3 and S4 to improve tracking performance.

S6) Repeat from S1 for $n+1 \leftarrow n$.

Similar to all decision-directed algorithms, convergence to the optimum is not guaranteed and error propagation occurs especially when the SNR is low. A heuristic remedy to mitigate error propagation at medium-low SNR is to insert pilot symbols periodically. The period is chosen depending on how much bandwidth one is willing to give up versus how reliable tracking performance is desired to be.

11.4.2.3 Numerical Examples To illustrate the KF-based algorithm numerically, we simulate an EDGE system with carrier frequency 1.9 GHz, terminal speed 250 km/hour and transmission rate 144 kbps. With CFO equal to f_o =1,000 Hz and the AR model in (11.60) used to generate the taps (α is known at the receiver), we compare one realization of the true and estimated channels. To avoid divergence of the KF tracker, we insert one pilot symbol every 12 symbols, which incurs an 8% bandwidth efficiency loss. Figure 11.9 shows that the estimated channels (dashed curves) track the true channels (solid curves) well.

We also compare BER performance with and without KF-based channel tracking. Here the SNR is defined as bit energy versus noise power. For fairness, when no channel tracking is used, the receiver assumes knowledge of α and updates the channels via $h_\mu(n+1) = \alpha h_\mu(n)$, with $h_\mu(n)$'s reset to their correct values every 12 symbols. Figure 11.10 confirms that channel tracking improves the BER performance, especially at high SNR.

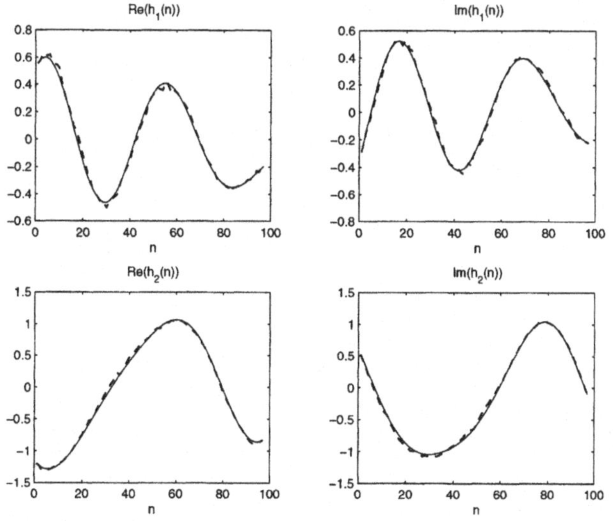

Figure 11.9 True (solid curves) and KF-based estimate (dashed curves) of channel variations

11.5 CARRIER SYNCHRONIZATION

In previous sections we presented various algorithms for acquisition and tracking of MIMO channels. All of them assume, however, that timing and frequency synchronization have been acquired perfectly. In this section we consider channel estimation and frequency synchronization of multi-antenna transmissions over frequency-selective MIMO channels. Timing synchronization amounts to estimating and compensating for the timing offset caused by the propagation delay and the mismatch between transmitter-receiver clocks. When the MIMO channel is flat fading, timing offset estimators and their MSE performance have been studied in [300, 301]. The Cramer-Rao bound (CRB) achieved by the ML time-delay estimator is inversely proportional to the number of receive-antennas but does not depend on the number of transmit-antennas [300,301]. Frame synchronization for the single-antenna setup has been considered in [238, 302], and joint frame synchronization and frequency offset estimation have been explored in [141].

Because timing offset can be viewed as a pure delay channel, its estimation can be considered jointly with the impulse response of the propagation channel using any of the MIMO channel estimation algorithms we described earlier in the chapter. Hence, it suffices to deal with CFO acquisition and estimation of the aggregate MIMO channel. We have already mentioned that CFO affects error performance of multi-carrier systems more severely. For this reason, our focus in this section is on frequency synchronization of MIMO OFDM transmissions over frequency-selective channels, keeping in mind that frequency-flat channels are subsumed as a special case and time-selective MIMO channels can be acquired similarly based on the duality principle.

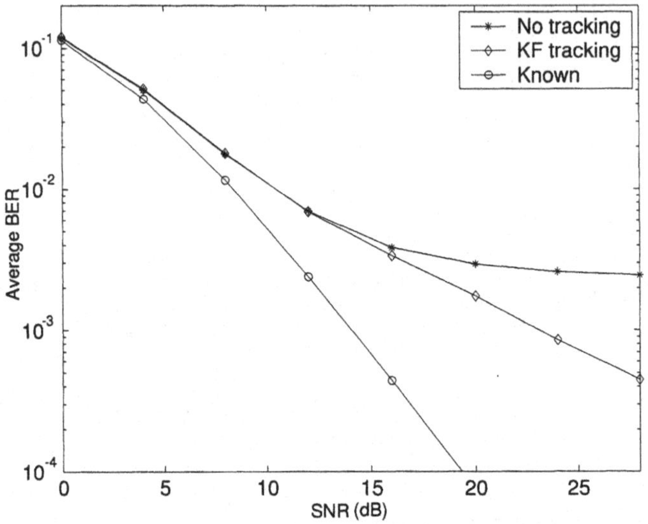

Figure 11.10 BER improvement with KF-based channel tracking

A number of approaches have dealt with CFO estimation in a single-input single-output OFDM setup [89, 155, 185, 190, 198, 199, 267]. Some rely on training blocks [198, 226], while others just take advantage of the standardized transmission format (e.g., the presence of null subcarriers is exploited in [88, 155, 190]). CFO estimation algorithms for MIMO OFDM systems can be found in [189, 190, 197, 265, 316]. Among those, we outline here the null-subcarrier-based scheme in [189] and the blind algorithm of [316].

11.5.1 Hopping Pilot-Based CFO Estimation

The input-output relationship of a MIMO OFDM system is described by (8.13) or (11.42) when training is accounted for explicitly. The channel from the μth transmit-antenna to the νth receive-antenna is defined as $\boldsymbol{h}_{\nu\mu} := [h_{\nu\mu}(0), \ldots, h_{\nu\mu}(L)]^T$. With T_s denoting the sampling period, the input-output relationship corresponding to the kth block of size P at the νth receive-antenna with the CFO $\omega_o := 2\pi f_o T_s$ present is

$$\boldsymbol{y}_\nu(k) = e^{j\omega_o(kP+L)} \boldsymbol{D}_N(\omega_o) \sum_{\mu=1}^{N_t} \tilde{\boldsymbol{H}}_{\nu\mu} \boldsymbol{F}_N^{\mathcal{H}} \bar{\boldsymbol{u}}_\mu(k) + \boldsymbol{v}_\nu(k), \tag{11.72}$$

where $D_N(\omega_o) := \text{diag}[1, e^{j\omega_o}, \ldots, e^{j\omega_o(N-1)}]$, $\tilde{H}_{\nu\mu}$ denotes as before the circulant channel matrix generated by $[h_{\nu\mu}, 0_{N-L-1}]^T$, and $\bar{u}_\mu(k)$ is the OFDM block transmitted through the μth antenna. Notice that even though we deal with a multi-antenna system, only a single CFO is considered. This is because only one oscillator is present at the transmitter and similarly at the receiver. Furthermore, even if CFO arises due to Doppler effects, typical values of the relative speed and the small distance between antenna elements justifies well the presence of a single CFO in most cases of practical interest.

As in Section 11.2.2, pilot symbols are inserted for channel estimation. Specifically, the training subblock $b_\mu(k)$ is inserted into the information-bearing subblock $c_\mu(k)$ per transmit-antenna, as follows (see also Figure 11.1):

$$u_\mu(k) = P_A c_\mu(k) + P_b b_\mu(k), \tag{11.73}$$

where the two permutation matrices P_A and P_b have sizes $K \times N_c$ and $K \times N_b$, respectively, and are selected to be mutually orthogonal: $P_A^T P_b = 0_{N_c \times N_b}$ according to Lemma 11.3. Note that $N_c + N_b = K$ and $K < N$. One example of such matrices is to form P_A using the last N_c column of $I_{N_b+N_c}$, and P_b with the first N_b columns of $I_{N_b+N_c}$:

$$P_A = [e_{N_b} \ \cdots \ e_{K-1}] \text{ and } P_b = [e_0 \ \cdots \ e_{N_b-1}], \quad \mu \in [1, N_t], \tag{11.74}$$

where e_n denotes the nth column of I_K.

To facilitate CFO estimation, we insert a certain number of zeros in every block, which in the MIMO OFDM transmission format will manifest themselves as null subcarriers. Later we will show that hopping null subcarriers from block to block will enable the CFO acquisition range in the full range $[-\pi, \pi)$.

Specifically, we insert $N - K$ zeros in the block $u_\mu(k)$ in (11.73) to obtain $\bar{u}_\mu(k)$. This insertion can be described analytically by left-multiplying $u_\mu(k)$ with a null subcarrier insertion matrix, which is defined as

$$T_{\text{sc}}(k) := \left[e_{q_k (\text{mod } N)}, \ldots, e_{q_k + K - 1 (\text{mod } N)} \right], \tag{11.75}$$

where $q_k := k\lfloor N/(L+1) \rceil$, and $\lfloor \cdot \rceil$ denotes the rounding operation. We call each subcarrier corresponding to a zero symbol as a null subcarrier. Dependence of the null subcarrier insertion matrix $T_{\text{sc}}(k)$ on the block index k, implies that the position of the inserted zero varies from block to block. In other words, (11.75) implements a null subcarrier hopping operation from block to block. Plugging (11.75) and (11.73) into (11.72), the samples received at the νth antenna can be expressed as

$$y_\nu(k) = e^{j\omega_o(kP+L)} D_N(\omega_o) F_N^{\mathcal{H}} \sum_{\mu=1}^{N_t} D_N(\tilde{h}_{\nu\mu}) T_{\text{sc}}(k) \left[P_A c_\mu(k) + P_b b_\mu(k) \right] + v_\nu(k). \tag{11.76}$$

We have described the insertion of two types of training symbols: zero and nonzero. We reiterate that the null pilot is hopping to a different position from block to block.

This idea of hopping pilots is instrumental in establishing the identifiability of the CFO estimator while offering a transmission pattern that minimizes the MMSE in channel estimation.

If the CFO was absent [$\omega_o \equiv 0$ in (11.76)], then similar to Section 11.2.2, we could isolate the part corresponding to the training symbols from the block received and by collecting a sufficient number of blocks (enough number of pilots), we could estimate the MIMO channel reliably. However, CFO destroys the orthogonality among subcarriers, and the training information is "mingled" with the unknown symbols and channels. This motivates acquiring the CFO first and estimating the channels afterward as described in [189].

The CFO estimation algorithm in [189] starts with a "dehopping" operation, implemented on a per block basis using the diagonal dehopping matrix

$$\boldsymbol{D}_N^{\mathcal{H}}(k) := \text{diag}[1, e^{-j2\pi q_k/N}, \ldots, e^{-j2\pi q_k(N-1)/N}]. \tag{11.77}$$

Because $\boldsymbol{T}_{\text{sc}}(k)$ in (11.75) is a permutation matrix and $\boldsymbol{D}_N(\tilde{\boldsymbol{h}}_{\nu\mu})$ is a diagonal matrix, it is not difficult to verify that $\boldsymbol{D}_N(\tilde{\boldsymbol{h}}_{\nu\mu})\boldsymbol{T}_{\text{sc}}(k) = \boldsymbol{T}_{\text{sc}}(k)\boldsymbol{D}_K(\tilde{\boldsymbol{h}}_{\nu\mu}(k))$, where $\tilde{\boldsymbol{h}}_{\nu\mu}(k)$ is formed by permuting the entries of $\tilde{\boldsymbol{h}}_{\nu\mu}$ as dictated by $\boldsymbol{T}_{\text{sc}}(k)$. Using the well-designed dehopping matrix in (11.77) and the fact that $\boldsymbol{T}_{\text{sc}}(0) = \boldsymbol{T}_{\text{zp}}$, it is easy to establish the identity

$$\boldsymbol{D}_N^{\mathcal{H}}(k)\boldsymbol{F}_N^{\mathcal{H}}\boldsymbol{T}_{\text{sc}}(k) = \boldsymbol{F}_N^{\mathcal{H}}\boldsymbol{T}_{\text{zp}}, \tag{11.78}$$

where $\boldsymbol{T}_{\text{zp}} := [\boldsymbol{I}_K \; \boldsymbol{0}_{K\times(N-K)}]^T$ is a zero-padding operator. Multiplying (11.76) by the dehopping matrix and using (11.78), we obtain

$$\bar{\boldsymbol{y}}_\nu(k) = \boldsymbol{D}_N^{\mathcal{H}}(k)\boldsymbol{y}_\nu(k) = e^{j\omega_o(kP+L)}\boldsymbol{D}_N(\omega_o)\boldsymbol{F}_N^{\mathcal{H}}\boldsymbol{T}_{\text{zp}}\boldsymbol{g}_\nu(k) + \bar{\boldsymbol{v}}_\nu(k), \tag{11.79}$$

where $\boldsymbol{g}_\nu(k) := \sum_{\mu=1}^{N_t} \boldsymbol{D}_K(\tilde{\boldsymbol{h}}_{\nu\mu}(k))\tilde{\boldsymbol{u}}_\mu(k)$ and $\bar{\boldsymbol{v}}_\nu(k) := \boldsymbol{D}_N^{\mathcal{H}}(k)\boldsymbol{v}_\nu(k)$. Since $\boldsymbol{T}_{\text{zp}}$ does not depend on the block index k, we deduce from (11.79) that after dehopping, null subcarriers in different blocks are at the same location. The system model (11.79) is similar to the one used in [155, 190] for a single-antenna OFDM system. This observation suggests that we can generalize the method of [155, 190] to estimate the CFO for MIMO OFDM systems.

To this end, we consider the covariance matrix of $\bar{\boldsymbol{y}}_\nu(k)$ [cf. (11.79)]:

$$\boldsymbol{R}_{\bar{y}_\nu} = \boldsymbol{D}_N(\omega_o)\boldsymbol{F}_N^{\mathcal{H}}\boldsymbol{T}_{\text{zp}}E\left[\boldsymbol{g}_\nu(k)\boldsymbol{g}_\nu^{\mathcal{H}}(k)\right]\boldsymbol{T}_{\text{zp}}^{\mathcal{H}}\boldsymbol{F}_N\boldsymbol{D}_N^{\mathcal{H}}(\omega_o) + \sigma^2\boldsymbol{I}_N, \tag{11.80}$$

where the noise $\bar{\boldsymbol{v}}_\nu(k)$ has covariance matrix $\sigma^2\boldsymbol{I}_N$. In practice, supposing that the channels remain time invariant over M blocks, we replace the ensemble correlation matrix $\boldsymbol{R}_{\bar{y}_\nu}$ by its sample estimate formed by averaging across $M \geq K$ blocks:

$$\hat{\boldsymbol{R}}_{\bar{y}_\nu} = \frac{1}{M}\sum_{k=0}^{M-1} \bar{\boldsymbol{y}}_\nu(k)\bar{\boldsymbol{y}}_\nu^{\mathcal{H}}(k). \tag{11.81}$$

The column space of $\boldsymbol{R}_{\bar{y}_\nu}$ consists of two parts: the signal subspace and the null subspace. In the absence of CFO, if $E\left[\boldsymbol{g}_\nu(k)\boldsymbol{g}_\nu^{\mathcal{H}}(k)\right]$ has full rank, the null space of

$\boldsymbol{R}_{\bar{\boldsymbol{y}}_\nu}$ is spanned by the missing columns (the location of the null subcarriers) of the FFT matrix. The presence of CFO introduces a shift in the null space. Similar to [190], a cost function can be built to measure this CFO-induced shift for the MIMO OFDM setup. With ω denoting the candidate CFO, this cost function can be written as

$$J(\omega) := \sum_{n=K}^{N-1} \boldsymbol{f}_N^{\mathcal{H}}\left(\frac{2\pi n}{N}\right) \boldsymbol{D}_N^{-1}(\omega) \left(\sum_{\nu=1}^{N_r} \boldsymbol{R}_{\bar{\boldsymbol{y}}_\nu}\right) \boldsymbol{D}_N(\omega) \boldsymbol{f}_N\left(\frac{2\pi n}{N}\right), \qquad (11.82)$$

where $\sum_{\nu=1}^{N_r} \boldsymbol{R}_{\bar{\boldsymbol{y}}_\nu} = \boldsymbol{D}_N(\omega_o)\boldsymbol{F}_N^{\mathcal{H}}\boldsymbol{T}_{\text{zp}}\left\{\sum_{\nu=1}^{N_r} E[\boldsymbol{g}_\nu(k)\boldsymbol{g}_\nu^{\mathcal{H}}(k)]\right\}\boldsymbol{T}_{\text{zp}}^{\mathcal{H}}\boldsymbol{F}_N\boldsymbol{D}_N^{\mathcal{H}}(\omega_o)$.

The cost function in (11.82) quantifies the signal energy that "leaks" to the null-subcarrier frequency bins. Clearly, if $\omega = \omega_o$, then $\boldsymbol{D}_N(\omega_o - \omega) = \boldsymbol{I}_N$. Next, recall that the matrix $\boldsymbol{F}_N^{\mathcal{H}}\boldsymbol{T}_{\text{zp}}$ is orthogonal to $\{\boldsymbol{f}_N(2\pi n/N)\}_{n=K}^{N-1}$. Hence, if $\omega = \omega_o$, the cost function $J(\omega_o)$ is zero in the absence of noise. However, we have to confirm that ω_o is the unique minimum of $J(\omega)$. The next proposition establishes this uniqueness [189]:

Proposition 11.1 *If $E[\boldsymbol{b}_\mu(k)\boldsymbol{b}_\mu^{\mathcal{H}}(k)]$ is diagonal, $\sum_{\mu=1}^{N_t} E[\boldsymbol{b}_\mu(k)\boldsymbol{b}_\mu^{\mathcal{H}}(k)]$ has full rank, $E[\boldsymbol{c}_{\mu_1}(k)\boldsymbol{c}_{\mu_2}^{\mathcal{H}}(k)] = \boldsymbol{0}$ and $E[\boldsymbol{b}_{\mu_1}(k)\boldsymbol{b}_{\mu_2}^{\mathcal{H}}(k)] = \boldsymbol{0}$, $\forall \mu_1 \neq \mu_2$, then $\sum_{\nu=1}^{N_r} E[\boldsymbol{g}_\nu(k)\, \boldsymbol{g}_\nu^{\mathcal{H}}(k)]$ has full rank. The latter implies that $J(\omega) \geq J(\omega_o)$, where the equality holds if and only if $\omega = \omega_o$.*

Proposition 11.1 shows that the CFO estimate can be found uniquely as

$$\hat{\omega}_o = \arg\min_{\omega} J(\omega). \qquad (11.83)$$

Thanks to subcarrier hopping, $J(\omega)$ has a unique minimum in $[-\pi, \pi)$ regardless of the position of channel nulls. This establishes the consistency of $\hat{\omega}_o$ and shows that the acquisition range of the CFO estimator in (11.83) is $[-\pi, \pi)$, which is the full range.

Having estimated the CFO as in (11.83), we can remove it and resort to the algorithms of Section 11.2.2 to estimate the $N_t \times N_r$ MIMO channel. For most schemes (e.g., [155, 190]), the CFO and channel estimation process ends here. However, since no CFO estimator is perfect, even residual CFO effects can degrade the BER severely with time, because the complex exponential factor in (11.72) increases with k. To cope with residual CFO effects, we introduce next a phase estimation algorithm.

After CFO compensation using the estimate in (11.83), the block received can be written as [cf. (11.79)]

$$\tilde{\boldsymbol{y}}_\nu(k) = e^{j(\omega_o - \hat{\omega}_o)(kP+L)}\boldsymbol{D}_N(\omega_o - \hat{\omega}_o)\boldsymbol{F}_N^{\mathcal{H}}\boldsymbol{T}_{\text{zp}}\boldsymbol{g}_\nu(k) + \boldsymbol{\zeta}_\nu(k), \qquad (11.84)$$

where $\hat{\omega}_o - \omega_o$ is the residual CFO and $\boldsymbol{\zeta}_\nu(k) := e^{-j\hat{\omega}_o(kP+L)}\boldsymbol{D}_N^{-1}(\hat{\omega}_o)\bar{\boldsymbol{v}}_\nu(k)$. We observe from (11.84) that when the CFO estimate is accurate enough, the matrix $\boldsymbol{D}_N(\hat{\omega}_o - \omega_o)$ can be approximated well by an identity matrix. However, the phase term $(\hat{\omega}_o - \omega_o)(kP + L)$ becomes increasingly large as the block index k increases.

Without mitigating it, the phase distortion degrades not only the performance of the channel estimator but also the BER performance over time.

To enhance BER performance, it is suggested in [189] that the phase distortion per block be estimated using nonzero training symbols, which are already present for training-based channel estimation. Suppose that for the kth block, we estimate the channel perfectly using (11.84). After equalizing the channel at the νth antenna, and assuming that $\boldsymbol{D}_N(\hat{\omega}_o - \omega_o) \approx \boldsymbol{I}_N$ holds true, the μth entry of the equalizer output block $\boldsymbol{z}_{\nu,b}(k)$ can be expressed as

$$\phi_\nu(k) = e^{j(\hat{\omega}_o - \omega_o)(kP+L)} b + w_\nu, \tag{11.85}$$

where $\phi_\nu(k) := [\boldsymbol{z}_{\nu,b}(k)]_\mu / [\tilde{\boldsymbol{h}}_b^{(\nu,\mu)}(k)]_\mu$, b is the nonzero training symbol, and w_ν is the equivalent noise term after removing the channel. Since b is known, the phase $(\hat{\omega}_o - \omega_o)(kP+L)$ can be estimated based on observations from N_r receive-antennas on a per block basis. It is worth stressing that this phase estimation step does not require any additional pilot symbol on top of that used for channel estimation, and the extra complexity it introduces is negligible. Simulations verify that phase estimation improves performance considerably.

Three remarks are now in order on CFO and channel estimation for MIMO OFDM systems:

Remark 11.1 The algorithms presented in this chapter for estimating the single common CFO and the MIMO channel in the *single-user* setup involving N_t transmit-antennas and N_r receive-antennas, can easily be modified to estimate CFOs and channels in a *multiuser* downlink scenario, where the access point deploys N_t transmit-antennas to broadcast OFDM-based transmissions to N_r mobile stations, each of which is equipped with a single antenna. In this case, there are N_r distinct CFOs and $N_t N_r$ frequency-selective channels to estimate. However, each mobile station can still apply the CFO estimator in (11.83) using the cost function in (11.82) with $N_r = 1$. In addition, based on the orthogonal training designs, it is easy to verify that the LMMSE channel estimator can be decoupled to estimate on a per receive-antenna basis the N_t channel impulse responses contained in $\boldsymbol{h}_\nu$ for $\nu = 1, \ldots, N_r$.

Remark 11.2 Since it depends on the driving noise, the CFO generated by an oscillator's circuit is random in nature (see, e.g., [103, Appendix B] and references therein). Furthermore, the power of the phase noise in an oscillator is inversely proportional to the CFO range. This implies that a CFO estimation algorithm offering a larger acquisition range is useful on two counts: (i) the outage probability of the CFO error is lower; and (ii) for the same oscillator hardware, the phase noise power is lower.

Remark 11.3 Recalling that the null subcarrier insertion matrix does not depend on the number of antennas implies that the CFO estimator for MIMO OFDM systems applies identically to a single-antenna OFDM setup. The decoupling of CFO estimation can also be implemented by training one antenna at a time. After compensating for the CFO, the same approach can be followed to estimate the MIMO channel using multiple single-antenna channel estimators. However, using existing training-based single-antenna CFO and separate single-antenna channel estimators to realize this one-antenna-at-a-time approach has the following limitations relative to the MIMO approach pursued in this chapter: (i) The need emerges to switch between training- and information-transmission modes per antenna, which is more difficult to implement relative to the single-mode MIMO transmission format presented here. (ii) Available single-antenna CFO estimators have a smaller acquisition range and require a coarse (in addition to a fine) CFO estimation step which costs in terms of both complexity and spectral efficiency. The resulting bandwidth loss becomes severe in high-mobility applications, where training has to be increasingly frequent. Furthermore, if a coarse estimation module is invoked to bring the CFO to a range manageable by existing algorithms (such as [199]), the full-range CFO estimator can also benefit from it to reduce the complexity of the nonlinear search required by (11.83) while still leading to improved error performance. (iii) Both limitations in complexity and spectral inefficiency are clearly illustrated in the broadcast-OFDM setup discussed in Remark 11.1, where the one-antenna-at-a-time approach is evidently inferior to the MIMO approach by a factor proportional to the number of active users.

11.5.2 Blind CFO Estimation

In this subsection we introduce a low-complexity blind CFO estimator for MIMO OFDM systems which relies on a kurtosis-based cost function to identify the CFO uniquely within the range of half subcarrier spacing [316]. Similar to all blind schemes, blind CFO estimation does not sacrifice bandwidth and is suitable for commercial systems where the traffic is continuous (as in, e.g., DVB-T broadcasting). Furthermore, in noncooperative (e.g., tactical) links, blind estimators are the only option since training-based ones cannot be implemented.

Consider again the input-output relationship in (11.72) describing a MIMO OFDM system in the presence of CFO. By stacking the blocks received from different receive-antennas to form $\boldsymbol{y}(k) := [\boldsymbol{y}_1^T(k)\cdots\boldsymbol{y}_{N_r}^T(k)]^T$, and redefining $\boldsymbol{s}(k) := [\boldsymbol{s}_1^T(k)\cdots\boldsymbol{s}_{N_t}^T(k)]^T$ and $\boldsymbol{w}(k) := [\boldsymbol{w}_1^T(k)\cdots\boldsymbol{w}_{N_r}^T(k)]^T$, the kth block at the receiver output after CP removal can be written as

$$\boldsymbol{y}(k) = e^{jk(N+L)\omega_o}\left(\boldsymbol{I}_{N_r}\otimes\boldsymbol{D}_N(\omega_o)\right)\boldsymbol{H}\left(\boldsymbol{I}_{N_t}\otimes\boldsymbol{F}_N^{\mathcal{H}}\right)\boldsymbol{s}(k)+\boldsymbol{w}(k), \qquad (11.86)$$

where the channel matrix $\boldsymbol{H}$ consists of $N_r \times N_t$ subblocks

$$\boldsymbol{H} = \begin{bmatrix} \boldsymbol{H}_{1,1} & \cdots & \boldsymbol{H}_{1,N_t} \\ \vdots & \ddots & \vdots \\ \boldsymbol{H}_{N_r,1} & \cdots & \boldsymbol{H}_{N_r,N_t} \end{bmatrix}, \qquad (11.87)$$

and the subblock $\boldsymbol{H}_{\nu,\mu}$ is the $N\times N$ circulant matrix representing the channel between the μth transmit-antenna and the νth receive-antenna.

Diagonalizing the circulant matrices using (I)FFT operations, we can rewrite (11.86) as

$$\boldsymbol{y}(k) = e^{jk(N+L)\omega_o}(\boldsymbol{I}_{N_r} \otimes \boldsymbol{D}_N(\omega_o)\boldsymbol{F}_N^{\mathcal{H}})\bar{\boldsymbol{H}}\boldsymbol{s}(k) + \boldsymbol{w}(k), \tag{11.88}$$

where

$$\bar{\boldsymbol{H}} := \begin{bmatrix} \boldsymbol{D}_H^{1,1} & \cdots & \boldsymbol{D}_H^{1,N_t} \\ \vdots & \ddots & \vdots \\ \boldsymbol{D}_H^{N_r,1} & \cdots & \boldsymbol{D}_H^{N_r,N_t} \end{bmatrix},$$

with subblocks $\boldsymbol{D}_H^{\nu\mu} := \text{diag}(H_{\nu\mu}(0), \ldots, H_{\nu\mu}(N-1))$ being diagonal matrices with diagonal entries $H_{\nu\mu}(n) := \sum_{l=0}^{L} h_{\nu\mu}(l) \exp(-j2\pi nl/N)$.

If the CFO is absent, the noise-free entries of $\boldsymbol{y}(k)$ will equal (within a scale) the source symbols; otherwise, they will be linear combinations of the source symbols. Furthermore, it is well known that a linear combination of independent random variables is closer to Gaussian than the original random variables, unless these random variables are Gaussian or the combination is trivial (only one nonzero weight) [61]. As symbols transmitted on different subcarriers are independent and adhere to a finite alphabet, the distribution of $\boldsymbol{y}(k)$ is more non-Gaussian when CFO is absent than when CFO is present.

Motivated by this observation, the basic idea in [316] is to perform blind CFO estimation based on a criterion that measures distance from non-Gaussianity. Pertinent cost functions could be devised using Fisher's information, Kullback-Leibler divergence, or kurtosis-based metrics. The criterion chosen in [316] is the normalized kurtosis, which for a random variable z is defined as

$$\text{kurtosis} := \frac{E[|z|^4] - 2(E[|z|^2])^2 - |E[z^2]|^2}{(E[|z|^2])^2}. \tag{11.89}$$

The normalized kurtosis has been used for blind channel estimation and equalization (see, e.g., [264] and references therein). Interestingly, when used for blind CFO acquisition, it yields a surprisingly simple closed-form solution, lending itself to a very low-complexity estimator. To derive this estimator, let $\hat{\omega}$ denote a candidate CFO, $\tilde{\omega}_o := \omega_o - \hat{\omega}_o$ the corresponding error, and consider the vector

$$\boldsymbol{z}(k) = (\boldsymbol{I}_{N_r} \otimes \boldsymbol{F}_N \boldsymbol{D}_N(-\hat{\omega}_o))\, \boldsymbol{y}(k),$$

with entries $\boldsymbol{z}(k) := [z_{1,0}(k) \cdots z_{1,N-1}(k) \cdots z_{N_t,0}(k) \cdots z_{N_t,N-1}(k)]^T$.

Using M blocks $\boldsymbol{y}(k)$ from (11.88), we can estimate using sample averaging the normalized kurtosis of $\boldsymbol{z}(k)$, which yields the cost function

$$J(\tilde{\omega}_o) = \frac{\sum_{k=0}^{M-1} \sum_{\nu=1}^{N_r} \sum_{n=0}^{N-1} |z_{\nu,n}(k)|^4}{\left(\sum_{k=0}^{M-1} \sum_{\nu=1}^{N_r} \sum_{n=0}^{N-1} |z_{\nu,n}(k)|^2\right)^2}. \tag{11.90}$$

It has been shown in [316] that if $\omega_o \in [-\pi/N, \pi/N]$, the estimate of ω_o can be obtained as the unique minimizer of $J(\tilde{\omega}_o)$. In fact, this claim is true even when the CP is less than the delay spread [316].

Although minimization of $J(\tilde{\omega}_o)$ can be performed using line search, the regularity of $J(\tilde{\omega}_o)$ makes it possible to design estimators with lower complexity. For example, it is possible to use a steepest-descent approach to update $\hat{\omega}_o$ iteratively or use stochastic gradient descent alternatives to develop symbol-by-symbol adaptive algorithms. Surprisingly, a closed-form solution is also possible using curve fitting.

For M sufficiently large, the cost function in (11.90) can be written as [316]

$$J(\tilde{\omega}_o) \cong A \cdot g(\boldsymbol{H}, \kappa_s) \cos \tilde{\omega}_o N + B, \tag{11.91}$$

where A and B are constants independent of $\tilde{\omega}_o$ and $g(\boldsymbol{H}, \kappa_s)$ depends on $\boldsymbol{H}$ with $\kappa_s := E[|s_n(k)|^4]/(E[|s_n(k)|^2])^2$. When $g(\boldsymbol{H}, \kappa_s) > 0$, it follows from (11.91) that $\tilde{\omega}_o = 0$ is a unique global maximum of J; similarly, when $g(\boldsymbol{H}, \kappa_s) < 0$, $\tilde{\omega}_o = 0$ is a unique global minimum [316].

An interesting special case occurs when the symbols are Gaussian. It is well known that for a blind source separation problem to be solvable, at most one source (here a symbol) can be Gaussian [39]. However, the analysis in [316, Equation (5)] suggests that blind CFO estimation is possible even when all sources are Gaussian as long as the channel is frequency selective. In this case, one no longer relies on the source non-Gaussianity to retrieve the CFO; instead, the method in [316] exploits the power difference between pairs of subcarriers.

To derive a low-complexity version of the kurtosis-based CFO estimator, it is possible to evaluate $J(\tilde{\omega}_o)$ on several points to find the values of $A \cdot g(\boldsymbol{H}, \kappa_s)$, B and ω_o that satisfy (11.91). For example, evaluating $J(\tilde{\omega}_o)$ at $\tilde{\omega}_o = -\pi/2N, 0, \pi/2N$, it is easy to see from (11.91) that an estimate $\hat{\omega}_o$ can be obtained as

$$\hat{\omega}_o = \begin{cases} \dfrac{1}{N} \sin^{-1} b, & \text{if } a \geq 0, \\ \dfrac{\pi}{N} - \dfrac{1}{N} \sin^{-1} b, & \text{if } a < 0 \text{ and } b \geq 0, \\ -\dfrac{\pi}{N} - \dfrac{1}{N} \sin^{-1} b, & \text{if } a < 0 \text{ and } b \leq 0, \end{cases} \tag{11.92}$$

where $a := [J(-\pi/2N) + J(\pi/2N)]/2 - J(0)$ and $b := [J(-\pi/2N) - J(\pi/2N)]/2$. Compared with an exhaustive line search, this algorithm exhibits much lower complexity. Numerical examples verify that it achieves almost identical performance to that of the more expensive line search.

11.5.3 Numerical Examples

In this section we test the CFO estimation algorithms with simulations. The SNR is defined as the average received symbol power/noise-ratio per receive-antenna.

Example 11.3 (Training-based CFO estimation) We consider system configurations with $(N_t, N_r) = (1, 1)$, $(N_t, N_r) = (1, 2)$, $(N_t, N_r) = (2, 1)$, and $(N_t, N_r) = (2, 2)$ and rely on $M = N$ blocks. For $(N_t, N_r) = (1, 1)$ and $(N_t, N_r) = (1, 2)$, we use four nonzero pilot symbols and one null subcarrier per OFDM block. Figure 11.11 depicts the normalized MSE of the CFO estimator as a function of the SNR for

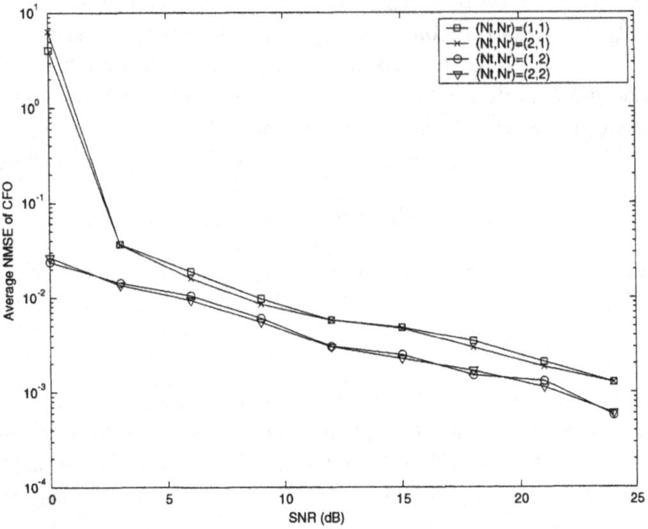

Figure 11.11 Average CFO NMSE

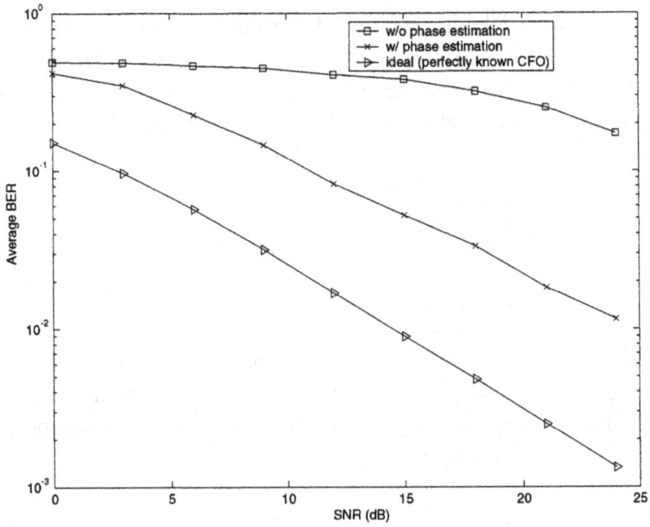

Figure 11.12 BER $(N_t, N_r) = (2, 2),\ M = N$

a variable number of transmit-receive antennas. We observe that as the number of receive-antennas increases, the performance of CFO estimation improves, thanks to the receive-diversity gains. Figure 11.12 depicts BER performance after mitigating the phase distortion and corroborates the claim that phase estimation improves BER performance considerably.

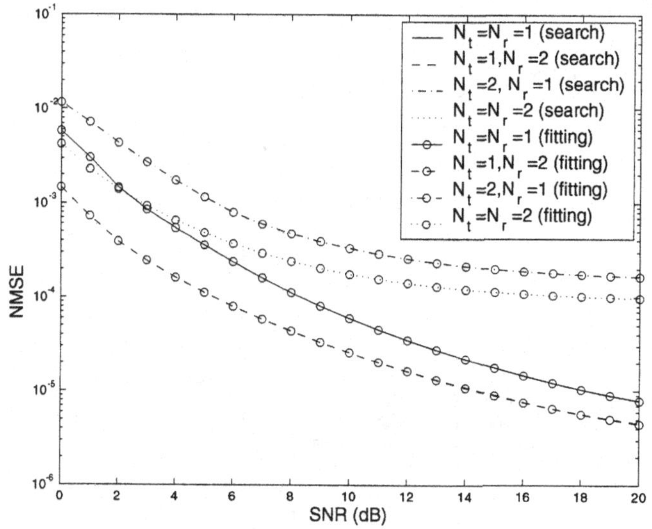

Figure 11.13 Performance of Kurtosis-based CFO estimator in MIMO OFDM

Example 11.4 (Blind CFO estimation) Here we simulate a MIMO OFDM setup where the channel has five uncorrelated Rayleigh-fading taps with exponentially decaying powers; in particular, we set $E[|h(l)|^2] = e^{-l/3}/\sum_{l=0}^{4} e^{-l/3}$, $l = 0, \ldots, 4$. We assume that the channel fading is slow enough so that the channel does not vary rapidly while CFO estimation is performed and the channels from different antennas are independent. The size of the OFDM block and the length of the cyclic prefix are 128 and 4, respectively. In each OFDM block, 128 randomly drawn QPSK symbols are transmitted. Figure 11.13 depicts normalized MSE (NMSE) performance of the kurtosis-based CFO estimator in MIMO OFDM systems with different numbers of transmit- and receive-antennas based on $M = 10$ blocks. Channels between different antenna pairs are independent. As expected, multiple receive-antennas improve CFO estimation performance thanks to receive-diversity gains. When there are multiple transmit-antennas, the NMSE performance of the CFO estimator exhibits an error floor, because the coupling between different transmit-antennas brings the symbol distribution closer to Gaussian. Hence, one must exercise caution in applying this method to systems with many (e.g., $N_t > 10$) transmit-antennas and ST coding[3]. We also compare the performance of the line search with the curve-fitting algorithm [based on (11.91)]. We observe that they have almost identical performance, suggesting that the low-complexity curve-fitting algorithm can also be applied to MIMO systems.

[3]Systems using transmit beamforming do not face this limitation because the same data symbols are transmitted by every antenna element.

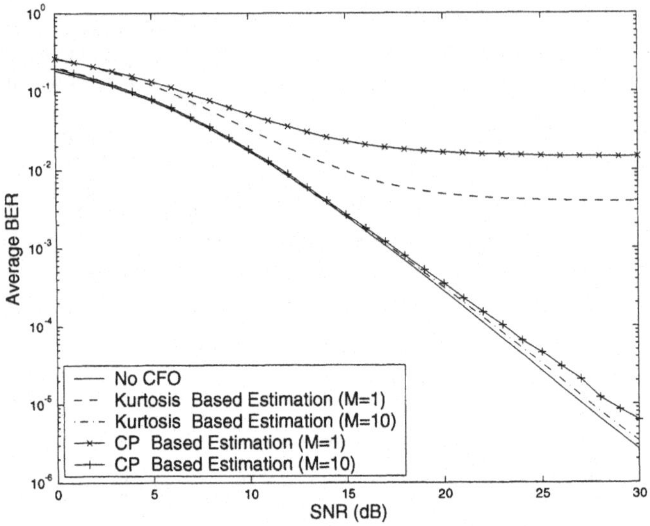

Figure 11.14 BER Performance of blind CFO estimator in MIMO OFDM

The kurtosis-based CFO estimator is applicable to MIMO OFDM systems *regardless* of the underlying ST code that may be employed. We simulate the BER performance of a MIMO OFDM system ($N_t = 2, N_r = 1$) when there is no CFO and when $M = 1$ and $M = 10$ OFDM blocks are used for CFO estimation in the presence of CFO. The kurtosis-based curve-fitting algorithm is used for CFO estimation, and Alamouti's code in (3.15) is employed to effect diversity gains. From Figure 11.14, we observe that significant BER performance loss is incurred if only one OFDM block is used for CFO estimation. However, we also observe that when $M = 10$, the BER performance loss caused by imperfect CFO estimation is negligible. So although kurtosis-based CFO estimation exhibits a floor in NMSE when there are multiple transmit-antennas, this NMSE floor does not translate to a BER floor. For comparison, we also plot the performance of the CP-based estimator ($M = 1$ and $M = 10$) and that of the modified CP-based algorithm in [185], using a sufficiently long CP of length 5. The kurtosis-based CFO estimator outperforms CP-based estimators.

11.6 CLOSING COMMENTS

In this chapter we presented carrier synchronization and MIMO channel estimation algorithms. A preamble-based approach was discussed first, which relied on training symbols to obtain LS or LMMSE optimal estimators of frequency-selective MIMO channels. Recognizing that preamble-based schemes do not optimize estimation performance jointly with transmitter resources (power and bandwidth), the placement and power allocation was optimized between training and information symbols per transmitted block. Optimality amounted to minimizing the MMSE in estimation, which for sufficiently high SNR turned out to be equivalent to maximizing a lower bound on average capacity when channel estimation is taken into account. For reduced complexity and flexibility, the optimal schemes were designed to decouple channel estimation from symbol detection and be universally applicable to all coherent ST coded systems.

In zero-padded block transmissions (which apply to the single-carrier systems in Chapter 7), the optimal design of each block per transmit-antenna turned out to have (i) the information-bearing subblock separated from the training subblock with an all-zero guard of size equal to the channel order; (ii) only one nonzero symbol in the training subblock placed in a position to ensure orthogonality among training subblocks across all other transmit-antennas; and (iii) generally unequal power allocated between the training subblock and the information subblock, both of which must contain equipowered symbols. In a dual fashion, for cyclic-prefixed block transmissions (which fit the MIMO OFDM systems in Chapter 8), the optimal training scheme also demanded separation of the information from training subblocks as well as orthogonality of the training subblocks across antennas, but now in the subcarrier domain. These optimal training based estimators of frequency-selective MIMO channels were specialized to frequency-flat and single-antenna systems. By appealing to the duality established in Chapter 9, they can also be used with necessary but minor modifications to estimate optimally time- and doubly selective MIMO channels.

Blind, semi-blind and decision-directed options were also mentioned for joint channel estimation and demodulation. The line between blind and non-blind estimators is often "gray" and the designer's choice is typically dictated by application-specific trade-offs. Decision-directed schemes offer a vital option when channel variations are slow and the SNR is sufficiently high to prevent error propagation. For faster time variations modeled as an autoregressive process, Kalman filtering can be used efficiently to track the channel in an orthogonal ST block coded system. When transmissions are encoded with the Alamouti ST code, decoding turned out to be insensitive to CFO.

Establishing carrier synchronization in all coherent ST codes is necessary and of paramount importance in MIMO OFDM systems. To this end, a pilot-symbol-assisted modulation was presented which consists of null subcarrier and nonzero training symbols inserted per block in each transmit-antenna. Again, training patterns were designed orthogonal across transmit-antennas to reduce complexity and decouple CFO from channel estimation. Hopping the position of a null subcarrier across blocks ensured the maximum possible CFO acquisition range. For acquisition ranges not exceeding half-subcarrier spacing, a low-complexity blind CFO estimator was developed for MIMO OFDM systems based on a kurtosis-type criterion.

In this chapter we dealt with MIMO channel estimation at the receiver in order to decode ST coded multi-antenna transmissions coherently. Interestingly, knowing the MIMO channel also at the transmitter can be very beneficial to the performance of wireless multi-antenna systems. The means and type of CSI that can be acquired pragmatically at the transmitter as well the improved ST designs that become available with this knowledge are the themes of the next two chapters.

12

ST Codes with Partial Channel Knowledge: Statistical CSI

Except for the number of degrees of freedom in space, time, multipath or Doppler dimensions, the multi-antenna transmitters designed in Chapters 3 to 9 are basically unaware of the underlying MIMO channels. Indeed, the only channel knowledge used by the ST codes introduced so far includes the channel order L for frequency-selective channels or the number of basis functions $Q + 1$ for time-selective channels. With such minimal knowledge available, optimization of ST-coded systems inevitably has to rely on conservative criteria so that transmitters are designed able to cope with worst-case propagation effects.

At the other extreme, information theory suggests that perfect knowledge of the intended channel realizations at the transmitter can boost error and rate performance even with single-antenna systems. Channel-adaptive transmission systems adjust parameters such as power levels, constellation sizes, transmission rates, error control coding schemes, and modulation types, depending on the channel state information (CSI), which can be made available to the transmitter (e.g., via feedback from the receiver as depicted in Figure 12.1). The merits of these closed-loop systems relative to their open-loop counterparts designed so far have been well appreciated in wireline and slowly fading wireless links [7, 97, 217, 263]. The capacity-achieving discrete multitone system that has been standardized for digital subscriber line modems is an example of a closed-loop adaptive transceiver [46].

Since no-CSI leads to robust but pessimistic designs and perfect-CSI is probably a utopia for most wireless links, modeling and exploitation of *partial CSI* at the transmitter promise to have great practical value for three reasons: (i) due to errors arising from channel estimation, prediction, quantization or feedback delays, only partial CSI can be pragmatically made available to the transmitter; (ii) transmissions

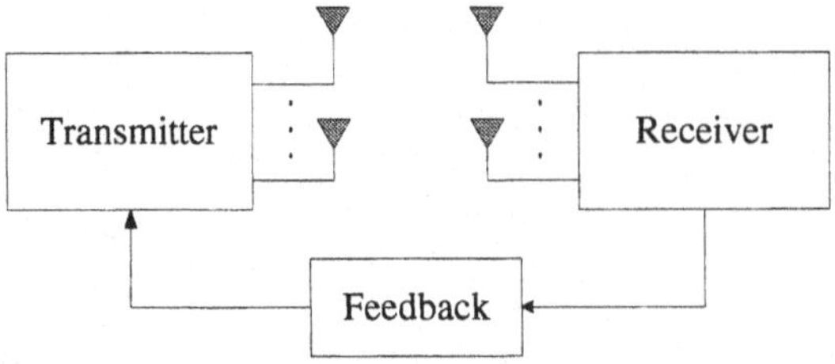

Figure 12.1 Closed-loop multi-antenna system with channel feedback

adapted to partial CSI can offer the "jack of both trades": improved error performance as well as enhanced data rates; and (iii) partial CSI based designs subsume in the limit the no-CSI and perfect-CSI extreme paradigms. Prompted by these considerations, the scope of the present chapter and the next is to explore ST-coded systems taking advantage of various forms of partial CSI that can become available to the multi-antenna transmitter. Partial CSI in Chapter 13 takes the form of a finite number of (typically, a few) bits fed back from the receiver to the transmitter.

In this chapter, partial CSI can come from the feedback channel in the form of either a mean vector or a covariance matrix, or it can be deciphered from the physics of the fading process and channel geometry when the latter does not change fast. In both cases it takes the form of a virtual complex Gaussian channel adopted by the transmitter to model statistically its uncertainty regarding the true MIMO channel. Using this *statistical CSI* model, we first select the ST matrix of a low-rate spread-spectrum multi-antenna system to optimize average error performance metrics. ST spreading is viewed as a transmit-beamforming operation and the optimal transmit-beamformer benchmarks the performance of designs based on the mean or the covariance of the statistical CSI model. The ST spread-spectrum system lends itself naturally to a cascade combination of transmit-beamforming with orthogonal ST block coding to increase the rate up to 1 symbol per channel use. Even though we have seen in Section 3.3 that OSTBCs lose rate with $N_t > 2$, one such coder-beamformer combination surprisingly guarantees 1 symbol per channel use for any number of transmit antennas, while ensuring near-optimal error performance at low complexity. In fact, rate performance can be improved further because the OSTBC coder-beamformer orthogonalizes the MIMO channel, thus allowing application of adaptive modulation algorithms developed for single-antenna systems to MIMO channels. Finally, we complement the error-performance-optimal ST designs based on partial CSI with their counterparts optimizing the average capacity when partial CSI is available at the multi-antenna transmitter.

12.1 PARTIAL CSI MODELS

Consider the closed-loop system in Figure 12.1 with N_t transmit-antennas and N_r receive-antennas, operating over a frequency-flat MIMO channel. Let $h_{\nu\mu}$ denote the channel coefficient from the μth transmit-antenna to the νth receive-antenna. Corresponding to each receive-antenna ν, we define the $N_t \times 1$ vector $\boldsymbol{h}_\nu := [h_{\nu 1}, \ldots, h_{\nu N_t}]^T$; and for future use, we concatenate all $\boldsymbol{h}_\nu$ vectors to form the longer composite channel vector

$$\boldsymbol{h} := \begin{bmatrix} \boldsymbol{h}_1 \\ \vdots \\ \boldsymbol{h}_{N_r} \end{bmatrix} \in \mathbb{C}^{N_t N_r \times 1}. \tag{12.1}$$

Given a sufficient number of training symbols, the algorithms presented in Chapter 11 for MIMO channel estimation can yield (nearly) perfect channel estimates at the receiver. However, assuming that perfect CSI is available at the transmitter is less realistic for most wireless links unless fading is extremely slow. Indeed, beyond estimation errors, feedback delay, quantization effects, and feedback errors prevent the transmitter from knowing perfectly each and every realization of the random vector $\boldsymbol{h}$. On the other hand, information-theoretic and array processing considerations suggest that modeling and exploiting even imperfect or partial CSI at the transmitter is very important since capacity can be enhanced and error performance can be improved through the use of transmit-beamforming. Critical to the design of these systems is judicious modeling of the partial CSI.

Partial CSI can be modeled in different ways. In this chapter we deal with what is termed *statistical CSI*, whereby the transmitter accounts for the uncertainty in the MIMO channel through, for example, the mean or the covariance of a virtual random channel. Chapter 13 will take into account the bandwidth-constrained nature of the feedback link, and the partial CSI will comprise a finite number of (often a few) bits characterizing the underlying channel. We next specify possible forms of statistical CSI models and how they can be acquired at the transmitter.

12.1.1 Statistical CSI

In lieu of deterministic knowledge of $\boldsymbol{h}$, the transmitter views the MIMO channel as a random vector[1] $\check{\boldsymbol{h}}$ and utilizes feedback information to characterize its multivariate probability density function (pdf) via, for example, the pair $(\bar{\boldsymbol{h}}, \boldsymbol{\Sigma}_h)$ of its first- and second-order moments. For analytical tractability, the virtual channel $\check{\boldsymbol{h}}$ perceived at the transmitter is modeled as complex Gaussian distributed; that is,

$$\check{\boldsymbol{h}} \sim \mathcal{CN}(\bar{\boldsymbol{h}}, \boldsymbol{\Sigma}_h). \tag{12.2}$$

[1] We use $\check{\boldsymbol{h}}$ to differentiate the channel perceived at the transmitter from the true channel $\boldsymbol{h}$. Notice that the statistics of $\check{\boldsymbol{h}}$ change every time updated feedback information becomes available.

Clearly, the Gaussian assumption in (12.2) is not always satisfied. However, it will greatly simplify the transmitter design, and the resulting solutions will provide valuable insight on transmitter optimization based on partial channel knowledge.

The perceived channel $\check{\boldsymbol{h}}$ in (12.2) can be represented equivalently by a *nominal-plus-perturbation* channel model as

$$\check{\boldsymbol{h}} = \overline{\boldsymbol{h}} + \boldsymbol{\epsilon}, \tag{12.3}$$

where $\overline{\boldsymbol{h}}$ is deterministic and $\boldsymbol{\epsilon}$ is a random perturbation vector the transmitter uses to model the uncertainty it has about the true channel [i.e., the partial CSI here includes the "nominal" channel $\overline{\boldsymbol{h}}$ and the statistical description of the "perturbation" error: $\boldsymbol{\epsilon} \sim \mathcal{CN}(\mathbf{0}_{N_t N_r \times 1}, \boldsymbol{\Sigma}_h)$]. When partial CSI is acquired through feedback, $\overline{\boldsymbol{h}}$ and $\boldsymbol{\Sigma}_h$ are updated each time feedback information from the receiver becomes available to the transmitter.

Either (12.2) or (12.3) provides a general statistical CSI model where imperfect knowledge about the underlying MIMO channel at the transmitter is captured through a multivariate Gaussian random vector. We next specify two simplified forms of statistical CSI and scenarios under which these become available at the transmitter using mean feedback, covariance feedback, or statistical a priori knowledge about the MIMO channel [273].

12.1.1.1 *Mean feedback* In the mean-feedback model, all entries of $\boldsymbol{\epsilon}$ are assumed to be i.i.d. with covariance σ_ϵ^2; and thus

$$\check{\boldsymbol{h}} \sim \mathcal{CN}(\overline{\boldsymbol{h}}, \sigma_\epsilon^2 \boldsymbol{I}_{N_t N_r}), \tag{12.4}$$

which implies identical uncertainty about all channel coefficients. Parameters $\overline{\boldsymbol{h}}$ and σ_ϵ^2 in (12.4) are estimated with arbitrarily high accuracy at the receiver, finely quantized, and sent back to the transmitter over the feedback channel. Presumably, sufficiently powerful error control codes are used in the low-rate feedback channel to ensure error-free reception. Furthermore, with the true channel remaining stationary over long time intervals, feedback updates can be infrequent. Estimation and quantization at the receiver can be sufficiently accurate for these parameters to be safely assumed perfectly known at the transmitter.

Let us now consider possible scenarios where feeding back the channel mean can be realized in practice.

Example 12.1 (Ricean fading channels) Suppose that a line-of-sight (LOS) path is present along with non-LOS diffuse paths between each transmit–receive antenna pair. In this case, the true channel itself is Ricean distributed. We assume further that the diffuse components of the channel coefficients are uncorrelated, all with identical variance $\sigma_{\text{non-LOS}}^2$; that is,

$$\boldsymbol{h} \sim \mathcal{CN}(\boldsymbol{\mu}_h, \sigma_{\text{non-LOS}}^2 \boldsymbol{I}_{N_t N_r}), \tag{12.5}$$

where $\boldsymbol{\mu}_h$ contains the channel gain corresponding to the LOS components.

The receiver feeds back to the transmitter instantaneous values of the LOS paths and the variance of the diffuse components without errors. We then have a mean-feedback model as in (12.4) with parameters

$$\bar{\boldsymbol{h}} = \boldsymbol{\mu}_h, \qquad \sigma_\epsilon^2 = \sigma_{\text{non-LOS}}^2. \tag{12.6}$$

Example 12.2 (Delayed feedback) Consider a setup where:

1. The antennas are well separated and the channel coefficients are i.i.d. complex Gaussian [i.e., $\boldsymbol{h} \sim \mathcal{CN}(\boldsymbol{0}_{N_tN_r\times 1}, \sigma_h^2 \boldsymbol{I}_{N_tN_r})$].
2. The channel coefficients are slowly time-varying according to the Jakes model with maximum Doppler frequency f_{max}.
3. The channel is acquired perfectly at the receiver and fed back to the transmitter via a noiseless channel that introduces delay τ.

Let $\boldsymbol{h}_\tau$ denote the delayed version of the channel available to the transmitter through feedback. Clearly, $\boldsymbol{h}$ and $\boldsymbol{h}_\tau$ are jointly Gaussian vectors with $E\{\boldsymbol{h}_\tau \boldsymbol{h}^{\mathcal{H}}\} = \rho\sigma_h^2 \boldsymbol{I}_{N_tN_r}$, where

$$\rho := J_0(2\pi f_{\text{max}}\tau) \tag{12.7}$$

denotes the correlation coefficient between $\boldsymbol{h}$ and $\boldsymbol{h}_\tau$ and $J_0(\cdot)$ is the zeroth-order Bessel function of the first kind. The minimum mean-square error (MMSE) estimator of $\boldsymbol{h}$ based on $\boldsymbol{h}_\tau$ is given by $E\{\boldsymbol{h}|\boldsymbol{h}_\tau\} = \rho \boldsymbol{h}_\tau$, with MMSE covariance matrix $\sigma_h^2(1 - |\rho|^2)\boldsymbol{I}_{N_tN_r}$. Thus, for each realization $\boldsymbol{h}_{\tau,0}$ of $\boldsymbol{h}_\tau$, the transmitter adheres to a mean-feedback model as in (12.4) with

$$\bar{\boldsymbol{h}} = \rho \boldsymbol{h}_{\tau,0}, \qquad \sigma_\epsilon^2 = \sigma_h^2(1 - |\rho|^2). \tag{12.8}$$

The values of $\bar{\boldsymbol{h}}$ are updated each time new feedback information becomes available.

In the delayed-feedback model of Example 12.2, a single parameter ρ quantifies the feedback quality. When $\rho = 0$, the feedback delay is so long that the channel at the receiver becomes outdated to the extent that it becomes irrelevant to the true channel and the instantaneous CSI is rendered useless; when $\rho = 1$, it corresponds to the other extreme where the transmitter has perfect CSI. Values of ρ between these two extreme cases ($\rho = 0$ and $\rho = 1$) model various degrees of partial CSI. We adopt this delayed-feedback model in our simulated examples in Section 12.6.1 to illustrate error performance under the mean-feedback model.

Because in both Examples 12.1 and 12.2 the nominal channel is assumed to be available to the transmitter error-free, mean feedback is more suitable for very slowly fading channels. A better fit for faster-fading MIMO channels is the paradigm of covariance feedback considered next.

12.1.1.2 Covariance Feedback If the channel $\boldsymbol{h}$ varies too rapidly for the transmitter to track its instantaneous value, feeding back the covariance of $\boldsymbol{h}$ is well motivated. The channel mean is usually set to zero and the relative geometry of the propagation environment manifests itself in a general non-diagonal covariance matrix Σ_h. Specifically, (12.2) in this case reduces to

$$\check{h} \sim \mathcal{CN}(\mathbf{0}_{N_t N_r \times 1}, \Sigma_h). \tag{12.9}$$

Once again, we suppose that sufficient training symbols have been used so that if Σ_h needs to be estimated at the receiver through sample averaging, the estimate is for all practical purposes perfect. Furthermore, notice that since no nominal channel realization is needed at the transmitter, feeding back Σ_h does not require frequent updates as long as the (even rapidly fading) channel process remains stationary.

Besides estimating it at the receiver and feeding it back to the transmitter, Σ_h can become available to the transmitter through field measurements, ray-tracing simulations, or physical channel models which do not require feedback. For certain applications (e.g., with immobile wireless stations) the channel covariance matrix can be determined from physical parameters such as antenna spacing, antenna arrangement, angle of arrival, and angle spread. In applications involving time-division duplex (TDD) protocols, the transmitter can also obtain channel statistics directly since the forward and reverse links share the same physical (and statistically invariant) channel characteristics, even when the time separation between the forward and the backward link is long enough to render the deterministic instantaneous channel estimates outdated. In frequency-division duplex (FDD) systems with small angle spread, the downlink channel covariance estimates can also be obtained accurately from the uplink channel covariance through proper frequency calibration processing [153]. The bottom line is that the covariance-feedback model in (12.9) does not necessarily require feedback since knowledge of the covariance matrix required can be acquired through other means.

The matrix Σ_h used in (12.9) to describe the form of partial CSI in the covariance-feedback model is allowed to have a general structure. In a number of applications, however, this matrix can be highly structured. Two such cases are described in the ensuing examples.

Example 12.3 (Transmit-correlation model) Consider an application scenario where the base station (BS) is unobstructed and the subscriber unit (SU) is surrounded by a large number of local scatterers. Relative to the BS, antenna spacing at the SU is much smaller (by one or two orders of magnitude), which renders the channels among different antenna elements uncorrelated. In this case the MIMO channel has zero mean, the waveforms at the receive-antennas can be assumed uncorrelated, and the transmit correlation across all receive-antennas is the same; that is,

$$\Sigma_t = E\{\boldsymbol{h}_\nu \boldsymbol{h}_\nu^{\mathcal{H}}\}, \quad \forall \nu, \tag{12.10}$$

where Σ_t is an arbitrary Hermitian matrix. In this simplified scenario we have

$$\Sigma_h = \boldsymbol{I}_{N_r} \otimes \Sigma_t. \tag{12.11}$$

Example 12.4 (Transmit-receive correlation model) Consider now an application setting where:

1. Corresponding to each receive-antenna, the correlation matrix of N_t transmit-antennas is the same (i.e., $\boldsymbol{\Sigma}_t = E\{\boldsymbol{h}_\nu \boldsymbol{h}_\nu^{\mathcal{H}}\}, \forall \nu$).
2. Corresponding to each transmit-antenna, the correlation matrix of N_r receive-antennas is the same (i.e., $\boldsymbol{\Sigma}_r = E\{\grave{\boldsymbol{h}}_\mu \grave{\boldsymbol{h}}_\mu^{\mathcal{H}}\}, \forall \mu$, where $\grave{\boldsymbol{h}}_\mu := [h_{1\mu}, \ldots, h_{N_r\mu}]^T$).

It turns out that in this case, $\boldsymbol{\Sigma}_h$ is structured and can be expressed as the Kronecker product ($\otimes$) of the transmit- and receive-correlation matrices; that is,

$$\boldsymbol{\Sigma}_h = \boldsymbol{\Sigma}_r \otimes \boldsymbol{\Sigma}_t. \tag{12.12}$$

The model in (12.12) has been used for analytical studies and has also been verified through field measurements in some practical MIMO settings (see, e.g., [87, 136]). The limitations of this Kronecker-product model were pointed out in [207]. An improved model may be found in [293].

12.2 ST SPREADING

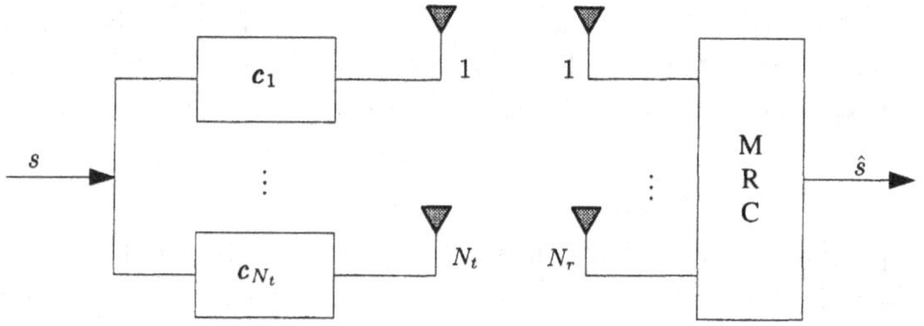

Figure 12.2 ST spread-spectrum system

Let us now consider utilizing the partial CSI models of Section 12.1 in designing the ST spread-spectrum system depicted in Figure 12.2. In each transmit-antenna μ, an incoming information-bearing symbol s is spread by a spreading code of length P denoted by $\boldsymbol{c}_\mu := [c_\mu(0), \ldots, c_\mu(P-1)]^T$. A different spreading code is used per transmit-antenna. Hence, the transmit-antenna array spreads s over both space and time. The pertinent ST code matrix of size $N_t \times P$ is

$$\boldsymbol{X} = \underbrace{\begin{bmatrix} c_1(0) & \cdots & c_1(P-1) \\ \vdots & \cdots & \vdots \\ c_{N_t}(0) & \cdots & c_{N_t}(P-1) \end{bmatrix}}_{:=\boldsymbol{C}} s. \tag{12.13}$$

Since only one information symbol is transmitted in P time slots, the spread-spectrum system in Figure 12.2 is by design a low-rate multi-antenna system. As we explain later in this section, the matrix $\boldsymbol{X}$ in (12.13) can also be interpreted as a transmit-beamforming matrix or as a linear precoding matrix. We first optimize the design of this low-rate system and then make up for the rate loss in Section 12.3.

With $\boldsymbol{X}$ in (12.13) transmitted over the MIMO channel, let us collect the received samples in an $N_r \times P$ matrix $\boldsymbol{Y}$ to obtain the input-output relationship

$$\boldsymbol{Y} = \sqrt{\bar{\gamma}}\boldsymbol{H}\boldsymbol{C}s + \boldsymbol{W}, \tag{12.14}$$

where $\boldsymbol{H}$ is the $N_r \times N_t$ channel matrix with (ν, μ)th entry $[\boldsymbol{H}]_{\nu\mu} = h_{\nu\mu}$ and $\boldsymbol{W}$ contains AWGN with each entry having zero mean and unit variance. To control the transmit-power through the average SNR parameter $\bar{\gamma}$, we constrain $\boldsymbol{C}$ to satisfy

$$\text{tr}\{\boldsymbol{C}^{\mathcal{H}}\boldsymbol{C}\} = \text{tr}\{\boldsymbol{C}^*\boldsymbol{C}^T\} = 1. \tag{12.15}$$

If s is drawn from a constellation with normalized energy $E[|s|^2] = 1$, then $\bar{\gamma}$ represents the ratio of the average transmitted energy per information symbol to noise variance.

Notice that $\boldsymbol{H}$ in (12.14) is related to the vector $\boldsymbol{h}$ in (12.1) via

$$\boldsymbol{h} := \text{vec}(\boldsymbol{H}^T), \tag{12.16}$$

where the vec($\cdot$) operator stacks all columns of a matrix vertically, one on top of the other. Correspondingly, we define $\boldsymbol{y} := \text{vec}(\boldsymbol{Y}^T)$, $\boldsymbol{w} := \text{vec}(\boldsymbol{W}^T)$ and use a property of Kronecker products to rewrite (12.14) as

$$\boldsymbol{y} = \sqrt{\bar{\gamma}}(\boldsymbol{I}_{N_r} \otimes \boldsymbol{C}^T)\boldsymbol{h}s + \boldsymbol{w}. \tag{12.17}$$

Our objective in this section is to find $\boldsymbol{C}$ so that a suitably chosen criterion is optimized. Before specifying criteria for this optimal transmit-beamformer design problem, let us note that for any given $\boldsymbol{C}$ the optimal receiver (in the ML sense) for recovering s from $\boldsymbol{y}$ in (12.17) amounts to maximum ratio combining (MRC). The latter is implemented using a filter $\boldsymbol{h}^{\mathcal{H}}(\boldsymbol{I}_{N_r} \otimes \boldsymbol{C}^T)^{\mathcal{H}}$ matched to the equivalent channel $(\boldsymbol{I}_{N_r} \otimes \boldsymbol{C}^T)\boldsymbol{h}$ in (12.17). The ML decision variable at the MRC output is

$$\hat{s} = \boldsymbol{h}^{\mathcal{H}}(\boldsymbol{I}_{N_r} \otimes \boldsymbol{C}^T)^{\mathcal{H}}\boldsymbol{y}. \tag{12.18}$$

The corresponding instantaneous SNR can easily be expressed as

$$\begin{aligned}\gamma &= \bar{\gamma}\boldsymbol{h}^{\mathcal{H}}(\boldsymbol{I}_{N_r} \otimes \boldsymbol{C}^T)^{\mathcal{H}}(\boldsymbol{I}_{N_r} \otimes \boldsymbol{C}^T)\boldsymbol{h} \\ &= \bar{\gamma}\boldsymbol{h}^{\mathcal{H}}\left[\boldsymbol{I}_{N_r} \otimes (\boldsymbol{C}^*\boldsymbol{C}^T)\right]\boldsymbol{h}.\end{aligned} \tag{12.19}$$

For notational convenience, we define

$$\boldsymbol{Z} = \bar{\gamma}\boldsymbol{I}_{N_r} \otimes (\boldsymbol{C}^*\boldsymbol{C}^T) \tag{12.20}$$

and simplify (12.19) as

$$\gamma = \boldsymbol{h}^{\mathcal{H}}\boldsymbol{Z}\boldsymbol{h}. \tag{12.21}$$

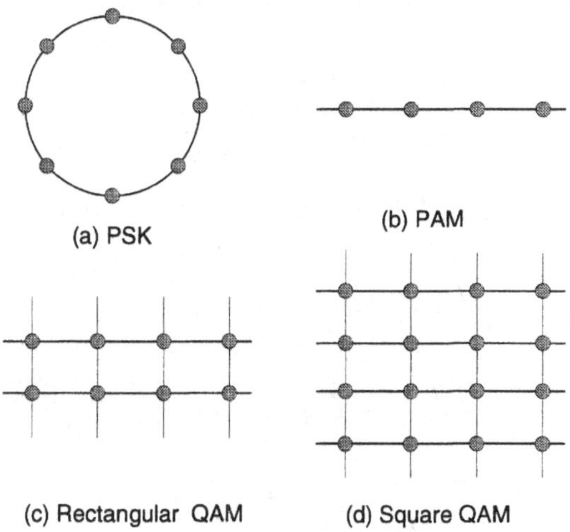

Figure 12.3 Signal constellations

Even though with sufficient number of training symbols the receiver can acquire $\boldsymbol{h}$ to perform MRC perfectly for all practical purposes, using feedback information or other means, the transmitter can never have exact knowledge of $\boldsymbol{h}$ realizations. To account for this inevitable uncertainty about the perfect CSI, the transmitter adopts a virtual channel $\check{\boldsymbol{h}}$ modeled as a random vector distributed according to (12.2). In agreement with this model, the SNR $\gamma = \check{\boldsymbol{h}}^{\mathcal{H}} \boldsymbol{Z} \check{\boldsymbol{h}}$ is viewed at the transmitter as a random variable.

The symbol error rate averaged over this random SNR variable will form our initial criterion, which we wish to optimize with respect to the ST spreading (or beamforming) matrix $\boldsymbol{C}$. To this end, we first need to express this average error performance criterion under the partial CSI model in terms of $\boldsymbol{C}$.

12.2.1 Average Error Performance

Let us suppose that s is drawn from commonly used constellations, including phase shift keying (PSK), pulse amplitude modulation (PAM), and rectangular and square quadrature amplitude modulation (QAM); see also Figure 12.3. As usual, the rectangular M-ary QAM constellation can be decomposed into two independent PAMs, one on the I-branch with size $\sqrt{2M}$ and the other on the Q-branch with size $\sqrt{M/2}$.

Relying on the finite-integral representation of the Gaussian $Q(\cdot)$ function

$$Q(x) := \int_x^\infty \frac{1}{\sqrt{2\pi}} e^{-\frac{t^2}{2}} dt = \frac{1}{\pi} \int_0^{\frac{\pi}{2}} \exp\left(-\frac{x^2}{2\sin^2\theta}\right) d\theta, \tag{12.22}$$

the instantaneous SER for commonly used constellations can be expressed as [236]

$$P_{s,\text{PSK}}(\gamma) = \frac{1}{\pi}\int_0^{\frac{(M-1)\pi}{M}} \exp\left(-\frac{g_{\text{PSK}}\gamma}{\sin^2\theta}\right) d\theta, \tag{12.23}$$

$$P_{s,\text{PAM}}(\gamma) = \frac{2}{\pi}\left(\frac{M-1}{M}\right)\int_0^{\frac{\pi}{2}} \exp\left(-\frac{g_{\text{PAM}}\gamma}{\sin^2\theta}\right) d\theta, \tag{12.24}$$

$$\begin{aligned} P_{s,\text{QAM}}^{\text{rect}}(\gamma) = {} & \frac{4}{\pi}\left(\frac{3}{\sqrt{8M}} - \frac{1}{M}\right)\int_0^{\frac{\pi}{4}} \exp\left(-\frac{g_{\text{QAM}}^{\text{rect}}\gamma}{\sin^2\theta}\right) d\theta \\ & + \frac{4}{\pi}\left(1 - \frac{3}{\sqrt{8M}}\right)\int_{\frac{\pi}{4}}^{\frac{\pi}{2}} \exp\left(-\frac{g_{\text{QAM}}^{\text{rect}}\gamma}{\sin^2\theta}\right) d\theta, \end{aligned} \tag{12.25}$$

$$\begin{aligned} P_{s,\text{QAM}}^{\text{square}}(\gamma) = {} & \frac{4}{\pi}\left(\frac{1}{\sqrt{M}} - \frac{1}{M}\right)\int_0^{\frac{\pi}{4}} \exp\left(-\frac{g_{\text{QAM}}^{\text{square}}\gamma}{\sin^2\theta}\right) d\theta \\ & + \frac{4}{\pi}\left(1 - \frac{1}{\sqrt{M}}\right)\int_{\frac{\pi}{4}}^{\frac{\pi}{2}} \exp\left(-\frac{g_{\text{QAM}}^{\text{square}}\gamma}{\sin^2\theta}\right) d\theta, \end{aligned} \tag{12.26}$$

where the constellation-specific constant g is defined as

$$g_{\text{PSK}} = \sin^2\left(\frac{\pi}{M}\right) \quad \text{for } M\text{-ary PSK}, \tag{12.27}$$

$$g_{\text{PAM}} = \frac{3}{M^2-1} \quad \text{for } M\text{-ary PAM}, \tag{12.28}$$

$$g_{\text{QAM}}^{\text{rect}} = \frac{6}{5M-4} \quad \text{for } M\text{-ary rectangular QAM} \tag{12.29}$$

$$g_{\text{QAM}}^{\text{square}} = \frac{3}{2(M-1)} \quad \text{for } M\text{-ary square QAM}. \tag{12.30}$$

To evaluate the average performance, we use the identity [247]

$$E_{\boldsymbol{z}}\{\exp(-\boldsymbol{z}^{\mathcal{H}}\boldsymbol{A}\boldsymbol{z})\} = \frac{\exp(-\boldsymbol{\mu}^{\mathcal{H}}\boldsymbol{A}(\boldsymbol{I}+\boldsymbol{\Sigma}\boldsymbol{A})^{-1}\boldsymbol{\mu})}{|\boldsymbol{I}+\boldsymbol{\Sigma}\boldsymbol{A}|}, \tag{12.31}$$

which holds for $\boldsymbol{z} \sim \mathcal{CN}(\boldsymbol{\mu}, \boldsymbol{\Sigma})$ and $\boldsymbol{A}$ an arbitrary Hermitian matrix. Recalling that $\check{\boldsymbol{h}}$ is distributed as in (12.2) and taking expectations of the instantaneous SER expressions in (12.24)–(12.26) over $\gamma = \check{\boldsymbol{h}}^{\mathcal{H}}\boldsymbol{Z}\check{\boldsymbol{h}}$, we can obtain the average SER for the commonly used constellations. This average SER can be written in compact form after using the identity in (12.31); for example, for PSK and square QAM, we have

$$P_{s,\text{PSK}} = \frac{1}{\pi}\int_0^{\frac{(M-1)\pi}{M}} \frac{\exp\left(-\bar{\boldsymbol{h}}^{\mathcal{H}} g\boldsymbol{Z}\left[\boldsymbol{I}\sin^2\theta + g\boldsymbol{\Sigma}_h\boldsymbol{Z}\right]^{-1}\bar{\boldsymbol{h}}\right)}{|\boldsymbol{I} + g\boldsymbol{\Sigma}_h\boldsymbol{Z}/\sin^2\theta|}\,d\theta, \tag{12.32}$$

$$\begin{aligned} P_{s,\text{QAM}}^{\text{square}} &= \frac{4}{\pi}\left(\frac{1}{\sqrt{M}} - \frac{1}{M}\right)\int_0^{\frac{\pi}{4}} \frac{\exp\left(-\bar{\boldsymbol{h}}^{\mathcal{H}} g\boldsymbol{Z}\left[\boldsymbol{I}\sin^2\theta + g\boldsymbol{\Sigma}_h\boldsymbol{Z}\right]^{-1}\bar{\boldsymbol{h}}\right)}{|\boldsymbol{I} + g\boldsymbol{\Sigma}_h\boldsymbol{Z}/\sin^2\theta|}\,d\theta \\ &+ \frac{4}{\pi}\left(1 - \frac{1}{\sqrt{M}}\right)\int_{\frac{\pi}{4}}^{\frac{\pi}{2}} \frac{\exp\left(-\bar{\boldsymbol{h}}^{\mathcal{H}} g\boldsymbol{Z}\left[\boldsymbol{I}\sin^2\theta + g\boldsymbol{\Sigma}_h\boldsymbol{Z}\right]^{-1}\bar{\boldsymbol{h}}\right)}{|\boldsymbol{I} + g\boldsymbol{\Sigma}_h\boldsymbol{Z}/\sin^2\theta|}\,d\theta, \end{aligned} \tag{12.33}$$

where g takes values as in (12.27) and (12.30), respectively, and the identity matrix $\boldsymbol{I}$ has size N_tN_r. The average SER for PAM and rectangular QAM can be obtained similarly.

We use (12.32) and (12.33) to evaluate the exact SER performance in the numerical examples of Section 12.4. However, for our stated goal to minimize the average SER over $\boldsymbol{C}$ [which shows up in $\boldsymbol{Z}$ according to (12.20)], these exact expressions are not amenable to optimization, due to their integral form. Furthermore, separate optimization is needed for each constellation. These considerations motivate the derivation of a unified upper bound on the average SER and its minimization with respect to $\boldsymbol{C}$, treating all signal constellations at once.

12.2.2 Optimization Based on Average SER Bound

It is not difficult to verify that the integrands in (12.23)–(12.26) peak at $\theta = \pi/2$. Based on this observation, it is possible to upper-bound the instantaneous SER in (12.23)–(12.26) using [333]

$$P_{s,\text{bound}}(\gamma) = \alpha\exp(-g\gamma), \tag{12.34}$$

where $\alpha := (M-1)/M$ and g is a constant that depends on the underlying constellation as in (12.27)–(12.30). Averaging (12.34) over γ and using (12.31) leads to the following unified upper bound on the average SER in (12.32)–(12.33)

$$P_{s,\text{bound}} = \alpha\exp\left(-\bar{\boldsymbol{h}}^{\mathcal{H}} g\boldsymbol{Z}\left[\boldsymbol{I} + g\boldsymbol{\Sigma}_h\boldsymbol{Z}\right]^{-1}\bar{\boldsymbol{h}}\right)|\boldsymbol{I} + g\boldsymbol{\Sigma}_h\boldsymbol{Z}|^{-1}. \tag{12.35}$$

We underscore that this bound applies to all constellations under consideration.

Together with the power constraint in (12.15), the ST spreading optimization problem can now be formulated as follows:

$$\boldsymbol{C}_{\text{opt}} = \underset{\boldsymbol{C}:\,\text{tr}\{\boldsymbol{C}^*\boldsymbol{C}^T\}=1}{\arg\min}\; P_{s,\text{bound}}. \tag{12.36}$$

Notice that $P_{s,\text{bound}}$ depends on $\boldsymbol{C}$ only through $\boldsymbol{Z}$, which is a function of $\boldsymbol{C}^*\boldsymbol{C}^T$ [cf. (12.20)]. Using the singular value decomposition (SVD), we can factor $\boldsymbol{C}$ as

$$\boldsymbol{C} = \boldsymbol{U}_c^*\boldsymbol{\Delta}^{\frac{1}{2}}\boldsymbol{\Phi}, \tag{12.37}$$

where $\boldsymbol{U}_c$ is an $N_t \times N_t$ unitary matrix, $\boldsymbol{\Phi}$ is also unitary of size $N_t \times P$, and $\boldsymbol{\Delta}$ is a diagonal matrix denoted as

$$\boldsymbol{\Delta} := \text{diag}(\delta_1, \ldots, \delta_{N_t}). \tag{12.38}$$

As long as $P \geq N_t$ and thus $\boldsymbol{\Phi}\boldsymbol{\Phi}^{\mathcal{H}} = \boldsymbol{I}_{N_t}$, we have

$$\boldsymbol{C}^*\boldsymbol{C}^T = \boldsymbol{U}_c\boldsymbol{\Delta}\boldsymbol{U}_c^{\mathcal{H}}, \tag{12.39}$$

which implies that the choice of $\boldsymbol{\Phi}$ does not affect the average SER. Without loss of generality, we can therefore assume that $P = N_t$, $\boldsymbol{\Phi} = \boldsymbol{I}_{N_t}$ and look for the optimal $\boldsymbol{U}_c$ and $\boldsymbol{\Delta}$. The optimization in (12.36) is then equivalent to

$$(\boldsymbol{U}_c, \boldsymbol{\Delta})_{\text{opt}} = \underset{\boldsymbol{U}_c, \boldsymbol{\Delta}:\ \text{tr}\{\boldsymbol{\Delta}\}=1}{\arg\min} \quad P_{s,\text{bound}}. \tag{12.40}$$

Solving (12.40) in closed form is generally impossible, and one has to resort to numerical nonlinear programming methods. Interestingly, we will see next that when partial CSI is available in the form of mean or covariance feedback, closed-form solutions are possible.

12.2.3 Mean Feedback

Setting $\boldsymbol{\Sigma}_h = \sigma_\epsilon^2 \boldsymbol{I}_{N_t N_r}$ in the mean-feedback model of (12.4), we deduce from (12.20) and (12.39) that

$$g\boldsymbol{Z}(\boldsymbol{I} + g\boldsymbol{\Sigma}_h\boldsymbol{Z})^{-1} = \boldsymbol{I}_{N_r} \otimes \left[\frac{\beta}{\sigma_\epsilon^2}\boldsymbol{U}_c\boldsymbol{\Delta}\boldsymbol{U}_c^{\mathcal{H}}\left(\boldsymbol{I} + \beta\boldsymbol{U}_c\boldsymbol{\Delta}\boldsymbol{U}_c^{\mathcal{H}}\right)^{-1}\right], \tag{12.41}$$

where the constant β is given by

$$\beta := g\sigma_\epsilon^2\bar{\gamma}. \tag{12.42}$$

Based on (12.3), we can collect the nominal channels in $\overline{\boldsymbol{H}}$, corresponding to the $N_r \times N_t$ channel $\boldsymbol{H}$, to form

$$\overline{\boldsymbol{H}}^T := [\bar{\boldsymbol{h}}_1, \bar{\boldsymbol{h}}_2, \ldots, \bar{\boldsymbol{h}}_{N_r}] \in \mathbb{C}^{N_t \times N_r}. \tag{12.43}$$

After straightforward manipulations, and using (12.41) and (12.43), we can express the bound in (12.35) as

$$P_{s,\text{bound}} = \frac{\alpha}{|\boldsymbol{I}_{N_t} + \beta\boldsymbol{\Delta}|^{N_r}} \exp\left(-\frac{1}{\sigma_\epsilon^2}\text{tr}\left[\boldsymbol{U}_c^{\mathcal{H}}\overline{\boldsymbol{H}}^T\overline{\boldsymbol{H}}^*\boldsymbol{U}_c\beta\boldsymbol{\Delta}(\boldsymbol{I}_{N_t} + \beta\boldsymbol{\Delta})^{-1}\right]\right). \tag{12.44}$$

The simplified average SER bound in (12.44) will allow us to find the optimal $\boldsymbol{U}_c$ and $\boldsymbol{\Delta}$ matrices in closed form. The next proposition addresses the optimization over $\boldsymbol{U}_c$.

Proposition 12.1 *Consider the eigenvalue decomposition*

$$\overline{\boldsymbol{H}}^T \overline{\boldsymbol{H}}^* = \boldsymbol{U}_H \boldsymbol{\Lambda} \boldsymbol{U}_H^{\mathcal{H}}, \tag{12.45}$$

where $\boldsymbol{U}_H$ contains eigenvectors and the diagonal matrix

$$\boldsymbol{\Lambda} := \text{diag}(\lambda_1, \lambda_2, \ldots, \lambda_{N_t}) \tag{12.46}$$

contains the eigenvalues arranged in a non-increasing order: $\lambda_1 \geq \cdots \geq \lambda_{N_t}$. The optimal $\boldsymbol{U}_c$ in (12.36) is then given by

$$\boldsymbol{U}_{c,\text{opt}} = \boldsymbol{U}_H. \tag{12.47}$$

Proof: For each fixed $\boldsymbol{\Delta}$, the optimal $\boldsymbol{U}_c$ shall maximize

$$\begin{aligned} &\text{tr}\left[\boldsymbol{U}_c^{\mathcal{H}} \overline{\boldsymbol{H}}^T \overline{\boldsymbol{H}}^* \boldsymbol{U}_c \beta \boldsymbol{\Delta} (\boldsymbol{I}_{N_t} + \beta \boldsymbol{\Delta})^{-1}\right] \\ &\qquad = \text{tr}(\boldsymbol{\Lambda}) - \text{tr}\left[\boldsymbol{U}_c^{\mathcal{H}} \boldsymbol{U}_H \boldsymbol{\Lambda} \boldsymbol{U}_H^{\mathcal{H}} \boldsymbol{U}_c (\boldsymbol{I}_{N_t} + \beta \boldsymbol{\Delta})^{-1}\right]. \end{aligned} \tag{12.48}$$

To proceed, we will need the following lemma, which is proved in [124, Equation (19)]:

Lemma 12.1 *Consider the $N \times N$ positive semi-definite matrix $\boldsymbol{\Lambda}_Q$ with diagonal entries $[\boldsymbol{\Lambda}_Q]_{1,1} \geq \cdots \geq [\boldsymbol{\Lambda}_Q]_{N,N}$; and the $N \times N$ positive definite matrix $\boldsymbol{\Lambda}_A$ with diagonal entries arranged in non-increasing order. If $\boldsymbol{Q} = \boldsymbol{U}_Q \boldsymbol{\Lambda}_Q \boldsymbol{U}_Q^{\mathcal{H}}$ denotes any unitary matrix, then*

$$\text{tr}(\boldsymbol{Q}\boldsymbol{\Lambda}_A^{-1}) \geq \text{tr}(\boldsymbol{\Lambda}_Q \boldsymbol{\Lambda}_A^{-1}), \tag{12.49}$$

where the equality holds for any $\boldsymbol{\Lambda}_A \neq \boldsymbol{I}_N$ only when $\boldsymbol{U}_Q = \boldsymbol{I}_N$.

Recall that for each $\boldsymbol{\Delta}$, we can arrange the diagonal entries of $\boldsymbol{\Delta}$ in a non-increasing order by reordering the eigenvectors in $\boldsymbol{U}_c$. Applying Lemma 12.1 to (12.48), we infer readily that $\boldsymbol{U}_{c,\text{opt}} = \boldsymbol{U}_H$. □

The SER bound in (12.44) with the optimal $\boldsymbol{U}_c$ in (12.47) can be written as

$$P_{s,\text{bound}} = \alpha \left[\prod_{\mu=1}^{N_t} \frac{1}{1+\delta_\mu \beta} \exp\left(\frac{-\mathcal{K}_\mu \delta_\mu \beta}{1+\delta_\mu \beta}\right)\right]^{N_r}, \tag{12.50}$$

where

$$\mathcal{K}_\mu = \frac{\lambda_\mu}{N_r \sigma_\epsilon^2}. \tag{12.51}$$

We next proceed to find the diagonal matrix $\boldsymbol{\Delta}$ solving the optimization in (12.36).

12.2.3.1 Exact solution for $N_r = 1$ The optimal Δ can be found in closed form only when $N_r = 1$, where only one eigenvalue λ_1 is non-zero. Since $\ln(\cdot)$ is a monotonically increasing function, let

$$\mathcal{E}_1 := \ln P_{s,\text{bound}} = \ln \alpha - \sum_{\mu=1}^{N_t} \ln(1+\delta_\mu \beta) - \frac{\lambda_1 \delta_1 \beta}{\sigma_\epsilon^2 (1+\delta_1 \beta)}, \tag{12.52}$$

and recast (12.36) in the equivalent constrained optimization form

$$\Delta_{\text{opt}} = \underset{\Delta \geq 0;\ \text{tr}(\Delta)=1}{\arg\min} \quad \mathcal{E}_1. \tag{12.53}$$

Using the Lagrange multiplier method, one can solve (12.53) as follows (see, e.g., [330] for details). Upon defining the constants

$$a := \left(1 + \frac{N_t}{\beta}\right)^2, \qquad c := N_t(N_t - 1),$$
$$b := \frac{\lambda}{\beta \sigma_\epsilon^2} + \left(1 + \frac{N_t}{\beta}\right)(2N_t - 1), \tag{12.54}$$

the optimal power-loading solution is given by

$$\delta_2 = \cdots = \delta_{N_t} = \left[\frac{2a}{b + \sqrt{b^2 - 4ac}} - \frac{1}{\beta}\right]_+,$$
$$\delta_1 = 1 - (N_t - 1)\delta_2, \tag{12.55}$$

where the notation $[x]_+$ stands for $\max(x, 0)$.

12.2.3.2 Approximate Solution Besides the closed-form solution which is available for $N_r = 1$ only, it is also possible to derive an approximate solution for Δ which is applicable to any N_r [330]. The key is to recognize that

$$E\{\exp(-\gamma)\} = \frac{1}{1+\delta_\mu \beta} \exp\left(-\frac{\mathcal{K}_\mu \delta_\mu \beta}{1+\delta_\mu \beta}\right) \tag{12.56}$$

when $\sqrt{\gamma}$ is Ricean distributed with Ricean factor $\mathcal{K}_\mu$ and power $(1+\mathcal{K}_\mu)\delta_\mu\beta$ [236]. It is well known that a Rice distribution with Ricean factor $\mathcal{K}_\mu$ can be approximated by a Nakagami-m distribution with m_μ given by [236]

$$m_\mu = \frac{(1+\mathcal{K}_\mu)^2}{1+2\mathcal{K}_\mu}. \tag{12.57}$$

Notice that a Ricean distribution with $\mathcal{K}_\mu = 0$ coincides with a Nakagami distribution having $m_\mu = 1$, and in this case both reduce to a Rayleigh distribution. Furthermore, for a Nakagami random variable $\sqrt{\gamma'}$ with power $(1+\mathcal{K}_\mu)\delta_\mu\beta$, we have

$$E\{\exp(-\gamma')\} = \left(1 + \frac{(1+\mathcal{K}_\mu)\delta_\mu\beta}{m_\mu}\right)^{-m_\mu}. \tag{12.58}$$

Approximating the Ricean distributed $\sqrt{\gamma}$ with a Nakagami distributed $\sqrt{\gamma'}$, we obtain [cf. (12.52)]

$$P_{s,\text{bound}} \approx \tilde{P}_{s,\text{bound}} = \alpha \left[\prod_{\mu=1}^{N_t} \left(1 + \frac{(1+\mathcal{K}_\mu)\delta_\mu \beta}{m_\mu}\right)^{-m_\mu} \right]^{N_r}. \tag{12.59}$$

Based on (12.59), we can adopt the objective function

$$\mathcal{E}_2 := \ln \tilde{P}_{s,\text{bound}} = \ln \alpha - N_r \left[\sum_{\mu=1}^{N_t} m_\mu \ln\left(1 + \frac{\delta_\mu (1+\mathcal{K}_\mu)\beta}{m_\mu}\right)\right] \tag{12.60}$$

and optimize it to obtain the approximate solution for the diagonal loading matrix as

$$\boldsymbol{\Delta}_{\text{opt}} = \underset{\boldsymbol{\Delta}\geq 0;\ \text{tr}(\boldsymbol{\Delta})=1}{\arg\min}\ \mathcal{E}_2. \tag{12.61}$$

Solving (12.61) using the Lagrange multiplier method, we arrive at

$$\delta_\mu = \left[\frac{m_\mu}{\zeta} - \frac{m_\mu}{(1+\mathcal{K}_\mu)\beta}\right]_+, \tag{12.62}$$

where ζ denotes the Lagrange multiplier. Based on (12.62), ζ is obtained by substituting (12.62) into the power constraint $\sum_{\mu=1}^{N_t} \delta_\mu = 1$.

Suppose that the final solution ends up with $\overline{N}_t$ non-zero eigenvalues in $\boldsymbol{\Delta}$. In this case we have $\delta_\mu = 0$ for $\mu \geq \overline{N}_t + 1$. For each $\mu = 1, \ldots, \overline{N}_t$, we can then find ζ using the power constraint, and obtain

$$\delta_\mu = \frac{m_\mu}{\sum_{l=1}^{\overline{N}_t} m_l}\left(1 + \sum_{l=1}^{\overline{N}_t} \frac{m_l}{(1+\mathcal{K}_l)\beta}\right) - \frac{m_\mu}{(1+\mathcal{K}_\mu)\beta}. \tag{12.63}$$

To ensure that $\delta_{\overline{N}_t} > 0$, the transmit-power (and hence the average SNR $\bar{\gamma}$) should be larger than a lower bound; that is,

$$\bar{\gamma} > \underbrace{\frac{1}{g\sigma_\epsilon^2} \sum_{l=1}^{\overline{N}_t - 1} \frac{(\lambda_l - \lambda_{\overline{N}_t})(N_r\sigma_\epsilon^2 + \lambda_l)}{(N_r\sigma_\epsilon^2 + \lambda_{\overline{N}_t})(N_r\sigma_\epsilon^2 + 2\lambda_l)}}_{:=\gamma_{\text{th},\overline{N}_t}}. \tag{12.64}$$

From (12.63) and (12.64), a practical algorithm for calculating the optimal $\boldsymbol{\Delta}$ can be described in the following steps:

S1) For $r = 1, \ldots, N_t$, calculate $\gamma_{\text{th},r}$ from (12.64), based only on the first r eigenvalues of $\boldsymbol{\Lambda}$.

S2) With the given power budget ensuring that $\bar{\gamma}$ falls in the interval $[\gamma_{\text{th},r}, \gamma_{\text{th},r+1}]$, set $\delta_{r+1}, \ldots, \delta_{N_t} = 0$ and obtain $\delta_1, \ldots, \delta_r$ according to (12.63) with $\overline{N}_t = r$.

Using this algorithm, $\boldsymbol{\Delta}$ can be obtained in closed form. This approximate solution holds for any N_r.

As an example, it is instructive to specify the approximate solution for $N_r = 1$. In this case, we have only one non-zero eigenvalue λ_1, and all thresholds in (12.64) reduce to

$$\gamma_{\text{th}} = \frac{\lambda_1}{g\sigma_\epsilon^4}\left(\frac{\sigma_\epsilon^2 + \lambda_1}{\sigma_\epsilon^2 + 2\lambda_1}\right). \tag{12.65}$$

Hence, the approximate solution in (12.62) yields

$$\delta_2 = \cdots = \delta_{N_t} = \begin{cases} \delta_2^o & \bar{\gamma} > \gamma_{\text{th}} \\ 0 & \bar{\gamma} \leq \gamma_{\text{th}}, \end{cases}$$

$$\delta_1 = 1 - \delta_2(N_t - 1), \tag{12.66}$$

where δ_2^o is simplified from (12.63) as

$$\delta_2^o = \frac{\sigma_\epsilon^2(\sigma_\epsilon^2 + 2\lambda_1)}{N_t\sigma_\epsilon^2(\sigma_\epsilon^2 + 2\lambda_1) + \lambda_1^2}\left[1 + \frac{1}{\beta}\left(N_t - \frac{\lambda_1}{\sigma_\epsilon^2 + 2\lambda_1}\right)\right] - \frac{1}{\beta}. \tag{12.67}$$

For $N_r = 1$ we compare the approximate solution in (12.66) with the exact solution in (12.55) in Section 12.4.

12.2.4 Covariance Feedback

For the covariance-feedback form of the partial CSI model, we consider only the special case of (12.11), where closed-form solutions turn out to be available [331]. Recall also that the term *feedback* in this form of partial CSI is perhaps a misnomer when the covariance matrix required is obtained a priori based on geometric considerations or experimental data.

Substituting (12.11) into (12.35), we obtain

$$P_{s,\text{bound}} = \alpha\left|\boldsymbol{I} + \boldsymbol{\Sigma}_t\boldsymbol{C}^*\boldsymbol{C}^T g\bar{\gamma}\right|^{-N_r}, \tag{12.68}$$

which we wish to optimize again with respect to $\boldsymbol{U}_c$ and $\boldsymbol{\Delta}$. The next proposition yields the optimal $\boldsymbol{U}_c$ for any loading matrix $\boldsymbol{\Delta}$.

Proposition 12.2 *Consider the eigenvalue decomposition*

$$\boldsymbol{\Sigma}_t = \boldsymbol{U}_H\boldsymbol{\Lambda}\boldsymbol{U}_H^{\mathcal{H}}, \tag{12.69}$$

where $\boldsymbol{U}_H$ contains the eigenvectors, and the diagonal matrix

$$\boldsymbol{\Lambda} = \text{diag}(\lambda_1, \ldots, \lambda_{N_t}) \tag{12.70}$$

contains the eigenvalues of $\boldsymbol{\Sigma}_t$ in a non-increasing order $\lambda_1 \geq \cdots \geq \lambda_{N_t} \geq 0$. For any given $\boldsymbol{\Delta}$, the optimal $\boldsymbol{U}_c$ with covariance feedback adhering to (12.11) *is*

$$\boldsymbol{U}_{c,\text{opt}} = \boldsymbol{U}_H. \tag{12.71}$$

Proof: Substituting (12.39) and (12.69) into (12.68), we obtain

$$P_{s,\text{bound}} = \alpha \left| \boldsymbol{I} + \boldsymbol{\Lambda}^{\frac{1}{2}} \boldsymbol{U}_H^{\mathcal{H}} \boldsymbol{U}_c \boldsymbol{\Delta} \boldsymbol{U}_c^{\mathcal{H}} \boldsymbol{U}_H \boldsymbol{\Lambda}^{\frac{1}{2}} g\bar{\gamma} \right|^{-N_r}. \tag{12.72}$$

Based on the Hadamard inequality, $P_{s,\text{bound}}$ is minimized when $\boldsymbol{U}_c = \boldsymbol{U}_H$. □

Proceeding to find the optimal $\boldsymbol{\Delta}$, we substitute $\boldsymbol{U}_{c,\text{opt}}$ into (12.68) and look for the minimum of the logarithm of the resulting objective function

$$\begin{aligned} \mathcal{E}_3 := \ln P_{s,\text{bound}} &= \ln\alpha - N_r \ln |\boldsymbol{I}_{N_t} + \boldsymbol{\Lambda}\boldsymbol{\Delta} g\bar{\gamma}| \\ &= \ln\alpha - N_r \sum_{\mu=1}^{N_t} \ln\left(1 + \lambda_\mu \delta_\mu g\bar{\gamma}\right). \end{aligned} \tag{12.73}$$

The optimal $\boldsymbol{\Delta}$ is the solution of

$$\boldsymbol{\Delta}_{\text{opt}} = \underset{\boldsymbol{\Delta} \geq 0:\, \text{tr}(\boldsymbol{\Delta})=1}{\arg\min} \quad \mathcal{E}_3. \tag{12.74}$$

Solving (12.74) using the Lagrange multiplier method, we find that

$$\delta_\mu = \left[-\frac{1}{\zeta} - \frac{1}{\lambda_\mu g\bar{\gamma}} \right]_+, \tag{12.75}$$

where ζ denotes the Lagrange multiplier, which can be obtained as before using the power constraint $\text{tr}(\boldsymbol{\Delta}) = 1$.

Suppose that the given power budget supports $\overline{N}_t$ non-zero δ_μ's. Solving for ζ using the power constraint leads to

$$\delta_\mu = \left[\frac{1}{\overline{N}_t} + \frac{1}{g\bar{\gamma}} \left(\frac{1}{\overline{N}_t} \sum_{l=1}^{\overline{N}_t} \frac{1}{\lambda_l} - \frac{1}{\lambda_\mu} \right) \right], \quad \mu = 1, \ldots, \overline{N}_t. \tag{12.76}$$

To ensure that $\delta_{\overline{N}_t} > 0$, the transmit-power should satisfy the inequality [cf. (12.64)]

$$\bar{\gamma} > \underbrace{\frac{1}{g} \left(\frac{\overline{N}_t}{\lambda_{\overline{N}_t}} - \sum_{\mu=1}^{\overline{N}_t} \frac{1}{\lambda_\mu} \right)}_{:=\gamma_{\text{th},\overline{N}_t}}. \tag{12.77}$$

Based on (12.77) and (12.76), the practical algorithm for optimizing $\boldsymbol{\Delta}$ can be summarized in the following steps:

S1) For $r = 1, \ldots, N_t$, calculate $\gamma_{\text{th},r}$ from (12.77) based only on the first r eigenvalues in $\boldsymbol{\Delta}$.

S2) With the given power budget leading to $\bar{\gamma}$ in the interval $[\gamma_{\text{th},r}, \gamma_{\text{th},r+1})$, set $\delta_{r+1}, \ldots, \delta_{N_t} = 0$ and obtain $\delta_1, \ldots, \delta_r$ according to (12.76) with $\overline{N}_t = r$.

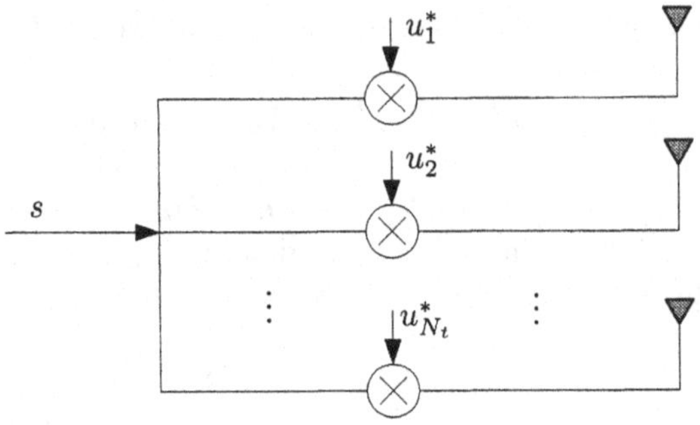

Figure 12.4 Transmit-beamforming operation

12.2.5 Beamforming Interpretation

We already mentioned that the ST spreading operation in (12.37) can be viewed as a transmit-beamformer. To recognize this, consider Figure 12.4, which depicts a transmit-beamformer. Each incoming information-bearing symbol s is weighted in every time slot by a coefficient u_μ^* per transmit-antenna μ. Vector $\boldsymbol{u} := [u_1^*, \ldots, u_{N_t}^*]^T$ is known as a *beamforming* or *beamsteering vector.*

Upon placing time slots (columns of $\boldsymbol{C}$) next to each other, we can view the precoder $\boldsymbol{C}$ in Figure 12.2 as a time-varying beamformer. In each time slot p, the pth column of $\boldsymbol{C}$ (denoted as $\tilde{\boldsymbol{c}}_p$) weighs s across N_t antennas and thus steers s toward a beam directed along $\tilde{\boldsymbol{c}}_p$. Since there are a total of P different beams used per information symbol, the time-varying beamformer is redundant in the sense that it repeats s in different directions.

When partial CSI is available at the transmitter in the form of mean or covariance feedback, the optimal $\boldsymbol{C}$ assumes a special structure. Indeed, if we suppose that $\boldsymbol{\Phi} = \boldsymbol{I}_{N_t}$, the antenna-steering vector at the pth time slot is

$$\tilde{\boldsymbol{c}}_p = \sqrt{\delta_p}\, \boldsymbol{u}_{H,p}^*, \quad p = 0, 1, \ldots, P-1, \tag{12.78}$$

where $\boldsymbol{u}_{H,p}$ is the $(p+1)$st eigenvector of $\boldsymbol{U}_H$ in (12.45) for mean feedback or (12.69) for covariance feedback. To underscore this special beamformer structure, we henceforth term the eigenvectors $\{\tilde{\boldsymbol{c}}_p\}_{p=0}^{P-1}$ in (12.78) *eigenbeams*. Accordingly, we can view $\boldsymbol{\Delta}$ as a matrix loading power onto those eigenbeams. Hence, the optimal transmit-beamformer with mean or covariance feedback deploys the eigenbeams in successive time slots, with proper power allocation among them.

On the other hand, with a general $\boldsymbol{\Phi}$, each beam direction (column of $\boldsymbol{C}$) comprises a weighted multiplexing of N_t eigenbeams:

$$\tilde{\boldsymbol{c}}_p = \sum_{\mu=1}^{N_t} [\boldsymbol{\Phi}]_{\mu,p} (\sqrt{\delta_\mu}\, \boldsymbol{u}_{H,\mu}^*), \quad p = 0, 1, \ldots, P-1, \tag{12.79}$$

where the power loading on N_t eigenbeams is fixed to δ_μ.

In a nutshell, we can view the matrix $\boldsymbol{U}_c$ in $\boldsymbol{C}$ alternatively as a beamforming matrix containing N_t basis directions and $\boldsymbol{\Delta}$ as the power-loading matrix across different beams. With mean or covariance feedback, the basis beams are directed along the corresponding channel eigenvectors [cf. (12.79)]. This viewpoint is useful and will be appreciated in the next section, where beamforming is coupled with ST coding in order to improve the spectral efficiency of the low-rate ST spreading scheme we dealt with in this section.

12.3 COMBINING OSTBC WITH BEAMFORMING

As we mentioned in Section 12.2, the ST spreading scheme of (12.14) effects a low-rate transmission since only one symbol is transmitted every P time slots (chip periods). In this section we increase bandwidth efficiency through data multiplexing [i.e., by transmitting multiple symbols, say $(s_1, \ldots, s_K)$, simultaneously]. The rate will then increase from $1/P$ to K/P symbols per channel use. However, error performance with bandwidth efficient multiplexing will be lower bounded by the average SER of the low-rate ST spreading system. Indeed, when detecting one particular symbol s_k, the best scenario happens when all other symbols have been detected correctly and their effect on s_k has been canceled perfectly; in such a case, each symbol can be treated as if it were transmitted separately through an ST spreading scheme.

Despite the rate increase due to multiplexing, our ultimate goal here is to achieve optimal demodulation performance for each multiplexed symbol, as if each symbol were transmitted alone with ST spreading. Interestingly, this can be achieved by combining the OSTBC design in Section 3.3 with the beamforming operation introduced in Section 12.2. Specifically, instead of transmitting the spreading matrix codeword $\boldsymbol{X} = \boldsymbol{U}_c^* \boldsymbol{\Delta}^{\frac{1}{2}} \boldsymbol{\Phi} s$, the combined OSTBC–beamformer system transmits the ST code matrix

$$\boldsymbol{X} = \boldsymbol{U}_c^* \boldsymbol{\Delta}^{\frac{1}{2}} \mathcal{O}_{N_t}, \tag{12.80}$$

where $K = N_s$ symbols are encoded in the OSTBC matrix $\mathcal{O}_{N_t}$ over $P = N_x$ time slots as in (3.15). As a result, the transmission rate is increased from $1/P = 1/N_x$ to

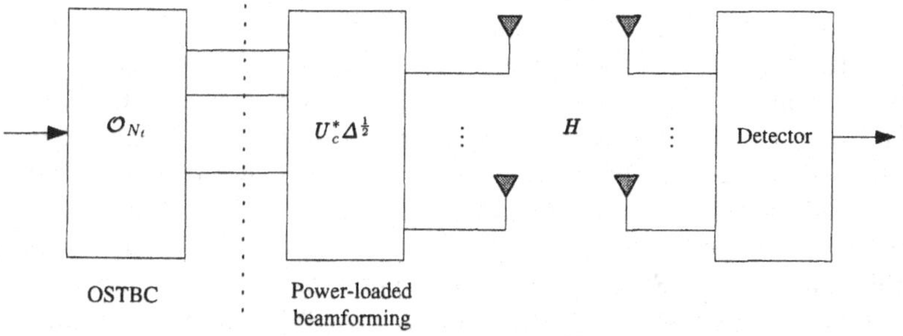

Figure 12.5 Multi-antenna system combining OSTBC with power-loaded beamforming

$K/P = N_s/N_x$, in accordance with the OSTBC transmission rate. [Recall (3.16)–(3.18), where (N_s, N_x) pairs are given as a function of the number of transmit-antennas.]

Let us now consider the received ST matrix

$$Y = \sqrt{\bar{\gamma}}HX + W = \sqrt{\bar{\gamma}}\,\underbrace{HU_c^*\Delta^{\frac{1}{2}}}_{:=H_{\rm eq}}\mathcal{O}_{N_t} + W, \tag{12.81}$$

where we reiterate that $\bar{\gamma}$ denotes the total average transmit-energy in X per information symbol s_k since we operate under the constraint $\mathrm{tr}(\Delta) = 1$. Equation (12.81) reveals that the OSTBC matrix $\mathcal{O}_{N_t}$ now "sees" an equivalent channel $H_{\rm eq}$ as depicted in Figure 12.5. Since OSTBC matrices are orthogonal by design, each symbol is equivalently passing through a scalar channel with input-output relationship

$$\hat{s}_k = \sqrt{\bar{\gamma}}\,\|HU_c^*\Delta^{\frac{1}{2}}\|_F s_k + w_k. \tag{12.82}$$

The instantaneous SNR in (12.82) is

$$\begin{aligned}\gamma_k &= \bar{\gamma}\,\|HU_c^*\Delta^{\frac{1}{2}}\|_F^2 \\ &= \bar{\gamma}\,h^{\mathcal{H}}(I_{N_r} \otimes U_c\Delta U_c^{\mathcal{H}})h.\end{aligned} \tag{12.83}$$

For a given pair of (U_c, Δ) matrices, the SNR in (12.83) coincides with that of a single symbol transmission as in (12.19) and (12.39). This implies that:

- The average SER of the OSTBC-beamformer system is the same as that of the low-rate system, which is based on ST spreading.
- Matrices U_c and Δ optimizing the average SER metrics for the OSTBC-beamformer system are identical to those found for the ST spreading system.

The idea of combining OSTBC with beamforming was introduced in [132] and [330]. With only linear increase in complexity, this combination optimizes the average SER while increasing the transmission rate relative to ST spreading up to 1 symbol per channel use — the limit imposed by the OSTBC designs in Section 3.3. With complex symbols in particular, the transmitter of (12.80) achieves optimal performance with no rate loss only when $N_t = 2$, and pays a rate penalty up to 50% when $N_t > 2$. To make up for this loss, the transmitter has to enlarge the constellation size, which necessitates more transmit-power to attain the same error performance.

At this point, one might be tempted to boost rates by replacing the OSTBC matrix with the ST code matrices we saw in Chapter 4 corresponding to layered (e.g., V-BLAST, D-BLAST, or FDFR) designs. However, we have to bear in mind that symbols multiplexed with these layered transmissions do not orthogonalize the MIMO channel and interfere with each other at the receiver side. This renders it impossible to achieve the average SER of the ST spreading system. In addition, coming up with an even approximate expression for the average SER, which is also amenable to optimization with respect to C, appears to be formidably challenging in the presence of interference.

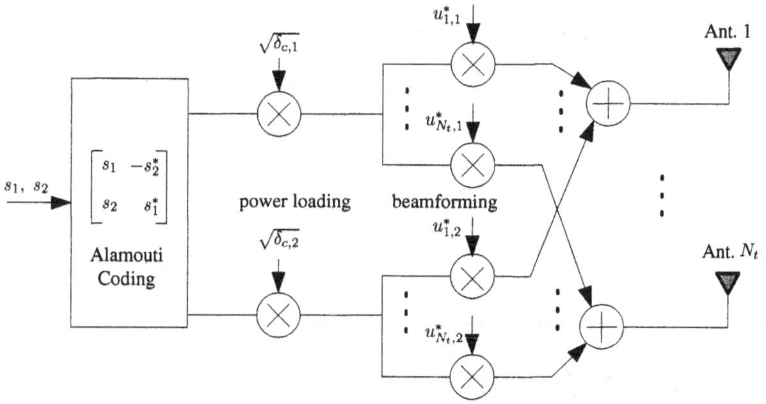

Figure 12.6 Two-dimensional (2D) coder-beamformer, $u_{p,q} := [\boldsymbol{U}_c]_{p,q}$

12.3.1 Two-Dimensional Coder-Beamformer

With the OSTBC-beamforming combination, it is also possible to trade off optimality in error performance for a guaranteed rate at 1 symbol per channel use, even if the number of transmit-antennas exceeds two. This surprising result can be accomplished by transmitting with the Alamouti code along the two strongest beam directions. Specifically, [330, 331] put forth a two-dimensional (2D) coder-beamformer which for any N_t transmits with the following ST code matrix of size $N_t \times 2$:

$$\boldsymbol{X} = \underbrace{\begin{bmatrix} \boldsymbol{u}_{c,1}^*, \boldsymbol{u}_{c,2}^* \end{bmatrix}}_{\boldsymbol{U}_c^*} \underbrace{\begin{bmatrix} \sqrt{\delta_1} & 0 \\ 0 & \sqrt{\delta_2} \end{bmatrix}}_{\boldsymbol{\Delta}^{1/2}} \underbrace{\begin{bmatrix} s_1 & -s_2^* \\ s_2 & s_1^* \end{bmatrix}}_{\mathcal{O}_2}. \tag{12.84}$$

The implementation of (12.84) is depicted in Figure 12.6.

Following the steps in Section 12.2, one can easily prove that:

Proposition 12.3 *When partial CSI available to the transmitter adheres to either the mean- or covariance-feedback model, the optimal beam directions are*

$$\boldsymbol{u}_{c,1} = \boldsymbol{u}_{H,1}, \qquad \boldsymbol{u}_{c,2} = \boldsymbol{u}_{H,2}, \tag{12.85}$$

where $(\boldsymbol{u}_{H,1}, \boldsymbol{u}_{H,2})$ *are the first two eigenvectors of* $\boldsymbol{U}_H$ *in* (12.45) *for mean feedback, or* $\boldsymbol{U}_H$ *in* (12.69) *for covariance feedback.*

The power-loading coefficients can be obtained following the steps in Section 12.2, but with only two eigenbeams. This corresponds to a virtual MIMO system with only two transmit-antennas and N_r receive-antennas. The resulting optimal power-loading solutions for mean and covariance feedback are summarized next.

Case 1: Mean feedback. In 2D beamforming, power is divided between two basis beams (δ_1, δ_2). The optimal solution is then

$$\delta_1 = 1 - \delta_2 \qquad \text{and} \qquad \delta_2 = \begin{cases} \delta_2^0 & \bar{\gamma} > \gamma_{\text{th},2} \\ 0 & \bar{\gamma} \leq \gamma_{\text{th},2}. \end{cases} \tag{12.86}$$

The threshold is obtained from (12.64) with two virtual antennas:

$$\gamma_{\text{th},2} = \frac{1}{g\sigma_\epsilon^2} \frac{(\lambda_1 - \lambda_2)(N_r\sigma_\epsilon^2 + \lambda_1)}{(N_r\sigma_\epsilon^2 + \lambda_2)(N_r\sigma_\epsilon^2 + 2\lambda_1)}, \tag{12.87}$$

where λ_1 and λ_2 are defined in (12.46) and δ_2^0 is obtained from (12.63) as

$$\delta_2^0 := \frac{1 + \dfrac{N_r\sigma_\epsilon^2 + \lambda_1}{(N_r\sigma_\epsilon^2 + 2\lambda_1)\beta} + \dfrac{N_r\sigma_\epsilon^2 + \lambda_2}{(N_r\sigma_\epsilon^2 + 2\lambda_2)\beta}}{1 + \dfrac{(N_r\sigma_\epsilon^2 + 2\lambda_2)(N_r\sigma_\epsilon^2 + \lambda_1)^2}{(N_r\sigma_\epsilon^2 + 2\lambda_1)(N_r\sigma_\epsilon^2 + \lambda_2)^2}} - \frac{N_r\sigma_\epsilon^2 + \lambda_2}{(N_r\sigma_\epsilon^2 + 2\lambda_2)\beta}. \tag{12.88}$$

The solution in (12.86) reduces to (12.66) if $N_t = 2$ and $N_r = 1$ (thus $\lambda_2 = 0$), as expected.

Case 2: Covariance feedback. The optimal 2D power loading is

$$\delta_1 = 1 - \delta_2 \qquad \text{and} \qquad \delta_2 = \begin{cases} \delta_2^0 & \bar{\gamma} > \gamma_{\text{th},2} \\ 0 & \bar{\gamma} \leq \gamma_{\text{th},2}. \end{cases} \tag{12.89}$$

The threshold is found from (12.77) with two virtual antennas as

$$\gamma_{\text{th},2} = \frac{1}{g}\left(\frac{1}{\lambda_2} - \frac{1}{\lambda_1}\right), \tag{12.90}$$

where λ_1 and λ_2 are defined in (12.70). The optimal δ_2 value in (12.89) is

$$\delta_2^0 = \frac{1}{2}\left[1 + \frac{1}{g\bar{\gamma}}\left(\frac{1}{\lambda_1} - \frac{1}{\lambda_2}\right)\right]. \tag{12.91}$$

Since only two beam directions are used in the 2D beamformer, only the first two eigenvalues λ_1 and λ_2 affect the power loading in both cases.

Notice that if we set $\delta_2 = 0$ in the 2D beamformer, (12.84) reduces to

$$\boldsymbol{X} = \left[\boldsymbol{u}_{c,1}^* s_1 \quad -\boldsymbol{u}_{c,1}^* s_2^*\right], \tag{12.92}$$

which corresponds to transmitting information symbols s_1 and $-s_2^*$ in two consecutive time slots using only one beamforming vector.

The transmission scheme in (12.92) amounts to conventional one-dimensional (1D) beamforming. How the latter compares with the 2D coder-beamformer is addressed in the next corollary, which summarizes pertinent results from [330, 331].

Corollary 12.1 *The 2D coder-beamformer subsumes the 1D beamformer as a special case and outperforms it uniformly, without rate reduction and without essential increase in complexity.*

Besides being more attractive than the conventional 1D beamformer, it is worth recalling that the 2D eigen-beamformer is optimal for systems with $N_t = 2$ transmit-antennas. Its ability to attain transmission rate up to 1 symbol per channel use for any N_t while achieving SER optimality over the practical SNR range at low complexity renders the 2D coder-beamformer an appealing candidate for closed-loop operation of multi-antenna systems with statistical CSI available at the transmitter. To improve the transmission rate further at a prescribed average SER, we consider in Section 12.5 a multi-antenna system that combines adaptive modulation with the 2D coder-beamformer.

12.4 NUMERICAL EXAMPLES

Using simulated examples, we illustrate in this section the performance improvement due to mean and covariance feedback for the ST spreading system in Section 12.2 and the combined OSTBC-beamforming system in Section 12.3. The average SNR in all plots refers to $\bar{\gamma}$ in decibels.

12.4.1 Performance with Mean Feedback

Here we consider the delayed feedback scenario in Example 12.2 with $\sigma_{\hat{h}}^2 = 1$ and correlation coefficient ρ. Two constellations, QPSK (4-PSK) and 16-QAM, are used; and simulated SER curves are averaged over 10,000 Monte Carlo feedback realizations.

Example 12.5 (Comparison between exact and approximate power loadings) We first compare optimal power loading based on the Ricean distribution (12.55) with that based on the Nakagami distribution (12.66). We consider a uniform linear array with $N_t = 4$ antennas at the transmitter and a single antenna at the receiver. Figure 12.7 verifies that for both distributions, the SER curves are almost identical. For this reason, we subsequently plot only the performance of power loading based on (12.66). Figure 12.7 also confirms that the SER bound is tight while exhibiting an approximately constant difference with the exact SER across the SNR range considered. This justifies well the approach in Section 12.2.2 of minimizing the bound to decrease the SER.

Example 12.6 (Comparison results with $N_r = 1$) Here we assume a uniform linear array with $N_t = 4$ antennas at the transmitter and a single antenna at the receiver. Figures 12.8 and 12.9 compare optimal power loading, equal power loading (that has the same performance as plain OSTBC without beamforming), and 1D and 2D beamforming for both QPSK and 16-QAM. When the feedback quality is low ($\rho =$

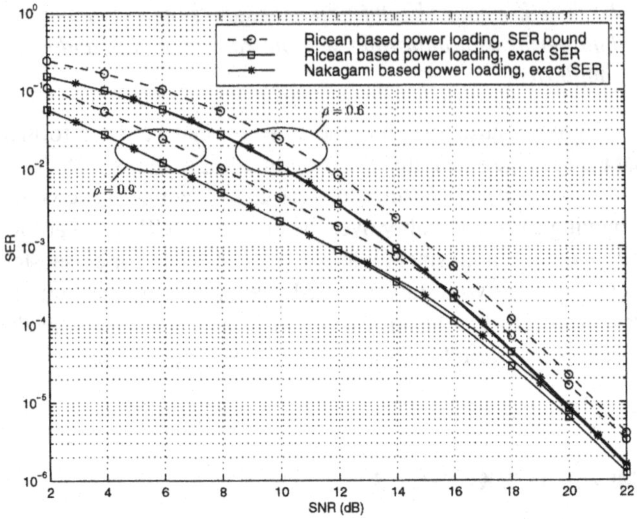

Figure 12.7 Power loading based on Ricean and Nakagami distributions (QPSK)

0.6), Figure 12.8 shows that optimal power loading performs close to equal power loading while it considerably outperforms conventional 1D beamforming. On the other hand, when the feedback quality improves to $\rho = 0.9$, equal power loading is highly suboptimal. The 1D beamforming performs close to the optimal power loading at low SNR, while it becomes inferior at sufficiently high SNRs. Notice that the 2D beamformer outperforms the 1D beamformer uniformly. When $\bar{\gamma} > \gamma_{\text{th}}$ for each feedback realization, although both 2D and 1D beamformers become suboptimal, the 2D beamformer benefits from the second-order diversity. Since $g_{\text{QPSK}}/g_{\text{16QAM}} = 5$, we infer that relative to QPSK, 7.0 dB higher power is required for 16-QAM to adopt N_t directions.

Example 12.7 (Comparisons with $N_r = 2$) We now test mean-feedback performance with multiple receive-antennas. Figures 12.10 and 12.11 are the counterparts of Figures 12.8 and 12.9, but with $N_r = 2$ receive-antennas. It can be seen that the performance of the 2D beamformer coincides with the optimal beamformer for a larger range of SNR than that of the 1D beamformer. This is different from the single receive-antenna case, where 2D and 1D beamformers deviate from the optimal beamformer since there is only one dominant direction.

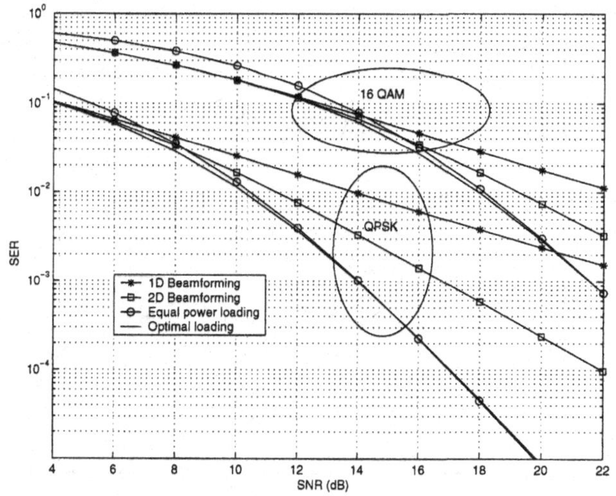

Figure 12.8 SER performance with $\rho = 0.6$, $N_t = 4$, $N_r = 1$

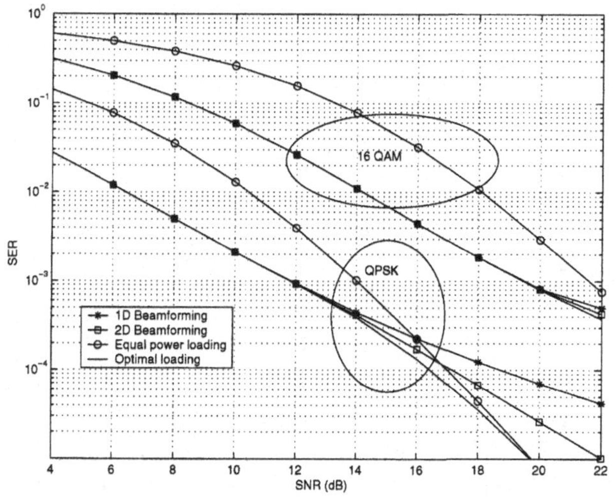

Figure 12.9 SER performance with $\rho = 0.9$, $N_t = 4$, $N_r = 1$

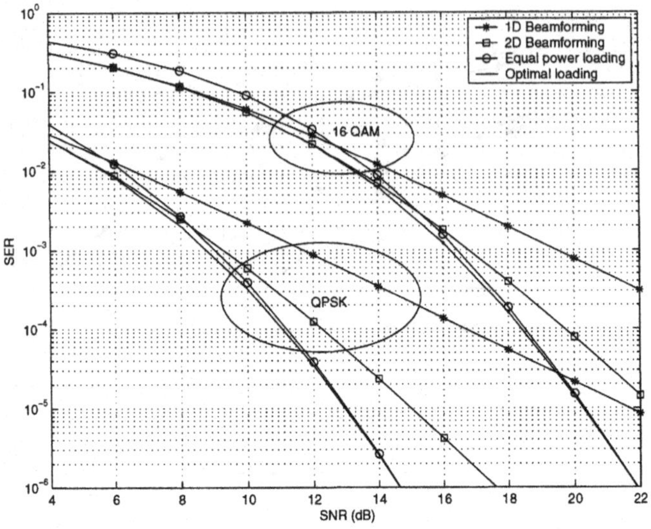

Figure 12.10 SER performance with $\rho = 0.6$, $N_t = 4$, $N_r = 2$

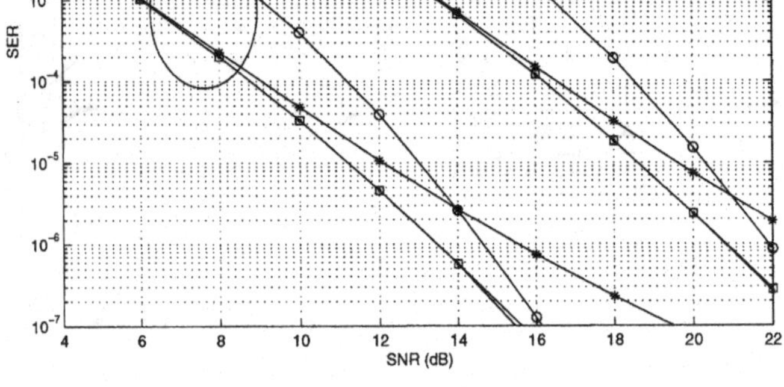

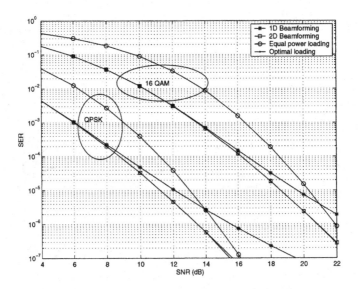

Figure 12.11 SER performance with $\rho = 0.9$, $N_t = 4$, $N_r = 2$

12.4.2 Performance with Covariance Feedback

Let us now consider a uniform linear array with $N_t = 4$ antennas at the transmitter and a single antenna at the receiver. We assume that the side information, including the distance between the transmitter and the receiver, the angle of arrival, and the angle spread are all available to the transmitter. Let λ be the wavelength of a narrowband signal, d_t the antenna spacing, and Δ the angle spread. We suppose further that the angle of arrival is perpendicular to the transmitter antenna array ("broadside" as in [235]). Using the result of [235, Equation (6)] for small angle spread, we can express the correlation coefficient between the pth and the qth transmit-antenna as

$$[\boldsymbol{\Sigma}_t]_{p,q} \approx \frac{1}{2\pi}\int_0^{2\pi} \exp\left[-j2\pi(p-q)\Delta\frac{d_t}{\lambda}\sin\theta\right] d\theta. \tag{12.93}$$

Using (12.93), we will test two channels: Channel 1 has $d_t = 0.5\lambda$ and $\Delta = 5°$; while channel 2 has lower spatial correlations with $d_t = 0.5\lambda$ and $\Delta = 25°$. Plugging these parameters into (12.93), we specify the correlation matrix and its eigenvalues for these two channels as:

Channel 1:

$$\boldsymbol{\Sigma}_t^{(1)} = \begin{bmatrix} 1.00 & 0.98 & 0.93 & 0.84 \\ 0.98 & 1.00 & 0.98 & 0.93 \\ 0.93 & 0.98 & 1.00 & 0.98 \\ 0.84 & 0.93 & 0.98 & 1.00 \end{bmatrix}, \tag{12.94}$$

$$\boldsymbol{\Lambda}^{(1)} = \text{diag}(3.81, 0.18, 0.007, 0.00). \tag{12.95}$$

Channel 2:

$$\boldsymbol{\Sigma}_t^{(2)} = \begin{bmatrix} 1.00 & 0.58 & -0.16 & -0.39 \\ 0.58 & 1.00 & 0.58 & -0.16 \\ -0.16 & 0.58 & 1.00 & 0.58 \\ -0.39 & -0.16 & 0.58 & 1.00 \end{bmatrix}, \tag{12.96}$$

$$\boldsymbol{\Lambda}^{(2)} = \text{diag}(1.79, 1.74, 0.45, 0.02). \tag{12.97}$$

Simple inspection of the eigenvalues in (12.95) and (12.97) reveals that channel 2 is less correlated than channel 1.

We next present simulations for two constellations: QPSK and 16-QAM.

Example 12.8 (Optimal power allocation) Figures 12.12 and 12.13 depict the optimal power allocation among different beams for channels 1 and 2, for both QPSK and QAM constellations. At low SNRs, the transmitter prefers to shut off certain beams, while it approximately equates power to all beams at sufficiently high SNRs to benefit from diversity. Notice that the choice of how many beams are retained depends on the constellation-specific SNR thresholds. For QPSK, we can verify that $\gamma_{\text{th},2} = 10.2$ dB and $\gamma_{\text{th},3} = 37.5$ dB for channel 1, while $\gamma_{\text{th},2} = -15.0$ dB and $\gamma_{\text{th},3} = 8.1$ dB for channel 2. Since $g_{\text{QPSK}}/g_{16\text{QAM}} = 5$, the threshold $\gamma_{\text{th},r}$ for 16-QAM is $10\log_{10}(5) = 7$ dB higher for QPSK; we observe that 7 dB higher power is required for 16-QAM before switching to as many beams are used for QPSK.

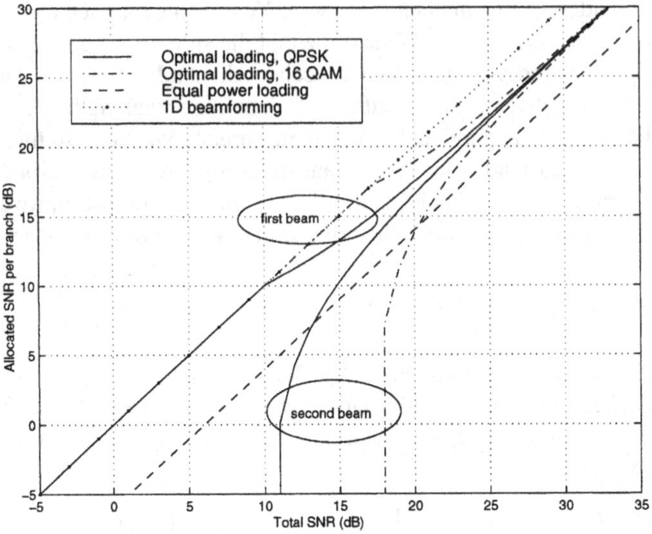

Figure 12.12 Optimal vs. equal power loading: channel 1

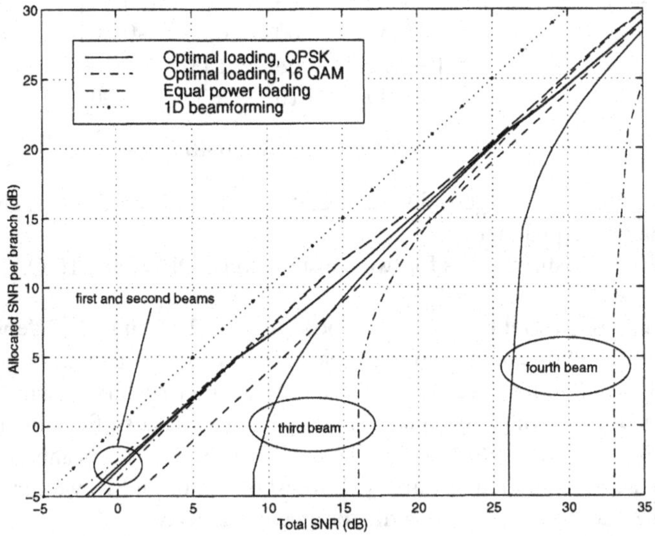

Figure 12.13 Optimal vs. equal power loading: channel 2

Example 12.9 (Error performance for channel 1) With channel 1, Figures 12.14 and 12.15 depict the exact SER along with the SER upper bound for optimal power loading, equal power loading (that has the same performance as plain OSTBC without beamforming), and 1D beamforming. Since channel 1 is highly correlated, only $r = 2$ beams are used for optimal loading in the SNR range considered. Therefore, the 2D eigen-beamformer is overall optimal for channel 1 in the SNR range considered, and its performance curves coincide with those of the optimal loading. Figures 12.14 and 12.15 confirm that the optimal allocation outperforms both the equal power allocation and the 1D beamforming. The difference between optimal loading and equal power loading is about 3 dB as SNR increases, since two out of four beams are so weak that the power allocated to them is wasted. The differences between the upper bound and the exact SER in Figures 12.14 and 12.15 justify the approach based on the SER bound.

Example 12.10 (Error performance for channel 2) Notice that channel 2 is less correlated than channel 1 and all four beams are used at high SNR. Equal power loading approaches the optimal loading when SNR is sufficiently high, but is inferior to both 2D eigen-beamforming and optimal loading in the low–medium SNR range, as confirmed by Figures 12.16 and 12.17. We can also verify that 2D beamforming outperforms 1D beamforming uniformly and that the difference is quite significant at moderate-to-high SNR. Notice that the first two eigenvalues of channel 2 in (12.97) are not disparate enough and the 1D beamformer is optimal only when $\bar{\gamma} \leq \gamma_{\mathrm{th},2} = -8.0$ dB for 16-QAM. On the other hand, the 2D eigen-beamformer achieves optimality up to $\bar{\gamma} = \gamma_{\mathrm{th},3} = 15.1$ dB for 16-QAM, as we verify from Figure 12.17. This observation corroborates the advantage of 2D beamforming relative to 1D beamforming.

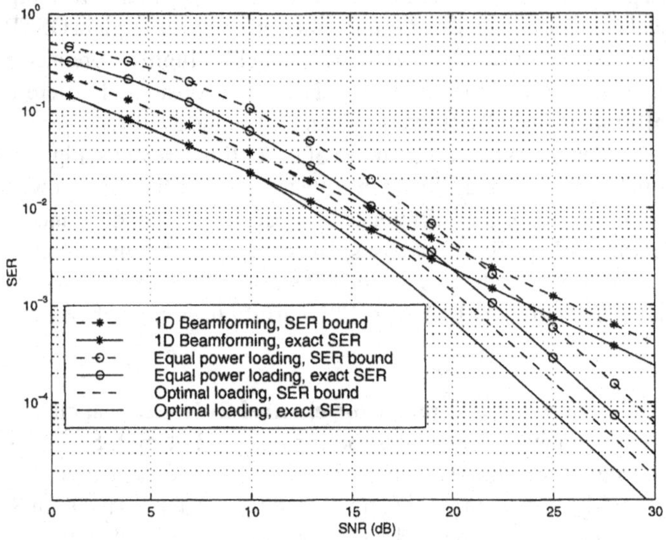

Figure 12.14 SER performance for channel 1, QPSK

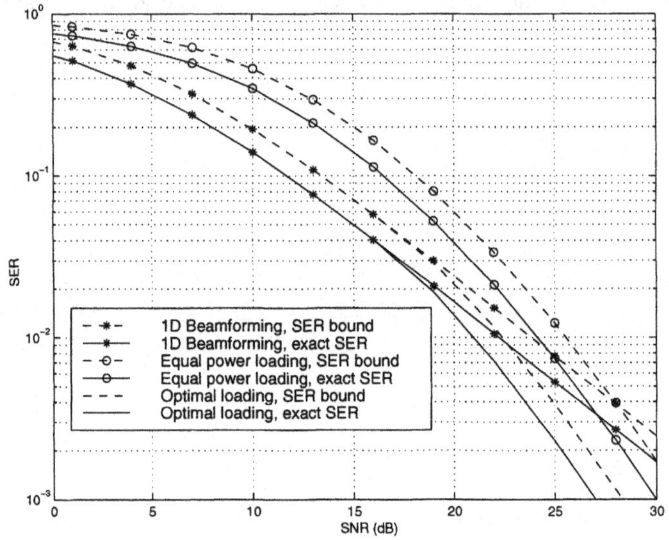

Figure 12.15 SER performance for channel 1, 16-QAM

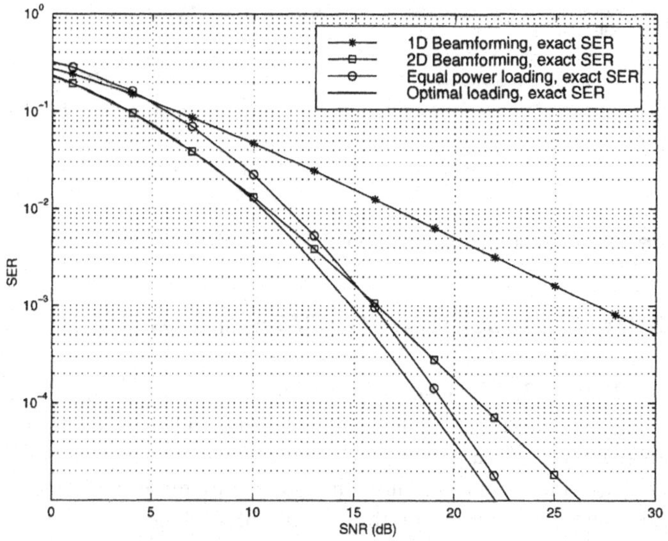

Figure 12.16 SER performance for channel 2, QPSK

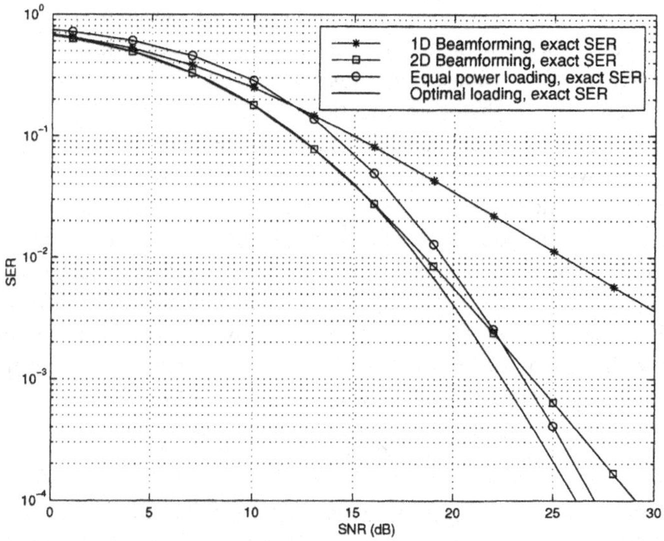

Figure 12.17 SER performance for channel 2, 16-QAM

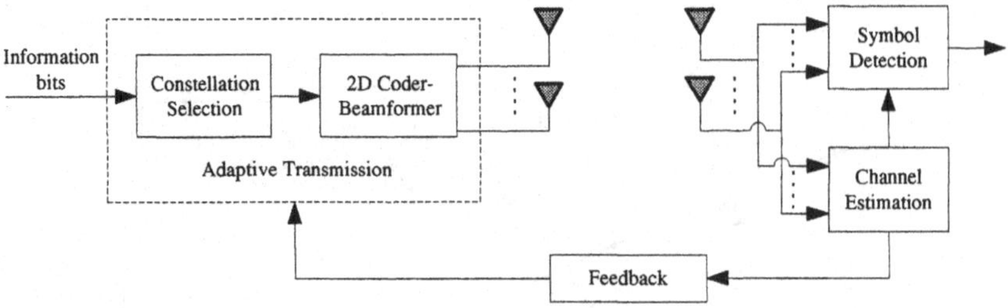

Figure 12.18 Adaptive modulation combined with 2D coding-beamforming

12.5 ADAPTIVE MODULATION FOR RATE IMPROVEMENT

In Sections 12.2 and 12.3, the transmitter is optimized based on partial CSI to improve the average SER performance with a fixed modulation. On the other hand, capacity considerations suggest that a transmitter should adjust its parameters dynamically in accordance with the intended channel in order to increase the transmission rate, subject to a prescribed constraint on the average SER. This design objective is accomplished through the use of adaptive modulation [97]. In this section we rely on adaptive modulation in conjunction with the ST coder-beamformer architecture to attain transmission rates up to 1 symbol per channel use.

Let us consider the multi-antenna transmitter depicted in Figure 12.18, which is equipped with an adaptive modulator feeding the 2D coder-beamformer introduced in previous sections [333]. The information bits are mapped to symbols drawn from a suitable constellation which changes depending on the partial CSI available at the transmitter. The symbol stream $\{s(n)\}$ is then fed to the 2D coder-beamformer and transmitted from the N_t antennas. For a prescribed transmission power, the adjustable transmitter parameters are the beam directions ($\boldsymbol{u}_{c,1}$ and $\boldsymbol{u}_{c,2}$), the power allocation (δ_1 and δ_2), and the size of the constellation M. The objective is to maximize the transmission rate while maintaining a target average BER denoted by $P_{b,\text{target}}$.

Since the constellations are constantly adjusted in adaptive modulation, we need to use the average BER to assess error performance instead of the average SER we relied on in Sections 12.2 and 12.3. Assuming Gray mapping of information bits, the average BER can be expressed in closed form for the general class of QAM constellations. But to facilitate adaptive modulation, we rely on the following approximate BER expression, which is simple enough to be used for optimally selecting transmitter parameters and can be applied to all constellations [97, 333]:

$$P_b \approx 0.2 \exp(-g\gamma), \tag{12.98}$$

where g is the constellation-specific constant taking the specific forms listed in (12.27)–(12.30).

Example 12.11 (BER approximation for QAM) In Figure 12.19, we compare the exact BERs reported in [278] against the approximate BERs of (12.98) for QAM

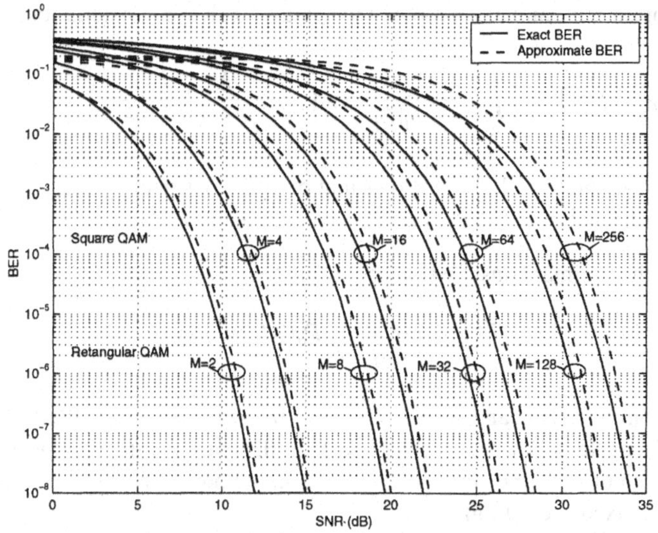

Figure 12.19 BER approximation for M-QAM constellations

constellations with $M = 2^i, i \in [1, 8]$ (BPSK is clearly a special case of rectangular QAM with $M = 2$). Figure 12.19 confirms that the approximation is within 2 dB, for all constellations at $P_b \leq 10^{-2}$.

We next describe the steps of the transmitter adaptation process with channel mean feedback, based on the closed-form solution obtained in Section 12.3.1. Suppose that the transmitter can choose among N different QAM constellations with sizes

$$M_i = 2^i, \qquad i = 1, 2, \ldots N. \tag{12.99}$$

For each constellation M_i, we denote the constant g as g_i. The value of g_i is evaluated from (12.29) or (12.30), depending on the constellation M_i. When the channel experiences deep fades, we allow our adaptive design to suspend data transmission by setting $M_0 = 0$.

Transmitter adaptation to mean feedback

S1) Collect the mean-feedback parameters $(\bar{\boldsymbol{h}}, \sigma_\epsilon^2)$ and the operating SNR $\bar{\gamma}$.

S2) Fix the two beamforming vectors $\boldsymbol{u}_{c,1}$ and $\boldsymbol{u}_{c,2}$ as in (12.85).

S3) For each constellation M_i, determine the optimal power-splitting factors δ_1 and δ_2 based on (12.86)–(12.88). Notice that the constant β in (12.88) is now replaced by

$$\beta_i = g_i \sigma_\epsilon^2 \bar{\gamma}. \tag{12.100}$$

With optimal δ_1 and δ_2, compute the average BER for constellation M_i as

$$P_b(M_i) \approx 0.2 \prod_{\mu=1}^{2} \left[\frac{1}{1+\delta_\mu \beta_i} \exp\left(-\frac{\lambda_\mu \delta_\mu \beta_i}{N_r \sigma_\epsilon^2 (1+\delta_\mu \beta_i)} \right) \right]^{N_r}. \tag{12.101}$$

The derivation of (12.101) follows that of (12.50), since $P_{s,\text{bound}}(\gamma)$ in (12.34) is proportional to $P_b(\gamma)$ in (12.98) and $\delta_\mu = 0, \forall \mu > 2$.

S4) Select the optimal constellation according to

$$M = \underset{M \in \{M_i\}_{i=0}^{N}}{\arg\max} \; P_b(M) \leq P_{b,\text{target}}. \tag{12.102}$$

Equation (12.102) can be solved simply by trial and error: We start with the largest constellation $M_i = M_N$ and then decrease i until we find the optimal M_i. With the optimal constellation chosen, the power-splitting factors in step 3 are then fixed accordingly.

In this adaptation procedure, although $\bar{\boldsymbol{h}}$ has $N_t N_r$ entries, the constellation selection depends only on two eigenvalues, λ_1 and λ_2. We can thus partition the 2D space of (λ_1, λ_2) into $N+1$ disjoint fading regions $\{D_i\}_{i=0}^{N}$, each associated with one constellation. Specifically, we choose

$$M = M_i \qquad \text{when } (\lambda_1, \lambda_2) \in D_i, \; \forall i = 0, 1, \ldots, N. \tag{12.103}$$

The rate achieved by the adaptive system in Figure 12.18 is therefore

$$R = \sum_{i=1}^{N} \log_2(M_i) \iint_{D_i} p(\lambda_1, \lambda_2)\, d\lambda_1\, d\lambda_2, \tag{12.104}$$

where $p(\lambda_1, \lambda_2)$ is the joint pdf of λ_1 and λ_2. The outage probability, which corresponds to the event of no data transmission, is clearly

$$P_{\text{out}} = \iint_{D_0} p(\lambda_1, \lambda_2)\, d\lambda_1\, d\lambda_2. \tag{12.105}$$

In general, Monte Carlo simulations are needed to evaluate the rate in (12.104) and the outage probability in (12.105). Analytical solutions are available only for special cases, one of which we present in the next example.

Example 12.12 Consider a system with multiple transmit-antennas and a single receive-antenna ($N_r = 1$). For this setup we have $\lambda_2 = 0$. The fading region D_i reduces to an interval of the real line denoted as $D_i = [\alpha_i, \alpha_{i+1})$. When $\lambda_1 \in D_i$, the target BER is met with the constellation having size M_i.

Suppose that the true channel $\boldsymbol{h}$ is distributed as $\mathcal{CN}(\boldsymbol{0}, \boldsymbol{I}_{N_t})$. For the delayed-feedback partial CSI model considered in Example 12.2, we have

$$\lambda_1 = (|\rho|^2)\,\|\boldsymbol{h}\|^2 = |\rho|^2 \sum_{\mu=1}^{N_t} |h_{\mu 1}|^2, \tag{12.106}$$

which is Gamma distributed with parameter N_t and mean $E\{\lambda_1\} = |\rho|^2 N_t$. In this case, the cumulative distribution function (cdf) of λ_1 is [236]

$$F(x) = \int_0^x p(\lambda_1)\,d\lambda_1 = 1 - e^{-\frac{x}{|\rho|^2}} \sum_{j=0}^{N_t-1} \frac{1}{j!}\left(\frac{x}{|\rho|^2}\right)^j, \quad x \geq 0. \tag{12.107}$$

Substituting (12.107) into (12.104) and (12.105), we obtain the rate [cf. (12.104)]

$$R = \sum_{i=1}^{N} \log_2(M_i)[F(\alpha_{i+1}) - F(\alpha_i)] \tag{12.108}$$

and the outage probability [cf. (12.105)]

$$P_{\text{out}} = F(\alpha_1). \tag{12.109}$$

The closed-loop multi-antenna system of this section uses adaptive modulation in conjunction with ST coding-beamforming over frequency-flat MIMO channels. Extension to adaptive MIMO-OFDM systems for frequency-selective MIMO channels may be found in [308], where joint power and bit loadings across OFDM subcarriers are combined with adaptive 2D coding-beamforming per subcarrier.

12.5.1 Numerical Examples

Here we adopt the channel setup of Example 12.2 with $\sigma_h^2 = 1$ and set $P_{b,\text{target}} = 10^{-3}$. Recall that the feedback quality σ_ϵ^2 is captured by the correlation coefficient $\rho = J_0(2\pi f_{\max}\tau)$ via $\sigma_\epsilon^2 = 1 - |\rho|^2$. In all plots of this subsection we use

$$\text{average SNR} := (1 - P_{\text{out}})\bar{\gamma} \tag{12.110}$$

to accommodate variable probabilities of no-transmission for different setups.

Example 12.13 (Distance from capacity) Figure 12.20 plots the rate achieved by the adaptive transmitter with $N_t = 2$, $N_r = 1$, and $\rho = 1, 0.95, 0.9, 0.8, 0$. It is clear that the rate decreases fast as the feedback quality drops.

For comparison, we also plot the average channel capacity with mean feedback, as summarized in Proposition 12.4. Figure 12.20 shows that the capacity is less sensitive to channel imperfections. The capacity with perfect CSI is larger than the capacity with no CSI by about $\log_2(N_t) = 1$ bit at high SNR, as predicted in

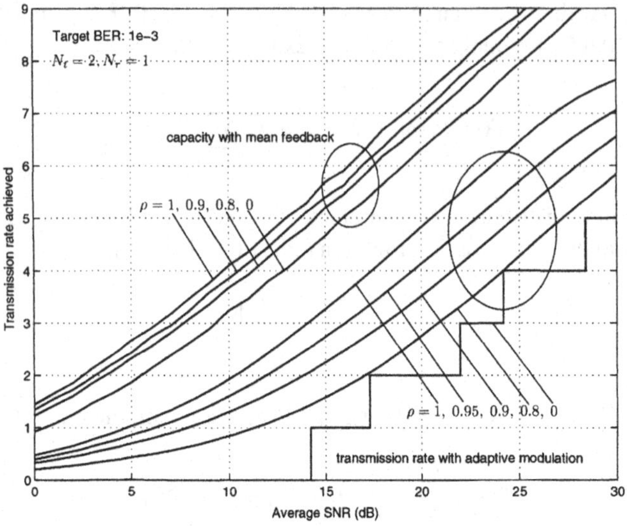

Figure 12.20 Transmission rates for variable feedback quality

[256]. With $\rho = 0.9$, the adaptive uncoded modulation is about 11 dB away from the average capacity curve. The gap can be decreased if adaptive modulation is used in conjunction with channel coding.

Example 12.14 (Rate improvement with multiple transmit-antennas) Here we test a system with a variable number of transmit-antennas, a single receive-antenna ($N_r = 1$), and correlation coefficient $\rho = 0.9$. Figure 12.21 demonstrates clearly that the transmission rate achieved increases as the number of transmit-antennas increases. The largest rate improvement occurs when N_t increases from one to two.

Example 12.15 (Trade-offs between feedback quality and hardware complexity) For single-antenna systems equipped with adaptive modulation, it has been argued that the CSI can be viewed as perfect when $f_{\max}\tau \leq 0.01$ ($\rho = 0.999$), where τ denotes the feedback delay [7]. In Figure 12.22 we verify that with two transmit-antennas, the rate achieved with $f_{\max}\tau = 0.1$ ($\rho = 0.904$) coincides with that attained when we deploy one transmit-antenna with perfect CSI. The rate with $N_t = 4$ and $f_{\max}\tau = 0.16$ ($\rho = 0.76$) is even better than that of $N_t = 1$ with perfect CSI. To achieve the same rate, the delay constraint imposed when we use a single-antenna can be relaxed if we deploy more transmit-antennas. This presents an interesting trade-off between feedback quality and hardware complexity. For example, to achieve the same rate with the same feedback delay τ, the system with two transmit-antennas and a single receive-antenna can accommodate a user that moves ten times faster than that in a single-antenna system. Figure 12.22 also reveals that the adaptive design becomes less sensitive to CSI imperfections when the number of transmit-antennas increases.

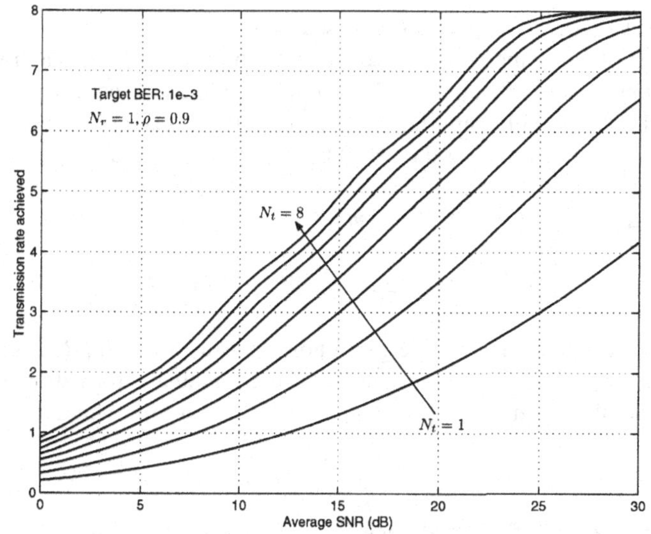

Figure 12.21 Rate improvement with multiple transmit-antennas

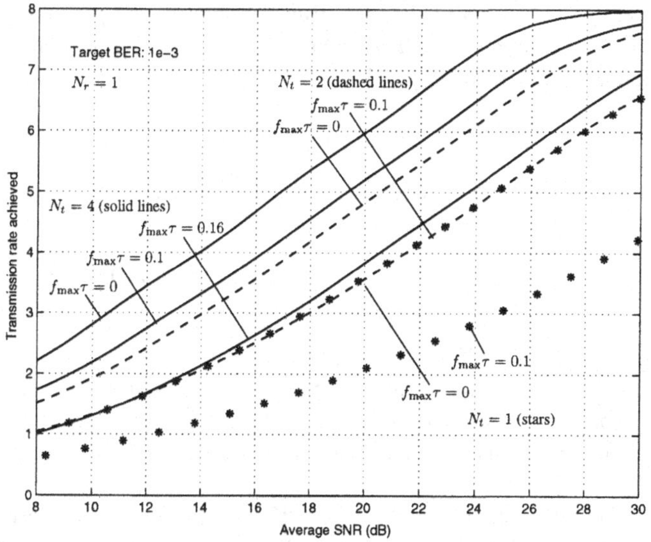

Figure 12.22 Trading-off feedback delay for hardware complexity

12.6 OPTIMIZING AVERAGE CAPACITY

So far, we have focused on partial CSI based transmitter designs optimizing average error rate performance metrics of closed-loop multi-antenna systems at transmission rates up to 1 symbol per channel use. Transmitter optimization based on information-theoretic criteria is equally important and has been studied extensively in the literature.

For a given realization $\boldsymbol{H}$ of the MIMO channel and Gaussian zero-mean channel input with covariance matrix $\boldsymbol{R}_x := E\{\boldsymbol{x}\boldsymbol{x}^{\mathcal{H}}\}$, the mutual information between the input $\boldsymbol{x}$ and the output $\boldsymbol{y} = \sqrt{\bar{\gamma}}\,\boldsymbol{H}\boldsymbol{x} + \boldsymbol{w}$ is

$$I(\boldsymbol{y};\boldsymbol{x}|\boldsymbol{H}) = \log_2 \det\left(\boldsymbol{I}_{N_r} + \bar{\gamma}\boldsymbol{H}\boldsymbol{R}_x\boldsymbol{H}^{\mathcal{H}}\right), \tag{12.111}$$

where $\bar{\gamma}$ is the average transmit-energy-to-noise-ratio as we set $\text{tr}\{\boldsymbol{R}_x\} = 1$. Aiming to select the optimal $\boldsymbol{R}_x$ that maximizes the average mutual information in (12.111), let us consider the eigen-decomposition of the Hermitian correlation matrix

$$\boldsymbol{R}_x = \boldsymbol{U}_x^*\boldsymbol{\Delta}_x\boldsymbol{U}_x^T, \tag{12.112}$$

where $\boldsymbol{U}_x$ contains the eigenvectors, and

$$\boldsymbol{\Delta}_x := \text{diag}(\delta_{x,1}, \ldots, \delta_{x,N_t}) \tag{12.113}$$

contains the non-negative eigenvalues.

With no CSI available at the transmitter, we saw in Chapter 2 that the input covariance $\boldsymbol{R}_x = (1/N_t)\boldsymbol{I}_{N_t}$ is optimal in the sense of maximizing the average (a.k.a. *ergodic*) capacity of the MIMO channel. But having available partial CSI modeled as in (12.2), the transmitter views $\boldsymbol{H}$ as a realization of the random channel $\breve{\boldsymbol{H}}$, which can be written in vector form as [cf. (12.16)]

$$\breve{\boldsymbol{h}} = \text{vec}(\breve{\boldsymbol{H}}^T). \tag{12.114}$$

Taking expectation over the channel $\breve{\boldsymbol{h}}$, transmitter optimization can be carried out based on the average capacity criterion to obtain

$$(\boldsymbol{U}_x, \boldsymbol{\Delta}_x)_{\text{opt}} = \underset{\boldsymbol{U}_x, \boldsymbol{\Delta}_x:\ \text{tr}\{\boldsymbol{\Delta}_x\}\leq 1}{\arg\max} \quad E_{\breve{\boldsymbol{h}}}\{I(\boldsymbol{y};\boldsymbol{x}|\breve{\boldsymbol{H}})\}. \tag{12.115}$$

Substituting the optimal solution from (12.115) into (12.112) and then on to the average of (12.111), we can obtain the maximum $E_{\breve{\boldsymbol{h}}}\{I(\boldsymbol{y};\boldsymbol{x}|\breve{\boldsymbol{H}})\}$, which yields the average capacity conditioned on partial CSI. This average capacity dictates the theoretical upper bound on the transmission rate of any adaptive transmitter based on the same partial CSI knowledge.

It turns out that the analytical solution of (12.115) for the general case is difficult to obtain. In multi-input single-output (MISO) systems, where $N_r = 1$, the optimal $\boldsymbol{R}_x$ for the mean- and covariance-feedback forms of partial CSI have been derived in [273]. The results of [273] can be summarized as follows:

Proposition 12.4 (MISO Mean Feedback) *With mean feedback* $\check{\boldsymbol{h}} \sim \mathcal{CN}(\overline{\boldsymbol{h}}, \sigma_\epsilon^2 \boldsymbol{I}_{N_t})$ *and the eigen-decomposition*

$$\overline{\boldsymbol{h}}\,\overline{\boldsymbol{h}}^{\mathcal{H}} = \boldsymbol{U}_h \,\mathrm{diag}(\|\overline{\boldsymbol{h}}\|, 0, \ldots, 0) \boldsymbol{U}_h^{\mathcal{H}}, \tag{12.116}$$

the optimal input covariance matrix satisfies

$$\boldsymbol{U}_{x,\mathrm{opt}} = \boldsymbol{U}_h, \tag{12.117}$$

$$\delta_{x,2} = \cdots = \delta_{x,N_t} = \delta_o, \qquad \delta_{x,1} = 1 - (N_t - 1)\delta_o, \tag{12.118}$$

where the optimal value δ_o *is found by numerical search.*

Proposition 12.5 (MISO Covariance Feedback) *Given covariance feedback* $\check{\boldsymbol{h}} \sim \mathcal{CN}(\boldsymbol{0}_{N_t \times 1}, \boldsymbol{\Sigma}_h)$ *and the eigen-decomposition*

$$\boldsymbol{\Sigma}_h = \boldsymbol{U}_h \,\mathrm{diag}(\lambda_1, \ldots, \lambda_{N_t}) \boldsymbol{U}_h^{\mathcal{H}}, \tag{12.119}$$

the optimal input covariance matrix satisfies

$$\boldsymbol{U}_{x,\mathrm{opt}} = \boldsymbol{U}_h, \tag{12.120}$$

and the optimal $\boldsymbol{\Delta}_x$ *can be obtained using numerical search.*

The optimization problem in (12.115) has been solved analytically for certain MIMO channels whose covariance matrices possess special structure for the mean- and covariance-feedback scenarios [122, 124, 237]. Conditions under which transmit-beamforming (where $\boldsymbol{\Sigma}_h$ has rank 1) is optimal in the sense of maximizing average capacity were provided in [22, 98, 121]. Besides average capacity, optimization of multi-antenna transmissions has also been carried out with respect to the outage capacity, and pertinent results can be found in [200, 237].

12.7 CLOSING COMMENTS

In this chapter we presented ST transmitter designs based on partial CSI. To account for the uncertainty about the true channel, partial CSI was modeled as a multivariate complex Gaussian vector, which represents the rendition of the true channel as perceived by the transmitter. The mean or covariance matrix of this virtual channel used to design multi-antenna systems can become available to the transmitter either through feedback from the receiver, during the reverse link (in TDD and FDD operation), or even a priori through sounding experiments of the propagation environment. When feedback is used, the updates are presumed to be infrequent in this statistical CSI model, which is more suitable for a fixed wireless MIMO system. In this setup the partial CSI parameters can be made available to the transmitter even beforehand without being necessary to rely on feedback.

Using the mean or the covariance matrix of the partial CSI model, we first optimized a low-rate spread-spectrum system which relies on transmit-beamforming and evaluated its average symbol error rate. Optimization relied on a unified SER upper bound and optimal beamformers were obtained analytically in approximate and, in some special cases, exact closed forms. Transmit-beamformers were further combined serially with OSTBCs to improve data rates based on partial CSI. From the suite of these coder-beamformers, the "hotspot combination" emerges when Alamouti's code is cascaded with transmit-beamforming. The resulting 2D coder-beamformer offers a number of attractive features: It outperforms conventional beamforming uniformly across the SNR range; it is SER optimal over the practical SNR range; it can afford low-complexity decoding; and it attains transmission rates up to 1 symbol per channel use for any number of transmit-antennas. We saw further that OSTBC-beamforming combinations facilitate adaptive modulation to be used in partial CSI based multi-antenna systems. These adaptive ST-coded systems improve the transmission rate, or for the same rate, they can afford an order-of-magnitude higher mobility, as we verified with a numerical example.

We finally outlined how the capacity of multi-antenna systems can be optimized when statistical CSI is available at the transmitter. This study can delineate achievable-rate regions over which transmit-beamforming alone is capacity optimal as well as regions over which ST multiplexing is also required. Results are available in analytical form for the mean- or covariance-only forms of the partial CSI model. Average error performance as well as average or outage capacity expressions for the partial CSI model with a non-zero mean vector *and* general covariance matrices remain largely open problems, especially in forms amenable to optimizing the design of multi-antenna transmitters. Other interesting directions include the use of duality detailed in Chapter 9 to map MIMO OFDM coder-beamformer designs [308] to their counterparts for time- and doubly selective MIMO channels, as well as FDFR-based coder-beamformer combinations based on mean or covariance feedback.

Although we have seen applications where feedback is not necessary, use of the feedback channel offers a pragmatic means for the transmitter to acquire partial CSI when dealing with mobile wireless systems. Updates can then be frequent, but they should always adhere to the bandwidth constraints of the (typically low-rate) feedback link. For such cases, this chapter's mean- or covariance-based model of exploiting partial CSI at the transmitter is not as appealing. However, the finite-rate CSI mode of conveying to the transmitter partial knowledge about the MIMO channel with a few feedback bits is practically feasible. Optimization of ST transmitters and their performance analysis based on finite-rate feedback forms of partial CSI is the theme of the next chapter.

13

ST Codes with Partial Channel Knowledge: Finite-Rate CSI

In Chapter 12, ST coded multi-antenna systems were optimized based on partial channel knowledge, which is available to the transmitter in the form of statistical CSI. The latter may be acquired through feedback from the receiver in a closed-loop system or a priori from the physics of the MIMO channel in an open-loop configuration. In this chapter, closed-loop designs are presented where partial CSI becomes available to the transmitter only through feedback from the receiver in the form of a finite number of bits. Since the feedback channel is subject to a (typically, stringent) bandwidth constraint, this type of partial CSI is referred to as *finite-rate CSI*.

Designing a multi-antenna system based on finite-rate CSI amounts to deciding how a MIMO channel can be described "best" with a prescribed number of (typically, few) bits and what criteria can be adopted to optimize the design based on these bits. The first question will be cast as a vector quantizer design problem using distance metrics which are unique to MIMO communications. With regard to the second question, optimality criteria will include error probability, average capacity, or transmission rate when adaptive modulation is used. Whatever the criterion, the optimal designs will depend on the structure of the multi-antenna transmitter. Using error probability criteria, we optimize multi-antenna systems equipped with transmit-beamforming, or precoded ST multiplexing, as well as systems in which precoding is combined serially with an orthogonal ST block code (OSTBC).

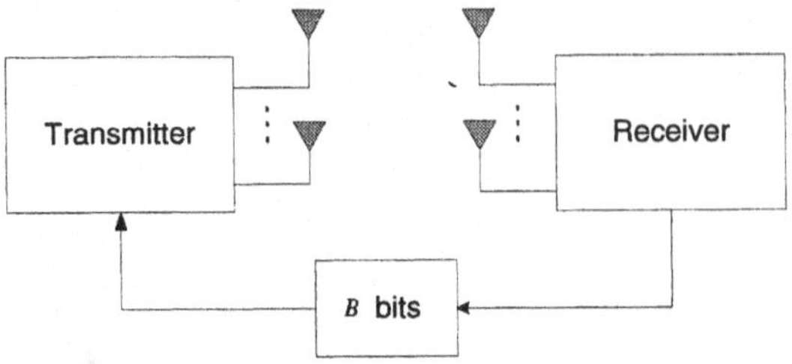

Figure 13.1 Closed-loop MIMO system with finite-rate feedback

13.1 GENERAL PROBLEM FORMULATION

Consider a closed-loop MIMO system where the receiver is allowed to feed B bits back to the transmitter, as depicted in Figure 13.1. To isolate the impact of the bandwidth constraint on the feedback link from other sources of feedback imperfection, such as feedback error and delay, we assume that:

A13.1) The finite-rate feedback link is error- and delay-free.

With B bits, the receiver can feed back up to $N = 2^B$ different messages to inform the transmitter about N possible states of the random channel matrix $\boldsymbol{H}$, which belongs to an $N_t N_r$-dimensional space. This suggests partitioning the channel space into N disjoint regions $\{\mathcal{R}_i\}_{i=1}^N$ and using as B-bit message the *index* of the region into which the current channel realization falls. In vector quantization terms, each $\mathcal{R}_i$ is a quantization region (or class) represented by the codeword $\boldsymbol{H}_{\mathcal{R}_i}$, which could be the centroid of $\mathcal{R}_i$ for example. The set $\{\boldsymbol{H}_{\mathcal{R}_i}\}_{i=1}^N$ of region representatives comprises the channel codebook.

Each time the transmitter receives one of these B-bit messages, it adapts to one of N transmission modes, where each transmission mode specifies parameters such as transmit power, modulation type, and ST coding format. Intuitively speaking, based on only N possible choices for the partially known MIMO channel, the transmitter adapts to the representative of the corresponding channel region indexed by the B-bit message fed back from the receiver. If we have a rule to select a unique transmission mode for each channel, we clearly have a one-to-one mapping between channel codewords and transmission-mode codewords. Since our goal is to optimize multi-antenna transmissions based on finite-rate feedback, we will let $\mathcal{M}_i$ denote the representative of the ith transmission mode and work with the codebook of transmission modes:

$$\{\mathcal{M}_1, \mathcal{M}_2, \ldots, \mathcal{M}_N\}, \tag{13.1}$$

bearing in mind that the latter corresponds one-to-one with the channel codebook $\{\boldsymbol{H}_{\mathcal{R}_i}\}_{i=1}^N$.

The codebook $\{\mathcal{M}_i\}_{i=1}^N$ is constructed *offline* and is available to both the transmitter and the receiver. In *online* operation, the receiver first forms an estimate $\hat{\boldsymbol{H}}$ of the current channel realization $\boldsymbol{H}$. By comparing $\hat{\boldsymbol{H}}$ with all $N = 2^B$ channel codewords $\{\boldsymbol{H}_{\mathcal{R}_i}\}_{i=1}^N$, the receiver selects the "closest" one, call it $\boldsymbol{H}_{\mathcal{R}_{\text{opt}}}$, and feeds back the B-bit index of the codeword $\mathcal{M}_{\text{opt}}$ for the transmitter to switch to the corresponding mode from the transmission-mode codebook $\{\mathcal{M}_i\}_{i=1}^N$. Recall that in the mean-feedback system of Chapter 12, for example, a quantized version of $\hat{\boldsymbol{H}}$ is fed back to the transmitter to form the nominal $\bar{\boldsymbol{H}}$. Here, only the binary index of the transmission-mode codeword $\mathcal{M}_{\text{opt}}$ that matches $\hat{\boldsymbol{H}}$ best is fed back.

For a MIMO system with finite-rate feedback, naturally the following two important issues need to be addressed:

- *Transmission-mode selection*: Given a channel, how do we select uniquely the mode a multi-antenna transmitter can adapt to optimally? And which are the pertinent optimality criteria?

- *Codebook construction*: Given the distribution of the random MIMO channel, how can we construct an optimal channel (and thus transmission mode) codebook? What are proper performance metrics?

For the transmission-mode selection, we need to specify a cost function that could be related to the capacity, error, or transmission rate of the MIMO system. If $f(\boldsymbol{H}, \mathcal{M})$ denotes such a cost function, which should clearly depend on the candidate transmission mode $\mathcal{M}$ and the current channel realization $\boldsymbol{H}$, the optimal mode is selected as

$$\mathcal{M}_{\text{opt}} = \underset{\mathcal{M} \in \{\mathcal{M}_i\}_{i=1}^N}{\arg\max}\; f(\boldsymbol{H}, \mathcal{M}), \tag{13.2}$$

where we assume that maximizing $f(\boldsymbol{H}, \mathcal{M})$ improves the system performance. Mode selection is performed online at the receiver, often by searching exhaustively over the transmission-mode codebook that has been designed offline.

When the estimated channel $\hat{\boldsymbol{H}}$ falls into a region $\mathcal{R}_i$, the corresponding transmission mode $\mathcal{M}_i$ is selected via (13.2) after replacing $\boldsymbol{H}$ with $\boldsymbol{H}_{\mathcal{R}_i}$; and the B-bit index of the optimal transmission mode (corresponding to the "closest" channel) codeword is fed back to the transmitter.[1]

As mentioned earlier, codebook construction is naturally linked to the design of a vector quantizer [86], here with input $\boldsymbol{H}$ and output one of the codewords $\{\mathcal{M}_i\}_{i=1}^N$. Similar to vector quantization, the channel codebook $\{\boldsymbol{H}_{\mathcal{R}_i}\}_{i=1}^N$ must depend on the channel distribution. However, designing the codebook $\{\mathcal{M}_i\}_{i=1}^N$ that we are ultimately interested in must also take into account the transmission-mode selection rule in (13.2). In fact, by mapping channels to transmission modes, this rule renders explicit specification of the channel regions $\{\mathcal{R}_i\}_{i=1}^N$ and their representatives $\{\boldsymbol{H}_{\mathcal{R}_i}\}_{i=1}^N$ unnecessary, since those have been taken into account [via (13.2)] when designing the transmission-mode codebook $\{\mathcal{M}_i\}_{i=1}^N$.

Another difference with respect to vector quantization lies in the performance metric. The typical one in vector quantization is the Euclidean distance between the

[1] This B-bit index of the channel codeword explains why this partial channel knowledge is referred to as finite-rate CSI.

quantizer's input and output, whereas performance metrics more suitable for MIMO communications with finite-rate feedback should relate to error probability, transmission rate, or mutual information [168]. In the ensuing sections we focus on those metrics minimizing error probability when the transmission rate is fixed (Sections 13.2 to 13.4), or maximizing the transmission rate subject to a target error probability (Section 13.6), or optimizing information-theoretic metrics (Section 13.5). In a practical closed-loop multi-antenna system, only a subset of transmission parameters may be adjustable based on finite-rate CSI. Using error performance as a criterion, we design these parameters optimally for multi-antenna transmitters based on transmit-beamforming, precoded spatial multiplexing, and precoded OSTBC systems.

13.2 FINITE-RATE BEAMFORMING

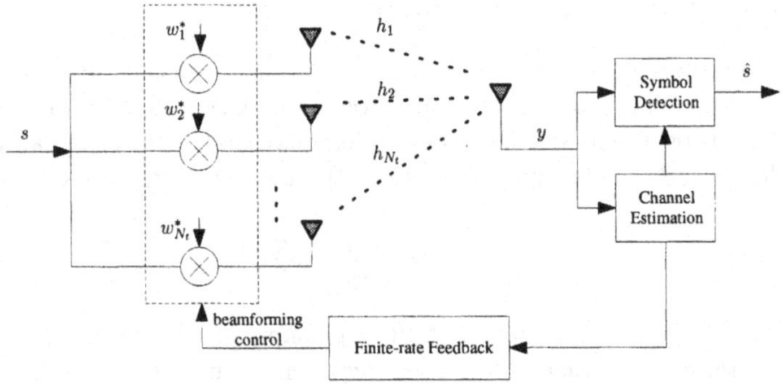

Figure 13.2 Transmit-beamforming MISO system based on finite-rate feedback

In this section we deal with one-dimensional transmit-beamforming in a MISO ($N_r = 1$) setting with finite-rate feedback [169,202,340]. As depicted in Figure 13.2, each information symbol s is multiplied by a beamforming vector $\boldsymbol{w}^*$, where $\boldsymbol{w} := [w_1, w_2, \ldots, w_{N_t}]^T$ has unit norm $\|\boldsymbol{w}\| = 1$. The N_t entries of the vector $\boldsymbol{w}^* s$ are transmitted simultaneously from N_t antennas over the vector flat fading channel $\boldsymbol{h} := [h_1, h_2, \ldots, h_{N_t}]^T$. The sample received can be expressed as

$$y = \sqrt{\bar{\gamma}}\, \boldsymbol{w}^{\mathcal{H}} \boldsymbol{h} s + v, \tag{13.3}$$

where v is AWGN with zero mean and unit variance. Having normalized the symbol and channel energies to 1 (i.e., $E\{|s|^2\} = 1$ and $E\{\|\boldsymbol{h}\|^2\} = 1$), $\bar{\gamma}$ represents the average SNR at the receiver.

With the signal constellation fixed, each transmission mode is characterized by the corresponding beamforming vector $\boldsymbol{w}$. We collect the set of candidate beamforming vectors $\{\boldsymbol{w}_n\}_{n=1}^N$ in the matrix

$$\boldsymbol{W} := [\boldsymbol{w}_1, \boldsymbol{w}_2, \ldots, \boldsymbol{w}_N], \tag{13.4}$$

which forms the codebook of beamforming vectors. In a finite-rate beamforming system, the $N = 2^B$ possible ST transmission mode codewords are clearly

$$\{\boldsymbol{w}_i^* s\}_{i=1}^N. \tag{13.5}$$

13.2.1 Beamformer Selection

Our objective is to improve error performance by adapting the beamforming directions based on finite-rate feedback. Given the codebook $\boldsymbol{W}$, the beamformer selection at the receiver is straightforward. The instantaneous SNR in (13.3) is

$$\gamma = |\boldsymbol{w}^{\mathcal{H}}\boldsymbol{h}|^2\bar{\gamma}, \tag{13.6}$$

where $\boldsymbol{w}$ denotes the beamformer and $\boldsymbol{h}$ the realization of the fading channel. To optimize system performance, the receiver selects the beamforming vector to maximize the instantaneous SNR; that is,

$$\boldsymbol{w}_{\text{opt}} = \underset{\boldsymbol{w}\in\{\boldsymbol{w}_i\}_{i=1}^N}{\arg\max} |\boldsymbol{w}^{\mathcal{H}}\boldsymbol{h}|^2. \tag{13.7}$$

With the beamformer in (13.7) specifying the generic transmission-mode selection rule in (13.2), the instantaneous SNR in (13.6) is

$$\gamma = \max_{1\leq i\leq N} |\boldsymbol{w}_i^{\mathcal{H}}\boldsymbol{h}|^2\bar{\gamma}, \tag{13.8}$$

which shows that SNR-based performance at the receiver depends critically on the design of the beamforming vectors $\{\boldsymbol{w}_i\}_{i=1}^N$. Since the latter are chosen from the codebook $\boldsymbol{W}$, selecting $\boldsymbol{W}$ judiciously plays a major role in the design of the finite-rate beamforming system.

13.2.2 Beamformer Codebook Design

Let us first consider a closed-loop system with infinite-rate feedback ($B = \infty$). For this asymptotic setup where the channel $\boldsymbol{h}$ is perfectly known at the transmitter, it is well known that (see, e.g., [203]):

Proposition 13.1 *If $B = \infty$, the optimal beamforming vector in the sense of maximizing the instantaneous SNR at the receiver is $\boldsymbol{w}_{\text{opt}} = \boldsymbol{h}/\|\boldsymbol{h}\|$.*

To select the beamforming codewords in the finite-rate case, let us suppose for simplicity that the fading channel coefficients are zero-mean complex Gaussian and i.i.d.; that is,

$$\boldsymbol{h} \sim \mathcal{CN}(\mathbf{0}, \boldsymbol{I}_{N_t}). \tag{13.9}$$

Codebook design for correlated channels is treated in [306]. Based on performance metrics such as average receive-SNR [169], outage probability [202], or, symbol error rate [340], a "good" beamformer codebook should minimize the maximum correlation between any pair of beamforming codewords; that is,

$$\boldsymbol{W}_{\text{opt}} = \underset{\boldsymbol{W} \in \mathbb{C}^{N_t \times N}}{\arg\min} \max_{1 \le i < j \le N} |\boldsymbol{w}_i^{\mathcal{H}} \boldsymbol{w}_j|. \tag{13.10}$$

Besides maximum correlation, the codebook constructed as in (13.10) minimizes an upper bound on the receive-SNR loss (and also on the capacity loss) when B is finite relative to the asymptotic case with $B = \infty$ [169].

In [169], this beamformer design problem is further linked to the Grassmannian line-packing problem in [48]. Specifically, entries of $\boldsymbol{w}_i$ are viewed as coordinates of a point on the surface of a hypersphere with unit radius centered at the origin. This point defines a straight line in the complex space $\mathbb{C}^{N_t}$ that passes through the origin. The chordal distance between any two lines generated by $\boldsymbol{w}_i$ and $\boldsymbol{w}_j$ is defined as

$$d_{\mathrm{c}}(\boldsymbol{w}_i, \boldsymbol{w}_j) := \sin\theta_{i,j} = \sqrt{1 - |\boldsymbol{w}_i^{\mathcal{H}} \boldsymbol{w}_j|^2}, \tag{13.11}$$

where $\theta_{i,j}$ denotes the angle between these two lines [48]. The beamformer design in (13.10) is equivalent to a line-packing problem where one wants to pack a finite number of lines through the origin with the minimum chordal distance between any two lines maximized; that is,

$$\boldsymbol{W}_{\text{opt}} = \underset{\boldsymbol{W} \in \mathbb{C}^{N_t \times N}}{\arg\max} \min_{1 \le i < j \le N} d_{\mathrm{c}}(\boldsymbol{w}_i, \boldsymbol{w}_j). \tag{13.12}$$

Having specified the beamformer codebook optimization metric and interpreted it as a line-packing problem, we proceed to find the solution that will yield the desired codebook.

13.2.2.1 Theoretical Lower Bounds For future use, let us define the maximum cross-correlation between beamforming vectors in the codebook $\boldsymbol{W}$ as

$$\rho_{\max}(\boldsymbol{W}) = \max_{1 \le i < j \le N} |\boldsymbol{w}_i^{\mathcal{H}} \boldsymbol{w}_j|. \tag{13.13}$$

The codebook design in (13.10) corresponds to minimizing $\rho_{\max}(\boldsymbol{W})$. A theoretical limit on the minimal $\rho_{\max}(\boldsymbol{W})$ is characterized by Welch's lower bound, specified in the following lemma [295].

Lemma 13.1 *For any codebook* $\boldsymbol{W}$ *with* $N \ge N_t$, *it holds that*

$$\rho_{\max}(\boldsymbol{W}) \ge \sqrt{\frac{N - N_t}{(N-1)N_t}}, \tag{13.14}$$

with the equality satisfied if and only if

$$|\boldsymbol{w}_\ell^{\mathcal{H}} \boldsymbol{w}_{\ell'}| = \sqrt{\frac{N - N_t}{(N-1)N_t}}, \quad \forall \ell \ne \ell'. \tag{13.15}$$

A codebook that meets this Welch bound on $\rho_{\max}(\boldsymbol{W})$ is referred to as a *maximum-Welch-bound-equality* (MWBE) *codebook.* Apparently, if a codebook $\boldsymbol{W}$ is MWBE, it solves the problem in (13.10) automatically.

Unfortunately, Welch's lower bound on $\rho_{\max}(\boldsymbol{W})$ is tight only for relatively small values of N, but becomes quite loose for large N, as we confirm later in Figure 13.3. A composite bound which can be tighter than Welch's bound for certain (N, N_t) pairs may be found in [309] and is given in the next lemma.

Lemma 13.2 *For any $N_t \times N$ codebook $\boldsymbol{W}$, the maximum cross-correlation is lower-bounded by*

$$\rho_{\max}(\boldsymbol{W}) \geq \max\left(\sqrt{\frac{N-N_t}{(N-1)N_t}},\ 1-2N^{-\frac{1}{N_t-1}}\right). \tag{13.16}$$

Obviously, the first half of the composite bound in (13.16) corresponds to the Welch bound. The new lower bound,

$$\rho_{\max}(\boldsymbol{W}) \geq 1-2N^{-\frac{1}{N_t-1}}, \tag{13.17}$$

can be derived using geometric arguments [202].

Having benchmarked $\rho_{\max}$, we proceed to solve the optimization problem in (13.10). The solution is obtained through numerical search for most (N, N_t) pairs and analytically in closed form for certain cases.

13.2.2.2 Codebook Construction in Closed Form

Closed-form constructions of MWBE codebooks are limited. We summarize next some known cases:

- When $N = N_t$, the optimal codebook can be any square unitary matrix. For example, one can set $\boldsymbol{W} = \boldsymbol{I}_{N_t}$. Finite-rate beamforming in this case corresponds to selecting the antenna with the largest channel amplitude. When $N < N_t$, the optimal codebook consists simply of any N columns of a square unitary matrix.

- When $N = N_t + 1$, the optimal codebook corresponds to a simplex signal set, which has $\boldsymbol{w}_i^{\mathcal{H}}\boldsymbol{w}_j = -1/(N-1), \forall i \neq j$. For example, one can take $\boldsymbol{W}$ to be proportional to the first $N-1$ rows of an $N \times N$ FFT matrix.

- When $N = 2N_t = 2^{d+1}$ with d denoting a positive integer, one can construct complex MWBE codebooks based on conference matrices [241]. When $N = 2N_t = p^d + 1$ where p is a prime number and d is a positive integer, one can construct real MWBE codebooks based on conference matrices [241].

- When $N \geq N_t$, MWBE codebooks can be constructed in closed form based on FFT matrices [241, 309]. Specifically, $\boldsymbol{W}$ then contains carefully selected

N_t rows of a scaled $N \times N$ FFT matrix:

$$\boldsymbol{W} = \frac{1}{\sqrt{N_t}} \begin{bmatrix} 1 & e^{j\frac{2\pi}{N}u_1} & \dots & e^{j\frac{2\pi}{N}u_1(N-1)} \\ 1 & e^{j\frac{2\pi}{N}u_2} & \dots & e^{j\frac{2\pi}{N}u_2(N-1)} \\ \vdots & \vdots & \vdots & \vdots \\ 1 & e^{j\frac{2\pi}{N}u_{N_t}} & \dots & e^{j\frac{2\pi}{N}u_{N_t}(N-1)} \end{bmatrix}, \tag{13.18}$$

where the row indices $u_k \in \{0, 1, \dots, N-1\}$, $1 \le k \le N_t$ are distinct (i.e., $u_k \ne u_{k'}$ for $k \ne k'$). Selecting the row indices u_k in this case relies on the notion of difference sets, which have been well studied in combinatorics [240] and are defined as follows.

Definition: A subset $\boldsymbol{u} = \{u_1, \dots, u_{N_t}\}$ of $\{0, 1, \dots, N-1\}$ is called a (N, N_t, λ) *difference set* if the $N_t(N_t-1)$ differences $(u_k - u_\ell) \mod N, k \ne \ell$, take all possible non-zero values $1, 2, \dots, N-1$, with each value taken exactly λ times.

Difference sets are known to exist for certain (N, N_t) pairs; and for those pairs it is possible to prove that [309]:

Lemma 13.3 *The $N_t \times N$ codebook $\boldsymbol{W}$ in* (13.18) *is MWBE if and only if $\boldsymbol{u}$ is an (N, K, λ) difference set.*

In what follows we provide several codebook design examples where $\boldsymbol{W}$ is available in closed form:

Example 13.1 For $N_t = 2$ and $N = 2N_t = 4$, $\boldsymbol{W}$ is constructed in [169] as

$$\boldsymbol{W}^T = \begin{bmatrix} -0.1612 - 0.7348j & -0.5135 - 0.4128j \\ -0.0787 - 0.3192j & -0.2506 + 0.9106j \\ -0.2399 + 0.5985j & -0.7641 - 0.0212j \\ -0.9541 & 0.2996 \end{bmatrix}. \tag{13.19}$$

The maximum correlation achieved is $\max_{1 \le i < j \le N} |\boldsymbol{w}_i^{\mathcal{H}} \boldsymbol{w}_j| = 0.57735$, which reaches Welch's lower bound. The minimum chordal distance in this case is $\min_{1 \le i < j \le N} d_c(\boldsymbol{w}_i, \boldsymbol{w}_j) = \sqrt{0.6713} = \sin(55.02°)$.

Example 13.2 Table 13.1 lists a number of MWBE codebooks from [309] using construction based on difference sets.

13.2.2.3 Codebook Construction via Lloyd's Algorithm Since analytical codebook constructions are limited to certain (N, N_t) pairs, for general N and N_t we have to resort to numerical approaches. As feedback messages consist of bits, we are interested primarily in cases where N is a power of 2.

Good codebooks can be constructed numerically using Lloyd's algorithm, which has been applied extensively and studied thoroughly in the context of vector quantization [86]. We will see how Lloyd's algorithm can be used to obtain the more

Table 13.1 MWBE CODEBOOKS BASED ON DIFFERENCE SETS

N	N_t	FFT Row Indexes $\{u_1, \ldots, u_{N_t}\}$
7	3	$\{1, 2, 4\}$
7	4	$\{0, 3, 5, 6\}$
13	4	$\{0, 1, 3, 9\}$
11	5	$\{1, 3, 4, 5, 9\}$
21	5	$\{3, 6, 7, 12, 14\}$
11	6	$\{0, 2, 6, 7, 8, 10\}$
31	6	$\{1, 5, 11, 24, 25, 27\}$
15	7	$\{0, 1, 2, 4, 5, 8, 10\}$
15	8	$\{3, 6, 7, 9, 11, 12, 13, 14\}$
57	8	$\{1, 6, 7, 9, 19, 38, 42, 49\}$
13	9	$\{2, 4, 5, 6, 7, 8, 10, 11, 12\}$
37	9	$\{1, 7, 9, 10, 12, 16, 26, 33, 34\}$
73	9	$\{1, 2, 4, 8, 16, 32, 37, 55, 64\}$
40	13	$\{0, 1, 3, 5, 9, 15, 22, 25, 26, 27, 34, 35, 38\}$

general class of precoder codebooks in Section 13.3.3.1, from which the construction of beamforming codebooks follows as a special case. Codebooks obtained through the use of Lloyd's algorithm may be found in [305]; codebooks constructed through random search are available in [164].

Example 13.3 For $N_t = 2$ and $N = 8$, the codebook $\boldsymbol{W}$ obtained by using Lloyd's algorithm is [305]

$$\boldsymbol{W}^T = \begin{bmatrix} 0.3918 + 0.4725j & 0.7894 \\ 0.3086 - 0.2204j & 0.9253 \\ -0.2472 + 0.0029j & 0.9690 \\ 0.9932 & -0.0126 - 0.1156j \\ 0.8489 & 0.5065 + 0.1510j \\ 0.7915 & -0.5965 + 0.1328j \\ 0.7123 & -0.3405 - 0.6137j \\ 0.7580 & -0.0739 + 0.6480j \end{bmatrix}. \tag{13.20}$$

The maximum cross-correlation achieved is $\max_{1 \le i < j \le N} |\boldsymbol{w}_i^{\mathcal{H}} \boldsymbol{w}_j| = 0.82161$, which in this case is higher than the composite lower bound (0.75). The minimum chordal distance achieved is $\min_{1 \le i < j \le N} d_c(\boldsymbol{w}_i, \boldsymbol{w}_j) = \sqrt{0.3250} = \sin(34.75°)$.

Example 13.4 In this example we compare the maximum cross-correlation among codewords of available codebooks against the lower bounds in (13.14) and (13.16). For $N_t = 2$ and different values of N, Figure 13.3 plots various bounds on $\rho_{\max}(\boldsymbol{W})$ together with the minimum $\rho_{\max}(\boldsymbol{W})$ for codebooks constructed in closed form or

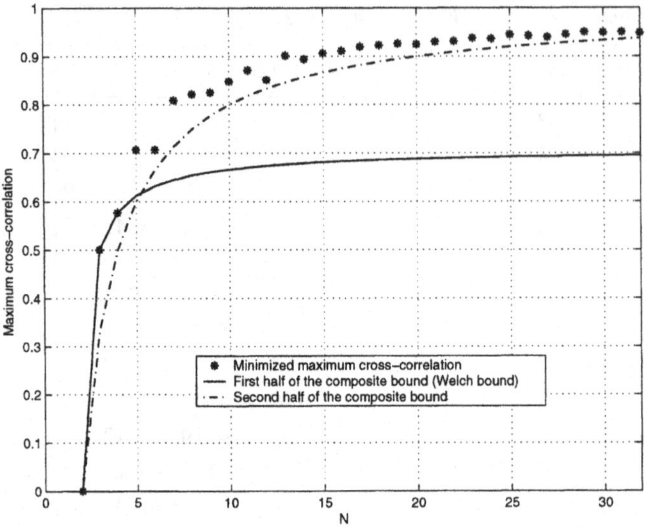

Figure 13.3 Lower bounds and minimum achieved values of $\rho_{\max}(\boldsymbol{W})$ for $N_t = 2$.

through the use of Lloyd's search algorithm [305]. As N increases, the Welch bound becomes looser while the bound in (13.17) gets tighter. The composite bound in (13.16) is effective throughout the range of N. Figure 13.3 demonstrates that most codebooks obtained by numerical search yield $\rho_{\max}(\boldsymbol{W})$ close to the bound, which speaks for the efficiency of Lloyd's algorithm.

13.2.3 Quantifying the Power Loss

How many feedback bits are needed to achieve a decent performance (quantified by the receive-SNR) with transmit-beamforming based on finite-rate CSI? To address this question, we need to evaluate the receive-SNR with finite-rate feedback. Although in this section the SNR evaluation assumes symbols s drawn from a PSK constellation, similar derivations can be carried out with other constellations [340].

Conditioned on the instantaneous SNR γ, the symbol error rate (SER) for M-ary PSK is available in (12.23) and replicated here for convenience:

$$\mathrm{SER}(\gamma) = \frac{1}{\pi} \int_0^{\frac{(M-1)\pi}{M}} \exp\left(-\frac{g_{\mathrm{PSK}}\gamma}{\sin^2\theta} \right) d\theta, \tag{13.21}$$

where $g_{\mathrm{PSK}} = \sin^2(\pi/M)$ as in (12.27). Since $\boldsymbol{h}$ is a random vector, the average SER is expressed as

$$\overline{\mathrm{SER}} = E_{\boldsymbol{h}}\{\mathrm{SER}(\gamma)\}. \tag{13.22}$$

Because γ in (13.6) depends on the specific beamformer design, the $\overline{\mathrm{SER}}$ in (13.22) is also beamformer dependent. The exact expression for the average SER turns out to be difficult to obtain in closed form. However, with $\boldsymbol{h}$ as in (13.9), a lower bound on $\overline{\text{SER}}$ is available [340]:

$$\overline{\text{SER}}_{\text{lb}}(N_t, N, \bar{\gamma}) = \frac{1}{\pi} \int_0^{\frac{(M-1)\pi}{M}} \left(1 + \frac{g_{\text{PSK}}\bar{\gamma}}{\sin^2\theta}\right)^{-1} \left[1 + \left[1 - \left(\frac{1}{N}\right)^{\frac{1}{N_t-1}}\right] \frac{g_{\text{PSK}}\bar{\gamma}}{\sin^2\theta}\right]^{1-N_t} d\theta. \tag{13.23}$$

Notice that $\overline{\text{SER}}_{\text{lb}}$ is a function of N_t, N, and $\bar{\gamma}$, but does not depend on the beamformer. A good beamformer should come as close as possible to this lower bound. This turns out to be consistent with the design guidelines in (13.10) and (13.12).

As N grows large, performance of beamformed transmissions with finite-rate feedback should approach the ideal case with perfect channel knowledge at the transmitter. With perfect CSI, the system is equivalent to a diversity system with maximum ratio combing (MRC) of N_t diversity branches. The average SER at the MRC output with PSK modulation is given by [236]

$$\overline{\text{SER}}_{\text{perfect CSI}}(N_t, \bar{\gamma}) = \frac{1}{\pi} \int_0^{\frac{(M-1)\pi}{M}} \left(1 + \frac{g_{\text{PSK}}\bar{\gamma}}{\sin^2\theta}\right)^{-N_t} d\theta. \tag{13.24}$$

Comparing (13.23) with (13.24), we deduce that the lower bound in (13.23) must become tight as N increases since it should coincide with (13.24) when $N = \infty$.

If the lower bound is achievable by a certain beamformer design, the distance between the two curves corresponding to (13.23) and (13.24) quantifies the performance loss due to the finite-rate constraint, as asserted by the following proposition [340].

Proposition 13.2 *Assume that the SER lower bound in* (13.23) *can be achieved by some practical beamformer designs. To compensate for the performance loss due to the finite-rate constraint, we can increase the transmission power from* $\bar{\gamma}$ *to* $\bar{\gamma}_0$ *so that performance with perfect CSI is achieved:*

$$\overline{\text{SER}}_{\text{lb}}(N_t, N, \bar{\gamma}_0) = \overline{\text{SER}}_{\text{perfect CSI}}(N_t, \bar{\gamma}). \tag{13.25}$$

The difference between $\bar{\gamma}_0$ *and* $\bar{\gamma}$ *representing the power loss (in decibels) due to the finite-rate constraint is*

$$L(N_t, N, \bar{\gamma}) = 10\log_{10}\bar{\gamma}_0 - 10\log_{10}\bar{\gamma} \tag{13.26}$$

and satisfies

$$L(N_t, N, \bar{\gamma}) \le L(N_t, N, \infty), \tag{13.27}$$

$$L(N_t, N, \infty) = 10\log_{10}\left[1 - \left(\frac{1}{N}\right)^{\frac{1}{N_t-1}}\right]^{\frac{1}{N_t}-1}. \tag{13.28}$$

Although derived for PSK constellations, the power loss in (13.26) holds true for any two-dimensional constellation [340]. Notice that the power loss $L(N_t, N, \bar{\gamma})$ is a function of $\bar{\gamma}$ and thus varies over the entire SNR range. However, Proposition 13.2 asserts that the power loss across the entire SNR range is less than or equal to $L(N_t, N, \infty)$, which depends only on N and N_t.

When $N = 2^B$ is large, we can use the approximation $\ln(1+x) \approx x$ for x small enough, to obtain

$$L(N_t, N, \infty) \approx \frac{10}{\ln 10}\left(1 - \frac{1}{N_t}\right) 2^{-\frac{B}{N_t - 1}}. \tag{13.29}$$

Equation (13.29) shows that the power loss (in decibels) due to the finite-rate constraint decays exponentially with the number of feedback bits B for large B. This implies that B does not need to be very large for the system to approach optimal error performance. This is also confirmed by the ensuing numerical results.

On the other hand, if a system requires the power loss to be within L_0 decibels relative to the perfect CSI case, we can easily identify from (13.28) the least number of beamforming vectors needed as

$$N \geq \left(1 - 10^{\frac{L_0}{10} \cdot \frac{N_t}{1 - N_t}}\right)^{-(N_t - 1)}. \tag{13.30}$$

13.2.4 Numerical Examples

Using QPSK, we compute the power loss in (13.28) for various cases and list the results in Table 13.2, where we are interested in non-trivial configurations with $N \geq N_t$.

Table 13.2 POWER LOSS (dB) DUE TO FINITE-RATE CONSTRAINT (B BITS)

	$B = 1$	$B = 2$	$B = 3$	$B = 4$	$B = 5$	$B = 6$	$B = 7$
$N_t = 2$	1.51	0.62	0.29	0.14	0.07	0.03	0.02
$N_t = 3$	—	2.01	1.26	0.83	0.56	0.39	0.27
$N_t = 4$	—	3.24	2.26	1.65	1.23	0.94	0.72
$N_t = 6$	—	—	3.90	3.09	2.51	2.07	1.72
$N_t = 8$	—	—	5.16	4.25	3.57	3.05	2.63

Table 13.3 MINIMUM NUMBER OF FEEDBACK BITS B FOR POWER LOSS WITHIN 1 dB

N_t	2	3	4	5	6	7	8
$N \geq$	3	12	55	256	1220	5851	28170
$B = \lceil \log_2 N \rceil$	2	4	6	8	11	13	15

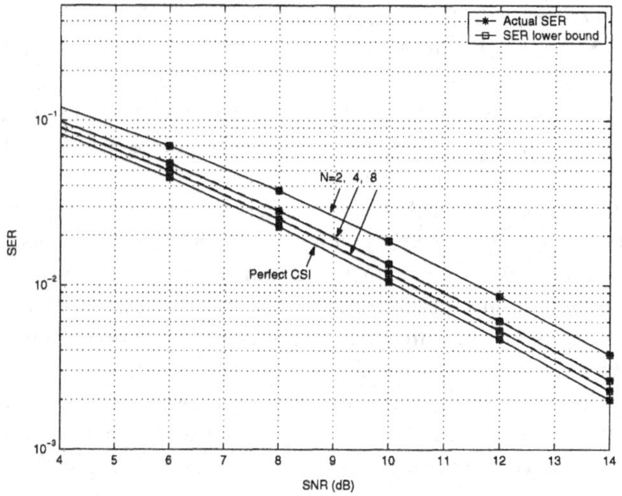

Figure 13.4 Actual SER vs. the lower bound ($N_t = 2$)

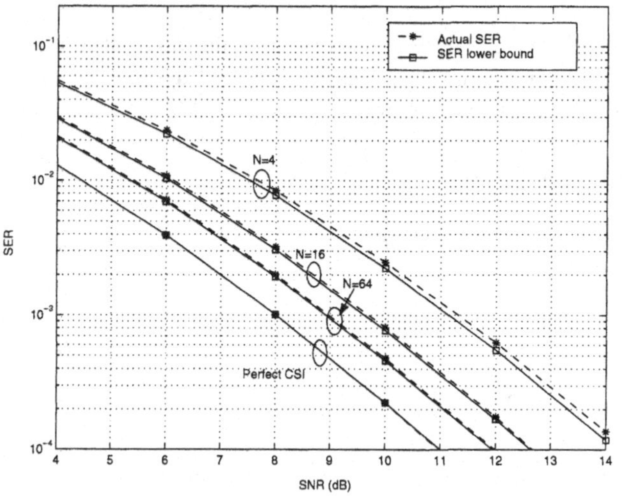

Figure 13.5 Actual SER versus the lower bound ($N_t = 4$)

On the other hand, if we want to limit the power loss to be within 1 dB relative to the perfect CSI case, the minimum number of beamforming vectors N (corresponding to the minimum number of feedback bits B) can be computed from (13.30). From the results listed in Table 13.3, we see clearly that as N_t increases, the required number of feedback bits increases considerably in order to bring the performance close to optimal.

We further compare the SER lower bound in (13.23) with the actual SER in (13.22) obtained via Monte Carlo simulations. Figure 13.4 depicts the results with $N_t = 2$. The beamformer codebooks for $N = 4, 8$ are listed in Examples 13.1 and 13.3. The codebook with $N = 2$ is $\boldsymbol{W} = \boldsymbol{I}_2$, which corresponds to the selection diversity scheme. As shown in Figure 13.4, the SER lower bound for $N_t = 2$ is almost identical to the actual SER even for small values of N.

We finally compare the SER lower bound with the actual SER for $N_t = 4$. We use the beamformers listed in [305] with $N = 16$ and 64. The codebook with $N = 4$ is $\boldsymbol{W} = \boldsymbol{I}_4$, corresponding to selection combining. As shown in Figure 13.5, the bound is also tight for those beamformers. This suggests that the lower bound, which does not depend on the beamformer design, is a reliable performance indicator for carefully constructed beamformers.

Figures 13.4 and 13.5 show that the distance between the SER lower bound and the SER with perfect CSI increases only slightly as the average SNR increases, yet it always stays bounded by the maximum power loss listed in Table 13.2. This confirms the validity of (13.27).

13.3 FINITE-RATE PRECODED SPATIAL MULTIPLEXING

As with the statistical CSI designs in Chapter 12, the finite-rate transmit-beamformers of Section 13.2 can attain a transmission rate of 1 symbol per channel use. Since multi-antenna transmitters with spatial multiplexing can achieve rates as high as N_t symbols per channel use even with open-loop operation, it is natural to explore such transmitter designs based on finite-rate feedback.

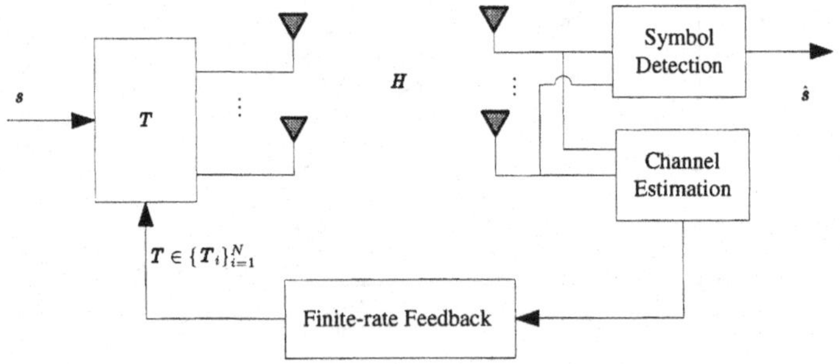

Figure 13.6 Precoded spatial multiplexing MIMO system with finite-rate feedback

To this end, consider the precoded spatial multiplexing closed-loop system depicted in Figure 13.6 with N_t transmit-antennas and N_r receive-antennas. The information symbol block $\boldsymbol{s} := [s_1, \ldots, s_K]^T$ is precoded by an $N_t \times K$ matrix $\boldsymbol{T}$ to obtain the precoded block $\boldsymbol{Ts}$, whose N_t entries are then transmitted from the N_t antennas simultaneously. As discussed in Section 12.2.5, *time-varying beamformer* is a synonymous term for the precoder $\boldsymbol{T}$. The reason we call it *precoder* here is twofold: (i) beamforming is a term traditionally used for the time-invariant vector $\boldsymbol{w}$ we dealt with in Section 13.2; and (ii) we want to distinguish the precoded transmissions of this chapter where the transmit-power is divided uniformly across data streams, from the coder-beamformers of Chapter 12, which also optimize power loading.

The samples received by the N_r receive-antennas are collected in the vector $\boldsymbol{y}$, which obeys the input-output relationship

$$\boldsymbol{y} = \sqrt{\frac{\bar{\gamma}}{K}} \boldsymbol{H}\boldsymbol{T}\boldsymbol{s} + \boldsymbol{v}, \tag{13.31}$$

where $\boldsymbol{H}$ is the $N_r \times N_t$ channel matrix and $\boldsymbol{v}$ is AWGN with each entry having zero mean and unit variance. We select the constellation to have unit energy and normalize the precoder so that $\mathrm{tr}(\boldsymbol{T}^{\mathcal{H}}\boldsymbol{T}) = K$. Therefore, $\bar{\gamma}$ can be interpreted as the average SNR per receive-antenna.

With the signal constellation fixed, transmission modes in this case are characterized by precoding matrices. Based on B feedback bits, a total of $N = 2^B$ candidate precoding matrices must be constructed. Let $\boldsymbol{T}_1, \boldsymbol{T}_2, \ldots, \boldsymbol{T}_N$ denote these matrices comprising the codebook

$$\mathcal{T} := \{\boldsymbol{T}_1, \ldots, \boldsymbol{T}_N\}. \tag{13.32}$$

The N transmission-mode codewords are now

$$\{\boldsymbol{T}_i\boldsymbol{s}\}_{i=1}^{N}. \tag{13.33}$$

We assume further that power is allocated uniformly across the K information symbols. This corresponds to constraining the precoder columns to be orthonormal, which means that the precoders are unitary and satisfy [165, 166]

$$\boldsymbol{T}_i^{\mathcal{H}}\boldsymbol{T}_i = \boldsymbol{I}_K, \quad i = 1, \ldots, N. \tag{13.34}$$

The receiver decides which precoder from the codebook $\mathcal{T}$ lies closest to the current channel realization and informs the transmitter to switch to that precoder by feeding back its B-bit codeword index. Clearly, the precoder selection rule as well as the codebook design metric depend on the receiver structure deployed. Notice that for the block transmission model in (13.31), one can apply various decoders, including ML, linear ZF/MMSE, nonlinear DF receivers, or the (near-)ML SDA described in Chapter 5. We next detail selection criteria for various receiver structures.

13.3.1 Precoder Selection Criteria

With precoders $\{\boldsymbol{T}_i\}_{i=1}^{N}$ satisfying (13.34), a number of precoder selection criteria have been introduced in [165, 166]:

- For an ML receiver, the precoder is chosen to either maximize the minimum distance (MD) among symbol vectors received (MD criterion), or to maximize the instantaneous capacity (capacity criterion).

- For a linear ZF receiver, the precoder is chosen to maximize the minimum singular value (SV) of $\boldsymbol{H}\boldsymbol{T}_i$ (SV criterion).

- For a linear MMSE receiver, the precoder is chosen to either minimize the trace of the MSE matrix (MMSE-trace criterion), or to minimize the determinant of the MSE matrix (MMSE-det criterion).

We focus on a selection criterion proposed by [336], which is based on the exact BER for linear ZF and MMSE receivers. Let $\overline{\text{BER}}(\boldsymbol{H}, \boldsymbol{T})$ denote the BER averaged over K data streams over which the channel realization is $\boldsymbol{H}$ and the candidate precoder is denoted by $\boldsymbol{T}$. The BER-based selection rule is then

$$\boldsymbol{T}_{\text{opt}} = \arg\min_{\boldsymbol{T}\in\mathcal{T}} \overline{\text{BER}}(\boldsymbol{H}, \boldsymbol{T}), \tag{13.35}$$

whose index is fed back to the transmitter.

For any given codebook, the BER-based precoder selection criterion outperforms all other criteria in terms of uncoded BER performance if the receivers adopted are linear. The BER corresponding to an ML receiver is not available in closed form, which prevents one from carrying out the optimization in (13.35) analytically with ML reception. However, this is possible for linear receivers, as we will see next.

13.3.1.1 Linear ZF Receiver Processing the vector $\boldsymbol{y}$ in (13.31) through the linear ZF receiver

$$\boldsymbol{G}^{\text{zf}} = \sqrt{\frac{K}{\bar{\gamma}}}\left(\boldsymbol{T}^{\mathcal{H}}\boldsymbol{H}^{\mathcal{H}}\boldsymbol{H}\boldsymbol{T}\right)^{-1}\boldsymbol{T}^{\mathcal{H}}\boldsymbol{H}^{\mathcal{H}}, \tag{13.36}$$

we obtain at its output

$$\boldsymbol{z} = \boldsymbol{G}^{\text{zf}}\boldsymbol{y} = \boldsymbol{s} + \boldsymbol{G}^{\text{zf}}\boldsymbol{v}, \tag{13.37}$$

where the noise has covariance matrix $\boldsymbol{G}^{\text{zf}}(\boldsymbol{G}^{\text{zf}})^{\mathcal{H}} = (K/\bar{\gamma})[\boldsymbol{T}^{\mathcal{H}}\boldsymbol{H}^{\mathcal{H}}\boldsymbol{H}\boldsymbol{T}]^{-1}$. Using $[\ ]_{k,l}$ to denote the (k,l)th entry of a matrix, the SNR for the kth data stream is therefore

$$\gamma_k^{\text{zf}} = \frac{\bar{\gamma}}{K\left[\left(\boldsymbol{T}^{\mathcal{H}}\boldsymbol{H}^{\mathcal{H}}\boldsymbol{H}\boldsymbol{T}\right)^{-1}\right]_{k,k}}. \tag{13.38}$$

Let $\text{BER}(\gamma)$ denote the BER in an AWGN channel at SNR γ. A closed-form expression for $\text{BER}(\gamma)$ with M-ary square QAM is [42]

$$\text{BER}(\gamma) = \frac{1}{\log_2\sqrt{M}}\sum_{k=1}^{\log_2\sqrt{M}}\frac{1}{\sqrt{M}}\sum_{i=0}^{(1-2^{-k})\sqrt{M}-1} \times\left\{(-1)^{\left\lfloor\frac{i\cdot 2^{k-1}}{\sqrt{M}}\right\rfloor}\left(2^{k-1}-\left\lfloor\frac{i\cdot 2^{k-1}}{\sqrt{M}}+\frac{1}{2}\right\rfloor\right)\cdot 2Q\left((2i+1)\sqrt{\frac{3}{M-1}\gamma}\right)\right\}, \tag{13.39}$$

where $Q(\cdot)$ denotes the Gaussian-Q function. For pulse amplitude modulation (PAM) and rectangular QAMs, closed-form BER expressions are also available [42]. When 4-QAM is used, (13.39) simplifies to

$$\text{BER}(\gamma) = Q(\sqrt{\gamma}). \tag{13.40}$$

Alternatively, one can compute $\mathrm{BER}(\gamma)$ easily using the recursive algorithm in [278]. The average BER over K data streams can then be expressed as

$$\overline{\mathrm{BER}}^{\mathrm{zf}}(\boldsymbol{H},\boldsymbol{T}) = \frac{1}{K}\sum_{k=1}^{K}\mathrm{BER}(\gamma_k^{\mathrm{zf}}). \tag{13.41}$$

13.3.1.2 Linear MMSE Receiver Let us now consider the linear MMSE receiver

$$\boldsymbol{G}^{\mathrm{mmse}} = \sqrt{\frac{\bar{\gamma}}{K}}\left(\frac{\bar{\gamma}}{K}\boldsymbol{T}^{\mathcal{H}}\boldsymbol{H}^{\mathcal{H}}\boldsymbol{H}\boldsymbol{T}+\boldsymbol{I}_K\right)^{-1}\boldsymbol{T}^{\mathcal{H}}\boldsymbol{H}^{\mathcal{H}}. \tag{13.42}$$

The signal-to-interference-plus-noise ratio (SINR) after MMSE equalization is

$$\gamma_k^{\mathrm{mmse}} = \frac{1}{\left[\left(\frac{\bar{\gamma}}{K}\boldsymbol{T}^{\mathcal{H}}\boldsymbol{H}^{\mathcal{H}}\boldsymbol{H}\boldsymbol{T}+\boldsymbol{I}_K\right)^{-1}\right]_{k,k}} - 1. \tag{13.43}$$

Furthermore, the residual interference-plus-noise term at the MMSE equalizer output can be well approximated by a Gaussian random variable [213]. Using this approximation, we can compute the average BER as

$$\overline{\mathrm{BER}}^{\mathrm{mmse}}(\boldsymbol{H},\boldsymbol{T}) = \frac{1}{K}\sum_{k=1}^{K}\mathrm{BER}(\gamma_k^{\mathrm{mmse}}). \tag{13.44}$$

13.3.2 Codebook Construction: Infinite Rate

We first consider the codebook construction with $B=\infty$. In this case we assume that the transmitter has full knowledge of the channel $\boldsymbol{H}$ and selects $\boldsymbol{T}$ directly based on $\boldsymbol{H}$. Consider the eigen-decomposition of the channel matrix

$$\boldsymbol{H}^{\mathcal{H}}\boldsymbol{H} = \boldsymbol{V}_H\boldsymbol{\Lambda}_H\boldsymbol{V}_H^{\mathcal{H}}, \tag{13.45}$$

where $\boldsymbol{\Lambda}_H = \mathrm{diag}(\lambda_1,\ldots,\lambda_{N_t})$ contains on its diagonal the eigenvalues arranged in non-increasing order (i.e., $\lambda_1 \geq \cdots \geq \lambda_{N_t}$).

Let $\overline{\boldsymbol{V}}_H$ denote the matrix containing the first K columns of $\boldsymbol{V}_H$ and $\boldsymbol{\Theta}_{\mathrm{cm}}$ denote a $K\times K$ unitary matrix with each entry having constant modulus $1/\sqrt{K}$. For example, $\boldsymbol{\Theta}_{\mathrm{cm}}$ could be a normalized FFT or a Hadamard matrix. As the BER in (13.39) depends on the constellation, the optimal precoders will be provided for three popular constellations: 4-QAM, 16-QAM, and 64-QAM; additional constellations are considered in [336], where the following results have been established:

Proposition 13.3 *With linear ZF reception and using the BER-based selection criterion, the optimal column-orthonormal precoder is given by*

$$\boldsymbol{T}_{\mathrm{opt}} = \begin{cases} \overline{\boldsymbol{V}}_H & \text{when } \frac{\lambda_1\bar{\gamma}}{K} \leq \Gamma_{\mathrm{th}} \\ \overline{\boldsymbol{V}}_H\boldsymbol{\Theta}_{\mathrm{cm}} & \text{when } \frac{\lambda_K\bar{\gamma}}{K} \geq \Gamma_{\mathrm{th}} \\ \text{unclear} & \text{otherwise,} \end{cases} \tag{13.46}$$

where the threshold $\Gamma_{\rm th}$ *is defined depending on the constellation as*

$$\Gamma_{\rm th} = \begin{cases} 3 & \text{for 4-QAM} \\ 14.9 & \text{for 16-QAM} \\ 62.5 & \text{for 64-QAM,} \end{cases} \tag{13.47}$$

$\lambda_1, \ldots, \lambda_K$ *are the* K *largest eigenvalues of* $\boldsymbol{H}^{\mathcal{H}}\boldsymbol{H}$ *in a descending order, and* $\overline{\boldsymbol{V}}_H$ *contains the corresponding eigenvectors* [*cf.* (13.45)].

Proposition 13.4 *With linear MMSE reception and using the BER-based selection criterion, the optimal column-orthonormal precoder is given by*

1. *For 4-QAM, the optimal precoder is always*

$$\boldsymbol{T}_{\rm opt} = \overline{\boldsymbol{V}}_H \boldsymbol{\Theta}_{\rm cm}. \tag{13.48}$$

2. *For 16-QAM and 64-QAM, the optimal precoder is*

$$\boldsymbol{T}_{\rm opt} = \begin{cases} \overline{\boldsymbol{V}}_H & \text{when } \Gamma_{\rm th,l} \le \frac{\lambda_K \bar{\gamma}}{K} \le \cdots \le \frac{\lambda_1 \bar{\gamma}}{K} \le \Gamma_{\rm th,h} \\ \overline{\boldsymbol{V}}_H \boldsymbol{\Theta}_{\rm cm} & \text{when } \frac{\lambda_K \bar{\gamma}}{K} \ge \Gamma_{\rm th,h} \text{ or } \frac{\lambda_1 \bar{\gamma}}{K} \le \Gamma_{\rm th,l} \\ \text{unclear} & \text{otherwise,} \end{cases} \tag{13.49}$$

where the constellation-specific thresholds $(\Gamma_{\rm th,l}, \Gamma_{\rm th,h})$ *are given by*

$$(\Gamma_{\rm th,l}, \Gamma_{\rm th,h}) = \begin{cases} (0.28, 13.7) & \text{for 16-QAM} \\ (0.26, 21.5) & \text{for 64-QAM,} \end{cases} \tag{13.50}$$

$\lambda_1, \ldots, \lambda_K$ *are the* K *largest eigenvalues of* $\boldsymbol{H}^{\mathcal{H}}\boldsymbol{H}$ *in a descending order, and* $\overline{\boldsymbol{V}}_H$ *contains the corresponding eigenvectors* [*cf.* (13.45)].

We now investigate the difference between the precoder choices $\overline{\boldsymbol{V}}_H$ and $\overline{\boldsymbol{V}}_H \boldsymbol{\Theta}_{\rm cm}$. When $\boldsymbol{T} = \overline{\boldsymbol{V}}_H$, we have

$$\gamma_k^{\rm zf} = \gamma_k^{\rm mmse} = \frac{\lambda_k \bar{\gamma}}{K}, \quad \forall k = 1, \ldots, K. \tag{13.51}$$

When $\boldsymbol{T} = \overline{\boldsymbol{V}}_H \boldsymbol{\Theta}_{\rm cm}$, we have

$$\gamma_1^{\rm zf} = \cdots = \gamma_K^{\rm zf} = \frac{1}{\frac{1}{\lambda_1 \bar{\gamma}} + \ldots + \frac{1}{\lambda_K \bar{\gamma}}}, \tag{13.52}$$

$$\gamma_1^{\rm mmse} = \cdots = \gamma_K^{\rm mmse} = \frac{1}{\frac{1}{1+\lambda_1 \bar{\gamma}} + \ldots + \frac{1}{1+\lambda_K \bar{\gamma}}} - 1. \tag{13.53}$$

Notice that when $\boldsymbol{T} = \overline{\boldsymbol{V}}_H \boldsymbol{\Theta}_{\rm cm}$, the SNRs are balanced (i.e., they are equal) across all information symbols.

To gain more insight as to why and when SNRs should be balanced, let us consider first the ZF receiver. Intuitively, if the subchannel SNRs of all K data streams are sufficiently high, the worst subchannel dominates the overall performance. Hence, the optimal precoder should balance the SNRs of all data streams to achieve the best performance in this scenario. This confirms Proposition 13.3 at high SNR, where $\boldsymbol{T} = \overline{\boldsymbol{V}}_H \boldsymbol{\Theta}_{\text{cm}}$. This high-SNR behavior of the ZF receiver was observed in [57] only for 2-QAM and 4-QAM. Notice that the common SNR for the ZF receiver in (13.52) is proportional to the *harmonic mean* of the K largest eigenvalues of $\boldsymbol{H}^{\mathcal{H}}\boldsymbol{H}$. On the other hand, when all subchannel SNRs are sufficiently low, they should not be balanced as suggested by (13.51).

Turning our attention to the MMSE receiver, it is well known that at high SNR it reduces to a ZF receiver. Hence, the behavior of the MMSE receiver should be similar to the ZF receiver at high SNR. This is reflected in Proposition 13.4, which asserts that the subchannel SNRs should be balanced when they exceed a certain threshold.

Different from the ZF receiver, however, the subchannel SNRs in the MMSE receiver should also be balanced when they are extremely low. The intuition behind this is that the MMSE receiver reduces to a matched filter when the SNR is very low, where balancing the SNRs is the optimal thing to do [336]. In short, the MMSE receiver behaves as a matched filter at low SNR and as a ZF receiver at high SNR. At both ends, the subchannel SNRs should be balanced, but for different reasons.

Between low and high SNR, there exists a range where the subchannel SNRs should not be balanced with MMSE reception, as established by Proposition 13.4. Interestingly, the 2-QAM and 4-QAM are exceptions where the subchannels should always be balanced throughout the entire SNR range.

Since the optimal precoder is not available for all scenarios, a practical solution with infinite-rate feedback is summarized in the following proposition.

Proposition 13.5 *With linear ZF or MMSE receivers, for each channel realization* $\boldsymbol{H}$*, the transmitter should select the precoder as either* $\overline{\boldsymbol{V}}_H$ *or* $\overline{\boldsymbol{V}}_H \boldsymbol{\Theta}_{\text{cm}}$*, depending on which one yields better BER performance.*

Notice that when precoders are designed in accordance with Proposition 13.5, there is no need to compare the channel eigenvalues with constellation-specific thresholds.

13.3.3 Codebook Construction: Finite Rate

Precoder codebook construction with finite-rate CSI turns out to be closely related to the Grassmannian subspace packing problem [48], as pointed out in [165, 166]. Based on each precoder $\boldsymbol{T}_i$, one can define a subspace spanned by the K columns of $\boldsymbol{T}_i$. For any two subspaces generated by $\boldsymbol{T}_i$ and $\boldsymbol{T}_j$, various distances can be defined [11]. These include the chordal distance $d_{\text{c}}(\boldsymbol{T}_i, \boldsymbol{T}_j)$, the Fubini-Study distance $d_{\text{FS}}(\boldsymbol{T}_i, \boldsymbol{T}_j)$, or the projection two-norm distance $d_{\text{p2}}(\boldsymbol{T}_i, \boldsymbol{T}_j)$, which are defined, respectively, as

$$d_c(\boldsymbol{T}_i, \boldsymbol{T}_j) = \frac{1}{\sqrt{2}} \|\boldsymbol{T}_i \boldsymbol{T}_i^{\mathcal{H}} - \boldsymbol{T}_j \boldsymbol{T}_j^{\mathcal{H}}\|_F, \tag{13.54}$$

$$d_{\text{FS}}(\boldsymbol{T}_i, \boldsymbol{T}_j) = \arccos |\det(\boldsymbol{T}_i^{\mathcal{H}} \boldsymbol{T}_j)|, \tag{13.55}$$

$$d_{\text{p2}}(\boldsymbol{T}_i, \boldsymbol{T}_j) = \|\boldsymbol{T}_i \boldsymbol{T}_i^{\mathcal{H}} - \boldsymbol{T}_j \boldsymbol{T}_j^{\mathcal{H}}\|_2. \tag{13.56}$$

Similar to (13.12), if the entries of $\boldsymbol{H}$ are i.i.d. complex Gaussian, the objective is to optimize the codebook so that the minimal distance between any two codewords is maximized; that is,

$$\mathcal{T}_{\text{opt}} = \arg\max_{\mathcal{T}} \min_{1 \le i < j \le N} d(\boldsymbol{T}_i, \boldsymbol{T}_j), \tag{13.57}$$

where $d(\boldsymbol{T}_i, \boldsymbol{T}_j)$ could be any of the distances in (13.54)–(13.56). For example:

- $d_{\text{FS}}(\boldsymbol{T}_i, \boldsymbol{T}_j)$ should be used in (13.57) for codebook construction when MMSE-det or capacity selection criteria are adopted [165, 166].
- $d_{\text{p2}}(\boldsymbol{T}_i, \boldsymbol{T}_j)$ should be used in (13.57) for codebook construction when MMSE-trace, SV, or MD criteria are employed [165, 166].

It is generally difficult to obtain a codebook maximizing the minimum distance among codewords as in (13.57). Construction methods based on FFT matrices are available in [113] and codebooks in [164] are obtained via modified versions of the algorithm in [113]. In the ensuing subsection we present a codebook construction scheme based on the Lloyd algorithm [336].

13.3.3.1 Codebook Construction Based on Lloyd's Algorithm Since Lloyd's algorithm has been derived for designing codebooks of vector quantizers, it is important to clarify the link between vector quantization and precoder codebook construction in the present context. Consider an isotropically distributed random matrix $\boldsymbol{V}$ of size $N_t \times K$ which we wish to quantize in order to obtain a finite number of codewords comprising the codebook $\mathcal{T}$. Adopting the chordal distance as the distance metric, we define the average distortion function

$$J = E\left\{\min_{1 \le i \le N} d_c^2(\boldsymbol{V}, \boldsymbol{T}_i)\right\} = \sum_{i=1}^{N} E_{\mathcal{R}_i}\{d_c^2(\boldsymbol{V}, \boldsymbol{T}_i)\} \Pr(\boldsymbol{V} \in \mathcal{R}_i), \tag{13.58}$$

where $E\{\cdot\}$ stands for expectation and $\Pr(\boldsymbol{V} \in \mathcal{R}_i)$ is the probability that $\boldsymbol{V}$ belongs to the region

$$\mathcal{R}_i := \{\boldsymbol{V} \mid d_c(\boldsymbol{V}, \boldsymbol{T}_i) < d_c(\boldsymbol{V}, \boldsymbol{T}_j), \quad \forall j \ne i\}. \tag{13.59}$$

Lloyd's algorithm guarantees that J in (13.58) is decreasing per iteration. The reason for choosing the chordal distance is that it can be expressed in terms of the trace as

$$d_c^2(\boldsymbol{T}_i, \boldsymbol{T}_j) = \text{tr}(\boldsymbol{I}_K - \boldsymbol{T}_j^{\mathcal{H}} \boldsymbol{T}_i \boldsymbol{T}_i^{\mathcal{H}} \boldsymbol{T}_j), \tag{13.60}$$

which allows for analytical simplifications in each iteration of Lloyd's algorithm.

The codebook design steps are as follows:

S1) Replace the expectation operator in (13.58) using the Monte Carlo (sample averaging) approach, which is based on a training set with N_{tr} samples $\{\boldsymbol{V}_n\}_{n=1}^{N_{\text{tr}}}$.

S2) Starting with an initial codebook (obtained via random computer search or using the best codebook currently available), carry out the following two substeps iteratively:

- Nearest-neighbor rule [86]: assign $\boldsymbol{V}_n$ to one of the regions using the rule

$$\boldsymbol{V}_n \in \mathcal{R}_i \quad \text{if} \quad d_c(\boldsymbol{V}_n, \boldsymbol{T}_i) < d_c(\boldsymbol{V}_n, \boldsymbol{T}_j), \ \forall j \neq i. \tag{13.61}$$

- Centroid condition [86]: For each region $\mathcal{R}_i$, find the optimal codebook as

$$\begin{aligned}\boldsymbol{T}_i^{\text{opt}} &= \arg\min_{\boldsymbol{T}} \frac{1}{N_{\text{tr}}} \sum_{\boldsymbol{V}_n \in \mathcal{R}_i} d_c^2(\boldsymbol{V}_n, \boldsymbol{T}) \\ &= \arg\min_{\boldsymbol{T}} \frac{1}{N_{\text{tr}}} \sum_{\boldsymbol{V}_n \in \mathcal{R}_i} \text{tr}(\boldsymbol{I}_K - \boldsymbol{T}^{\mathcal{H}} \boldsymbol{V} \boldsymbol{V}^{\mathcal{H}} \boldsymbol{T}) \\ &= \arg\max_{\boldsymbol{T}} \text{tr}(\boldsymbol{T}^{\mathcal{H}} \boldsymbol{R} \boldsymbol{T}),\end{aligned} \tag{13.62}$$

  where $\boldsymbol{R}$ is defined as

$$\boldsymbol{R} = \frac{1}{N_{\text{tr}}} \sum_{\boldsymbol{V}_n \in \mathcal{R}_i} \boldsymbol{V}_n \boldsymbol{V}_n^{\mathcal{H}}. \tag{13.63}$$

  Perform the eigen-decomposition

$$\boldsymbol{R} = \boldsymbol{U}_R \boldsymbol{\Delta}_R \boldsymbol{U}_R^{\mathcal{H}}, \tag{13.64}$$

  and construct the optimal $\boldsymbol{T}_i^{\text{opt}}$ in (13.62) using the K eigenvectors of $\boldsymbol{R}$ corresponding to the K largest eigenvalues.

Since J is monotonically decreasing, Lloyd's algorithm converges at least to a local optimum. But this does not mean that the minimum distance of the codebook is monotonically improving, as observed in [306, 309]. During each iteration, we examine the tentative codebook and record whether its minimum distance is larger than the currently best. This is applicable to any of the distances d_c, d_{FS}, and d_{p2}.

S3) Go back to S1 to generate another training set and rerun Lloyd's algorithm in S2. Stop if no further improvement on the minimum distance is observed.

If we set $K = 1$ in Lloyd's algorithm, we will obtain a finite-rate beamforming codebook. In this case, the multiple distance options are not necessary, and using any of them will lead to the same codebook. Notice that the chordal distance in (13.11)

Table 13.4 MINIMUM DISTANCES OF CODEBOOKS OBTAINED VIA LLOYD'S ALGORITHM

Feedback Bits	Setup: $N_t = 4, K = 2$ (d_c, d_{FS}, d_{p2})	Setup: $N_t = 6, K = 3$ (d_c, d_{FS}, d_{p2})
$B = 2$	$(\mathbf{1.1089}, 1.3461, 0.9534)$	$(\mathbf{1.4116}, 1.5232, 0.9786)$
	$(1.0003, \mathbf{1.5409}, \mathbf{0.9992})$	$(1.3972, \mathbf{1.5613}, \mathbf{0.9986})$
$B = 3$	$(\mathbf{1.0282}, 1.1770, 0.8717)$	$(\mathbf{1.2899}, 1.3819, 0.9371)$
	$(1.0164, \mathbf{1.2615}, 0.9171)$	$(1.2466, \mathbf{1.4404}, \mathbf{0.9699})$
	$(1.0189, 1.2425, \mathbf{0.9250})$	
$B = 4$	$(\mathbf{0.9755}, 1.0966, 0.7845)$	$(\mathbf{1.2281}, 1.3259, 0.8881)$
	$0.9703, \mathbf{1.1205}, 0.7948)$	$(1.1740, \mathbf{1.3548}, 0.9100)$
	$(0.9528, 1.0686, \mathbf{0.8355})$	$(1.1782, 1.3218, \mathbf{0.9314})$
$B = 5$	$(\mathbf{0.8751}, 0.9917, 0.7210)$	$(\mathbf{1.1539}, 1.2500, 0.8636)$
	$(0.8587, \mathbf{1.0030}, 0.7450)$	$(1.1424, \mathbf{1.2730}, 0.8748)$
	$(0.8583, 1.0016, \mathbf{0.7760})$	$(1.1401, 1.2635, \mathbf{0.8830})$
$B = 6$	$(\mathbf{0.8002}, 0.8610, 0.6444)$	$(\mathbf{1.0625}, 1.1004, 0.7637)$
	$(0.7906, \mathbf{0.8669}, 0.6464)$	$(1.0216, \mathbf{1.1710}, 0.7813)$
	$(0.7670, 0.8309, \mathbf{0.6734})$	$(1.0242, 1.1529, \mathbf{0.8239})$

is a special case of (13.60) with $K = 1$. Codebooks obtained based on Lloyd's algorithm are reported in [328].

Example 13.5 Table 13.4 lists the minimum distances of codebooks constructed for $(N_t, K) = (4, 2)$ and $(N_t, K) = (6, 3)$. When multiple codebooks are listed for one configuration, boldface fonts highlight the maximized minimum distances d_c, d_{FS}, or d_{p2}. The codebooks obtained using Lloyd's algorithm exhibit improved minimum distances relative to those reported in [164].

13.3.4 Numerical Examples

In this subsection we test two different configurations: $(N_t, N_r, K) = (4, 2, 2)$ and $(N_t, N_r, K) = (6, 3, 3)$. In all plots, the average SNR per receive antenna refers to $\bar{\gamma}$. The constellation used is 4-QAM unless specified otherwise. Each BER curve is averaged over 10^4 channel realizations.

Example 13.6 (Infinite-rate feedback) For $(N_t, N_r, K) = (6, 3, 3)$, we first validate the theoretical analysis in Section 13.3.2 where $B = \infty$. With 4-QAM, we observe from Figure 13.7 that the ZF receiver with the precoder $\boldsymbol{T} = \overline{\boldsymbol{V}}_H$ outperforms that with $\boldsymbol{T} = \overline{\boldsymbol{V}}_H \boldsymbol{\Theta}_{\text{cm}}$ at low SNR, but the converse holds true at high SNR. The solution suggested by Proposition 13.5 outperforms both since it always chooses

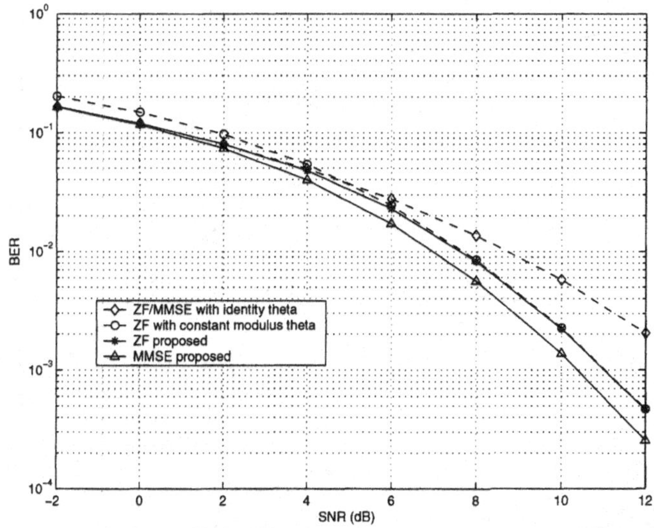

Figure 13.7 Error performance with 4-QAM and infinite-rate feedback for $(N_t, N_r, K) = (6, 3, 3)$

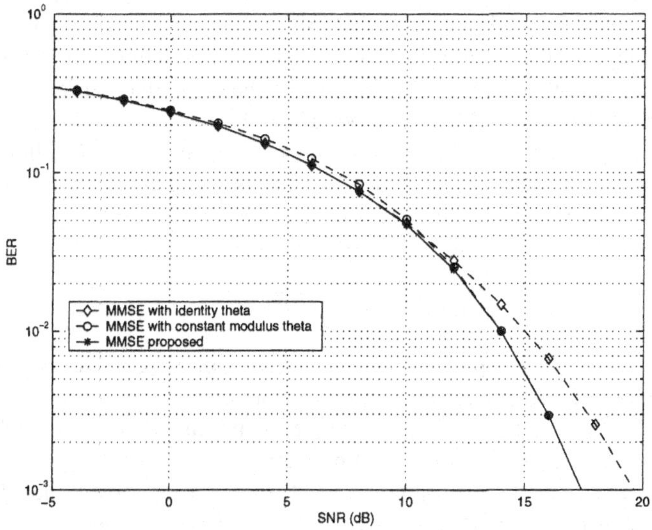

Figure 13.8 Error performance with 16-QAM and infinite-rate feedback for $(N_t, N_r, K) = (6, 3, 3)$

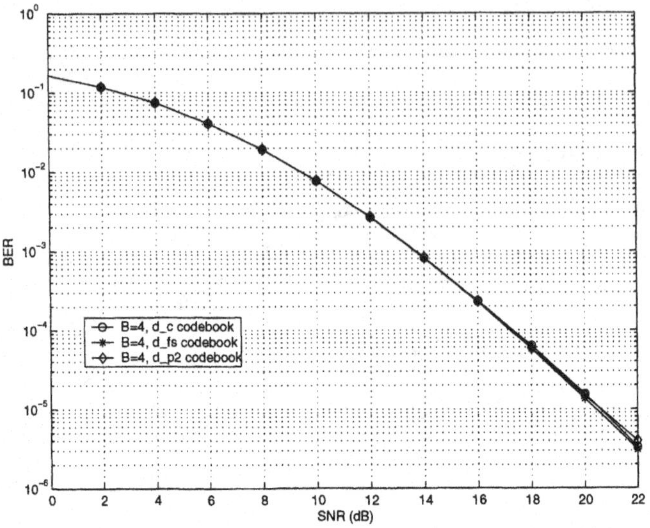

Figure 13.9 Codebooks optimized based on different distances for $(N_t, N_r, K) = (4, 2, 2)$

the better one for each channel realization. Figure 13.7 corroborates the theoretical analysis in Proposition 13.3 for ZF receivers.

From Figure 13.7, we verify that the MMSE receiver with precoder $\boldsymbol{T} = \overline{\boldsymbol{V}}_H \boldsymbol{\Theta}_{\rm cm}$ always outperforms that with precoder $\boldsymbol{T} = \overline{\boldsymbol{V}}_H$ when 4-QAM is used. This is not the case with 16-QAM, where the MMSE receiver with precoder $\boldsymbol{T} = \overline{\boldsymbol{V}}_H \boldsymbol{\Theta}_{\rm cm}$ performs better than that with $\boldsymbol{T} = \overline{\boldsymbol{V}}_H$ when the SNR is at either the high or the low end (the curves at low SNR coincide, but this can be verified numerically). However, it leads to inferior performance in the moderate SNR range, as verified by Figure 13.8.

Summarizing, Figures 13.7 and 13.8 confirm the analytical claims in Proposition 13.4 for MMSE receivers. Proposition 13.5 offers a practical solution for both ZF and MMSE receivers when infinite-rate feedback is available.

Example 13.7 (BER comparison for codebooks optimized under different distance metrics) Table 13.4 lists various codebooks which are optimized based on different definitions of the subspace distance. A pertinent question is which codebook one should use for improved BER performance.

With $(N_t, N_r, K) = (4, 2, 2)$ and $B = 4$, Figure 13.9 depicts the BER for the codebooks with maximized minimum $d_{\rm c}$, $d_{\rm FS}$, and $d_{\rm p2}$. Differences when using different codebooks are not discernible. The same observation applies to all other configurations. Hence, *sticking to the codebook with any distance optimized will be equally good* in terms of BER performance. Although individual codebooks entail optimization of different minimum distances, the impact on the overall error

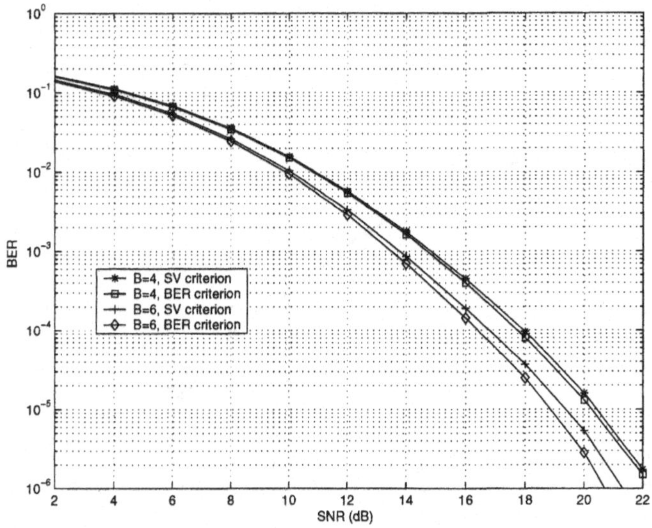

Figure 13.10 ZF receiver: SV and BER criteria for $(N_t, N_r, K) = (6, 3, 3)$

performance is negligible. This has to do with the fact that unlike the maximized minimum distance, BER is averaged over different channel realizations.

Example 13.8 (Comparison among different selection criteria) Here we compare BER-based criteria against those in [166]. Figure 13.10 compares SV against BER selection criteria for ZF receivers. We observe that the SV criterion yields performance quite close to that of the BER-based criterion. This is reasonable because the SV criterion aims at improving the worst SNR for K data streams, implicitly enforcing a certain degree of averaging across subchannel SNRs. On the other hand, Figure 13.11 compares MMSE-trace and MMSE-det with the BER-based criterion for MMSE reception. We observe that MMSE-det performs worse than the MMSE-trace criterion, and both of them are inferior to the BER-based criterion.

Recall also that computing BER is straightforward either through closed-form expressions [42] or using the recursive algorithm in [278]. On the other hand, SV and MMSE-trace do not require knowledge of the underlying signal constellation.

Example 13.9 (Performance improvement with the number of feedback bits) Here we test the performance improvement as a function of the number of feedback bits. Figure 13.12 illustrates this function for $(N_t, N_r, K) = (4, 2, 2)$ with MMSE reception; whereas Figure 13.13 depicts the same function for $(N_t, N_r, K) = (6, 3, 3)$ with ZF reception. The $B = 0$ case corresponds to an open-loop $N_r \times N_r$ system with linear ZF or MMSE reception. The $B = 1$ case corresponds to antenna subset selection with N_t antennas partitioned in two sets, each with K elements. The asymptotic case with $B = \infty$ is also included in order to benchmark performance.

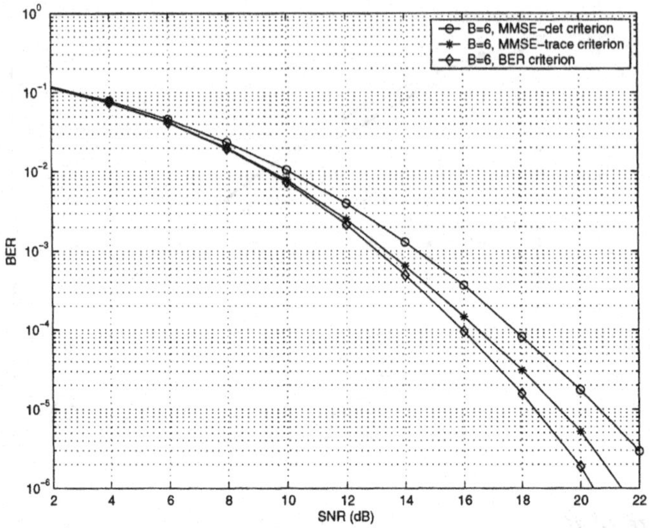

Figure 13.11 MMSE receiver with different selection criteria for $(N_t, N_r, K) = (6, 3, 3)$

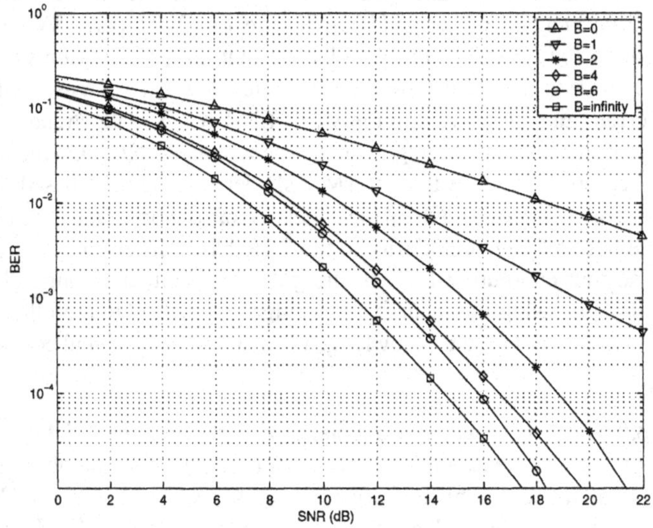

Figure 13.12 Impact of the number of feedback bits for $(N_t, N_r, K) = (4, 2, 2)$; MMSE receiver

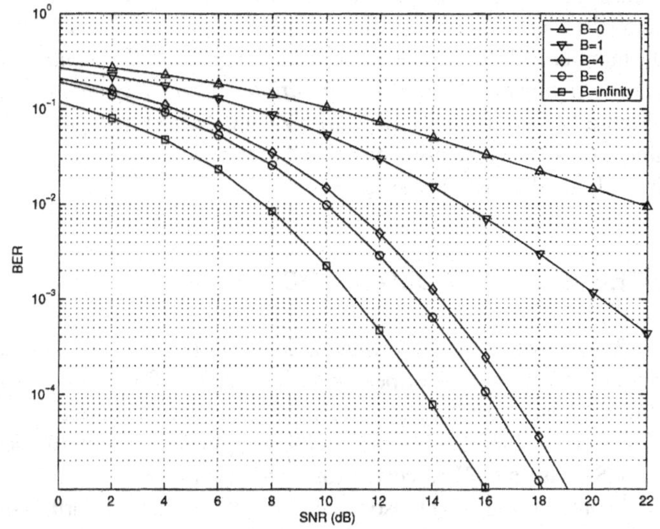

Figure 13.13 Impact of the number of feedback bits for $(N_t, N_r, K) = (6, 3, 3)$; ZF receiver

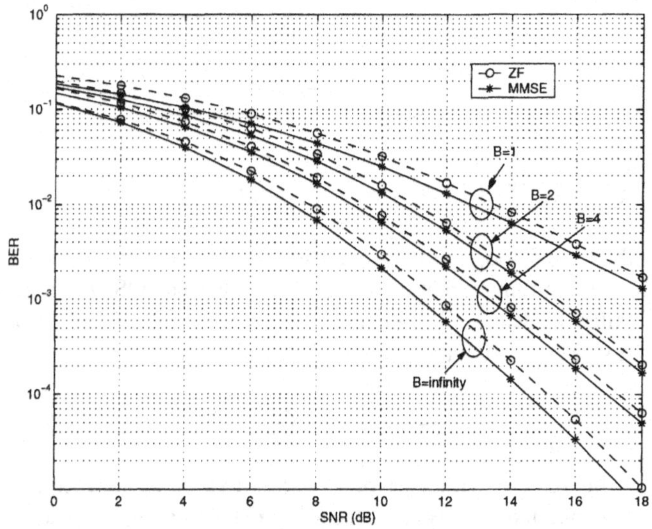

Figure 13.14 Comparison of ZF and MMSE receivers for $(N_t, N_r, K) = (4, 2, 2)$

We observe that: (i) the feedback link improves system performance markedly; (ii) the performance gain exhibits diminishing returns as the number of feedback bits increases; and (iii) a large portion of the feedback gain is achieved with only a relatively small number of bits ($B = 6$ in both cases). Hence, *the number of feedback bits required in practice for closed-loop operation does not have to be large.*

Finally, Figure 13.14 depicts the BER with ZF and MMSE reception. The MMSE receiver outperforms the ZF receiver by a small margin, as expected, since the former incorporates additional SNR knowledge.

13.4 FINITE-RATE PRECODED OSTBC

In Section 13.3, we saw that for a multi-antenna transmitter that relies on precoded ST multiplexing, optimization of error performance based on finite-rate CSI is possible provided that one resorts to suboptimum linear receiver structures. On the other hand, recall that in Section 12.3 it was feasible to optimize error performance based on statistical CSI with an ML receiver for a multi-antenna transmitter combining OSTBC with beamforming. The reason is that an OSTBC system orthogonalizes the MIMO channel and allows for ML reception with a linear receiver. These features also motivate the precoded OSTBC we present in this section based on finite-rate CSI.

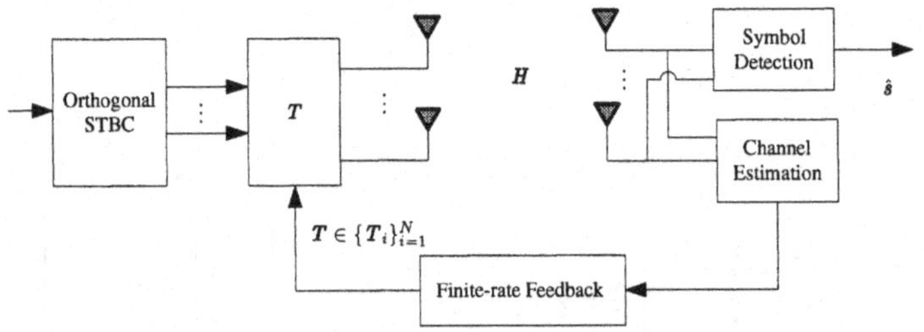

Figure 13.15 OSTBC-precoder system with finite-rate feedback

Consider the closed-loop multi-antenna system depicted in Figure 13.15 with an OSTBC-precoder combination at the transmitter side. The OSTBC matrix $\mathcal{O}_K$, designed as detailed in Section 3.3, encodes incoming symbols and feeds them to a precoder $\boldsymbol{T}$ of size $N_t \times K$ with $K < N_t$. The received ST matrix is

$$\boldsymbol{Y} = \sqrt{\frac{\bar{\gamma}}{N_t}}\boldsymbol{H}\boldsymbol{T}\mathcal{O}_K + \boldsymbol{V}, \tag{13.65}$$

where $\boldsymbol{V}$ contains AWGN with zero mean and unit variance per entry.

With the ST code fixed, only the precoding matrix can be adjusted to adapt the transmission mode depending on the finite-rate CSI. With B feedback bits, the precoder codebook comprising $N = 2^B$ codewords is

$$\mathcal{T} := \{\boldsymbol{T}_1, \ldots, \boldsymbol{T}_N\}, \tag{13.66}$$

and correspondingly, the N aggregate ST code matrices are

$$\{T_i \mathcal{O}_K\}_{i=1}^N. \tag{13.67}$$

13.4.1 Precoder Selection Criterion

Thanks to the nice structure of the OSTBC matrix, one can separately decode each information symbol without interference from other symbols using low-complexity linear receiver processing. Each information symbol is equivalently passing through a scalar channel with SNR [cf. (12.82) and (12.83)]

$$\gamma = \|HT\|_F^2 \bar{\gamma}. \tag{13.68}$$

Since SER and BER are monotonically decreasing as γ increases, maximizing the instantaneous SNR γ is optimal in terms of error performance. For this reason, a reasonable transmission-mode selection rule is

$$T_{\text{opt}} = \arg\max_{T \in \mathcal{T}} \|HT\|_F^2 \tag{13.69}$$

and includes (13.7) as a special case corresponding to $N_r = 1$ and $K = 1$.

13.4.2 Codebook Construction: Infinite Rate

As in the preceding two sections, let us first consider the optimal precoder with $B = \infty$, which corresponds to knowing the channel H perfectly at the transmitter.

Imposing the column-orthonormality constraint on T as in (13.34), the following result has been established in [167]:

Proposition 13.6 *In the asymptotic case with $B = \infty$, the precoder optimizing error performance subject to $T^{\mathcal{H}}T = I_K$ is $T_{\text{opt}} = \overline{V}_H$, where $\overline{V}_H$ is formed by the first K columns of V_H in* (13.45).

With orthonormal columns, the precoder T does not perform any power allocation across the K data streams. Figure 13.15 differs on this point from Figures 12.5 and 12.6, where we allowed for power loading. Subject to a total transmit-power constraint, optimal precoding with perfect CSI has already been established in Section 12.3. For comparison purposes we recall that:

Proposition 13.7 *In the asymptotic case with $B = \infty$, the precoder optimizing error performance subject to the trace constraint $\text{tr}(T^{\mathcal{H}}T) = K$ is $T_{\text{opt}} = \sqrt{K}[v_1, \mathbf{0}, \ldots, \mathbf{0}]$, where v_1 is the first eigenvector of $H^{\mathcal{H}}H$ in* (13.45).

The precoded OSTBC transmission in Proposition 13.7 then reduces to the one-dimensional beamformer-coder combination dealt with in Section 12.3.

With the column-orthonormal precoder T in Proposition 13.6, the instantaneous SNR in (13.68) is

$$\gamma = (\lambda_1 + \ldots + \lambda_K)\bar{\gamma}, \tag{13.70}$$

where $\lambda_1 \geq \cdots \geq \lambda_K$ are the eigenvalues of $\boldsymbol{H}^{\mathcal{H}}\boldsymbol{H}$. Without the column-orthonormality constraint, the optimal precoder $\boldsymbol{T}$ in Proposition 13.7 leads to the instantaneous SNR

$$\gamma = K\lambda_1\bar{\gamma}. \tag{13.71}$$

Notice that imposing column orthonormality entails certain performance loss. However, it enables simple codebook designs, as we will see next.

13.4.3 Codebook Construction: Finite Rate

Similar to (13.12) and (13.57), if the entries of $\boldsymbol{H}$ are i.i.d. complex Gaussian, the precoder codebook is constructed under the column-orthonormality constraint so that [167]

$$\mathcal{T}_{\text{opt}} = \arg\max_{\mathcal{T}} \min_{1\leq i<j\leq N} d_c(\boldsymbol{T}_i, \boldsymbol{T}_j), \tag{13.72}$$

where d_c is the chordal distance defined in (13.54). The optimization in (13.72) can be cast into minimization of an upper bound on the average SNR (or power) loss at the receiver [167]; and good precoder codebooks can be constructed via Lloyd's algorithm following the steps outlined in Section 13.3.3.1.

Systematic codebook construction *without* the column-orthonormality constraint on each precoder is an interesting open problem. For a MISO system, however, transmit-beamforming turns out to be optimal in attaining a SER lower bound when $N \geq N_t$ [341] (i.e., the optimal precoding matrices boil down to beamforming vectors when $N_r = 1$). This result agrees with the conclusion reached by [123] from an average capacity perspective. When $N < N_t$, optimal precoding matrices for a MISO system amount to antenna selection within the first N antennas and uniform power allocation across the remaining $N - N_t$ antennas [341].

13.4.4 Numerical Examples

In this subsection we present numerical examples for a closed-loop OSTBC-precoder system with $(N_t, N_r, K) = (4, 2, 2)$. As $K = 2$, the Alamouti code is adopted to avoid loss in transmission rate that larger OSTBC matrices entail with complex constellations. While varying the number of feedback bits, we will test the precoder codebook obtained using Lloyd's algorithm (see Table 13.4), where each precoder has orthonormal columns.

Figure 13.16 confirms that finite-rate feedback indeed improves error performance. With a small number of feedback bits, BER approaches the asymptotic performance attained with $B = \infty$. Diminishing gains are observed as the number of feedback bits grows large.

More important, Figure 13.16 demonstrates that the optimal precoder without the column-orthonormality constraint outperforms considerably its counterpart under the column-orthonormality constraint, when perfect CSI is available at the transmitter. This suggests that even in the finite-rate case, codebook construction with non-unitary precoders is worthwhile to pursue for further performance enhancement.

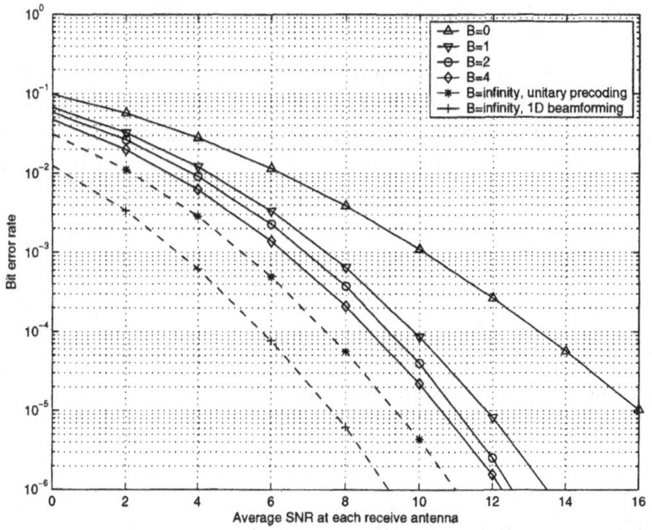

Figure 13.16 BER of a closed-loop OSTBC-precoder system with $(N_t, N_r, K) = (4, 2, 2)$

13.5 CAPACITY OPTIMIZATION WITH FINITE-RATE FEEDBACK

In this section we investigate transceiver optimization with finite-rate feedback from an information-theoretic perspective. With B feedback bits, the system needs to prepare a codebook comprising $N = 2^B$ transmission modes. Using capacity as a criterion, the transmission modes are characterized by covariance matrices of the channel input. Specifically, the mutual information $I(\boldsymbol{x}; \boldsymbol{y})$ for the input-output relationship

$$\boldsymbol{y} = \sqrt{\frac{\bar{\gamma}}{N_t}} \boldsymbol{H}\boldsymbol{x} + \boldsymbol{v}$$

is maximized when the input $\boldsymbol{x}$ is Gaussian distributed. If $\boldsymbol{R} := E\{\boldsymbol{x}\boldsymbol{x}^{\mathcal{H}}\}$, the transmission-mode codebook consists of N covariance matrices

$$\{\boldsymbol{R}_{x,i}\}_{i=1}^{N}. \tag{13.73}$$

13.5.1 Selection Criterion

Using capacity as a figure of merit, a natural choice for the input covariance matrix is to maximize the instantaneous mutual information [20, 146]; that is,

$$\boldsymbol{R}_{\text{opt}} = \underset{\boldsymbol{R} \in \{\boldsymbol{R}_{x,i}\}_{i=1}^{N}}{\arg\max} \; \log_2 \left| \boldsymbol{I}_{N_r} + \frac{\bar{\gamma}}{N_t} \boldsymbol{H}\boldsymbol{R}\boldsymbol{H}^{\mathcal{H}} \right|. \tag{13.74}$$

Matrix $\boldsymbol{R}_{\text{opt}}$ is obtained at the receiver, and its index is fed back to the transmitter.

13.5.2 Codebook Design

The codebook design can again be guided by related vector quantization principles [20, 146]. Supposing that the channel space is partitioned into N disjoint regions so that $\boldsymbol{H}$ falls into the ith region $\mathcal{R}_i$, the ith covariance matrix $\boldsymbol{R}_{x,i}$ is optimal according to the selection rule of (13.74). With $A_i := \mathrm{P}(\boldsymbol{H} \in \mathcal{R}_i)$ denoting the probability that $\boldsymbol{H}$ falls in the ith region, the mutual information averaged over all possible channel realizations is

$$E\left\{I(\boldsymbol{x};\boldsymbol{y}|\{\mathcal{R}_i,\boldsymbol{R}_{x,i}\}_{i=1}^N)\right\} = \sum_{i=1}^{N} E_{\boldsymbol{H}\in\mathcal{R}_i}\left\{\log_2\left|\boldsymbol{I}_{N_r} + \frac{\bar{\gamma}}{N_t}\boldsymbol{H}\boldsymbol{R}_{x,i}\boldsymbol{H}^{\mathcal{H}}\right|\right\} A_i. \tag{13.75}$$

The ergodic capacity with finite-rate feedback is thus

$$C = \max_{\{\mathcal{R}_i,\boldsymbol{R}_{x,i}\}} E\left\{I(\boldsymbol{x};\boldsymbol{y}|\{\mathcal{R}_i,\boldsymbol{R}_{x,i}\}_{i=1}^N)\right\}, \tag{13.76}$$

subject to either an instantaneous power constraint,

$$\mathrm{tr}[\boldsymbol{R}_{x,i}] = N_t, \;\; \forall i, \tag{13.77}$$

or an average power constraint,

$$\sum_{i=1}^{N} A_i \,\mathrm{tr}[\boldsymbol{R}_{x,i}] = N_t. \tag{13.78}$$

Capacity optimization with the instantaneous power constraint (13.77) is addressed in [20] and with the average power constraint (13.78) in [146].

Constraining the power to be constant over all transmission modes as in (13.77), the capacity optimization problem can be solved using Lloyd's algorithm [86]. Specifically, the generalized Lloyd's algorithm in this case implements the following two steps iteratively:

S1) Given transmission modes $\{\boldsymbol{R}_{x,i}\}_{i=1}^N$, partition the channels according to a *nearest-neighbor* rule [146]:

$$\boldsymbol{H} \in \mathcal{R}_i \quad \text{if} \quad d(\boldsymbol{H},\boldsymbol{R}_{x,i}) \le d(\boldsymbol{H},\boldsymbol{R}_{x,j}), \quad \forall j \ne i, \tag{13.79}$$

where $d(\boldsymbol{H},\boldsymbol{R}_x) = -\log_2|\boldsymbol{I}_{N_r} + (\bar{\gamma}/N_t)\boldsymbol{H}\boldsymbol{R}_x\boldsymbol{H}^{\mathcal{H}}|$ is a capacity-related distance metric replacing the Euclidean distance, which is typically adopted in vector quantization problems.

S2) Given channel space partitions $\{\mathcal{R}_i\}_{i=1}^N$, determine the optimal transmission modes $\{\boldsymbol{R}_{x,i}\}_{i=1}^N$ based on the *centroid condition*

$$\boldsymbol{R}_{x,i} = \arg\min_{\boldsymbol{R}_x;\mathrm{tr}\{\boldsymbol{R}_x\}=N_t} E_{\boldsymbol{H}\in\mathcal{R}_i} d(\boldsymbol{H},\boldsymbol{R}_x). \tag{13.80}$$

Although clearly formulated, the solution of (13.80) is in general difficult to obtain; and so far, only approximate solutions are available [20, 146].

On the other hand, capacity optimization with the average power constraint corresponds to a vector quantization problem with an additional entropy constraint on the codeword probabilities [45]. The generalized Lloyd's algorithm is still applicable in this case to maximize the corresponding objective function

$$E\{I(\boldsymbol{x};\boldsymbol{y}|\{\mathcal{R}_i, \boldsymbol{R}_{x,i}\}_{i=1}^N)\} + \lambda \sum_{i=1}^{N} A_i \operatorname{tr}\{\boldsymbol{R}_{x,i}\},$$

where λ is the Lagrange multiplier, as detailed in [146].

13.6 COMBINING ADAPTIVE MODULATION WITH BEAMFORMING

In previous sections we saw how finite-rate feedback can be used to improve BER and the capacity of closed-loop multi-antenna systems. Our aim in this section is to improve the transmission rate via adaptive modulation used in conjunction with transmit-beamforming in a MISO system ($N_r = 1$), subject to a target BER performance; see also [138], where adaptive modulation is combined with OSTBC. Different from Figure 13.2, we now allow both the transmission power and the signal constellation to vary, as depicted in Figure 13.17.

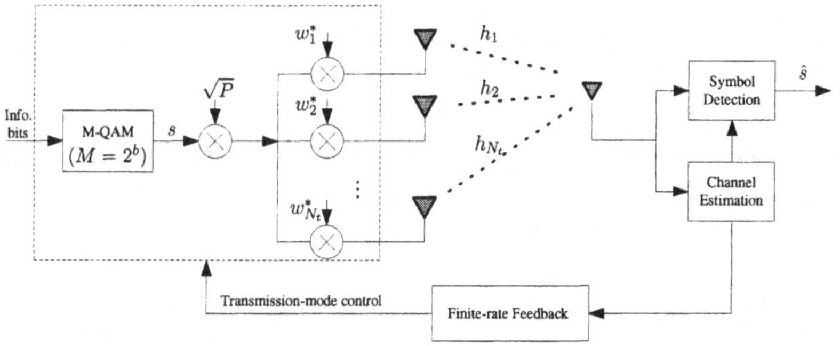

Figure 13.17 Closed-loop MISO system with adaptive modulation/beamforming

With P_t denoting transmit-power normalized by the noise variance, the information symbol s is drawn from a constellation of size M with average energy $E_s = 1$. Using the beamformer $\boldsymbol{w}$, the vector $\boldsymbol{w}^{\mathcal{H}}\sqrt{P_t}\, s$ is transmitted from $N_t > 1$ antennas. The symbol received is thus [cf. (13.3)]

$$y = \boldsymbol{w}^{\mathcal{H}}\boldsymbol{h}\sqrt{P_t}\, s + v, \tag{13.81}$$

where v is AWGN with zero mean and unit variance.

The set of candidate modulations will include rectangular and square QAM constellations of size $M = 2^b$, where b is a positive integer. The transmission mode we will adapt in this setup is then characterized by the triplet (constellation size, transmission power, beamforming vector), which we denote as

$$\mathcal{M} := (M, P, \boldsymbol{w}). \tag{13.82}$$

With B feedback bits, the transmitter needs to construct $N = 2^B$ transmission modes, with each mode defined as

$$\mathcal{M}_i := \left(M_i, P_i, \boldsymbol{w}_i\right). \tag{13.83}$$

Notice that different transmission modes may adopt the same signal constellation. The collection of all transmission modes comprises the codebook in (13.1).

13.6.1 Mode Selection

When the transmission mode selected is $\mathcal{M}_i$, it follows from (13.81) that the output SNR is

$$\gamma_i = \left|\boldsymbol{w}_i^{\mathcal{H}}\boldsymbol{h}\right|^2 P_i. \tag{13.84}$$

The BER can then be computed for each transmission in closed form as a function of the SNR and the signal constellation [cf. (13.39)]. Let $\text{BER}(\boldsymbol{h}, \mathcal{M}_i)$ denote the instantaneous BER. Aiming to improve the transmission rate subject to a prescribed instantaneous $\text{BER}_{\text{target}}$, we select the transmission mode as

$$\mathcal{M}_{\text{opt}} = \underset{\substack{\mathcal{M}\in\{\mathcal{M}_1,\ldots,\mathcal{M}_N\},\\ \text{BER}(\boldsymbol{h},\mathcal{M})\leq \text{BER}_{\text{target}}}}{\arg\max} \log_2 M. \tag{13.85}$$

If, on the other hand, we wish to improve the transmission rate subject to an average BER constraint, as formulated in the next subsection, the selection rule needs to be modified accordingly [310].

13.6.2 Codebook Design

At the receiver side, the channel space is partitioned in N non-overlapping regions, denoted by $\mathcal{R}_1, \ldots, \mathcal{R}_N$. The design of these fading regions and transmission modes determines the achievable data rate. Our ultimate goal is to design $\{\mathcal{R}_i\}_{i=1}^N$ and $\{\mathcal{M}_i\}_{i=1}^N$ *jointly* to maximize the system transmission rate, subject to certain average error performance and power constraints.

The BER averaged over the channels in the region $\mathcal{R}_i$ can be written as

$$\overline{\text{BER}}_i = E_{\boldsymbol{h}\in\mathcal{R}_i}\left[\text{BER}(\boldsymbol{h}, \mathcal{M}_i)\right]. \tag{13.86}$$

Let $b_i := \log_2 M_i$ denote the number of bits associated with a constellation of size M_i, $A_i := \text{P}(\boldsymbol{h} \in \mathcal{R}_i)$ denote the probability that the channel vector $\boldsymbol{h}$ lies in region $\mathcal{R}_i$, $\overline{P}_0$ stand for the average power (normalized by the noise variance), and $\overline{\text{BER}}_{\text{target}}$

be the prescribed average BER performance. By jointly designing $\{\mathcal{R}_i\}_{i=1}^N$ and $\{\mathcal{M}_i\}_{i=1}^N$, the ultimate objective is to:

$$\begin{aligned}
&\text{maximize} && \sum_{i=1}^{N} A_i b_i, \\
&\text{subject to} \quad \text{(i)} && \sum_{i=1}^{N} A_i P_i \le \overline{P}_0, \\
&\text{(ii)} && \|\boldsymbol{w}_i\| = 1, \quad \forall i, \\
&\text{(iii)} && b_i \in \{0,1,2,3,\ldots\}, \quad \forall i, \\
&\text{(iv)} && \text{an average BER constraint} \\
& && \overline{\text{BER}}_i \le \overline{\text{BER}}_{\text{target}}, \quad \forall i, \\
& && \text{or an instantaneous BER constraint} \\
& && \text{BER}(\boldsymbol{h}, \mathcal{M}_i) \le \text{BER}_{\text{target}}, \forall \boldsymbol{h} \in \mathcal{R}_i.
\end{aligned} \tag{13.87}$$

Equation (13.87) provides the general formulation for optimizing, based on finite-rate feedback, the design of multi-antenna transmitters using adaptive modulation and one-dimensional beamforming. Because carrying out this optimization problem appears intractable, suboptimal alternatives are well motivated.

One possible approach is to rely on the finite-rate beamformer of Section 13.2. For example, one can divide the B feedback bits into two parts, $B = B_1 + B_2$, with B_2 bits dedicated to beamforming vectors and the remaining B_1 bits conveying information about power loading and constellation selection. With the beamformers designed as in Section 13.2, the task is to exploit the finite-rate CSI in deriving optimal power and bit allocation schemes. The optimal distribution of B_1 and B_2 must also be specified in this approach.

Another approach is provided by [310], where the problem in (13.87) is reduced to a simpler vector quantization problem through a series of approximations. The iterative algorithm described in Section 13.5 is also modified and used to find an approximate solution, as detailed in [310].

13.7 FINITE-RATE FEEDBACK IN MIMO OFDM

So far in this chapter we delineated the role of finite-rate CSI in closed-loop multi-antenna systems signaling over flat fading MIMO channels. With symbol rates targeted by broadband wireless communication systems increasing, the frequency selectivity of the underlying MIMO channel becomes increasingly pronounced. In this context, the open-loop MIMO OFDM system we presented in Chapter 8 offers an attractive solution and motivates looking into closed-loop options based on finite-rate feedback to enhance its error and rate performance.

To this end, we can simply apply the transmission-mode selection rules and codebook constructions of Sections 13.2 to 13.4 using finite-rate CSI on a per subcarrier basis. Indeed, if each subcarrier is assigned B feedback bits, we can design a common codebook $\mathcal{T}$ of size $N = 2^B$ for all subcarriers. For a MIMO OFDM system with N_c subcarriers, the total number of feedback bits will then be $N_c B$.

Exploiting the fact that channel responses across OFDM subcarriers are correlated, the feedback overhead can be reduced considerably if we interpolate the finite-rate CSI. Such an interpolation-based scheme is described in [43, 44], where: (i) the subcarriers are divided into (say, N_g) groups, (ii) only the index of the optimal beamforming vector or precoding matrix corresponding to one subcarrier from each group is fed back, and (iii) the beamforming vectors or precoding matrices for the rest of the subcarriers are obtained by interpolating those specified by the feedback. The feedback overhead includes $N_g B$ bits plus the non-negligible number of bits needed to specify the interpolation parameters.

Viewing reduction in the feedback overhead as a compression problem for correlated sources, two methods are proposed in [337] to minimize the number of bits using tools from the vector quantization literature [86]. One is recursive feedback encoding, where optimal beamforming vectors or precoding matrices are selected sequentially across the subcarriers. A time-varying codebook of smaller size (say, 2^{B_2}) is then adopted per subcarrier, depending on prior decisions. This recursive approach is in principle similar to differential pulse-coded modulation (DPCM). Instead of coding per subcarrier the codeword index itself, the DPCM-based approach quantizes the relative difference with respect to the previous codeword by searching only over its neighbors.

The second method entails trellis-based feedback encoding where the optimal beamforming/precoding decisions for all subcarriers are made at once by searching through a trellis structure based on Viterbi's algorithm. Both methods require the same number of feedback bits (about $N_c B_2$). The trellis-based approach outperforms the recursive one at the price of increased receiver complexity.

13.8 CLOSING COMMENTS

In this chapter, closed-loop multi-antenna systems were designed using finite-rate CSI which becomes available to the transmitter through feedback from the receiver. MIMO transmitters were considered based on beamforming with and without adaptive modulation, on unitary-precoded ST multiplexing to further increase transmission rates, and on precoded OSTBCs, which can afford ML decoding at low complexity. All designs were cast in a vector quantization framework, which entailed offline construction of a codebook to specify a finite number of optimal transmission modes (beamformer or precoder codewords). One codeword per feedback cycle was selected from this codebook in the online operation to "best" adapt the transmitter to the quantized MIMO channel codeword conveyed by the feedback bits.

Average error probability was used as a figure of merit in both the codebook design and in the codeword selection rule. Capacity-based designs as well as MIMO OFDM extensions to frequency-selective channels were discussed briefly. Analysis and numerical examples confirmed that even a few bits which can be made available pragmatically through feedback improve considerably the performance of closed-loop MIMO systems relative to their open-loop counterparts. This justifies the popularity of finite-rate CSI based MIMO OFDM systems considered in the IEEE 802.11n WLAN standard, which aims at data rates exceeding 100 Mbps, as well as in the IEEE 802.16 WMAN, standard which targets high-speed broadband wireless access for integrated data, voice, and video services.

14

ST Codes in the Presence of Interference

The open- and closed-loop multi-antenna systems dealt with so far are designed to operate over fading MIMO channels with AWGN present at the receiver. In certain applications, however, wireless links are also affected by additional forms of intentional or unintentional interference. In a direct-sequence code-division multiple access (DS-CDMA) system, for example, multiuser interference (MUI) emerges naturally due to non-zero cross-correlation among spreading sequences. Although this type of MUI can be eliminated or controlled by designing spreading sequences judiciously, the *co-site MUI* can still arise when multiple systems operate simultaneously. In tactical communications, radios operate in hostile environments where intentional jamming is almost always present. The challenge is that unlike AWGN, interference may be narrowband or broadband, periodic or aperiodic, continuous or intermittent, and colored or white in space or/and time, depending on the source. Whatever the source, interference will adversely affect the reliability of wireless links, and ST codes must explicitly account for it in order to enhance the robustness of ST-coded multi-antenna systems.

As far as fundamental limits are concerned, the capacity of MIMO channels in the presence of interference was investigated in [19,201]. Receivers designed to cope with spatially correlated interference were reported in [143] for multi-antenna transmissions with orthogonal space-time codes (OSTBCs). For single-antenna transmissions, spreading sequences were also constructed in [254,255] to combat interference with known correlation structure in a spread-spectrum CDMA system. Intuition gained by these results suggests that exploiting knowledge of the interference properties at the transmitter side should also be beneficial for single-user multi-antenna systems. In fact, the closed-loop designs in Chapter 12 demonstrated that the performance of

ST-coded systems can be improved considerably when statistical channel knowledge is available at the transmitter. In the same spirit, multi-antenna transmitters in the present chapter will utilize the channel's spatial correlation and the interference's temporal correlation to robustify the operation of ST codes in the presence of interference. In particular, interference will be treated as zero-mean colored Gaussian noise with correlation perfectly known at both the transmitter and receiver.

A low-rate space-time spreading (STS) transmission scheme is presented first to establish that if one transmits along directions where the channels are "strong" and the interference is "weak," considerable performance gain can be achieved. STS-based systems will achieve the diversity provided by the MIMO channel and effectively suppress interference at the price of bandwidth expansion. To improve spectral efficiency without sacrificing diversity, serial STS-OSTBC combinations will be devised to increase the transmission rate up to 1 symbol per channel use. But as in all ST-coded transmissions, in order to collect the maximum available diversity, ML decoding is required. The latter can be prohibitively complex in the presence of interference and certainly requires estimation of the MIMO channel, which is also more challenging when interference is present. These considerations will guide the design of low-complexity suboptimal receivers and will suggest optimization of training approaches dealt with in Chapter 11 when knowledge of the interference color is available at the transmitter.

14.1 ST SPREADING

Consider a wireless communication system with N_t transmit-antennas and a single receive-antenna ($N_r = 1$) signaling over a flat fading MIMO channel. Extension to multiple receive-antennas is easy if what follows is applied on a per receive-antenna basis. As in the STS scheme in Section 12.2, each information symbol s is spread by a different length-P spreading code $\boldsymbol{c}_\mu := [c_\mu(0), \ldots, c_\mu(N_x)]^T$ per transmit-antenna. Hence, the transmit-antenna array spreads s over both space and time. The pertinent ST code matrix of size $N_t \times N_x$ is

$$\boldsymbol{X} = \boldsymbol{C}^T s, \tag{14.1}$$

where $\boldsymbol{C} := [\boldsymbol{c}_1, \ldots, \boldsymbol{c}_{N_t}]$ is the $N_x \times N_t$ STS code matrix. Since one symbol is transmitted every N_x time slots, this system has a low rate ($1/N_x$ symbol per channel use).

Let us collect the received samples in an $N_x \times 1$ block $\boldsymbol{y}$ to obtain the input-output relationship [cf. (2.13)]

$$\boldsymbol{y} = \sqrt{\bar{\gamma}}\, \boldsymbol{C}\boldsymbol{h}s + \sqrt{\bar{\gamma}_i}\, \boldsymbol{w}_i + \boldsymbol{w}_n, \tag{14.2}$$

where $\boldsymbol{h} := [h_1, \ldots, h_{N_t}]^T$ collects the channel coefficients between N_t transmit-antennas and the receive-antenna, $\boldsymbol{w}_i$ represents the interference, and $\boldsymbol{w}_n$ denotes AWGN with zero mean and covariance matrix $\boldsymbol{I}_{N_x}$. Upon constraining $\boldsymbol{C}$ to satisfy

$$\text{tr}(\boldsymbol{C}\boldsymbol{C}^{\mathcal{H}}) = 1$$

and normalizing s so that $E[|s|^2] = 1$, it can be verified that $\bar{\gamma}$ in (14.2) represents the ratio of the average energy transmitted per information symbol to noise variance. We model $\boldsymbol{w}_i$ as a Gaussian random vector with zero mean and covariance matrix $\boldsymbol{R}_i$ with all diagonal entries normalized to 1. Consequently, $\bar{\gamma}_i$ stands for the interference-to-noise ratio (INR). We consider quasi-static channels (i.e., $\boldsymbol{h}$ is allowed to vary from block to block but remains invariant during each block). We further assume that $\boldsymbol{h}$ is complex Gaussian distributed with zero mean and covariance $\boldsymbol{R}_h := E[\boldsymbol{h}\boldsymbol{h}^{\mathcal{H}}]$.

As argued in Section 12.1.1.2, although the channel itself may change rapidly, the channel correlation $\boldsymbol{R}_h$ varies slowly. The interference covariance matrix $\boldsymbol{R}_i$ is determined by the interference's power spectral density (PSD) or/and the interference's transmit-antenna correlation, and thus varies slowly with time. Since both $\boldsymbol{R}_h$ and $\boldsymbol{R}_i$ are slowly varying, they can be estimated at the receiver and fed back to the transmitter. Supposing that $\boldsymbol{R}_h$ and $\boldsymbol{R}_i$ are known at the transmitter, the objective in this section is to find an STS matrix $\boldsymbol{C}$ that optimizes error performance.

14.1.1 Maximizing the Average SINR

Let us define $\boldsymbol{w} := \sqrt{\bar{\gamma}_i}\,\boldsymbol{w}_i + \boldsymbol{w}_n$ and rewrite (14.2) as

$$\boldsymbol{y} = \sqrt{\bar{\gamma}}\,\boldsymbol{C}\boldsymbol{h}s + \boldsymbol{w}. \tag{14.3}$$

Equation (14.3) is a special form of (12.14) with $N_r = 1$, except that the vector $\boldsymbol{w}$ in (14.3) contains both interference and noise and is generally colored with covariance matrix $\boldsymbol{R}_w := E[\boldsymbol{w}\boldsymbol{w}^{\mathcal{H}}] = \bar{\gamma}_i\boldsymbol{R}_i + \boldsymbol{I}_{N_x}$.

Supposing that the channel $\boldsymbol{h}$ is perfectly known at the receiver, the ML optimal processor for detecting s from $\boldsymbol{y}$ entails a pre-whitening filter followed by a maximum ratio combining (MRC) module. The decision variable at the MRC output is

$$z = \sqrt{\bar{\gamma}}\,\boldsymbol{h}^{\mathcal{H}}\boldsymbol{C}^{\mathcal{H}}\boldsymbol{R}_w^{-1}\boldsymbol{C}\boldsymbol{h}s + \tilde{w}, \tag{14.4}$$

where $\tilde{w} := \boldsymbol{h}^{\mathcal{H}}\boldsymbol{C}^{\mathcal{H}}\boldsymbol{R}_w^{-1}\boldsymbol{w}$ has zero mean and variance $\boldsymbol{h}^{\mathcal{H}}\boldsymbol{C}^{\mathcal{H}}\boldsymbol{R}_w^{-1}\boldsymbol{C}\boldsymbol{h}$. The signal-to-interference-plus-noise-ratio (SINR) in (14.4) is given by

$$\gamma_z = \bar{\gamma}\boldsymbol{h}^{\mathcal{H}}\boldsymbol{C}^{\mathcal{H}}\boldsymbol{R}_w^{-1}\boldsymbol{C}\boldsymbol{h}. \tag{14.5}$$

Depending on the figure of merit chosen, the optimal STS matrix $\boldsymbol{C}$ may take different forms. The goal of this subsection is to find the $\boldsymbol{C}$ that maximizes the average SINR

$$\begin{aligned}\bar{\gamma}_z :&= E[\gamma_z] = \bar{\gamma}E[\mathrm{tr}(\boldsymbol{C}^{\mathcal{H}}\boldsymbol{R}_w^{-1}\boldsymbol{C}\boldsymbol{h}\boldsymbol{h}^{\mathcal{H}})]\\ &= \bar{\gamma}\,\mathrm{tr}(\boldsymbol{C}^{\mathcal{H}}\boldsymbol{R}_w^{-1}\boldsymbol{C}\boldsymbol{R}_h),\end{aligned} \tag{14.6}$$

under the constraint $\mathrm{tr}(\boldsymbol{C}\boldsymbol{C}^{\mathcal{H}}) = 1$.

Since $\boldsymbol{R}_h$ and $\boldsymbol{R}_w$ are both Hermitian, they can be eigen-decomposed as

$$\begin{aligned}\boldsymbol{R}_h &= \boldsymbol{U}_h\boldsymbol{\Lambda}_h\boldsymbol{U}_h^{\mathcal{H}},\\ \boldsymbol{R}_w &= \boldsymbol{U}_w\boldsymbol{\Lambda}_w\boldsymbol{U}_w^{\mathcal{H}},\end{aligned}$$

where $\boldsymbol{\Lambda}_h := \text{diag}(\lambda_{h,1}, \ldots \lambda_{h,N_t})$ and $\boldsymbol{\Lambda}_w := \text{diag}(\lambda_{w,1}, \ldots, \lambda_{w,N_x})$ are diagonal matrices containing the eigenvalues of $\boldsymbol{R}_h$ and $\boldsymbol{R}_w$, while the unitary matrices $\boldsymbol{U}_h := [\boldsymbol{u}_{h,1}, \ldots, \boldsymbol{u}_{h,N_t}]$ and $\boldsymbol{U}_w := [\boldsymbol{u}_{w,1}, \ldots, \boldsymbol{u}_{w,N_x}]$ comprise the eigenvectors of $\boldsymbol{R}_h$ and $\boldsymbol{R}_w$, respectively. For convenience and without loss of generality, we assume that the diagonal entries of $\boldsymbol{\Lambda}_h$ are arranged in a non-increasing order: $\lambda_{h,1} \geq \ldots \geq \lambda_{h,N_t}$, while those of $\boldsymbol{\Lambda}_w$ are arranged in a non-decreasing order: $\lambda_{w,1} \leq \ldots \leq \lambda_{w,N_x}$. The following proposition proved in [35] characterizes the SINR-maximizing $\boldsymbol{C}$:

Proposition 14.1 *The optimal STS matrix $\boldsymbol{C}$ maximizing the average SINR in* (14.6) *under the constraint* $\text{tr}(\boldsymbol{C}\boldsymbol{C}^{\mathcal{H}}) = 1$ *is given by the rank-one matrix* $\boldsymbol{C}_{opt} = \alpha \boldsymbol{u}_{w,1}\boldsymbol{u}_{h,1}^{\mathcal{H}}$, *where α is a constant satisfying $|\alpha| = 1$, $\boldsymbol{u}_{h,1}$ is the eigenvector of $\boldsymbol{R}_h$ corresponding to the largest eigenvalue $\lambda_{h,1}$, and $\boldsymbol{u}_{w,1}$ is the eigenvector of $\boldsymbol{R}_w$ corresponding to the smallest eigenvalue $\lambda_{w,1}$. The maximum average SINR at the MRC output is given by $\bar{\gamma}_{z,\max} = \bar{\gamma}\lambda_{h,1}/\lambda_{w,1}$.*

Proposition 14.1 reveals that the optimal STS matrix $\boldsymbol{C}$ should concentrate transmit-power on the strongest eigen-direction of the channel and the weakest eigen-direction of the interference to maximize the average SINR, which is intuitively reasonable. Note that Proposition 14.1 generalizes the SNR maximizing beamformer derived in [203] in the absence of interference. Note also that in proving Proposition 14.1, there is no need to assume that the interference is Gaussian distributed.

Although proportional to SINR, the ultimate metric at the physical layer of a communication system is error probability. In fact, maximizing the average SINR does not necessarily minimize error probability in fading channels, because error probability is not only determined by the average SINR but also is affected by the probability density function (pdf) of the SINR. We are thus motivated to look for the STS matrix $\boldsymbol{C}$, which optimizes error probability performance.

14.1.2 Minimizing the Average Error Bound

Assuming that the interference is Gaussian distributed, an exact symbol error rate (SER) expression for the STS system described by (14.2) can be found in [35]. However, it is not in a form that is amenable to optimization with respect to $\boldsymbol{C}$. For this reason, we resort to a high-SNR pairwise error probability (PEP) approximation of the SER, which will allow for optimizing $\boldsymbol{C}$. Let $P_2(s \to \tilde{s})$ denote the PEP (when s is erroneously decoded as $\tilde{s}$) averaged over all possible channel realizations. Following the steps in Section 2.4, $P_2(s \to \tilde{s})$ can be Chernoff-bounded by

$$\begin{aligned} P_2(s \to \tilde{s}) &< \det{}^{-1}\Big(\boldsymbol{I}_{N_t} + \frac{\bar{\gamma}|s-\tilde{s}|^2}{4}\boldsymbol{C}^{\mathcal{H}}\boldsymbol{R}_w^{-1}\boldsymbol{C}\boldsymbol{R}_h\Big) \\ &\leq \underbrace{\det{}^{-1}\Big(\boldsymbol{I}_{N_t} + \frac{\bar{\gamma}\Delta_{\min}^2}{4}\boldsymbol{C}^{\mathcal{H}}\boldsymbol{R}_w^{-1}\boldsymbol{C}\boldsymbol{R}_h\Big)}_{P_{\text{bound}}}, \end{aligned} \tag{14.7}$$

where $\Delta^2_{\min}$ is the minimum of $|s - \tilde{s}|^2$ over the set $\{(s, \tilde{s}) | s \neq \tilde{s}\}$. With proper scaling, the Chernoff bound P_{bound} can also serve as a lower bound on the average SER and the gap between the upper and lower bounds is relatively small [331]. Hence, to reduce the SER, it is reasonable to seek a $\boldsymbol{C}$ that minimizes P_{bound}. To this end, it would have been convenient if the product of matrices that P_{bound} depends on could be diagonalized. Before exploring whether this is feasible, it is useful to examine a couple of special cases for which such a diagonalization is possible.

Using the singular value decomposition (SVD), we can write $\boldsymbol{C} = \boldsymbol{U}_c \boldsymbol{D}_c \boldsymbol{V}_c^{\mathcal{H}}$, where matrices $\boldsymbol{U}_c$ and $\boldsymbol{V}_c$ are unitary and $\boldsymbol{D}_c$ contains the singular values of $\boldsymbol{C}$. In the absence of interference, the vector $\boldsymbol{w}$ is white with $\boldsymbol{R}_w = \boldsymbol{I}_{N_x}$ and the optimal $\boldsymbol{C}$ minimizing P_{bound} is provided by Proposition 12.1. The latter asserts that $\boldsymbol{V}_c = \boldsymbol{U}_h$, and $\boldsymbol{U}_c$ can be any unitary matrix, which allows the optimal $\boldsymbol{D}_c$ to be found analytically [cf. (12.63)]. If, on the other hand, the entries of $\boldsymbol{h}$ are independent and identically distributed (i.i.d.), we have $\boldsymbol{R}_h = \boldsymbol{I}_{N_t}$, and as we proved in Proposition 12.2 the optimal STS matrix must have $\boldsymbol{U}_c = \boldsymbol{U}_w$, whereas $\boldsymbol{V}_c$ can be any unitary matrix and the optimal $\boldsymbol{D}_c$ can be found in closed form [cf. (12.76)].

When both $\boldsymbol{R}_h$ and $\boldsymbol{R}_w$ are not proportional to the identity matrix, diagonalizing the product of matrices in P_{bound} is impossible. For this reason, motivated by the two special cases, we confine our search for the optimal STS matrix to the class described by

$$\boldsymbol{C} = \boldsymbol{U}_w \boldsymbol{P} \boldsymbol{D}_c \boldsymbol{U}_h^{\mathcal{H}}, \tag{14.8}$$

where $\boldsymbol{P}$ is a $N_x \times N_x$ permutation matrix. Letting $N_{\min} := \min(N_x, N_t)$, we can express the $N_x \times N_t$ matrix $\boldsymbol{D}_c$ as $\boldsymbol{D}_c := [\tilde{\boldsymbol{D}}_c, \boldsymbol{0}]^T$ if $N_x > N_t$, and $\boldsymbol{D}_c := [\tilde{\boldsymbol{D}}_c, \boldsymbol{0}]$ if $N_x < N_t$ with $\tilde{\boldsymbol{D}}_c$ being an $N_{\min} \times N_{\min}$ diagonal matrix. With $\boldsymbol{C}$ chosen as in (14.8), the error probability bound P_{bound} in (14.7) reduces to

$$P_{\text{bound}} = \det^{-1}\Big(\boldsymbol{I}_{N_t} + \frac{\bar{\gamma}\Delta^2_{\min}}{4} \boldsymbol{D}_c^{\mathcal{H}} \boldsymbol{P}^{\mathcal{H}} \boldsymbol{\Lambda}_w^{-1} \boldsymbol{P} \boldsymbol{D}_c \boldsymbol{\Lambda}_h\Big). \tag{14.9}$$

Within the class of matrices chosen a fortiori as in (14.8), it is possible to optimize with respect to $\boldsymbol{P}$ in order to obtain the STS matrix that minimizes P_{bound} in (14.9). This optimization has been carried out in [35], and the result can be summarized as follows:

Proposition 14.2 *With the diagonal entries of $\boldsymbol{\Lambda}_h$ arranged in a non-increasing order and those of $\boldsymbol{\Lambda}_w$ in a non-decreasing order, the bound P_{bound} in* (14.9) *is minimized when $\boldsymbol{P} = \boldsymbol{I}_{N_x}$.*

Similar to Section 12.2.5, we can interpret $\boldsymbol{C}$ in (14.8) as a time-varying or ST beamformer (STBF). Indeed, $\boldsymbol{C}$ in (14.8) steers transmission along the eigen-directions of the channel in the space domain and along the eigen-directions of the interference in the time domain. By choosing $\boldsymbol{P} = \boldsymbol{I}_{N_x}$, strong channel directions are "aligned" with weak interference directions in the optimal STBF, which improves error performance by minimizing P_{bound}. Diagonal entries of $\boldsymbol{D}_c$ determine the power loaded on different eigen-beams, which will be optimized later to further reduce P_{bound}. Notice that the average SINR-maximizing $\boldsymbol{C}$ in Proposition 14.1 is a special case of (14.8) with $|[\boldsymbol{D}_c]_{11}| = 1$ and $[\boldsymbol{D}_c]_{ii} = 0\ \forall i \neq 1$; thus, it cannot have better SER performance than an optimally power-loaded STBF. It further follows from the proof of Proposition 14.2 that when $\{\lambda_{w,i}\}_{i=1}^{N_{\min}}$ are distinct, the optimal $\boldsymbol{P}$ is unique (i.e., P_{bound} is minimized if and only if $\boldsymbol{P} = \boldsymbol{I}_{N_x}$) [35]. Hence, if we combine STBC with STBF to increase transmission rates, as we will do in the ensuing section, SER performance will degrade because it will turn out that each symbol in the ST code matrix is conveyed by a different permutation matrix.

With $\boldsymbol{P} = \boldsymbol{I}_{N_x}$, we can proceed to optimize the PEP bound with respect to the diagonal matrix $\boldsymbol{D}_c$. Toward this objective, we rewrite P_{bound} in (14.9) as

$$P_{\text{bound}} = \prod_{i=1}^{N_{\min}} \left(1 + \frac{\bar{\gamma}\Delta_{\min}^2}{4} \frac{\lambda_{h,i}[\boldsymbol{D}_c]_{ii}^2}{\lambda_{w,i}}\right)^{-1}. \tag{14.10}$$

Since $\ln(x)$ is a monotonically increasing function, minimizing P_{bound} is equivalent to minimizing $\ln(P_{\text{bound}})$ or maximizing $-\ln(P_{\text{bound}})$; hence, we can formulate the following optimization problem to find the optimal loading matrix:

$$\boldsymbol{D}_{c,\text{opt}} = \arg\max_{\boldsymbol{D}_c} \sum_{i=1}^{N_{\min}} \ln\left(1 + \frac{\bar{\gamma}\Delta_{\min}^2}{4} \frac{\lambda_{h,i}[\boldsymbol{D}_c]_{ii}^2}{\lambda_{w,i}}\right)$$

$$\text{subject to} \sum_{i=1}^{N_{\min}} [\boldsymbol{D}_c]_{ii}^2 = 1. \tag{14.11}$$

Using the Lagrange multiplier method, the optimal solution can be found in closed form as

$$[\boldsymbol{D}_{c,\text{opt}}]_{ii}^2 = \left[\frac{1}{\bar{N}} + \frac{\bar{\gamma}\Delta_{\min}^2}{4}\left(\frac{1}{\bar{N}}\sum_{j=1}^{\bar{N}} \frac{\lambda_{w,j}}{\lambda_{h,j}} - \frac{\lambda_{w,i}}{\lambda_{h,i}}\right)\right]_+, \tag{14.12}$$

where $\bar{N} \in [1, N_{\min}]$ denotes the number of non-zero $[\boldsymbol{D}_{c,\text{opt}}]_{ii}$'s and $[x]_+ := \max(x, 0)$.

14.2 COMBINING STS WITH OSTBC

Even though STS can effectively suppress interference while enabling channel-induced diversity, it is not spectrally efficient since it can only attain transmission rate $1/N_x$ symbols per channel use. To increase the transmission rate, in this section we mimic the approach of Section 12.3 and concatenate STS with OSTBC. Different from Section 12.3, however, the presence of interference will degrade the error performance as the rate increases. Nonetheless, the serial combination of STS with OSTBC will reduce error probability considerably compared to the case when OSTBC is employed alone.

Instead of using the ST code matrix in (14.1) with $\boldsymbol{C}$ specified in (14.8), let us consider transmission using

$$\boldsymbol{X}^T = \boldsymbol{U}_w \boldsymbol{P} \boldsymbol{X}_o^T(\boldsymbol{s}) \boldsymbol{D}_c \boldsymbol{U}_h^{\mathcal{H}}, \tag{14.13}$$

where $\boldsymbol{X}_o(\boldsymbol{s})$ is the $N_t \times N_x$ OSTBC matrix specified in Section 3.3.1. Because $\boldsymbol{X}_o(\boldsymbol{s})$ contains N_s information symbols in $\boldsymbol{s} := [s_1, \ldots, s_{N_s}]^T$, the transmission rate is N_s/N_x, which represents an N_s-fold increase relative to the low-rate STS transmission of Section 14.1. As before, the $N_t \times N_t$ diagonal matrix $\boldsymbol{D}_c$ and the $N_x \times N_x$ permutation matrix $\boldsymbol{P}$ in (14.13) will be chosen to reduce error probability.

ML decoding of $\boldsymbol{s}$ from $\boldsymbol{y}$ now yields

$$\hat{\boldsymbol{s}} = \arg\min_{\boldsymbol{s}} ||\boldsymbol{y} - \sqrt{\bar{\gamma}}\boldsymbol{X}^T\boldsymbol{h}||^2,$$

which for the ST code matrix in (14.13) can be written as

$$\hat{\boldsymbol{s}} = \arg\min_{\boldsymbol{s}} ||\boldsymbol{\Lambda}_w^{-1/2}(\tilde{\boldsymbol{y}} - \sqrt{\bar{\gamma}}\boldsymbol{P}\boldsymbol{X}_o^T(\boldsymbol{s})\boldsymbol{D}_c\boldsymbol{U}_h^{\mathcal{H}}\boldsymbol{h})||^2, \tag{14.14}$$

with $\tilde{\boldsymbol{y}} := \boldsymbol{U}_w^{\mathcal{H}}\boldsymbol{y}$. In the absence of interference, we have $\boldsymbol{R}_w = \boldsymbol{I}_{N_x}$ and $\boldsymbol{\Lambda}_w = \boldsymbol{I}_{N_x}$, and thanks to the OSTBC structure, the ML decoder in (14.14) reduces to N_s symbol-by-symbol linear decoders as in Section 3.3.2. When interference is present, however, exhaustive search or the low-complexity decoder we introduce in the next section, is required to find $\hat{\boldsymbol{s}}$.

Based on (14.2) and (14.14), the average PEP, $P_2(\boldsymbol{s} \to \tilde{\boldsymbol{s}})$, can be Chernoff-bounded as [230, Page 595]

$$P_2(\boldsymbol{s} \to \tilde{\boldsymbol{s}}) \tag{14.15}$$
$$< \det{}^{-1}\Big(\boldsymbol{I}_{N_t} + \frac{\bar{\gamma}}{4}\boldsymbol{D}_c[\boldsymbol{X}_o^*(\boldsymbol{s}) - \boldsymbol{X}_o^*(\tilde{\boldsymbol{s}})]\boldsymbol{P}^{\mathcal{H}}\boldsymbol{\Lambda}_w^{-1}\boldsymbol{P}[\boldsymbol{X}_o^T(\boldsymbol{s}) - \boldsymbol{X}_o^T(\tilde{\boldsymbol{s}})]\boldsymbol{D}_c\boldsymbol{\Lambda}_h\Big).$$

Without interference, we have $\boldsymbol{\Lambda}_w = \boldsymbol{I}_{N_x}$; and thus,

$$\Big[\boldsymbol{X}_o^*(\boldsymbol{s}) - \boldsymbol{X}_o^*(\tilde{\boldsymbol{s}})\Big]\boldsymbol{P}^{\mathcal{H}}\boldsymbol{\Lambda}_w^{-1}\boldsymbol{P}\Big[\boldsymbol{X}_o^T(\boldsymbol{s}) - \boldsymbol{X}_o^T(\tilde{\boldsymbol{s}})\Big] = \sum_{k=1}^{N_s}|e_k|^2,$$

where $e_k := s_k - \tilde{s}_k$. Hence, the PEP $P_2(\boldsymbol{s} \to \tilde{\boldsymbol{s}})$ can be bounded by a simple upper bound: $P_2(\boldsymbol{s} \to \tilde{\boldsymbol{s}}) < \left|\boldsymbol{I}_{N_t} + (1/4)\bar{\gamma}\Delta_{\min}^2\boldsymbol{D}_c^2\boldsymbol{\Lambda}_h\right|^{-1}$; and the optimal $\boldsymbol{D}_c$ can be found by minimizing this PEP bound as in Section 12.3.

In the presence of interference, however, such a simple upper bound on $P_2(\boldsymbol{s} \to \tilde{\boldsymbol{s}})$ cannot be obtained. For this reason, we consider the single-error PEP $P_{2,k} := P(\boldsymbol{s} \to \tilde{\boldsymbol{s}})$ whereby $\boldsymbol{s}$ and $\tilde{\boldsymbol{s}}$ differ only in their kth entry. Notice that the overall error probability is affected primarily by these single-error PEPs. The Chernoff bound on $P_{2,k}$ can now be found from (14.15) as

$$P_{k,\text{bound}} = \det{}^{-1}\Big(\boldsymbol{I}_{N_t} + \frac{\bar{\gamma}}{4}\boldsymbol{D}_c\boldsymbol{\Delta}_k^*\boldsymbol{P}^{\mathcal{H}}\boldsymbol{\Lambda}_w^{-1}\boldsymbol{P}\boldsymbol{\Delta}_k^T\boldsymbol{D}_c\boldsymbol{\Lambda}_h\Big), \tag{14.16}$$

where $\boldsymbol{\Delta}_k := \boldsymbol{X}_o(\boldsymbol{s}) - \boldsymbol{X}_o(\tilde{\boldsymbol{s}}) = \Re(e_k)\boldsymbol{\Phi}_k + j\Im(e_k)\boldsymbol{\Psi}_k$ [cf. (2.58)], and matrices $\boldsymbol{\Phi}_k$ and $\boldsymbol{\Psi}_k$ are specified in Section 3.3.1. Since $\boldsymbol{P}$ is a permutation matrix, $\boldsymbol{P}^{\mathcal{H}}\boldsymbol{\Lambda}_w^{-1}\boldsymbol{P}$ is a diagonal matrix whose diagonal entries are obtained by arranging those of $\boldsymbol{\Lambda}_w^{-1}$, possibly in a different order. Due to the special properties of $\boldsymbol{\Phi}_k$ and $\boldsymbol{\Psi}_k$ in (3.21), it can be verified that $\boldsymbol{\Phi}_k^* \boldsymbol{P}^{\mathcal{H}} \boldsymbol{\Lambda}_w^{-1} \boldsymbol{P} \boldsymbol{\Phi}_k^T = \boldsymbol{\Psi}_k^* \boldsymbol{P}^{\mathcal{H}} \boldsymbol{\Lambda}_w^{-1} \boldsymbol{P} \boldsymbol{\Psi}_k^T$ and $\boldsymbol{\Phi}_k^* \boldsymbol{P}^{\mathcal{H}} \boldsymbol{\Lambda}_w^{-1} \boldsymbol{P} \boldsymbol{\Psi}_k^T = \boldsymbol{\Psi}_k^* \boldsymbol{P}^{\mathcal{H}} \boldsymbol{\Lambda}_w^{-1} \boldsymbol{P} \boldsymbol{\Phi}_k^T$ are all diagonal. Based on the latter, $P_{k,\text{bound}}$ in (14.16) can be expressed in a simple analytical form.

In the STS scheme of Section 14.1, the choice $\boldsymbol{P} = \boldsymbol{I}_{N_x}$ was found to be optimal. In the combined STS-OSTBC system, though, (14.16) shows that selecting a matrix $\boldsymbol{P}$ that minimizes $P_{k,\text{bound}}$ $\forall k$ is not feasible, because it is impossible for the diagonal entries of $\boldsymbol{Q}_k := \boldsymbol{\Delta}_k^* \boldsymbol{P}^{\mathcal{H}} \boldsymbol{\Lambda}_w^{-1} \boldsymbol{P} \boldsymbol{\Delta}_k^T$ to appear in a non-increasing order $\forall k$. For this reason, we adopt the following heuristic rule to improve error probability: *Choose* $\boldsymbol{P}$ *so that the large diagonal entries of* $\boldsymbol{\Lambda}_h$ *coincide with the large diagonal entries of* $\boldsymbol{Q}_k$, $\forall k$.

Matrix $\boldsymbol{P}$, obeying this rule, can be determined offline since it depends only on the OSTBC structure, which is specified by $\{\boldsymbol{\Phi}_k, \boldsymbol{\Psi}_k\}_{k=1}^K$. Proceeding as before to optimize the power-loading matrix $\boldsymbol{D}_c$, we find

$$\boldsymbol{D}_{c,\text{opt}} = \arg\min_{\boldsymbol{D}_c} \sum_{k=1}^{N_s} P_{k,\text{bound}}, \qquad \text{subject to } \operatorname{tr}(\boldsymbol{D}_c^2) = 1. \tag{14.17}$$

Upon defining $x_i := [\boldsymbol{D}_c]_{ii}^2$, the error probability bound $P_{k,\text{bound}}$ can be expressed in terms of $\{x_i\}$ as $P_{k,\text{bound}} = \prod_{i=1}^{N_t}(1 + a_i x_i)$, where the positive constant a_i can be obtained from (14.16). Because $P_{k,\text{bound}}$ can be shown to be a convex function of $\{x_i\}$ [35], the cost function in (14.17) has a unique global minimum. Therefore, (14.17) can be solved efficiently using numerical search algorithms, for example, sequential quadratic programming (SQP) [67].

To illustrate how $\boldsymbol{P}$ and $\boldsymbol{D}_c$ are obtained, let us consider an example where the OSTBC is given by [152, Equation (62)]

$$\boldsymbol{X}_o = \begin{bmatrix} s_3 & 0 & s_2^* & s_1^* \\ 0 & s_3 & s_1 & -s_2 \\ s_2 & s_1^* & -s_3^* & 0 \\ s_1 & -s_2^* & 0 & -s_3^* \end{bmatrix}. \tag{14.18}$$

In this case we have $\boldsymbol{\Phi}_3^* \boldsymbol{P}^{\mathcal{H}} \boldsymbol{\Lambda}_w^{-1} \boldsymbol{P} \boldsymbol{\Phi}_3^T = \boldsymbol{\Psi}_3^* \boldsymbol{P}^{\mathcal{H}} \boldsymbol{\Lambda}_w^{-1} \boldsymbol{P} \boldsymbol{\Psi}_3^T = \boldsymbol{P}^{\mathcal{H}} \boldsymbol{\Lambda}_w^{-1} \boldsymbol{P}$, and thus $\boldsymbol{P} = \boldsymbol{I}_{N_x}$ is optimal for s_3. On the other hand, $\boldsymbol{P} = \boldsymbol{I}_{N_x}$ will cause the worst error performance when decoding s_1, because with $\boldsymbol{P} = \boldsymbol{I}_{N_x}$, we have $\boldsymbol{\Phi}_1^* \boldsymbol{\Lambda}_w^{-1} \boldsymbol{\Phi}_1^T = \boldsymbol{\Psi}_1^* \boldsymbol{\Lambda}_w^{-1} \boldsymbol{\Psi}_1^T = \operatorname{diag}(1/\lambda_{w,N_x}, \ldots, 1/\lambda_{w,1})$, which implies that the strongest channel coincides with the strongest interference. Based on the aforementioned heuristic rule, we choose $\boldsymbol{P}$ as

$$\boldsymbol{P} = \begin{bmatrix} 0 & 0 & 1 & 0 \\ 0 & 0 & 0 & 1 \\ 1 & 0 & 0 & 0 \\ 0 & 1 & 0 & 0 \end{bmatrix}. \tag{14.19}$$

Letting $d_i := [\boldsymbol{D}_c]_{ii}$ and $\rho := \bar{\gamma}\Delta^2_{\min}/4$, the upper bounds of the single-error PEP can be expressed as

$$P_{1,\text{bound}} = \left[\left(1+\frac{\rho d_1^2\lambda_{h,1}}{\lambda_{w,2}}\right)\left(1+\frac{\rho d_2^2\lambda_{h,2}}{\lambda_{w,1}}\right)\left(1+\frac{\rho d_3^2\lambda_{h,3}}{\lambda_{w,4}}\right)\left(1+\frac{\rho d_4^2\lambda_{h,4}}{\lambda_{w,3}}\right)\right]^{-1},$$

$$P_{2,\text{bound}} = \left[\left(1+\frac{\rho d_1^2\lambda_{h,1}}{\lambda_{w,1}}\right)\left(1+\frac{\rho d_2^2\lambda_{h,2}}{\lambda_{w,2}}\right)\left(1+\frac{\rho d_3^2\lambda_{h,3}}{\lambda_{w,3}}\right)\left(1+\frac{\rho d_4^2\lambda_{h,4}}{\lambda_{w,4}}\right)\right]^{-1},$$

$$P_{3,\text{bound}} = \left[\left(1+\frac{\rho d_1^2\lambda_{h,1}}{\lambda_{w,3}}\right)\left(1+\frac{\rho d_2^2\lambda_{h,2}}{\lambda_{w,4}}\right)\left(1+\frac{\rho d_3^2\lambda_{h,3}}{\lambda_{w,1}}\right)\left(1+\frac{\rho d_4^2\lambda_{h,4}}{\lambda_{w,2}}\right)\right]^{-1},$$

and the optimal $\boldsymbol{D}_c$ can be found numerically by minimizing $\sum_{k=1}^{3} P_{k,\text{bound}}$.

14.2.1 Low-Complexity Receivers

ML decoding in (14.14) requires high computational complexity (exponential in the dimension of $\boldsymbol{s}$), since it requires exhaustive search. In this subsection we show that the sphere decoding algorithm (SDA) described in Chapter 5 can be applied to decode $\boldsymbol{s}$ with average polynomial complexity which is cubic in N_s for the range of SNR and the N_s dimensions encountered in most applications.

Since $\boldsymbol{P}^T\boldsymbol{P} = \boldsymbol{I}_{N_x}$, we can think of (14.14) as implementing ML detection for the linear model

$$\boldsymbol{P}^T\tilde{\boldsymbol{y}} = \sqrt{\bar{\gamma}}\boldsymbol{X}_o^T(\boldsymbol{s})\boldsymbol{D}_c\boldsymbol{U}_h^{\mathcal{H}}\boldsymbol{h} + \tilde{\boldsymbol{w}}, \tag{14.20}$$

where $\tilde{\boldsymbol{w}} := \boldsymbol{P}^T\boldsymbol{U}_w\boldsymbol{w}$. Upon defining $\bar{\boldsymbol{y}} := [\Re(\boldsymbol{P}^T\tilde{\boldsymbol{y}})^T, \Im(\boldsymbol{P}^T\tilde{\boldsymbol{y}})^T]^T$, $\bar{\boldsymbol{s}} := [\Re(\boldsymbol{s})^T, \Im(\boldsymbol{s})^T]^T$, and $\bar{\boldsymbol{w}} := [\Re(\tilde{\boldsymbol{w}})^T, \Im(\tilde{\boldsymbol{w}})^T]^T$, the input-output relationship (14.20) can be rewritten as

$$\bar{\boldsymbol{y}} = \sqrt{\bar{\gamma}}\boldsymbol{H}\bar{\boldsymbol{s}} + \bar{\boldsymbol{w}}, \tag{14.21}$$

where $\boldsymbol{H}$ is a $2P \times 2N_t$ channel matrix satisfying $\boldsymbol{H}^{\mathcal{H}}\boldsymbol{H} = \sum_{\mu=1}^{N_t}|h_\mu|^2\boldsymbol{I}_{2N_t}$, and the covariance of $\bar{\boldsymbol{w}}$ is given by the block diagonal matrix $\boldsymbol{\Lambda}_{\bar{w}} = \text{diag}(\boldsymbol{P}^T\boldsymbol{\Lambda}_w\boldsymbol{P}/2,\ \boldsymbol{P}^T\boldsymbol{\Lambda}_w\boldsymbol{P}/2)$. Based on the linear model of (14.21), we can apply the SDA to decode $\boldsymbol{s}$ after whitening $\bar{\boldsymbol{w}}$.

Certainly, one can further reduce complexity by using a linear receiver [e.g., a matched filter (MF), an MMSE, or a ZF decoder] at the price of sacrificing performance. Since $\boldsymbol{H}^{\mathcal{H}}\boldsymbol{H} = \sum_{\mu=1}^{N_t}|h_\mu|^2\boldsymbol{I}_{2N_t}$, ZF and MF decoders have identical performance and they both achieve full diversity; however, they cannot effectively suppress interference and may degrade error performance considerably.

14.3 OPTIMAL TRAINING WITH INTERFERENCE

In previous sections we assumed that the channel is perfectly known at the receiver. In this section we consider channel estimation based on the signal model

$$\boldsymbol{y} = \sqrt{N_x\bar{\gamma}_{\text{tr}}}\boldsymbol{B}\boldsymbol{h} + \boldsymbol{w}, \tag{14.22}$$

where $\bar{\gamma}_{\text{tr}}$ controls the power spent on training, and the $N_x \times N_t$ ($N_x \geq N_t$) training matrix $\boldsymbol{B}$ is known at the receiver and satisfies $\text{tr}(\boldsymbol{B}\boldsymbol{B}^{\mathcal{H}}) = 1$. Notice that different from the preamble training approach in Section 11.1, $\boldsymbol{w}$ in (14.22) contains interference which is generally colored, but we assume its covariance matrix to be available at the transmitter. We will consider both least-squares (LS) and linear minimum-mean-square-error (LMMSE) channel estimators. Furthermore, we investigate how to construct $\boldsymbol{B}$ so that the mean-square error (MSE) of the resulting channel estimator is minimized. It will turn out that the optimal $\boldsymbol{B}$ depends only on the covariance of $\boldsymbol{w}$, without requiring $\boldsymbol{w}$ to be Gaussian distributed.

14.3.1 LS Channel Estimation

Letting $\tilde{\boldsymbol{B}} := \sqrt{N_x \bar{\gamma}_{\text{tr}}}\,\boldsymbol{B}$, the least-squares (LS) channel estimator is

$$\hat{\boldsymbol{h}} = (\tilde{\boldsymbol{B}}^{\mathcal{H}}\tilde{\boldsymbol{B}})^{-1}\tilde{\boldsymbol{B}}^{\mathcal{H}}\boldsymbol{y} = \boldsymbol{h} + (\tilde{\boldsymbol{B}}^{\mathcal{H}}\tilde{\boldsymbol{B}})^{-1}\tilde{\boldsymbol{B}}^{\mathcal{H}}\boldsymbol{w}, \tag{14.23}$$

and its covariance matrix is given by

$$\boldsymbol{\Sigma}_{\hat{h}} = \frac{1}{N_x \bar{\gamma}_{\text{tr}}}(\boldsymbol{B}^{\mathcal{H}}\boldsymbol{B})^{-1}\boldsymbol{B}^{\mathcal{H}}\boldsymbol{R}_w\boldsymbol{B}(\boldsymbol{B}^{\mathcal{H}}\boldsymbol{B})^{-1}. \tag{14.24}$$

The MSE of $\hat{\boldsymbol{h}}$ is then given by $\beta_{\text{LS}} = \text{tr}(\boldsymbol{\Sigma}_{\hat{h}})$. Using the SVD of $\boldsymbol{B} = \boldsymbol{U}_b\boldsymbol{D}_b\boldsymbol{V}_b$, it is possible to express $\boldsymbol{\Sigma}_{\hat{h}}$ as $\boldsymbol{\Sigma}_{\hat{h}} = (1/N_x\bar{\gamma}_{\text{tr}})\boldsymbol{V}_b\tilde{\boldsymbol{D}}_b^{-1}\boldsymbol{U}_b^{\mathcal{H}}\boldsymbol{R}_w\boldsymbol{U}_b\tilde{\boldsymbol{D}}_b^{-1}\boldsymbol{V}_b^{\mathcal{H}}$, where $\tilde{\boldsymbol{D}}_b$ is defined similar to $\tilde{\boldsymbol{D}}_c$. Thus, β_{LS} can be written as

$$\beta_{\text{LS}} = \frac{1}{N_x \bar{\gamma}_{\text{tr}}}\text{tr}(\tilde{\boldsymbol{U}}_b\boldsymbol{\Lambda}_w\tilde{\boldsymbol{U}}_b^{\mathcal{H}}\tilde{\boldsymbol{D}}_b^{-2}), \tag{14.25}$$

where $\tilde{\boldsymbol{U}}_b := \boldsymbol{U}_b^{\mathcal{H}}\boldsymbol{U}_w$. If the diagonal entries of $\tilde{\boldsymbol{D}}_b^2$ are arranged in a non-decreasing order, which we will verify later, then based on (14.25) and using Lemma 12.1, we deduce that β_{LS} is minimized when $\tilde{\boldsymbol{U}}_b = \boldsymbol{I}_{N_x}$, which implies that $\boldsymbol{U}_b = \boldsymbol{U}_w$. Thus, we have established the following proposition [35]:

Proposition 14.3 *To minimize β_{LS} in* (14.25), *the optimal training matrix is given by $\boldsymbol{B} = \boldsymbol{U}_w\boldsymbol{D}_b\boldsymbol{V}_b^{\mathcal{H}}$, where $\boldsymbol{V}_b$ is any unitary matrix.*

With $\boldsymbol{U}_b = \boldsymbol{U}_w$, the channel MSE β_{LS} in (14.25) reduces to

$$\beta_{\text{LS}} = \frac{1}{N_x \bar{\gamma}_{\text{tr}}}\sum_{i=1}^{N_t}\frac{\lambda_{w,i}}{[\boldsymbol{D}_b]_{ii}^2}. \tag{14.26}$$

The optimal $\boldsymbol{D}_b$ minimizing β_{LS} subject to the constraint $\text{tr}(\boldsymbol{D}_b\boldsymbol{D}_b^{\mathcal{H}}) = 1$ has diagonal entries

$$[\boldsymbol{D}_b]_{ii}^2 = \frac{\sqrt{\lambda_{w,i}}}{\sum_{j=1}^{N_t}\sqrt{\lambda_{w,j}}}, \tag{14.27}$$

which leads to

$$\beta_{\text{LS}} = \frac{1}{N_x \bar{\gamma}_{\text{tr}}} \Big(\sum_{i=1}^{N_t} \sqrt{\lambda_{w,i}} \Big)^2. \tag{14.28}$$

It is seen from (14.28) that the entries $\{[\boldsymbol{D}_b]_{ii}^2\}$ are arranged in a non-decreasing order because $\{\lambda_{w,i}\}$ are arranged in a non-decreasing order. Without interference, we have $\lambda_{w,1} = \ldots = \lambda_{w,N_x}$, and thus we obtain from (14.27) that $[\boldsymbol{D}_b]_{ii}^2 = 1/N_t$, $\forall i$. Hence, the optimal training matrix satisfies $\boldsymbol{B}^{\mathcal{H}}\boldsymbol{B} = \boldsymbol{I}_{N_t}/N_x$.

14.3.2 LMMSE Channel Estimation

The LS channel estimator does not exploit the statistics of the channel and the interference. This is accomplished, however, by the LMMSE estimator, which for the model in (14.22) is given by

$$\hat{\boldsymbol{h}} = \sqrt{N_x \bar{\gamma}_{\text{tr}}}\, \boldsymbol{R}_h \boldsymbol{B}^{\mathcal{H}} (N_x \bar{\gamma}_{\text{tr}} \boldsymbol{B} \boldsymbol{R}_h \boldsymbol{B} + \boldsymbol{R}_w)^{-1} \boldsymbol{y} \tag{14.29}$$

and has covariance matrix

$$\boldsymbol{\Sigma}_{\hat{h}} = \boldsymbol{R}_h - N_x \bar{\gamma}_{\text{tr}} \boldsymbol{R}_h \boldsymbol{B}^{\mathcal{H}} (\boldsymbol{R}_w + N_x \bar{\gamma}_{\text{tr}} \boldsymbol{B} \boldsymbol{R}_h \boldsymbol{B}^{\mathcal{H}})^{-1} \boldsymbol{B} \boldsymbol{R}_h. \tag{14.30}$$

Similar to (14.8), we choose the training matrix based on the STBF matrix $\boldsymbol{B} = \boldsymbol{U}_w \boldsymbol{D}_b \boldsymbol{U}_h^{\mathcal{H}}$. The corresponding channel MSE $\beta_{\text{LMMSE}} := \text{tr}(\boldsymbol{\Sigma}_{\hat{h}})$ is

$$\beta_{\text{LMMSE}} = \text{tr}\left[\boldsymbol{\Lambda}_h - N_x \bar{\gamma}_{\text{tr}} \boldsymbol{\Lambda}_h \boldsymbol{D}_b^{\mathcal{H}} (\boldsymbol{\Lambda}_w + N_x \bar{\gamma}_{\text{tr}} \boldsymbol{D}_b \boldsymbol{\Lambda}_h \boldsymbol{D}_b^{\mathcal{H}})^{-1} \boldsymbol{D}_b \boldsymbol{\Lambda}_h \right]. \tag{14.31}$$

Letting $N_h = \text{rank}(\boldsymbol{\Lambda}_h)$, we can also simplify (14.31) as

$$\beta_{\text{LMMSE}} = \sum_{i=1}^{N_h} \left(\frac{1}{\lambda_{h,i}} + N_x \bar{\gamma}_{\text{tr}} \frac{[\boldsymbol{D}_b]_{ii}^2}{\lambda_{w,i}} \right)^{-1}. \tag{14.32}$$

Minimizing β_{LMMSE} with respect to $\boldsymbol{D}_b$ under the constraint $\text{tr}(\boldsymbol{D}_b \boldsymbol{D}_b^{\mathcal{H}}) = 1$, we obtain [35]

$$[\boldsymbol{D}_b]_{ii}^2 = \left[\frac{1 + \frac{1}{N_x \bar{\gamma}_{\text{tr}}} \sum_{j \in \mathcal{I}} \frac{\lambda_{w,j}}{\lambda_{h,j}}}{\sum_{j \in \mathcal{I}} \sqrt{\lambda_{w,j}}} \sqrt{\lambda_{w,i}} - \frac{1}{N_x \bar{\gamma}_{\text{tr}}} \frac{\lambda_{w,i}}{\lambda_{h,i}} \right]_+, \tag{14.33}$$

where $\mathcal{I}$ denotes the set $\mathcal{I} := \{[\boldsymbol{D}_b]_{ii} | [\boldsymbol{D}_b]_{ii} \neq 0\}$. Note that the power loading in (14.33), which minimizes the channel estimation error, is different from that in (14.12), which minimizes the error probability bound.

14.4 NUMERICAL EXAMPLES

In this section we present simulations to test the performance of ST codes designed in the presence of interference. We consider a linear array with $N_t = 4$ transmit-antennas and a single receive-antenna. The STS matrix $\boldsymbol{C}$ has size $N_t \times N_t$ (i.e.,

$N_x = N_t$). The N_t transmit-antennas are equi-spaced by d. We assume that the direction of arrival is perpendicular to the transmit-antenna array; let λ denote the wavelength of the transmitted signal and φ denote the angle spread. When φ is small, the channel correlation can be calculated from the "one-ring" channel model as [235]

$$[\boldsymbol{R}_h]_{m,n} \approx \frac{1}{2\pi\lambda} \int_0^{2\pi} \exp[-j2\pi d(m-n)\varphi \sin\theta]\, d\theta.$$

In the ensuing examples, we consider two channels with different correlations. Channel 1 has $d = 0.5\lambda$ and $\varphi = 5°$, while channel 2 has $d = 0.5\lambda$ and $\varphi = 25°$. Channels are normalized so that $\text{tr}(\boldsymbol{R}_h) = N_t$. For channel 1, the eigenvalues of $\boldsymbol{R}_h$ are $\boldsymbol{\Lambda}_{h_1} = \text{diag}(3.81849, 0.18079, 0.00071, 0.00001)$; and for channel 2, we have $\boldsymbol{\Lambda}_{h_2} = \text{diag}(1.79, 1.741, 0.454, 0.015)$. While channel 1 is highly correlated, channel 2 is less correlated and provides more diversity. We consider partial band interference with a normalized power spectral density $S(f) = \sin(\pi W f)/(\pi W f)$, where W denotes the interference bandwidth. The center frequency of the interference coincides with the carrier frequency. We use two different values of W, $W = 0.1/T$ and $W = 0.4/T$, where T is the sampling period that is used to obtain the discrete signal model (14.2). When $W = 0.1/T$, the eigenvalue matrix is $\boldsymbol{\Lambda}_{i_1} = \text{diag}(3.8412, 0.1579, 0.0009, 0.0000)$; and $\boldsymbol{\Lambda}_{i_2} = \text{diag}(2.391, 1.394, 0.210, 0.005)$ when $W = 0.4/T$. Note that when $W = 0.1/T$, interference concentrates on two eigenvectors and is relatively easy to avoid. QPSK modulation is adopted in all simulations.

Example 14.1 (Theoretical BER comparison) BER performance here is calculated from the SINR in (14.5) using an expression derived in [236, Equation (9.11)]. Figures 14.1 and 14.2 depict BER of the low-rate STS system in Section 14.1 for channels 1 and 2. We plot BER curves without beamforming ($\boldsymbol{C} = \boldsymbol{I}_{N_t}$) and with optimally power-loaded STBF [$\boldsymbol{C}$ as in (14.8)]. When $W = 0.1/T$, performance of STBF in the presence of interference comes very close to its lower bound in the absence of interference. This is because most of the interference power lies in the weakest channel that contains no or very little signal power. When $W = 0.4/T$, the gap between STBF with interference and its lower bound increases because the interference power is spread over more eigen-directions. In both cases it is observed that optimally power-loaded STBF has considerably larger performance gain relative to the transmission without beamforming. Figure 14.3 depicts BER versus INR for STS transmissions. STBF can tolerate interference power more than 25 dB higher than transmissions without beamforming when $W = 0.1/T$ or more than 10 dB when $W = 0.4/T$.

Example 14.2 (Simulated BER comparison) Figures 14.4 to 14.7 depict BERs of STBC with ML and LMMSE receivers. The transmitted block $\boldsymbol{X}$ is given in (14.13) with STBF, and $\boldsymbol{X} = \boldsymbol{X}_o(\boldsymbol{s})$ without beamforming where the STBC matrix is given by (14.18). The ML receiver is implemented with the low-complexity SDA whose performance shows no noticeable difference relative to that of exhaustive search; for this reason, we do not plot the BER corresponding to the exhaustive

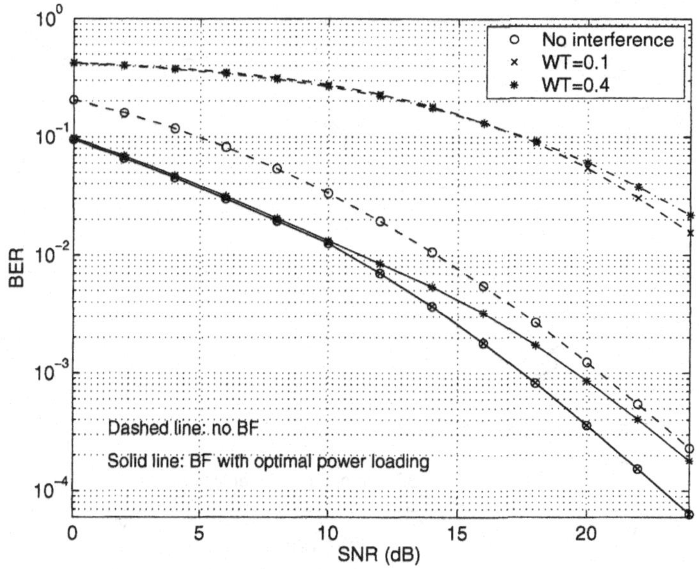

Figure 14.1 BER of STBF, channel 1, $\bar{\gamma}_i = 10$ dB

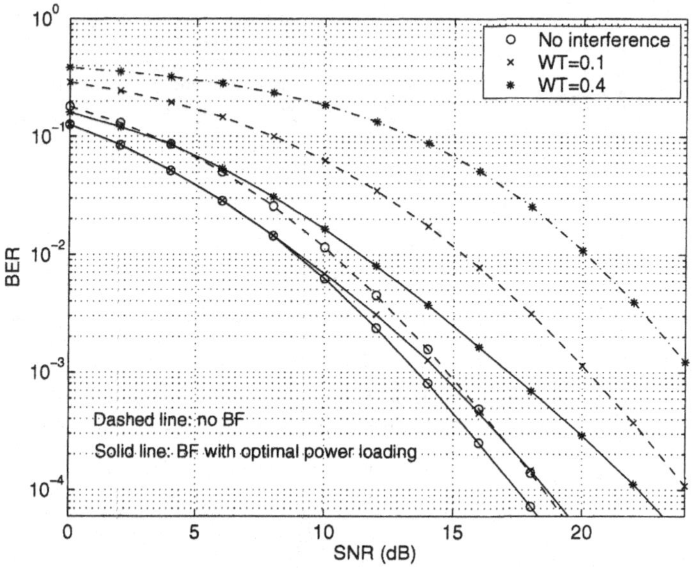

Figure 14.2 BER of STBF, channel 2, $\bar{\gamma}_i = 10$ dB

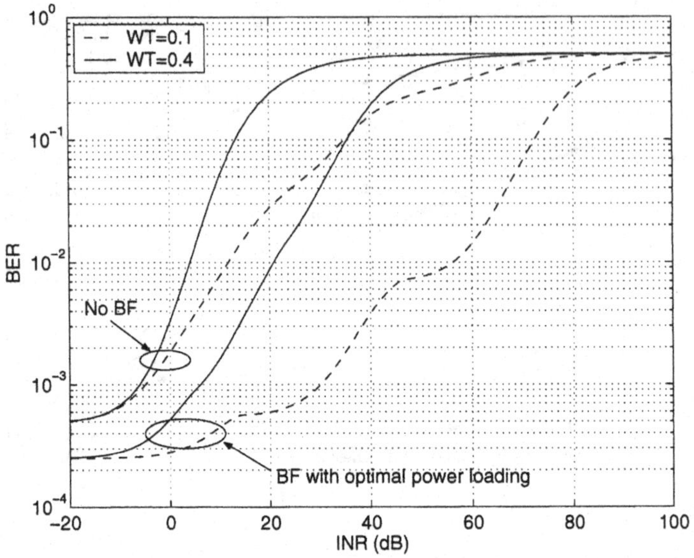

Figure 14.3 BER of STBF vs. INR, channel 2, $\bar{\gamma} = 16$ dB

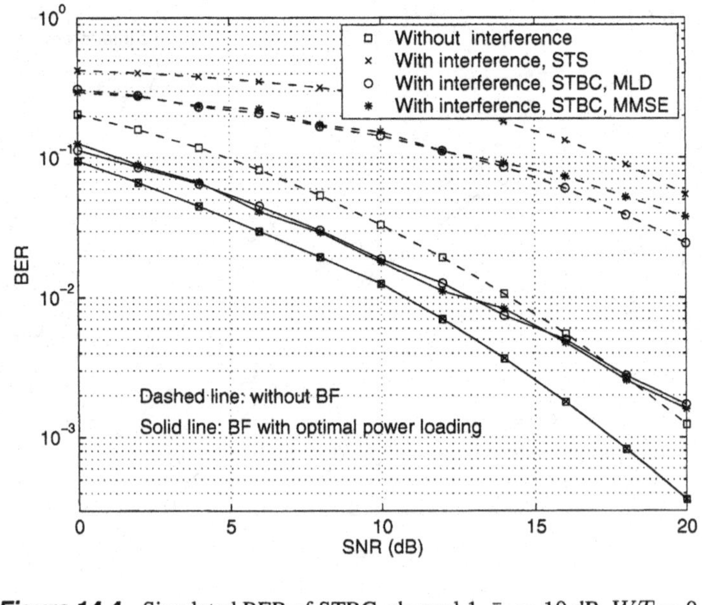

Figure 14.4 Simulated BER of STBC, channel 1, $\bar{\gamma}_i = 10$ dB, $WT = 0.1$

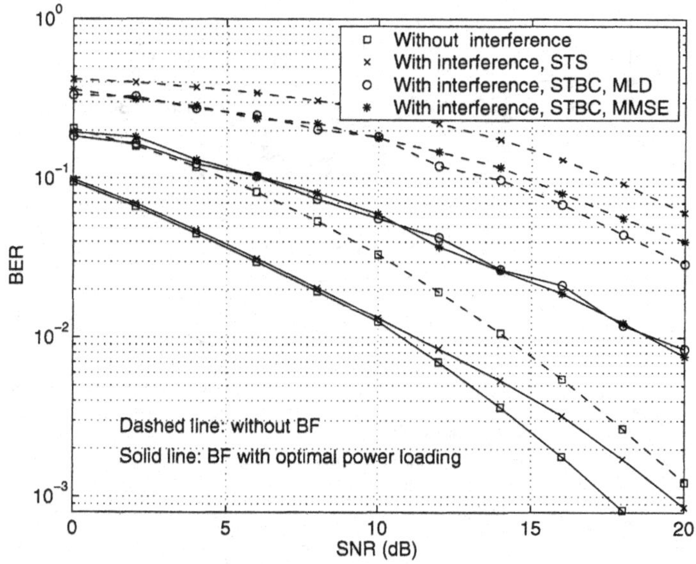

Figure 14.5 Simulated BER of STBC, channel 1, $\bar{\gamma}_i = 10$ dB, $WT = 0.4$

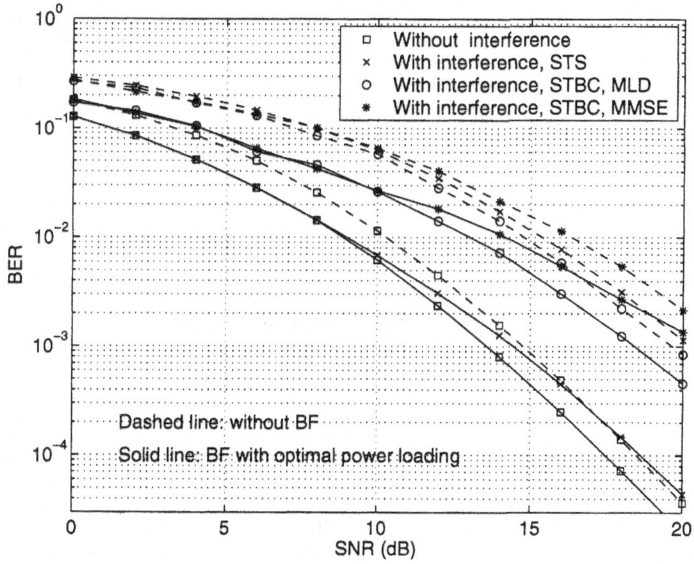

Figure 14.6 Simulated BER of STBC, channel 2, $\bar{\gamma}_i = 10$ dB, $WT = 0.1$

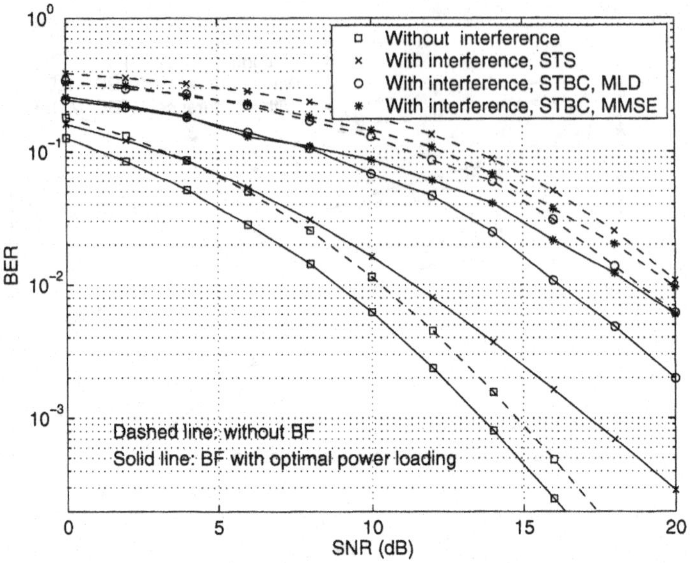

Figure 14.7 Simulated BER of STBC, channel 2, $\bar{\gamma}_i = 10$ dB, $WT = 0.4$

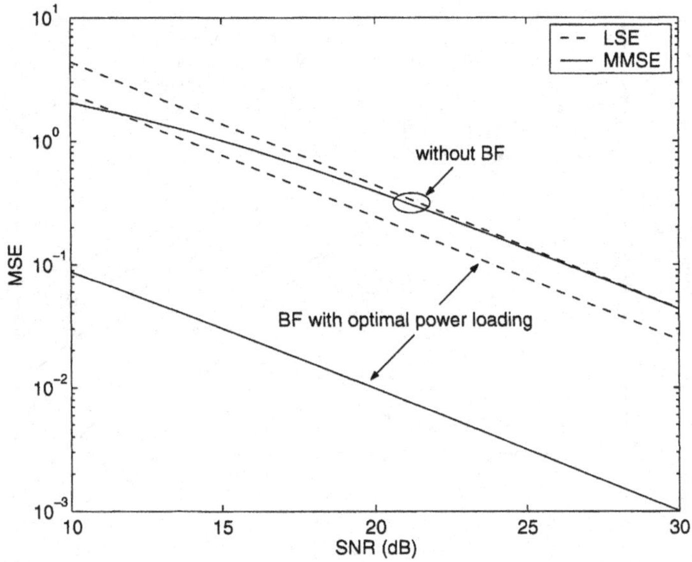

Figure 14.8 MSE of channel estimators, channel 1, $\bar{\gamma}_i = 10$ dB, $WT = 0.1$

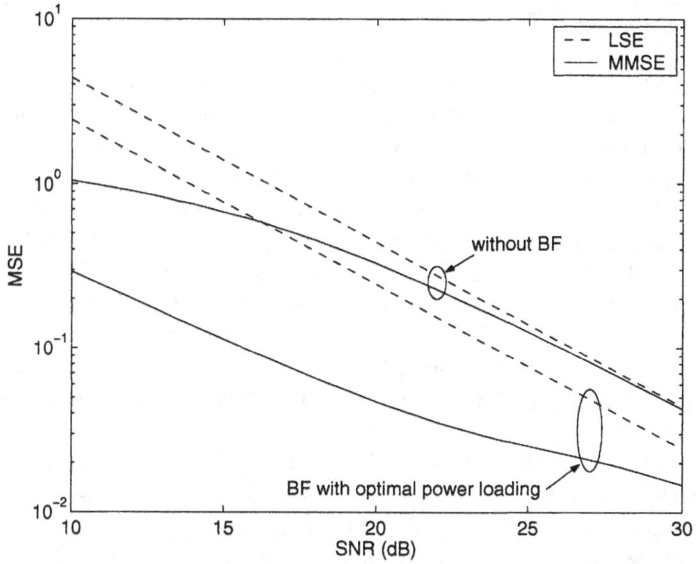

Figure 14.9 MSE of channel estimators, channel 2, $\bar{\gamma}_i = 10$ dB, $WT = 0.4$

search. As observed in Figures 14.4 and 14.5, STBF exhibits a large performance gain relative to the transmission without beamforming; although this performance gain reduces in channel 2, it is still considerably large as verified by Figures 14.6 and 14.7. Comparing Figure 14.4 with 14.5, and Figure 14.6 with 14.7, we see that when the interference bandwidth increases, BER degrades. One surprising observation is that without beamforming, BER of STS is worse than that of STBC. This is actually reasonable because each symbol in an ST block codeword corresponds to a different permutation matrix. Some of these permutation matrices can reduce the single-error PEP as analyzed in Section 14.2, and thereby decrease the overall error probability.

Example 14.3 (MSE of channel estimation) Figures 14.8 and 14.9 depict the MSE of the LS and LMMSE channel estimators. We plot the channel MSE for cases without beamforming ($\boldsymbol{C} = \boldsymbol{I}_{N_t}$) and with optimally power-loaded beamforming. Without beamforming, LS and LMMSE estimators have almost identical MSEs when $\bar{\gamma}_{\text{tr}}$ is moderately large. With optimal training, the LS channel estimator has a slightly better MSE than that without beamforming, because channel correlations are not exploited by the LS channel estimator. On the other hand, STBF considerably improves the performance of the LMMSE channel estimator.

14.5 CLOSING COMMENTS

In this chapter second-order spatial statistics of the MIMO channel and temporal statistics of the interference were exploited to design ST transceivers for multi-antenna wireless communication systems. Low-rate ST spreading systems were designed first to maximize the average SINR at the receiver by transmitting along the strongest eigen-direction of the channel and the weakest eigen-direction of the interference. Optimal power-loaded ST beamforming schemes were also derived capable of reducing error probability considerably. To increase transmission rates, OSTBC and STBF were combined and power-loading schemes were derived along with low-complexity near-ML receivers. As the latter require reliable channel estimators, training sequences were optimized for LS and LMMSE channel estimation in the presence of interference by exploiting knowledge of the interference covariance matrix at the transmitter.

What enabled this chapter's robust designs is the fact that correlation properties of the interference were available and accounted for explicitly, when designing the ST-coded transmitter. In multi-access communications, however, this is less likely to be available since multiuser interference changes on a constant basis as users drop in and out. Nonetheless, we will see in Chapter 15 that it is still possible to suppress multiuser interference while retaining the diversity and coding gains of ST transmissions, through judicious block spreading operations which orthogonalize the multi-access channel even in the presence of frequency-selective channels.

15

ST Codes for Orthogonal Multiple Access

In previous chapters we considered ST codes for single-user transmissions over a point-to-point MIMO channel. In this chapter we deal with ST codes in a multiple access setup. In particular, we focus on orthogonal multiple access systems where multiuser interference can be deterministically suppressed at the receiver so that the ST codes considered so far can be applied on a per user basis.

When the underlying MIMO channels are flat fading, orthogonality among multiple users can easily be achieved if users transmit synchronously and employ orthogonal spreading sequences as in a synchronous code-division multiple-access (CDMA) system [115] (Non-orthogonal spreading for flat fading MIMO channels has also been considered in, e.g., [149]). Multiple access over frequency-selective channels is challenging, however, because frequency-selectivity destroys the orthogonality of spreading sequences which gives rise to multiuser interference (MUI) in addition to inter-symbol interference (ISI).

In this chapter we see how properly designed block (as opposed to symbol) spreading code matrices can be coupled with ST matrices to enable orthogonal multiple access so that multiuser separation can be achieved at the receiver regardless of unknown frequency-selective MIMO channels. To cover the two popular transmission modalities, we consider both single- and multi-carrier systems. For each system, we outline one design paradigm.

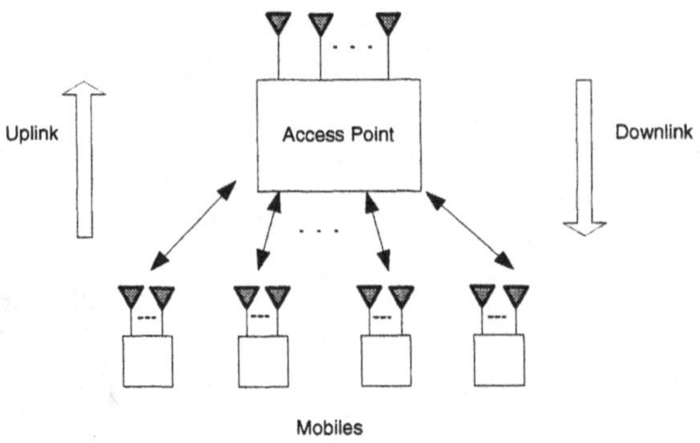

Figure 15.1 MIMO multiuser downlink and uplink transmissions

15.1 SYSTEM MODEL

With reference to Figure 15.1, we consider a cellular system where an access point (AP) serves multiple users simultaneously. As usual, *downlink* (a.k.a. *forward link*) refers to transmission from the AP to mobile users, while *uplink* (a.k.a. *reverse link*) refers to transmissions from mobile users to the AP. Both the AP and the mobile users can have multiple antennas deployed. Clearly, the AP can afford the cost and size of deploying more antennas than each mobile terminal, which in the case of cellular phones can deploy no more than two or three antennas with the current technology. As our system will be designed to support both narrowband and broadband applications, we consider frequency-selective MIMO channels. We describe next the downlink and uplink operations in more detail.

15.1.1 Synchronous Downlink

To reach multiple users in the downlink, the AP transmitter multiplexes signals intended for different users. The resulting aggregate waveform experiences a common propagation channel from the AP to each particular mobile user. To allow for user separation at each receiver, after ST coding, the information sequence of each user is spread by a user-specific spreading sequence before multiplexing. The resulting chip sequences of all users are added up synchronously.

Let $x_\mu^{(u)}(n)$ denote the sequence after ST coding and spreading intended for user u that is transmitted from the μth AP antenna over the symbol interval indexed by n. If N_u denotes the number of users served, the aggregate multiplexed sequence broadcast from the AP's μth antenna at time n can be written in a discrete-time equivalent baseband form as

$$\bar{x}_\mu(n) = \sum_{u=0}^{N_u-1} \sqrt{\frac{\overline{\gamma}_u}{N_t}}\, x_\mu^{(u)}(n), \tag{15.1}$$

where N_t is the number of antennas at the AP and $\overline{\gamma}_u$ is used to control the uth user's transmission power.

Each frequency-selective channel is modeled as a finite impulse response linear time-invariant filter (tapped delay line with a finite number of taps). With $\tau_{\max,s}$ denoting the maximum delay spread of the multipath channels (usually obtained through channel-sounding experiments), and supposing that each waveform received is sampled at the chip rate $1/T_c$, the maximum order of the discrete-time baseband equivalent channels is

$$L := \left\lceil \frac{\tau_{\max,s}}{T_c} \right\rceil, \tag{15.2}$$

where $\lceil \cdot \rceil$ denotes the ceiling operation and T_c is the chip period (equal to the sampling interval). Let $h_{\nu\mu}^{(u)}(l)$ denote the lth tap of the channel between the μth AP transmit-antenna and the νth receive-antenna of the uth user ($l = 0, 1, \ldots, L$).

The sequence received at the νth receive-antenna of the user of interest, say u_0, can be expressed as

$$y_\nu^{(u_0)}(n) = \sum_{\mu=1}^{N_t} \sum_{l=0}^{L} \bar{x}_\mu(n-l) h_{\nu\mu}^{(u_0)}(l) + w_\nu^{(u_0)}(n), \tag{15.3}$$

where $w_\nu^{(u_0)}(n)$ is AWGN with zero mean and unit variance. Substituting (15.1) into (15.3), we obtain

$$y_\nu^{(u_0)}(n) = \sum_{u=0}^{N_u-1} \sqrt{\frac{\overline{\gamma}_u}{N_t}} \sum_{\mu=1}^{N_t} \sum_{l=0}^{L} x_\mu^{(u)}(n-l) h_{\nu\mu}^{(u_0)}(l) + w_\nu^{(u_0)}(n). \tag{15.4}$$

When compared with (7.2), each user in (15.4) suffers from ISI, which is due to frequency-selective propagation, but also from MUI, which arises from the simultaneous transmissions intended to other users.

15.1.2 Quasi-synchronous Uplink

Uplink transmissions from the mobile users to the AP are naturally asynchronous. In a cellular system, however, the asynchronism can be rendered bounded because the AP typically transmits a pilot waveform on a separate channel based on which mobile users attempt to synchronize. Relative mobility and timing estimation errors only allow for coarse timing acquisition, causing each user's timing to be off by a few chip periods. If τ_u denotes the uth user's mistiming, the timing offset of the uth user's waveform received at the AP is $2\tau_u$. Clearly, the maximum asynchronism $\tau_{\max,a} := \max_{u \in [0, N_u-1]}(2\tau_u)$ arises between the nearest and farthest mobile users within the cell and can be predetermined from the radius of the cell. The important point, also illustrated in Figure 15.2, is that the amount of asynchronism is bounded to a few chips of a symbol and can thus be lumped into the channel's impulse response.

Under this bounded asynchronism among users, uplink transmissions are referred to as *quasi-synchronous*. A quasi-synchronous multiuser system can be modeled as

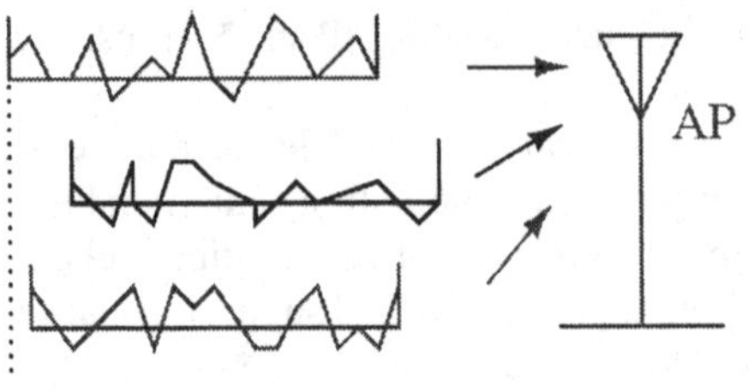

Figure 15.2 Quasi-synchronous multiple access

a synchronous system where the asynchronism is accounted by the zero taps of each user's channel. With asynchronism included, the maximum channel order becomes [cf. (15.2)]

$$L := \left\lceil \frac{\tau_{\max,s} + \tau_{\max,a}}{T_c} \right\rceil . \tag{15.5}$$

Hence, quasi-synchronous uplink transmissions can be viewed as synchronous transmissions experiencing a longer channel $h_{\nu\mu}^{(u)}(l)$ from the μth transmit-antenna of the uth user to the νth AP receive-antenna.

Let $x_{\mu}^{(u)}(n)$ denote the transmitted information symbol from the μth antenna of user u during the nth symbol interval. Corresponding to the single-user case treated in (7.2), the sequence received at the νth AP receive-antenna is now given by

$$y_{\nu}(n) = \sum_{u=0}^{N_u-1} \sqrt{\frac{\overline{\gamma}_u}{N_t}} \sum_{\mu=1}^{N_t} \sum_{l=0}^{L} x_{\mu}^{(u)}(n-l) h_{\nu\mu}^{(u)}(l) + w_{\nu}(n), \tag{15.6}$$

where $w_{\nu}(n)$ is zero-mean unit-variance AWGN and $\overline{\gamma}_u$ denotes the average receive-SNR corresponding to user u when the channel energy is normalized. Again, each user in (15.6) suffers from both ISI and MUI.

Let us now compare the downlink model in (15.4) with the uplink model in (15.6). Clearly, the downlink model can be subsumed by the uplink model if we set $h_{\nu\mu}^{(u)}(l) = h_{\nu\mu}^{(u_0)}(l)$, $\forall u \neq u_0$, when user u_0 is the user of interest. However, we should bear in mind two facts: (i) the downlink channel corresponding to each user of interest is different, and (ii) the maximum channel order in the synchronous downlink operation is usually much smaller than the order of the corresponding channel experienced during the quasi-synchronous uplink operation.

In the ensuing sections, we will rely on the uplink input-output relationship in (15.6) to describe both uplink and downlink operations. In either operation, the only channel knowledge assumed available at the transmitter is L (or an upper bound on it), which will be used explicitly in the system design. As depicted in Figure 15.3, the block diagram unifying both operations includes single-user ST coding and decoding modules. User-specific (de-)spreading modules are also employed for user separation. The challenge is to design (de-)spreading operations so that MUI can be suppressed deterministically at the receiver regardless of the frequency-selective

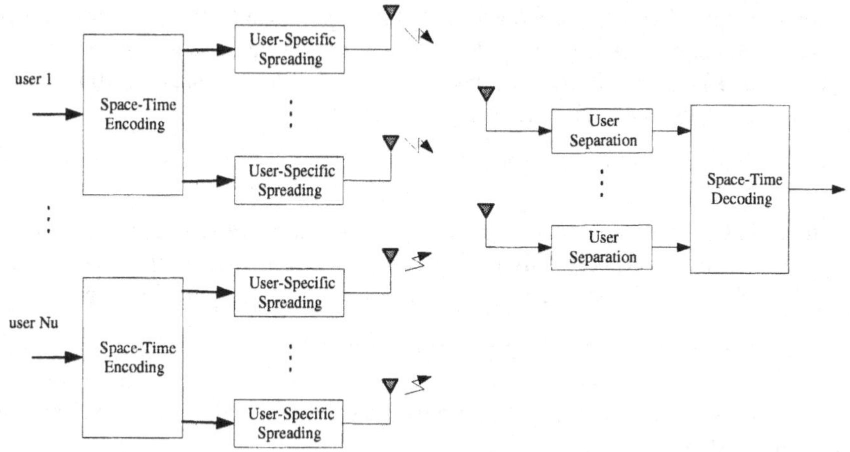

Figure 15.3 Orthogonal ST-coded multiple access

multipath propagation experienced. If this task can be accomplished, it will be possible to permeate through orthogonal channelization all the benefits of single-user ST coded systems to multi-access communications. We specify how this goal can be achieved with single- and multi-carrier systems.

15.2 SINGLE-CARRIER SYSTEMS: OSTBC-CIBS-CDMA

In this section we introduce a single-carrier multi-access system where each transmitter entails ST coding and spreading operations. Both operations are applied on a block-by-block basis. Specifically, the ST coding module relies on the time-reversal (TR) OSTBC developed in Chapter 7, while spreading is accomplished through the chip-interleaved block-spread (CIBS) CDMA scheme introduced by [334, 342] (an extension of CIBS-CDMA to doubly selective channels may be found in [148]). Unlike conventional symbol-periodic CDMA, where each user's signature vector spreads each symbol to obtain the chip sequence, each user in CIBS-CDMA relies on a signature matrix to spread each block of symbols in a way similar to multi-code CDMA. The combination of CIBS-CDMA followed by TR-OSTBC was pursued in [209] for downlink operation, where various receiver options were also considered. Relative to [209], the exposition here also covers the uplink, and the order of the modules in the combination is reversed (ST coding first followed by CIBS-CDMA).

15.2.1 CIBS-CDMA for User Separation

Suppose that the information-bearing symbols are parsed into blocks of size K. Let $\bar{\boldsymbol{s}}_{\mu}^{(u)}(i)$ denote the ith ST-coded symbol block of user u that is to be transmitted through antenna μ. How $\bar{\boldsymbol{s}}_{\mu}^{(u)}(i)$ is formed is detailed in the next section. Our concern in this section is to spread $\bar{\boldsymbol{s}}_{\mu}^{(u)}(i)$ judiciously so that after propagation through an unknown frequency-selective MIMO channel, MUI can be avoided at the receiver.

To this end, each user is first assigned a unique $M \times 1$ signature spreading vector $\boldsymbol{c}_u$ which is orthogonal to all other users' spreading vectors; that is,

$$\boldsymbol{c}_u^{\mathcal{H}} \boldsymbol{c}_{u'} = \delta[u - u'], \tag{15.7}$$

where $\delta[\cdot]$ is the Kronecker delta. The orthogonality among signature vectors implies that the maximum number of users that can be accommodated by this system is $N_u \leq M$. Based on $\boldsymbol{c}_u$, each user constructs a $P \times K$ block spreading matrix

$$\boldsymbol{C}_u = \boldsymbol{c}_u \otimes \boldsymbol{T}_{\text{zp}}, \tag{15.8}$$

where $P := M(K+L)$, $\otimes$ stands for the Kronecker product and $\boldsymbol{T}_{\text{zp}}$ is a zero-padding matrix defined as

$$\boldsymbol{T}_{\text{zp}} := \begin{bmatrix} \boldsymbol{I}_K \\ \boldsymbol{0}_{L \times K} \end{bmatrix}. \tag{15.9}$$

The symbol block $\bar{\boldsymbol{s}}_{\mu}^{(u)}(i)$ is then spread to obtain the $P \times 1$ chip block

$$\boldsymbol{x}_{\mu}^{(u)}(i) = \boldsymbol{C}_u \bar{\boldsymbol{s}}_{\mu}^{(u)}(i), \quad \forall \mu \in [1, N_t]. \tag{15.10}$$

Notice that each of the K symbols in $\boldsymbol{s}_{\mu}^{(u)}(i)$ is spread by a different sequence (column of $\boldsymbol{C}_u$) of length P. After parallel-to-serial conversion of $\boldsymbol{x}_{\mu}^{(u)}(i)$, the chip sequence $x_{\mu}^{(u)}(n)$ is transmitted through the μth antenna of user u.

The block spreading operation in (15.10) can be implemented conveniently by symbol spreading (each symbol is spread by the signature vector $\boldsymbol{c}_u$) followed by a chip interleaver with guards. As depicted in Figure 15.4, conventional direct-sequence (DS) CDMA corresponds to transmitting the interleaver entries row-wise, while the

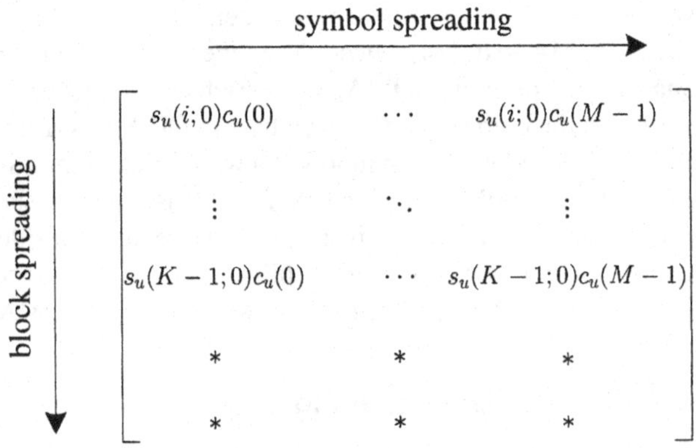

Figure 15.4 Redundant chip interleaver for the uth user ($*$ stands for guard zeros, which could be replaced by other known symbols)

CIBS-CDMA transmitter outputs the interleaver entries column-wise. This not only explains the acronym chip-interleaved block-spread CDMA, but also highlights how readily implementable (and backward-compatible) CIBS-CDMA is simply by cascading an existing DS-CDMA system with a chip interleaver that contains guard zeros.

Let us now reflect on the potential benefits of CIBS transmissions. Suppose that at the νth antenna output we collect received samples from (15.6) into blocks of size $P \times 1$, as $\boldsymbol{y}_\nu(i) := [y_\nu(iP), \ldots, y_\nu(iP+P-1)]^T$. Notice that the last L entries of $\boldsymbol{x}_\mu^{(u)}(i)$ are zeros, which prevents inter-block interference. As a result, the received symbol block $\boldsymbol{y}_\nu(i)$ depends only on the current transmitted blocks $\{\boldsymbol{x}_\mu^{(u)}(i)\}_{\mu=1,u=0}^{N_t,N_u-1}$ and can be expressed as

$$\begin{aligned}\boldsymbol{y}_\nu(i) &= \sum_{u=0}^{N_u-1}\sqrt{\frac{\overline{\gamma}_u}{N_t}}\sum_{\mu=1}^{N_t}\boldsymbol{H}_{\nu\mu}^{(u)}\boldsymbol{x}_\mu^{(u)}(i) + \boldsymbol{w}_\nu(i)\\ &= \sum_{u=0}^{N_u-1}\sqrt{\frac{\overline{\gamma}_u}{N_t}}\sum_{\mu=1}^{N_t}\boldsymbol{H}_{\nu\mu}^{(u)}\boldsymbol{C}_u\overline{\boldsymbol{s}}_\mu^{(u)}(i) + \boldsymbol{w}_\nu(i),\end{aligned} \tag{15.11}$$

where $\boldsymbol{w}_\nu(i)$ is AWGN defined in accordance with $\boldsymbol{y}_\nu(i)$ and $\boldsymbol{H}_{\nu\mu}^{(u)}$ is a $P \times P$ lower-triangular Toeplitz matrix with entries

$$[\boldsymbol{H}_{\nu\mu}^{(u)}]_{i,j} := h_{\nu\mu}^{(u)}((i-j) \bmod L). \tag{15.12}$$

To suppress MUI, the receiver of user u relies on the following despreading matrix:

$$\boldsymbol{D}_u = \boldsymbol{c}_u \otimes \boldsymbol{I}_{K+L}. \tag{15.13}$$

Due to the special structure of $\boldsymbol{C}_u$ in (15.8), the matrix $\boldsymbol{D}_u$ in (15.13) and the channel matrix $\boldsymbol{H}_{\nu\mu}^{(u)}$ in (15.12) can be shown to satisfy [334]

$$\boldsymbol{H}_{\nu\mu}^{(u)}\boldsymbol{C}_u = \boldsymbol{D}_u\overline{\boldsymbol{H}}_{\nu\mu}^{(u)}\boldsymbol{T}_{\text{zp}}, \tag{15.14}$$

where $\overline{\boldsymbol{H}}_{\nu\mu}^{(u)}$ is a $(K+L) \times (K+L)$ Toeplitz matrix with

$$[\overline{\boldsymbol{H}}_{\nu\mu}^{(u)}]_{i,j} := h_{\nu\mu}^{(u)}(((i-j))\text{mod } L). \tag{15.15}$$

Equation (15.14) reveals that the received block $\boldsymbol{H}_{\nu\mu}^{(u)}\boldsymbol{C}_u\overline{\boldsymbol{s}}_\mu^{(u)}(i)$ from user u belongs to the column space of the despreading matrix $\boldsymbol{D}_u$. However, the column spaces of the matrices $\boldsymbol{D}_u$ are mutually orthogonal since [cf. (15.13) and (15.7)]

$$\begin{aligned}\boldsymbol{D}_u^{\mathcal{H}}\boldsymbol{D}_{u'} &= (\boldsymbol{c}_u \otimes \boldsymbol{I}_{K+L})^{\mathcal{H}}(\boldsymbol{c}_{u'} \otimes \boldsymbol{I}_{K+L})\\ &= (\boldsymbol{c}_u^{\mathcal{H}}\boldsymbol{c}_{u'}) \otimes (\boldsymbol{I}_{K+L}^{\mathcal{H}}\boldsymbol{I}_{K+L})\\ &= \delta[u-u']\boldsymbol{I}_{K+L}.\end{aligned} \tag{15.16}$$

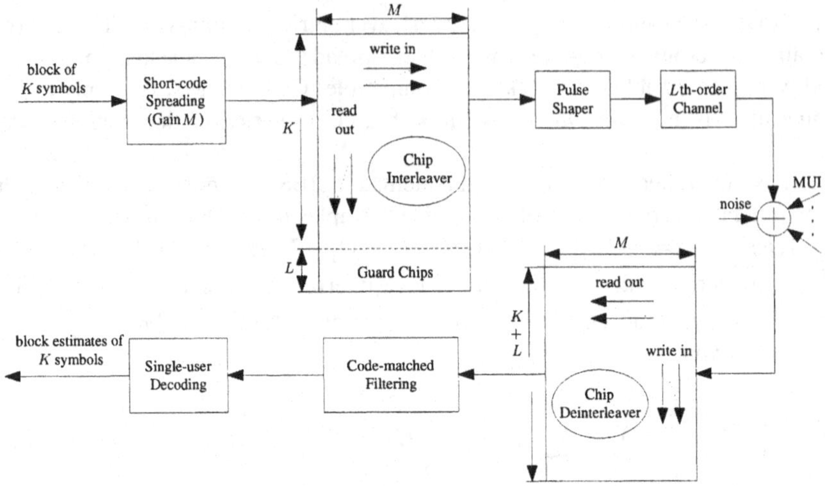

Figure 15.5 CIBS-CDMA transceiver of each user (only one transmit-receive antenna pair is shown)

Hence, MUI can be suppressed deterministically by projecting $\boldsymbol{y}_\nu(i)$ onto the corresponding subspaces to obtain

$$\bar{\boldsymbol{y}}_\nu^{(u)}(i) := \boldsymbol{D}_u^{\mathcal{H}} \boldsymbol{y}_\nu(i) = \sum_{\mu=1}^{N_t} \sum_{u'=0}^{N_u-1} \sqrt{\frac{\overline{\gamma}_{u'}}{N_t}} \boldsymbol{D}_u^{\mathcal{H}} \boldsymbol{H}_{\nu\mu}^{(u')} \boldsymbol{C}_{u'} \bar{\boldsymbol{s}}_\mu^{(u')}(i) + \boldsymbol{D}_u^{\mathcal{H}} \boldsymbol{w}_\nu(i)$$

$$= \sqrt{\frac{\overline{\gamma}_u}{N_t}} \sum_{\mu=1}^{N_t} \overline{\boldsymbol{H}}_{\nu\mu}^{(u)} \boldsymbol{T}_{\text{zp}} \bar{\boldsymbol{s}}_\mu^{(u)}(i) + \overline{\boldsymbol{w}}_\nu^{(u)}(i), \qquad (15.17)$$

where $\overline{\boldsymbol{w}}_\nu^{(u)}(i) := \boldsymbol{D}_u^{\mathcal{H}} \boldsymbol{w}_\nu(i)$ is the processed noise which remains white if $\boldsymbol{w}_\nu(i)$ is white since $\boldsymbol{D}_u^{\mathcal{H}} \boldsymbol{D}_u = \boldsymbol{I}_{K+L}$. Therefore, user separation by $\boldsymbol{D}_u$ does not degrade error performance in decoding the information symbols of user u.

Multiplying $\boldsymbol{D}_u^{\mathcal{H}}$ with $\boldsymbol{y}_\nu(i)$ amounts to passing $\boldsymbol{y}_\nu(i)$ through a deinterleaver (that is common to all users) and processing the output with a single user despreading filter matched to $\boldsymbol{c}_u$ to obtain each user's symbols. The block diagram of a transmit-receive antenna pair corresponding to each user is depicted in Figure 15.5. It entails symbol spreading (by the short signature code $\boldsymbol{c}_u$) and interleaving at the transmitter, together with deinterleaving and matched filtering at the receiver. Notice that processing steps to all other transmit-receive antenna pairs (not shown in Figure 15.5) are identical.

It is important to stress that deterministic user separation is accomplished without requiring any channel knowledge. This nice property is enabled by the judicious CIBS design, which converts a multiuser detection problem into a set of parallel single-user detection problems [cf. (15.17)]. Single-user ST coding schemes can then be applied to each user, as we explain next.

15.2.2 OSTBC Encoding and Decoding

Equation (15.17) describes the input-output relationship of a single-user ST-coded multi-antenna system signaling over a frequency-selective MIMO channel. ST codes for such systems have been presented in Chapter 7. Among other possibilities, we adopt here the OSTBC described in Section 7.3, which is specialized here for $N_t = 2$ for presentation clarity. (Extension to $N_t > 2$ antennas can be carried out similarly as detailed in Section 7.3.5.)

Each user (say, the uth one) partitions incoming information symbols $s_u(n)$ into blocks of length K, denoted as $\boldsymbol{s}_u(i) := [s_u(iK), s_u(iK+1), \ldots, s_u(iK+K-1)]^T$. For every two consecutive blocks $\boldsymbol{s}_u(2i)$ and $\boldsymbol{s}_u(2i+1)$ entering the ST encoder we obtain at its output two blocks $\bar{\boldsymbol{s}}_\mu^{(u)}(2i)$ and $\bar{\boldsymbol{s}}_\mu^{(u)}(2i+1)$ per antenna μ, where $\mu = 1, 2$. The input-output relationship of the ST encoder is [cf. (7.19)]

$$\boldsymbol{S}_u(i) := \begin{bmatrix} \bar{\boldsymbol{s}}_1^{(u)}(2i) & \bar{\boldsymbol{s}}_1^{(u)}(2i+1) \\ \bar{\boldsymbol{s}}_2^{(u)}(2i) & \bar{\boldsymbol{s}}_2^{(u)}(2i+1) \end{bmatrix} = \begin{bmatrix} \boldsymbol{s}_u(2i) & -\boldsymbol{P}_K^{(0)}\boldsymbol{s}_u^*(2i+1) \\ \boldsymbol{s}_u(2i+1) & \boldsymbol{P}_K^{(0)}\boldsymbol{s}_u^*(2i) \end{bmatrix}, \quad (15.18)$$

where the time-reversal permutation matrix $\boldsymbol{P}_K^{(0)}$ is defined as in (7.21).

At the receiver side, single-user decoding is applied individually on each user, after multiuser separation. Let us define a circulant matrix $\tilde{\boldsymbol{H}}_{\nu\mu}^{(u)}$ with size $K+L$ and entries $[\tilde{\boldsymbol{H}}_{\nu\mu}^{(u)}]_{i,j} := h_{\nu\mu}^{(u)}((i-j) \bmod (K+L))$. One can easily verify that

$$\bar{\boldsymbol{H}}_{\nu\mu}^{(u)}\boldsymbol{T}_{\text{zp}} = \tilde{\boldsymbol{H}}_{\nu\mu}^{(u)}\boldsymbol{T}_{\text{zp}}, \quad (15.19)$$

thanks to the fact that the last L rows in $\boldsymbol{T}_{\text{zp}}$ are all zeros. Upon replacing the Toeplitz matrix $\bar{\boldsymbol{H}}_{\nu\mu}^{(u)}$ by the circulant matrix $\tilde{\boldsymbol{H}}_{\nu\mu}^{(u)}$ in (15.17), we obtain

$$\bar{\boldsymbol{y}}_\nu^{(u)}(i) = \sqrt{\frac{\overline{\gamma}_u}{2}}\sum_{\mu=1}^{N_t}\tilde{\boldsymbol{H}}_{\nu\mu}^{(u)}\boldsymbol{T}_{\text{zp}}\bar{\boldsymbol{s}}_\mu^{(u)}(i) + \bar{\boldsymbol{w}}_\nu^{(u)}(i). \quad (15.20)$$

Adopting the ST code matrix in (15.18) and using the identity $\boldsymbol{T}_{\text{zp}}\boldsymbol{P}_K^{(0)} = \boldsymbol{P}_J^{(K)}\boldsymbol{T}_{\text{zp}}$, we can express the blocks received over two consecutive symbol block intervals as

$$\bar{\boldsymbol{y}}_\nu^{(u)}(2i) = \sqrt{\frac{\overline{\gamma}_u}{2}}\tilde{\boldsymbol{H}}_{\nu1}^{(u)}\boldsymbol{T}_{\text{zp}}\boldsymbol{s}_u(2i) + \sqrt{\frac{\overline{\gamma}_u}{2}}\tilde{\boldsymbol{H}}_{\nu2}^{(u)}\boldsymbol{T}_{\text{zp}}\boldsymbol{s}_u(2i+1) + \bar{\boldsymbol{w}}_\nu^{(u)}(2i), \quad (15.21)$$

$$\bar{\boldsymbol{y}}_\nu^{(u)}(2i+1) = -\sqrt{\frac{\overline{\gamma}_u}{2}}\tilde{\boldsymbol{H}}_{\nu1}^{(u)}\boldsymbol{P}_J^{(K)}\boldsymbol{T}_{\text{zp}}\boldsymbol{s}_u^*(2i+1) + \sqrt{\frac{\overline{\gamma}_u}{2}}\tilde{\boldsymbol{H}}_{\nu2}^{(u)}\boldsymbol{P}_J^{(K)}\boldsymbol{T}_{\text{zp}}\boldsymbol{s}_u^*(2i) + \bar{\boldsymbol{w}}_\nu^{(u)}(2i+1). \quad (15.22)$$

Comparing (15.21) and (15.22) with (7.30) and (7.31), we recognize that a single-user ZP-only transmission is in effect per user. This can also be verified if we let $N_u = 1$ and observe that the transmitted chip sequence in (15.10) coincides with the sequence in Figure 7.6. This coincidence implies that the receiver processing described in Section 7.3.2 can readily be applied to each user, and each information block can be decoded separately using any of the detectors discussed in Section 7.3.2.

Notice that only the time-reversal permutation matrix $\boldsymbol{P}_K^{(0)}$ is applied on each symbol block. Hence, combining OSTBC with CIBS-CDMA extends the single-user ZP-only scheme of Chapter 7 to a multiuser setting. Instead of ZP, it is possible to have CIBS-CDMA transmissions relying on a cyclic prefix (CP) guard [334]. This CP-based version of CIBS-CDMA allows for other permutation matrices but is not as spectrally efficient as the ZP-only version presented here.

15.2.3 Attractive Features of OSTBC-CIBS-CDMA

The OSTBC-CIBS-CDMA system comes with a number of attractive features which stem from corresponding properties that are present even in the single-antenna CIBS-CDMA system.

15.2.3.1 Deterministic Multiuser Separation Unlike DS-CDMA, the CIBS-CDMA system achieves deterministic multiuser separation regardless of the underlying frequency-selective multipath channel, which as we mentioned in Section 15.2.2 also includes bounded asynchronism. This is achieved via chip interleaving with guards at the transmitter and low-complexity code-matched filtering at the receiver.

More interestingly, since the despreading matrix $\boldsymbol{D}_u$ is unitary, we have

$$p\left(\boldsymbol{y}_\nu(i)\middle|\{\bar{\boldsymbol{s}}_\mu^{(u)}(i)\}_{u=0,\mu=1}^{N_u-1,N_t}\right) = \prod_{u=0}^{N_u-1} p\left(\bar{\boldsymbol{y}}_\nu^{(u)}(i)\middle|\{\bar{\boldsymbol{s}}_\mu^{(u)}(i)\}_{\mu=1}^{N_t}\right), \qquad (15.23)$$

where $p(\cdot|\cdot)$ is used to denote a conditional probability density function. Equation (15.23) shows that CIBS through $\boldsymbol{C}_u$ and multiuser separation via $\boldsymbol{D}_u$ has rendered a multiuser detection problem *equivalent* to a set of single-user detection problems. If the latter are ML decoded, CIBS-CDMA preserves maximum likelihood (ML) optimality at single-user complexity.

Multiuser separation is achieved for a very general class of codes as long as users construct their spreading matrices starting with orthogonal signature vectors [cf. (15.7) and (15.8)]. Choosing $\{\boldsymbol{c}_m\}_{m=0}^{M-1}$ as the canonical basis vectors (columns of the $M \times M$ identity matrix $\boldsymbol{I}_M$) yields a block-spread TDMA system. With $\boldsymbol{f}_m$ denoting the mth column of an FFT matrix, selecting $\boldsymbol{c}_m = \boldsymbol{f}_m$ corresponds to the generalized multi-carrier (MC)-CDMA of [93, 286], which subsumes existing MC-CDMA variates that separate users in the code or subcarrier domain.

15.2.3.2 Flexible Single-User Decoding Since detection in OSTBC-CIBS-CDMA requires no coordination among users, each mobile can choose its own equalizer, depending on the computational complexity it can afford without interfering with other users. This is in sharp contrast with the joint (and thus fully coordinated) multiuser detection needed for ML demodulation of (asynchronous) DS-CDMA transmissions over multipath channels. The basic distinction originates from the fundamental paradigm shift offered by CIBS, which capitalizes on carefully selecting the spreading at the transmitter in addition to the receiver. Indeed, different from DS-CDMA, which allows for fully uncoordinated transmissions at the cost of coordinated joint multiuser detection, MUI-free CIBS-CDMA relies on minimally coordinated transmissions and judicious design of (de-)spreading operations to enable low-complexity uncoordinated detection. This uncoordinated detection capability favors MUI-free transceivers because it enhances their robustness to imperfect system parameters, for example, imperfect channel estimation and imbalanced power control. (Intuitively speaking, poor channel estimation accuracy from one user does not affect other users' performance.)

With TR-OSTBC, decoding of ST-coded blocks is also decoupled, as detailed in Chapter 7. With (near-)optimal sphere decoding, each user achieves the maximum diversity $N_t N_r(L_h + 1)$ provided by the MIMO channel, which has order $L_h \leq L$ after excluding the number of zero taps capturing the asynchronism. Besides sphere decoding, suboptimal linear (ZF or MMSE) equalizers or nonlinear decision-feedback alternatives can also be used to trade-off error performance for reduced complexity.

15.2.3.3 Bandwidth Efficiency In every block consisting of $P = M(K + L)$ chips, each user $u \in [0, M - 1]$ can transmit K information-bearing chips. Taking into account both the loss due to zero padding in CIBS and the rate loss due to the TR-based OSTBC, the bandwidth efficiency of OSTBC-CIBS-CDMA turns out to be

$$\eta_{\text{ST-CIBS}} = \frac{K}{K + L}\,\eta_{\text{OSTBC}}. \tag{15.24}$$

To increase bandwidth efficiency, one has to increase the block size K. However, increasing K leads to increased decoding complexity, as larger blocks need to be processed; for further discussion of decoding complexity issues, see Chapter 5. Also, the channel coherence time may limit the maximum value of K, which in turn determines the highest-possible bandwidth efficiency. For example, consider a chip rate of $1/T_c = 3.84$ MHz and a time duration of $T_d = 2/3$ ms corresponding to one time slot in the time-division (TD) CDMA based UTRA (UMTS terrestrial radio access) system operating in time-division-duplex (TDD) mode [102]. Assuming that the channel only remains constant during one time slot, the OSTBC-CIBS-CDMA codeword can have a maximum of $T_d/T_c = 2560$ chips. Suppose that the transmitter has two antennas and the system hosts a maximum of $M = 16$ users. Then we shall have $2M(K + L) < 2560$, which limits $K + L$ to be less than 128. If $L = 32$, the factor $K/(K + L)$ in (15.24) would in turn be bounded by $(128 - 32)/128 = 75\%$.

For this reason, OSTBC-CIBS-CDMA is best suited for systems with a small channel order L. This is more likely the case in the synchronous downlink operation, where the channel order depends solely on the delay spread and L may be small. For channels with a long delay spread, one can also adopt OSTBC-CIBS-CDMA, provided that a channel-shortening scheme has been applied first. An interesting approach to MIMO channel shortening has been described in [10].

15.2.3.4 Power Efficiency In the uplink scenario, selecting $\boldsymbol{c}_m$ spreading vectors to form the CIBS matrices $\boldsymbol{C}_m$ in (15.8) with constant-modulus (C-M) entries leads to transmitted blocks in (15.10) which have C-M except for a guard interval of L samples. Although not a problem for the power amplifier, the use of guard zeros may require special care when switching between on and off states in a continuous-time analog transmitter implementation. However, with a digital signal processing (DSP) unit and a digital-to-analog (D/A) converter at the transmitter, inserting zeros is not a problem for software radio systems [219], for example. Nonetheless, the OSTBC-CIBS-CDMA transmitter can dispense with zero padding and achieve perfectly C-M transmissions during its uplink operation if one fills the guard zeros by known symbols drawn from the same alphabet of information symbols, without any change at the receiver [334].

15.2.3.5 Power Control Versus Delay Control Since mobile users are typically distributed in space, the power received from each user may exhibit a large dynamic range which necessitates stringent power control to mitigate MUI. However, power control requirements are alleviated in OSTBC-CIBS-CDMA since the signals received from multiple users are separable by design.

The key parameter in the OSTBC-CIBS-CDMA system design is the maximum channel order L. Large L implies large K to achieve high bandwidth efficiency, which increases decoding complexity and decoding delay. It is thus preferable to keep L as small as possible. As we discussed earlier, L depends on $\tau_{\max,a}$ and $\tau_{\max,s}$ [cf. (15.5)]. Notice that $\tau_{\max,s}$ is determined by the propagation environment and is thus not up to the designer's control. To maintain a small channel order L, CIBS-CDMA may rely on delay control to decrease $\tau_{\max,a}$ in uplink transmissions. In this case, each mobile user synchronizes with the AP's pilot waveform and roughly calculates the distance from the AP by, for example, power measurements. In uplink, the mobile advances its transmission by a corresponding amount to compensate for the two-way propagation time that is consumed between the AP and the mobile user. With such a delay control mechanism, the maximum $\tau_{\max,a}$ could be reduced considerably.

15.2.3.6 Limitations of OSTBC-CIBS-CDMA Compared with competing alternatives, OSTBC-CIBS-CDMA has certain limitations. First, it applies only to synchronous or quasi-synchronous but not completely asynchronous transmissions. DS-CDMA, on the other hand, can be used in an asynchronous setting. Second, OSTBC-CIBS-CDMA is less attractive for long channels, as we mentioned in Section 15.2.3.3. This is also the case, however, for other single-carrier approaches, including DS-CDMA. For long channels, multi-carrier systems should be preferred instead; one example is provided in Section 15.3. Third, OSTBC-CIBS-CDMA does not take advantage of system load variations; the receiver always deals with ISI among K information symbols by design. When a system is mostly operating at very low load, conventional DS and multi-carrier (MC) CDMA may have advantages in terms of error performance and complexity, as illustrated by numerical results.

15.2.4 Numerical Examples

To illustrate the merits of OSTBC-CIBS-CDMA, we compare its BER performance with that of DS-CDMA and MC-CDMA. We will test a system with two transmit-antennas and a single receive-antenna.

The DS-CDMA transmissions are generated as follows. Each user first encodes two information symbols using the Alamouti code

$$\begin{bmatrix} s_{u,1} & -s_{u,2}^* \\ s_{u,2} & s_{u,1}^* \end{bmatrix} \tag{15.25}$$

and forwards its rows to the two transmit-antennas. On each antenna, ST-coded symbols are spread with Walsh-Hadamard sequences of length 16, and three zeros are inserted after each spreading operation to avoid interblock interference. These Walsh-Hadamard spreading vectors are subsequently scrambled by a common pseudo-noise (PN) sequence. We assume that users have equal average received SNR $\overline{\gamma}$. The blocks received in two consecutive time slots are

$$\boldsymbol{y}_1 = \sum_{u=0}^{N_u-1} \sqrt{\frac{\overline{\gamma}}{2}} \left[\boldsymbol{H}_1^{(u)} \boldsymbol{c}_u s_{u,1} + \boldsymbol{H}_2^{(u)} \boldsymbol{c}_u s_{u,2} \right] + \boldsymbol{w}_1, \tag{15.26}$$

$$\boldsymbol{y}_2 = \sum_{u=0}^{N_u-1} \sqrt{\frac{\overline{\gamma}}{2}} \left[-\boldsymbol{H}_1^{(u)} \boldsymbol{c}_u s_{u,2} + \boldsymbol{H}_2^{(u)} \boldsymbol{c}_u s_{u,1}^* \right] + \boldsymbol{w}_2, \tag{15.27}$$

where $\boldsymbol{H}_\mu^{(u)}$ is a 19×16 Toeplitz matrix representing the channel corresponding to the μth antenna of user u. Rearranging (15.26) and (15.27) leads to

$$\begin{aligned} \begin{bmatrix} \boldsymbol{y}_1 \\ \boldsymbol{y}_2^* \end{bmatrix} &= \begin{bmatrix} \boldsymbol{w}_1 \\ \boldsymbol{w}_2^* \end{bmatrix} \\ &+ \begin{bmatrix} \boldsymbol{H}_1^{(0)} \boldsymbol{c}_0 & \boldsymbol{H}_2^{(0)} \boldsymbol{c}_0 & \cdots & \boldsymbol{H}_1^{(N_u-1)} \boldsymbol{c}_{N_u-1} & \boldsymbol{H}_2^{(N_u-1)} \boldsymbol{c}_{N_u-1} \\ [\boldsymbol{H}_2^{(0)} \boldsymbol{c}_0]^* & -[\boldsymbol{H}_1^{(0)} \boldsymbol{c}_0]^* & \cdots & [\boldsymbol{H}_2^{(N_u-1)} \boldsymbol{c}_{N_u-1}]^* & -[\boldsymbol{H}_1^{(N_u-1)} \boldsymbol{c}_{N_u-1}]^* \end{bmatrix} \\ &\times \sqrt{\frac{\overline{\gamma}}{2}} \begin{bmatrix} s_{0,1} \\ s_{0,2} \\ \vdots \\ s_{N_u-1,1} \\ s_{N_u-1,2} \end{bmatrix}. \end{aligned} \tag{15.28}$$

When $N_u = 1$, there is no MUI and the two symbols involved in Alamouti's ST code can be decoded separately. However, MUI will prevent such a separation when $N_u > 1$. This necessitates joint decoding of all $2N_u$ symbols.

MC-CDMA users also adopt the Alamouti code in (15.25). The information symbols on each antenna are spread in the subcarrier domain using Walsh-Hadamard sequences of length 16 and are transmitted after FFT processing and insertion of a CP of size $L = 3$. Following similar steps that led to (15.28), we obtain an equivalent input-output relationship in the subcarrier domain as

$$\begin{bmatrix} \boldsymbol{y}_1 \\ \boldsymbol{y}_2^* \end{bmatrix} = \begin{bmatrix} \boldsymbol{w}_1 \\ \boldsymbol{w}_2^* \end{bmatrix}$$
$$+ \begin{bmatrix} \boldsymbol{D}_1^{(0)} \boldsymbol{c}_0 & \boldsymbol{D}_2^{(0)} \boldsymbol{c}_0 & \cdots & \boldsymbol{D}_1^{(N_u-1)} \boldsymbol{c}_{N_u-1} & \boldsymbol{D}_2^{(N_u-1)} \boldsymbol{c}_{N_u-1} \\ [\boldsymbol{D}_2^{(0)} \boldsymbol{c}_0]^* & -[\boldsymbol{D}_1^{(0)} \boldsymbol{c}_0]^* & \cdots & [\boldsymbol{D}_2^{(N_u-1)} \boldsymbol{c}_{N_u-1}]^* & -[\boldsymbol{D}_1^{(N_u-1)} \boldsymbol{c}_{N_u-1}]^* \end{bmatrix}$$
$$\times \sqrt{\frac{\bar{\gamma}}{2}} \begin{bmatrix} s_{0,1} \\ s_{0,2} \\ \vdots \\ s_{N_u-1,1} \\ s_{N_u-1,2} \end{bmatrix}, \tag{15.29}$$

where the diagonal channel $\boldsymbol{D}_\mu^{(u)}$ contains on its diagonal entries the channel responses on the FFT grid corresponding to the μth antenna of user u. Due to MUI, we again have to decode $2N_u$ symbols together. Only when $N_u = 1$ can decoding of symbol pairs be performed separately.

For OSTBC-CIBS-CDMA, we test two scenarios. Parameters for the first are $K = 6$, $M = 12$ and for the second $K = 12$, $M = 16$. In both cases, a spreading gain of $P/K = M(K + L)/K \approx 19$ per information symbol is chosen, identical to what is used by DS-CDMA (counting the size of the ZP guard) and by MC-CDMA (counting also the size of the CP guard). For all three systems, QPSK symbols are transmitted and are detected using a linear MMSE equalizer. The complexity of both DS-CDMA and MC-CDMA is on the order of $O\left((2N_u)^3\right)$, which clearly grows as the number of users increases. The complexity of OSTBC-CIBS-CDMA, on the other hand, is $O(K^3)$, which does not depend on the number of users. Increasing the block size K allows for more MUI-free users but requires decoding of larger blocks.

We first consider the uplink operation described by (15.6) but with an artificially small channel order $L = 3$. The $L+1 = 4$ channel taps between each antenna pair are assumed Gaussian distributed with zero mean and variance $1/(L+1)$. The resulting BER curves are plotted in Figures 15.6 and 15.7. We also consider a downlink scenario with channel order $L = 3$, following the model in (15.4). The BER curves are depicted in Figures 15.8 and 15.9. We observe that for OSTBC-CIBS-CDMA, the BER remains unchanged when the number of active users $N_u \in [1, M]$ varies. However, for both DS-CDMA and MC-CDMA, the error performance of each user degrades with N_u increasing, as confirmed by Figures 15.6 to 15.9. When the system load is light (i.e., $N_u \ll M$), multiuser detection based receivers could exhibit improved BER since MUI is less severe in this case. When the system is moderately or heavily loaded (e.g., $N_u \approx M$), the MUI-free CIBS-CDMA system outperforms DS-CDMA equipped with multiuser detection. Figures 15.6 to 15.9 also corroborate that DS-CDMA and MC-CDMA have quite different characteristics in the presence of multipath; see also [335] for related results in the single-antenna setup. Finally, we underscore that advanced receivers based on, for example, the sphere decoding algorithm described in Chapter 5 can be used in OSTBC-CIBS-CDMA to improve each user's BER performance without requiring coordination among users.

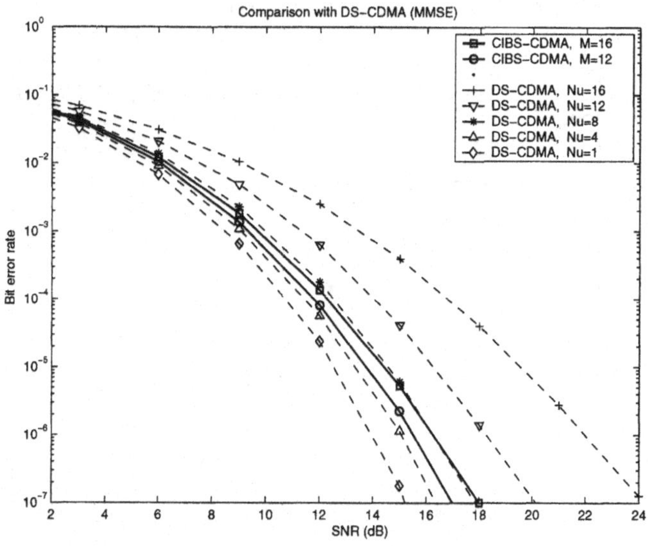

Figure 15.6 CIBS-CDMA vs. DS-CDMA in multi-antenna uplink

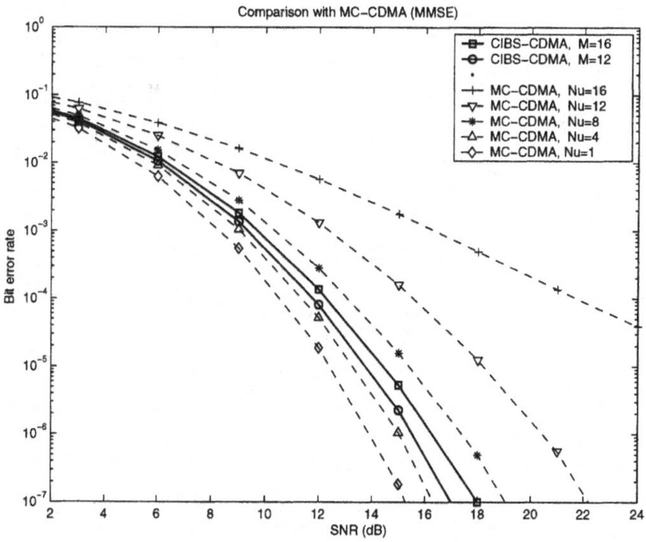

Figure 15.7 CIBS-CDMA vs. MC-CDMA in multi-antenna uplink

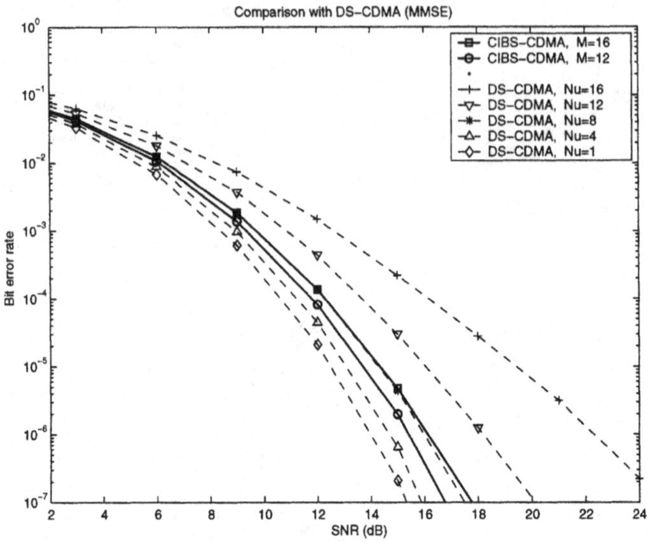

Figure 15.8 CIBS-CDMA vs. DS-CDMA in multi-antenna downlink

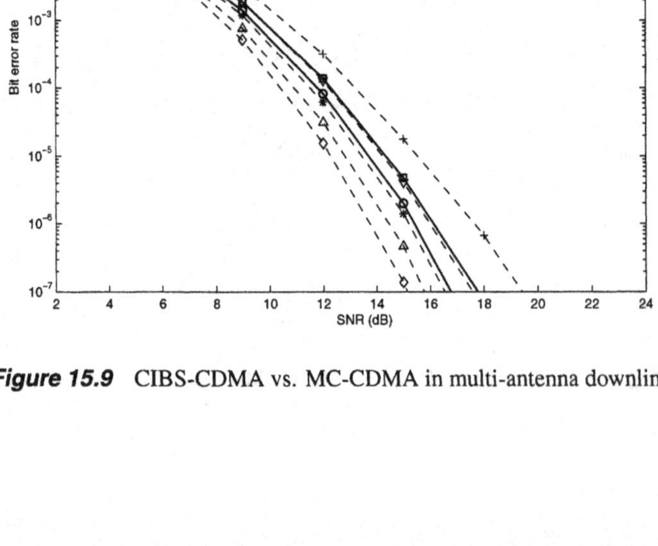

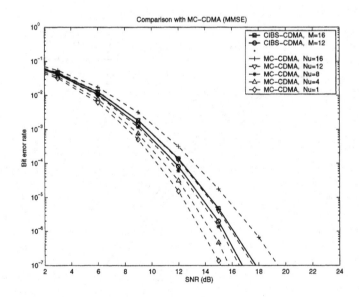

Figure 15.9 CIBS-CDMA vs. MC-CDMA in multi-antenna downlink

15.3 MULTI-CARRIER SYSTEMS: STF-OFDMA

In this section we consider an orthogonal multiple access system where users communicate with the AP over frequency-selective MIMO channels using multi-carrier ST-coded waveforms. Each user still employs single-user ST (de)coding, whereas the (de)spreading module in Figure 15.3 separates users in the subcarrier domain. The ST (de)coding module operates using the general framework described in Chapter 8 for single-user ST coded transmissions for MIMO-OFDM systems. Orthogonal frequency division multiple access (OFDMA) is a natural extension of OFDM to multiuser communications. OFDMA relies on the simple idea of allocating distinct subcarriers to each user, which clearly enables MUI-free access.

15.3.1 OFDMA for User Separation

Consider the scenario where all users have equal rates; extensions to multi-rate systems are possible along the lines of [287]. Suppose that the OFDMA system has a total number of N_c subcarriers. These N_c subcarriers are equally distributed to N_u users so that a group of $K := N_c/N_u$ subcarriers is assigned per user. Clearly, if K is non-integer, we use its integer part. ST coding is then performed across the assigned subcarriers per user.

We suppose that the K subcarriers assigned to each user are equispaced. Specifically, we let the generic user u be assigned with subcarriers having index

$$p_{u,k} = kN_c + u, \qquad k = 0, 1, \ldots, K-1, \quad u = 0, 1, \ldots, N_u - 1. \tag{15.30}$$

This equispaced subcarrier allocation is analogous to one used in the subcarrier grouping approach described in Section 8.3.1, with the group index replaced here by the user index.

Let $x_\mu^{(u)}(i; p_{u,k})$ denote the symbol transmitted on the subcarrier $p_{u,k}$ from the μth antenna of user u at the ith OFDMA interval. Correspondingly, $y_\nu^{(u)}(i; p_{u,k})$ is the received sample on subcarrier $p_{u,k}$ at the output of the νth antenna of user u over the ith interval. According to (8.7), the input-output relationship corresponding to subcarrier $p_{u,k}$ is

$$y_\nu^{(u)}(i; p_{u,k}) = \sqrt{\frac{\overline{\gamma}_u}{N_t}} \sum_{\mu=1}^{N_t} H_{\nu\mu}^{(u)}(p_{u,k}) x_\mu^{(u)}(i; p_{u,k}) + w_\nu^{(u)}(i; p_{u,k}), \tag{15.31}$$

where $H_{\nu\mu}^{(u)}(p_{u,k}) := \sum_{l=0}^{L} h_{\nu\mu}^{(u)}(l) e^{-j\frac{2\pi}{N_c} p_{u,k} l}$ and $w_\nu^{(u)}(i; p_{u,k})$ is AWGN. Suppose that the ST code entails N_x OFDM blocks. On each subcarrier, we collect transmitted and received symbols in matrices $\boldsymbol{X}^{(u)}(p_{u,k})$ and $\boldsymbol{Y}^{(u)}(p_{u,k})$, respectively, with corresponding entries

$$\left[\boldsymbol{X}^{(u)}(p_{u,k})\right]_{\mu,i} = x_\mu^{(u)}(i; p_{u,k}), \tag{15.32}$$

$$\left[\boldsymbol{Y}^{(u)}(p_{u,k})\right]_{\nu,i} = y_\nu^{(u)}(i; p_{u,k}), \tag{15.33}$$

where $i = 0, 1, \ldots, N_x - 1$. We can then rewrite (15.31) as

$$\boldsymbol{Y}^{(u)}(p_{u,k}) = \sqrt{\frac{\overline{\gamma}_u}{N_t}} \boldsymbol{H}^{(u)}(p_{u,k}) \boldsymbol{X}^{(u)}(p_{u,k}) + \boldsymbol{W}^{(u)}(p_{u,k}), \tag{15.34}$$

where

$$[\boldsymbol{H}^{(u)}(p_{u,k})]_{\nu\mu} := H_{\nu\mu}^{(u)}(p_{u,k}), \tag{15.35}$$

$$[\boldsymbol{W}^{(u)}(p_{u,k})]_{\nu,i} := w_{\nu}^{(u)}(i; p_{u,k}). \tag{15.36}$$

The block transmitted across all subcarriers

$$\boldsymbol{X}^{(u)} := [\boldsymbol{X}^{(u)}(p_{u,0}), \boldsymbol{X}^{(u)}(p_{u,1}), \ldots, \boldsymbol{X}^{(u)}(p_{u,K-1})] \tag{15.37}$$

is the STF codeword of user u. Notice the similarity with the STF code matrix used in the grouped MIMO-OFDM system of Section 8.3.

15.3.2 STF Block Codes

In Chapter 8 various classes of STF codes were presented for single-user MIMO-OFDM multi-antenna systems. Here we focus on the STF block codes constructed as in (8.28), where linear complex-field coding (LCFC) is combined with OSTBC designs. In this chapter's multi-access context, this system will be referred to henceforth as STF-OFDMA.

Following the argument used in Section 8.3 to analyze the diversity of GSTF block-coded MIMO OFDM, we can readily infer that with each ST-OFDMA user relying on K subcarriers to transmit over an Lth-order frequency-selective channel, the diversity order can be as high as $N_t N_r \min(K, L+1)$. When $K = 1$ and no precoding across subcarriers is performed, STF-OFDMA reduces to an OFDMA system with OSTBC applied on a per subcarrier basis.

15.3.3 Attractive Features of STF-OFDMA

Ensuring MUI-free access with deterministic user separation, STF-OFDMA shares several properties with the OSTBC-CIBS-CDMA system described in Section 15.2.3. In addition, we will see that STF-OFDMA can be bandwidth efficient even for frequency-selective MIMO channels with long impulse response. Furthermore, STF-OFDMA with equispaced subcarrier allocation in (15.30) will entail perfectly constant-modulus uplink transmission for each user. This means that STF-OFDMA enjoys high power efficiency in the uplink.

15.3.3.1 Bandwidth Efficiency To avoid interblock interference, each OFDMA transmission requires a CP of length at least equal to L. Also taking into account the rate loss due to the OSTBC, the bandwidth efficiency of the STF-OFDMA system is

$$\eta_{\text{STF-OFDMA}} = \frac{N_c}{N_c + L}\eta_{\text{OSTBC}} = \frac{K}{K + L/N_u}\eta_{\text{OSTBC}}. \tag{15.38}$$

As shown in Section 15.1.2, the channel order L increases by the number of zero taps added to account for the asynchronism among users. However, high bandwidth efficiency is still possible if the block size K is small when N_u increases, so that K is much larger than L/N_u. Hence, STF-OFDMA can be bandwidth efficient even for moderate or large L. This is particularly appealing for uplink operation.

15.3.3.2 Low Decoding Complexity Even in the presence of long channels, K can be made small without loss in bandwidth efficiency. Note that choosing a small K reduces the receiver complexity considerably. In a practical system with channel coding, iterative processing between channel decoding and STF demodulation proves to be an effective method to approach optimal reception. In the single-antenna setting, it has been shown that a small K will harvest most of the benefit of linear precoding when coupled with channel coding [292, 307]; see also Chapter 10. This suggests that only a small K (e.g., $K = 4$ or $K = 8$) is necessary for STF-OFDMA when combined with channel coding.

15.3.3.3 Power Efficiency As we discussed in Section 15.2.3.4, C-M transmissions are desirable at the power amplification stage. Even though we know from Section 8.1.1 that standard OFDM transmissions have non-constant modulus, we will see here that the equispaced subcarrier allocation in (15.30) ensures that each STF-OFDMA user's transmission is C-M. To establish this surprising result, suppose that PSK-modulated information symbols $s_u(n)$ constitute the C-M input of the STF module. We wish to show that the STF output is also C-M.

In a single-antenna OFDMA uplink operation, if only one subcarrier is assigned per user (i.e., $K = 1$), the sequence transmitted has C-M. On the other hand, if multiple subcarriers are assigned to each user, we will show that LCFC across subcarriers can guarantee C-M of the waveforms transmitted from all antennas in the uplink. For simplicity, we again consider the two-transmit-antenna case and let $\boldsymbol{s}_u(2i)$ and $\boldsymbol{s}_u(2i+1)$ denote two consecutive STF-coded symbol blocks. Rearranging the code matrix in (8.28) in a form similar to (15.18), we have

$$\boldsymbol{S}_u(i) = \begin{bmatrix} \boldsymbol{\Theta}\boldsymbol{s}_u(2i) & -\boldsymbol{\Theta}^*\boldsymbol{s}_u^*(2i{+}1) \\ \boldsymbol{\Theta}\boldsymbol{s}_u(2i{+}1) & \boldsymbol{\Theta}^*\boldsymbol{s}_u^*(2i) \end{bmatrix}. \tag{15.39}$$

Based on (15.39), the ST-coded block comprising K subcarriers per transmit-antenna can generally be written as $\boldsymbol{\Theta a}$ or $\boldsymbol{\Theta}^*\boldsymbol{a}^*$, where $\boldsymbol{a} = \boldsymbol{s}_u(i)$ contains C-M symbols.

Let the $N_c \times K$ matrix $\tilde{\boldsymbol{F}}_u^*$ contain the K digital subcarriers allocated to user u. According to the subcarrier allocation in (15.30), the $(k+1)$st column of $\tilde{\boldsymbol{F}}_u^*$ corresponds to the $(p_{u,k}+1)$st column of the IFFT matrix $\boldsymbol{F}_{N_c}^{\mathcal{H}}$. Hence, the $(i +$

$1, k+1)$st entry of $\tilde{\boldsymbol{F}}_u^*$ is

$$[\tilde{\boldsymbol{F}}_u^*]_{i,k} = \frac{1}{\sqrt{N_c}} e^{j\frac{2\pi}{N_c} i p_{u,k}} = \frac{1}{\sqrt{N_c}} e^{j\frac{2\pi}{N_c} i(kN_u+u)}. \tag{15.40}$$

Regardless of the value of K, we choose the precoder matrix as

$$\boldsymbol{\Theta} = \boldsymbol{F}_K^{\mathcal{H}} \operatorname{diag}(1, \alpha, \ldots, \alpha^{K-1}), \tag{15.41}$$

where $\alpha := \exp(j\pi/2K)$. It has been established in Section 3.5.5 that this choice is optimal when K is a power of 2. The transmitted block in the time domain with the CP excluded is either

$$\tilde{\boldsymbol{F}}_u^* \boldsymbol{\Theta} \boldsymbol{a} = \tilde{\boldsymbol{F}}_u^* \boldsymbol{F}_K^{\mathcal{H}} \operatorname{diag}\big(1, \ldots, \alpha^{K-1}\big) \boldsymbol{a}. \tag{15.42}$$

or

$$\tilde{\boldsymbol{F}}_u^* \boldsymbol{\Theta}^* \boldsymbol{a}^* = \tilde{\boldsymbol{F}}_u^* \boldsymbol{F}_K \operatorname{diag}\big(1, \ldots, (\alpha^*)^{K-1}\big) \boldsymbol{a}^*. \tag{15.43}$$

For each $p = 0, \ldots, N_u - 1$ and $q = 0, \ldots, K-1$, we can easily verify that the $(pK + q + 1)$st entries of $\tilde{\boldsymbol{F}}_u^* \boldsymbol{\Theta} \boldsymbol{a}$ and $\tilde{\boldsymbol{F}}_u^* \boldsymbol{\Theta}^* \boldsymbol{a}^*$ are

$$[\tilde{\boldsymbol{F}}_u^* \boldsymbol{\Theta} \boldsymbol{a}]_{pK+q} = \frac{1}{\sqrt{N_u}} e^{j\frac{2\pi}{N_c}(pK+q)u} \alpha^{f(q)} [\boldsymbol{a}]_{f(q)}, \tag{15.44}$$

where $f(q) := (K - q) \bmod K$, and

$$[\tilde{\boldsymbol{F}}_u^* \boldsymbol{\Theta}^* \boldsymbol{a}^*]_{pK+q} = \frac{1}{\sqrt{N_u}} e^{j\frac{2\pi}{N_c}(pK+q)u} (\alpha^*)^q [\boldsymbol{a}^*]_q. \tag{15.45}$$

Equation (15.45) shows that the STF output sequence corresponding to each transmit-antenna has C-M. This C-M property is achieved thanks to (i) the equispaced subcarrier allocation in (15.30), and (ii) the specific unitary precoder adopted in (15.41). In a single-antenna setup, the C-M property in (15.45) was originally established by [307].

With a large diversity gain as well as high bandwidth and power efficiency ensured by C-M transmissions, the STF-OFDMA system is an attractive choice for wireless multiuser communications, especially in the uplink operation.

15.3.3.4 Limitations of STF-OFDMA Chapter 8 has provided comparisons between single- and multi-carrier approaches. As a multi-carrier scheme, STF-OFDMA inherits certain properties of single-user OFDM. For example, STF-OFDMA is sensitive to subcarrier drifts arising due to Doppler and mismatch between transmit-receive oscillators. In the STF-OFDMA context, inter-carrier interference gives rise to multiuser interference.

15.3.4 Numerical Examples

We test an STF-OFDMA system with different configurations of transmit- and receive-antennas. The number of subcarriers per user varies as $K = 1, 4, 8, 16$. We

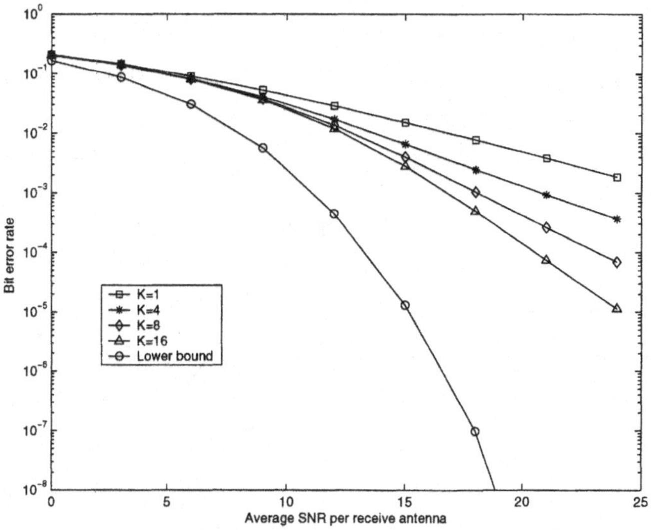

Figure 15.10 Performance of STF-OFDMA ($N_t = 1$, $N_r = 1$)

set $N_c = 64$, $L = 8$, and adopt a linear MMSE receiver for detection. Figures 15.10 to 15.13 depict the BER performance of STF-OFDMA with QPSK corresponding to $(N_t, N_r) = (1,1), (2,1), (2,2)$, and $(4,2)$, respectively. In Figures 15.10 to 15.13, we also plot the matched filter performance bound, which assumes that each of the $N_t N_r (L+1)$ channel taps contributes an independent signal copy for optimal diversity combining; for example, one can consider a low-rate scheme with one information symbol followed by L zeros transmitted per antenna and repeated sequentially across all antennas.

We observe that increasing K leads to improved BER performance. When K is small, the gap between the BER performance of STF-OFDMA with a linear receiver and the matched filter performance bound is large. This performance gap shrinks quickly when (i) the number of subcarriers per user increases, and (ii) more important, when N_t and/or N_r increase. When N_t and N_r are small, the optimal ML receiver, implemented using, for example, the sphere decoding algorithm of Chapter 5, can be adopted to reduce this performance gap. When N_t and/or N_r is large, each subcarrier enjoys considerable antenna diversity, of order $N_t N_r$, which renders each subchannel to behave more like a non-fading channel. In such a case, linear receiver processing approaches the optimal ML performance.

We underscore that the performance results herein pertain to uncoded BER performance. When channel coding is incorporated, the performance difference between different K's could change; see Chapter 10 for results with joint error control coding and precoding with iterative (turbo) decoding.

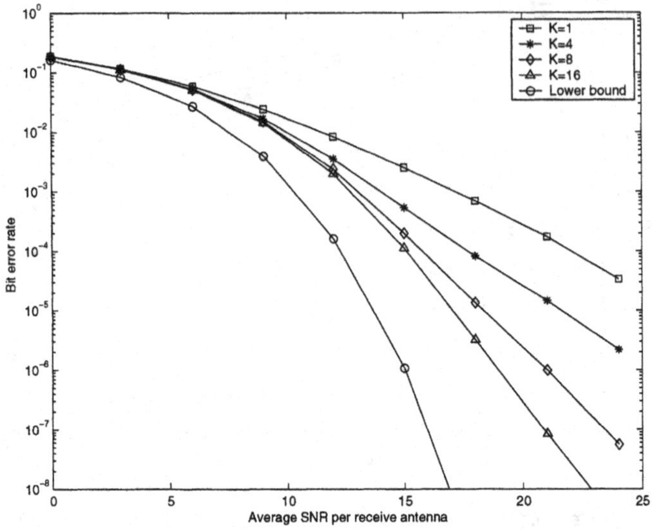

Figure 15.11 Performance of STF-OFDMA ($N_t = 2$, $N_r = 1$)

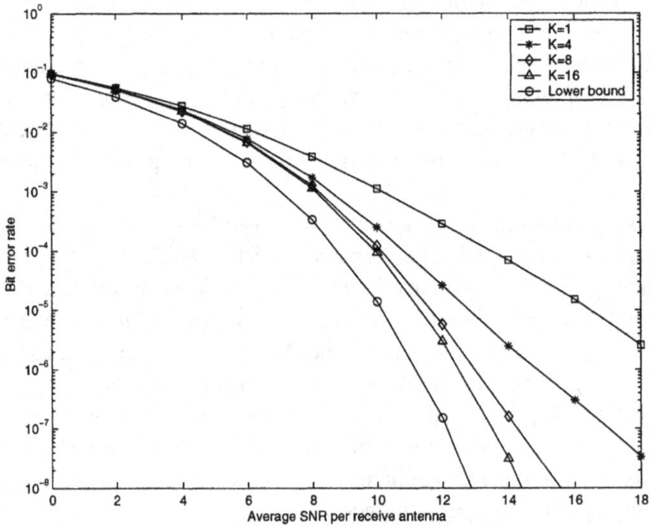

Figure 15.12 Performance of STF-OFDMA ($N_t = 2$, $N_r = 2$)

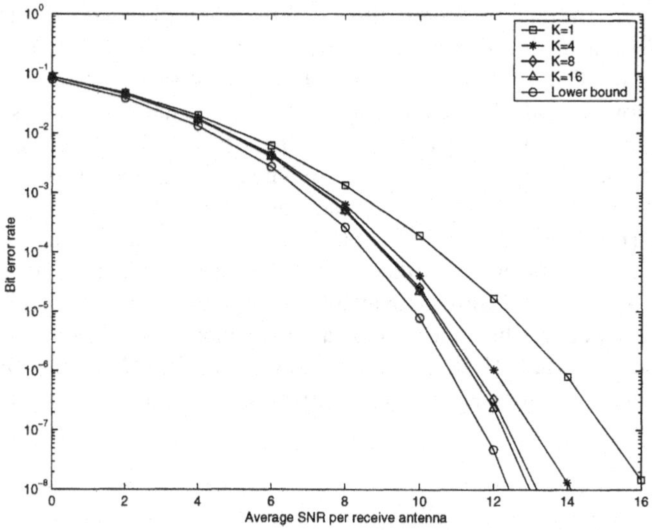

Figure 15.13 Performance of STF-OFDMA ($N_t = 4$, $N_r = 2$)

15.4 CLOSING COMMENTS

In this chapter we presented two orthogonal ST multiple access schemes in the presence of frequency-selective MIMO channels. The first one relies on single-carrier modulation and is based on chip interleaving and block spreading to suppress multiuser interference deterministically. The second one uses multicarrier modulation and is based on MIMO OFDM, where non-overlapping subcarriers are assigned to individual users. With deterministic multiuser separation, ST coding is performed on a per user basis. In the single-carrier approach, each user encodes with the time-reversed OSTBC of Chapter 7, while in the multi-carrier scheme, each user adopts the STF block code in Chapter 8.

Multiuser MIMO systems have received considerable attention lately. The uplink channel is also referred to as a MIMO multiple access channel (MAC), where each mobile transmits independent information symbols to the AP, and the downlink channel is also known as the MIMO broadcast channel (BC), where the AP transmits independent information to each mobile. Recent information-theoretic results assert that when non-orthogonal access schemes are used in the MIMO MAC or BC channels, large capacity gains are possible beyond those achieved by orthogonal access [36, 98, 272, 274, 294, 317]. In particular, the MIMO MAC sum capacity grows *linearly* with $\min(N_r, N_u N_t)$, where N_u is the number of users [98, 256]. Therefore, for systems with a large number of users, *linear* growth of the sum capacity can be achieved by increasing N_r at the AP while keeping N_t constant per user. For the MIMO BC channel, the achievable capacity region is obtained in [36, 317] using what is known as *dirty paper coding* (DPC) under the assumption of perfect CSI at the transmitter. The duality theorems in [272,274,294] further assert that the capacity of MIMO BC channels under a power constraint P coincides with the capacity of a dual MIMO MAC channel with sum power constraint P on all users. As a result, uplink and downlink MIMO channels are equally rich as far as capacity is concerned. The practical implication is that the great benefit of MIMO is effected even when each mobile user deploys a small number of antennas (e.g., one or two) in a cellular network by increasing the number of receive-antennas at the AP.

Although promising, these information-theoretic results do not translate directly to practical transceiver designs. Furthermore, techniques such as DPC assume perfect knowledge of all users' CSI to be available at the transmitter and require long error control coded blocks, which means increased complexity and decoding delay at the receiver side [36,272,274]. Practical transceivers for MIMO MAC and BC channels with only partial CSI are therefore well motivated and constitute interesting research topics.

References

1. S. Adireddy, L. Tong, and H. Viswanathan, "Optimal placement of training for unknown channels," *IEEE Transactions on Information Theory*, vol. 48, no. 8, pp. 2338–2353, Aug. 2002.

2. D. Agrawal, T. J. Richardson, and R. Urbanke, "Multiple-antenna signal constellations for fading channels," *IEEE Transactions on Information Theory*, vol. 47, pp. 2618–2626, Sept. 2001.

3. D. Agrawal, V. Tarokh, A. Naguib, and N. Seshadri, "Space-time coded OFDM for high data-rate wireless communication over wideband channels," in *Proc. of Vehicular Technology Conference*, Ottawa, Canada, 1998, vol. 3, pp. 2232–2236.

4. E. Agrell, T. Eriksson, A. Vardy, and K. Zeger, "Closest point search in lattices," *IEEE Transactions on Information Theory*, vol. 48, no. 8, pp. 2201–2214, Aug. 2002.

5. N. Al-Dhahir, "Single-carrier frequency-domain equalization for space-time block-coded transmissions over frequency-selective fading channels," *IEEE Communications Letters*, vol. 5, no. 7, pp. 304–306, July 2001.

6. S. M. Alamouti, "A simple transmit diversity technique for wireless communications," *IEEE Journal on Selected Areas in Communications*, vol. 16, no. 8, pp. 1451–1458, Oct. 1998.

7. M. S. Alouini and A. J. Goldsmith, "Adaptive modulation over Nakagami fading channels," *Kluwer Journal on Wireless Communications*, vol. 13, no. 1–2, pp. 119–143, May 2000.

8. S. Barbarossa, *Multiantenna Wireless Communication Systems*, Artech House, Norwood, MA, 2005.

9. S. Barbarossa and F. Cerquetti, "Simple space-time coded SS-CDMA systems capable of perfect MUI/ISI elimination," *IEEE Communications Letters*, vol. 5, no. 12, pp. 471–473, Dec. 2001.

10. S. Barbarossa, G. Scutari, and A. Swami, "MUI-free CDMA systems incorporating space-time coding and channel shortening," in *Proc. of International Conference on Acoustics, Speech, and Signal Processing*, Orlando, FL, May 2002, vol. 3, pp. 2213–2216.

11. A. Barg and D. Yu. Nogin, "Bounds on packings of spheres in the Grassmann manifold," *IEEE Transactions on Information Theory*, vol. 48, no. 9, pp. 2450–2454, Sept. 2002.

12. S. Baro, G. Bauch, and A. Hansmann, "Improved codes for space-time trellis-coded modulation," *IEEE Communications Letters*, vol. 4, no. 1, pp. 20–22, Jan. 2000.

13. S. Benedetto and E. Biglieri, *Principles of Digital Transmission with Wireless Applications*, Kluwer Academic, Norwell, MA/Plenum Publishers, New York, 1999.

14. S. Benedetto, G. Montorsi, D. Divsalar, and F. Pollara, "A soft-input soft-output maximum a posteriori (MAP) module to decode parallel and serial concatenated codes," *NASA JPL TDA Progress Report*, vol. 42-127, Nov. 1996.

15. C. Berrou, A. Glavieux, and P. Thitimajshima, "Near Shannon limit error-correcting coding and decoding: Turbo codes," in *Proc. of International Conference on Communications*, Geneva, Switzerland, May 1993, pp. 1064–1070.

16. H. Bessai, *MIMO Signals and Systems*, Springer-Verlag, New York, 2005.

17. E. Biglieri, J. Proakis, and S. Shamai, "Fading channels: Information-theoretic and communications aspects," *IEEE Transactions on Information Theory*, vol. 44, no. 6, pp. 2619–2692, Oct. 1998.

18. J. A. C. Bingham, "Multicarrier modulation for data transmission: An idea whose time has come," *IEEE Communications Magazine*, pp. 5–14, May 1990.

19. R. S. Blum, "MIMO capacity with interference," *IEEE Journal on Selected Areas in Communications*, vol. 21, no. 5, pp. 793 –801, June 2003.

20. R. S. Blum, "MIMO with limited feedback of channel state information," in *Proc. of International Conference on Acoustics, Speech, and Signal Processing*, Hong Kong, Apr. 2003, pp. 89–92.

21. A. Boariu and D. M. Ionescu, "A class of nonorthogonal rate-one space-time block codes with controlled interference," *IEEE Transactions on Wireless Communications*, vol. 2, no. 2, pp. 270–276, Mar. 2003.

22. H. Boche and E. A. Jorswieck, "Optimal power allocation for MISO systems and complete characterization of the impact of correlation on the capacity," in *Proc. of International Conference on Acoustics, Speech, and Signal Processing*, Hong Kong, Apr. 2003, vol. 4, pp. 373–376.

23. H. Bölcskei and A. J. Paulraj, "Performance of space-time codes in the presence of spatial fading correlation," in *Proc. of 34th Asilomar Conference on Signals, Systems, and Computers*, Pacific Grove, CA, Oct. 29–Nov. 1, 2000, vol. 1, pp. 687–693.

24. H. Bölcskei and A. J. Paulraj, "Space-frequency coded broadband OFDM systems," in *Proc. of Wireless Communications and Networking Conference*, Chicago, IL, Sept. 2000, vol. 1, pp. 1–6.

25. H. Bölcskei, R. W. Heath, Jr., and A. J. Paulraj, "Blind channel identification and equalization in OFDM-based multiantenna systems," *IEEE Transactions on Signal Processing*, vol. 50, no. 1, pp. 96–108, Jan. 2002.

26. M. Borgmann and H. Bölcskei, "Noncoherent space-frequency coded MIMO-OFDM," *IEEE Journal on Selected Areas in Communications*, vol. 23, no. 9, pp. 1799–1810, Sept. 2005.

27. G. E. Bottomley and S. Chennakeshu, "Unification of MLSE receivers and extension to time-varying channels," *IEEE Transactions on Communications*, vol. 46, no. 4, pp. 464–472, Apr. 1998.

28. K. Boullè and J.-C. Belfiore, "Modulation scheme designed for the Rayleigh fading channel," in *Proc. of Conference on Information Sciences and Systems*, Princeton, NJ, Mar. 1992, pp. 288–293.

29. J. Boutros and E. Viterbo, "Signal space diversity: A power- and bandwidth-efficient diversity technique for the Rayleigh fading channel," *IEEE Transactions on Information Theory*, vol. 44, no. 4, pp. 1453–1467, July 1998.

30. J. Boutros, E. Viterbo, C. Rastello, and J.-C. Belfiore, "Good lattice constellations for both Rayleigh and Gaussian channel," *IEEE Transactions on Information Theory*, vol. 42, no. 2, pp. 502–518, Mar. 1996.

31. M. Brehler and M. K. Varanasi, "Training-codes for the noncoherent multi-antenna block-Rayleigh-fading channel," in *Proc. of 37th Conference on Information Sciences and Systems*, Baltimore, MD, Mar. 12–14, 2003.

32. L. Brunel and J. Boutros, "Euclidean space lattice decoding for joint detection in CDMA systems," in *Proc. of Information Theory Workshop*, Kruger National Park, South Africa, June 1999.

33. A. Burg, M. Borgmann, M. Wenk, M. Zellweger, W. Fichtner, and H. Bölcskei, "VLSI implementation of MIMO detection using the sphere decoding algorithm," *IEEE Journal on Selected Areas in Communications*, vol. 40, no. 7, pp. 1566–1577, July 2005.

34. X. Cai and G. B. Giannakis, "Differential space-time modulation with transmit-beamforming for correlated MIMO fading channels," *IEEE Transactions on Signal Processing*, vol. 54, no. 4, pp. 1279–1288, Apr. 2006.

35. X. Cai, G. B. Giannakis, and M. D. Zoltowski, "Space-time spreading and block coding for correlated fading channels in the presence of interference," *IEEE Transactions on Communications*, vol. 53, no. 3, pp. 515–525, Mar. 2005.

36. G. Caire and S. Shamai, "On the achievable throughput of a multiantenna Gaussian broadcast channel," *IEEE Transactions on Information Theory*, vol. 49, no. 7, pp. 1691–1706, July 2003.

37. A. Cano, X. Ma, and G. B. Giannakis, "Block-differential modulation for doubly-selective wireless fading channels," *IEEE Transactions on Communications*, vol. 53, no. 12, pp. 2157–2166, Dec. 2005.

38. A. Cano, X. Ma, and G. B. Giannakis, "Space-time differential modulation using linear constellation precoding," in *Proc. of International Conference on Communications*, Istanbul, Turkey, June 2006.

39. J. F. Cardoso, "Blind signal separation: Statistical principles," *Proceedings of the IEEE*, vol. 86, no. 10, pp. 2009–2025, Oct. 1998.

40. J. K. Cavers, "An analysis of pilot symbol assisted modulation for Rayleigh fading channels," *IEEE Transactions on Vehicular Technology*, vol. 40, no. 4, pp. 686–693, Nov. 1991.

41. A. Chan and I. Lee, "A new reduced-complexity sphere decoder for multiple antenna systems," in *Proc. of International Conference on Communications*, New York, Apr. 28–May 2, 2002, vol. 1, pp. 460–464.

42. K. Cho and D. Yoon, "On the general BER expression of one- and two-dimensional amplitude modulations," *IEEE Transactions on Communications*, vol. 50, no. 7, pp. 1074–1080, July 2002.

43. J. Choi and R. W. Heath, Jr., "Interpolation based unitary precoding for spatial multiplexing MIMO-OFDM with limited feedback," in *Proc. of Global Telecommunications Conference*, Dallas, TX, Nov. 29–Dec. 3, 2004.

44. J. Choi and R. W. Heath, Jr., "Interpolation based transmit beamforming for MIMO-OFDM with limited feedback," *IEEE Transactions on Signal Processing*, vol. 53, no. 11, pp. 4125–4135, Nov. 2005.

45. P. A. Chou, T. Lookabaugh, and R. M. Gray, "Entropy-constrained vector quantization," *IEEE Transactions on Acoustics, Speech, and Signal Processing*, vol. 37, no. 1, pp. 31–42, Jan. 1989.

46. J. M. Cioffi, "A multicarrier primer," http://www-isl.stanford.edu/~cioffi.

47. G. Colavolpe and R. Raheli, "Noncoherent sequence detection in frequency nonselective slowly fading channels," *IEEE Journal on Selected Areas in Communications*, vol. 18, no. 11, pp. 2302–2322, Nov. 2000.

48. J. H. Conway, R. H. Hardin, and N. J. A. Sloane, "Packing lines, planes, etc.: Packings in Grassmannian space," *Experimental Mathematics*, no. 5, pp. 139–159, 1996.

49. T. M. Cover and J. A. Thomas, *Elements of Information Theory*, Wiley, New York, 1991.

50. M. O. Damen, K. Abed-Meraim, and J.-C. Belfiore, "Diagonal algebraic space-time block codes," *IEEE Transactions on Information Theory*, vol. 48, no. 3, pp. 628–636, Mar. 2002.

51. M. O. Damen, N. C. Beaulieu, and J.-C. Belfiore, "A number theory based dual transmit antennas space-time code," in *Proc. of GLOBECOM Conference*, San Antonio, TX, Nov. 2001, pp. 485–489.

52. M. O. Damen, H. El Gamal, and N. C. Beaulieu, "Linear threaded algebraic space-time constellations," *IEEE Transactions on Information Theory*, vol. 49, pp. 2372–2388, Oct. 2003.

53. M. O. Damen, A. Tewfik, and J.-C. Belfiore, "A construction of a space-time code based on number theory," *IEEE Transactions on Information Theory*, vol. 48, no. 3, pp. 753–760, Mar. 2002.

54. V. M. DaSilva and E. S. Sousa, "Fading-resistant modulation using several transmitter antennas," *IEEE Transactions on Communications*, vol. 45, no. 10, pp. 1236–1244, Oct. 1997.

55. P. Dayal, M. Brehler, and M. K. Varanasi, "Leveraging coherent space-time codes for noncoherent communication via training," *IEEE Transactions on Information Theory*, vol. 50, no. 9, pp. 2058–2080, Sept. 2004.

56. Y. L. C. de Jong and T. J. Willink, "Iterative tree search detection for MIMO wireless systems," *IEEE Transactions on Communications*, vol. 53, no. 6, pp. 930–935, June 2005.

57. Y. Ding, T. N. Davidson, Z.-Q. Luo, and K. M. Wong, "Minimum BER block precoders for zero-forcing equalization," *IEEE Transactions on Signal Processing*, vol. 51, no. 9, pp. 2410–2423, Sept. 2003.

58. D. Divsalar and M. K. Simon, "The design of trellis coded MPSK for fading channels: Performance criteria," *IEEE Transactions on Communications*, vol. 36, no. 9, pp. 1004–1021, Sept. 1988.

59. M. Dong and L. Tong, "Optimal design and placement of pilot symbols for channel estimation," *IEEE Transactions on Signal Processing*, vol. 50, no. 12, pp. 3055–3069, Dec. 2002.

60. M. Dong, L. Tong, and B. M. Sadler, "Optimal insertion of pilot symbols for transmissions over time-varying flat fading channels," *IEEE Transactions on Signal Processing*, vol. 52, no. 5, pp. 1403–1418, May 2004.

61. D. Donoho, "On minimum entropy deconvolution," *Applied Time-Series Analysis II*, pp. 569–609, 1981.

62. C. Douillard, C. B. Jezequel, A. Picart, P. Didier, and A. Glavieux, "Iterative correction of intersymbol interference: Turbo-equalization," *European Transactions on Telecommunications*, vol. 6, pp. 507–511, Sept. 1995.

63. ETSI Normalization Committee, "Channel models for HIPERLAN/2 in different indoor scenarios," Norme ETSI, document 3ERI085B, European Telecommunications Standards Institute, Sophia-Antipolis, Valbonne, France, 1998.

64. ETSI Normalization Committee, "Broadband radio access networks (BRAN); HIgh PErformance Radio Local Area Networks (HIPERLAN) Type 2; System overview," Norme ETSI, document ETR0230002, European Telecommunications Standards Institute, Sophia-Antipolis, Valbonne, France, 1999.

65. S. A. Fechtel and H. Meyr, "Optimal parametric feedforward estimation of frequency-selective fading radio channels," *IEEE Transactions on Communications*, vol. 42, no. 2/3/4, pp. 1639–1650, Feb./Mar./Apr. 1994.

66. U. Fincke and M. Pohst, "Improved methods for calculating vectors of short length in a lattice, including a complexity analysis," *Mathematics of Computation*, vol. 44, pp. 463–471, Apr. 1985.

67. R. Fletcher, *Practical Methods of Optimization*, Vol. 2, *Constrained Optimization*, Wiley, New York, 1980.

68. G. J. Foschini and M. J. Gans, "Layered space-time architecture for wireless communication in a fading environment when using multiple antennas," *Bell System Technical Journal*, vol. 1, no. 2, pp. 41–59, Autumn 1996.

69. G. J. Foschini and M. J. Gans, "On limits of wireless communication in a fading environment when using multiple antennas," *Wireless Personal Communications*, vol. 6, no. 3, pp. 311–335, Mar. 1998.

70. G. J. Foschini, G. D. Golden, R. A. Valenzuela, and P. W. Wolniansky, "Simplified processing for high spectral efficiency wireless communication employing multi-element arrays," *IEEE Journal on Selected Areas in Communications*, vol. 17, no. 11, pp. 1841–1852, Nov. 1999.

71. C. Fragouli, N. Al-Dhahir, and W. Turin, "Finite-alphabet constant-amplitude training sequence for multiple-antenna broadband transmissions," in *Proc. of International Conference on Communications*, New York, Apr. 2002, vol. 1, pp. 6–10.

72. C. Fragouli, N. Al-Dhahir, and W. Turin, "Reduced-complexity training schemes for multiple-antenna broadband transmissions," in *Proc. of Wireless Communications and Networking Conference*, Orlando, FL, Mar. 2002, vol. 1, pp. 78–83.

73. A. Furuskar, S. Mazur, F. Muller, and H. Olofsson, "EDGE: Enhanced data rates for GSM and TDMA/136 evolution," *IEEE Personal Communications*, vol. 6, no. 3, pp. 56–66, June 1999.

74. S. Galliou and J.-C. Belfiore, "A new family of full rate, fully diverse space-time codes based on Galois theory," in *Proc. of IEEE International Symposium on Information Theory*, Lausanne, Switzerland, June 30–July 5, 2002, p. 149.

75. H. El Gamal and A. R. Hammons, Jr., "On the design of algebraic space-time codes for MIMO block-fading channels," *IEEE Transactions on Information Theory*, vol. 49, no. 1, pp. 151–163, Jan. 2003.

76. H. El Gamal, D. Aktas, and M. O. Damen, "Noncoherent space-time coding: An algebraic perspective," *IEEE Transactions on Information Theory*, vol. 51, no. 7, pp. 2380–2390, July 2005.

77. H. El Gamal and M. O. Damen, "Linear space-time coding at full rate and full diversity," in *Proc. of IEEE International Symposium on Information Theory*, Lausanne, Switzerland, June 30–July 5, 2002, p. 132.

78. H. El Gamal and M. O. Damen, "Universal space-time coding," *IEEE Transactions on Information Theory*, vol. 49, no. 5, pp. 1097–1119, May 2003.

79. H. El Gamal and A. R. Hammons, Jr., "A new approach to layered space-time coding and signal processing," *IEEE Transactions on Information Theory*, vol. 47, pp. 2321–2334, Sept. 2001.

80. G. Ganesan and P. Stoica, "Space-time diversity," in *Signal Processing Advances in Wireless and Mobile Communications*, G. B. Giannakis, Y. Hua, P. Stoica, and L. Tong (Eds.), Prentice Hall, Upper Saddle River, NJ, 2000, vol. II, chap. 2.

81. G. Ganesan and P. Stoica, "Achieving optimum coded diversity with scalar codes," *IEEE Transactions on Information Theory*, vol. 47, no. 5, pp. 2078–2080, July 2001.

82. G. Ganesan and P. Stoica, "Differential modulation using space-time block codes," *IEEE Signal Processing Letters*, vol. 9, pp. 57–60, Feb. 2002.

83. D. Garrett, L. Davis, S. ten Brink, B. Hochwald, and G. Knagge, "Silicon complexity for maximum likelihood MIMO detection using spherical decoding," *IEEE Journal of Solid-State Circuits*, vol. 39, no. 9, pp. 1544–1552, Sept. 2004.

84. A. O. Gelfond, *Transcendental and Algebraic Numbers*, Dover, New York, 1960.

85. A. Gershman and N. Sidiropoulos (Eds.), *Space-Time Processing for MIMO Communications*, Wiley, Hoboken, NJ, 2005.

86. A. Gersho and R. M. Gray, *Vector Quantization and Signal Compression*, Kluwer Academic, Norwell, MA, 1992.

87. D. Gesbert, H. Bölcskei, D. A. Gore, and A. J. Paulraj, "Outdoor MIMO wireless channels: Models and performance prediction," *IEEE Transactions on Communications*, vol. 50, no. 12, pp. 1926–1934, Dec. 2002.

88. M. Ghogho and A. Swami, "Semi-blind frequency offset synchronization for OFDM," in *Proc. of International Conference on Acoustics, Speech, and Signal Processing*, Orlando, FL, May 2002, pp. 2333–2336.

89. M. Ghogho and A. Swami, "Carrier frequency synchronization for OFDM systems," in *Signal Processing for Wireless Communication Handbook*, vol. 40, no. 6, pp. 1012–1029, I. Ibnkahla and M. Ibnkahla (Eds.), CRC Press, Boca Raton, FL, 2004.

90. G. B. Giannakis, "Filterbanks for blind channel identification and equalization," *IEEE Signal Processing Letters*, vol. 4, pp. 184–187, June 1997.

91. G. B. Giannakis, X. Ma, G. Leus, and S. Zhou, "Space-time-Doppler coding over time-selective fading channels with maximum diversity and coding gains," in *Proc. of International Conference on Acoustics, Speech, and Signal Processing*, Orlando, FL, May 2002, pp. 2217–2220.

92. G. B. Giannakis and C. Tepedelenlioglŭ, "Basis expansion models and diversity techniques for blind identification and equalization of time-varying channels," *Proceedings of the IEEE*, vol. 86, no. 10, pp. 1969–1986, Nov. 1998.

93. G. B. Giannakis, Z. Wang, A. Scaglione, and S. Barbarossa, "AMOUR: Generalized multicarrier transceivers for blind CDMA irrespective of multipath," *IEEE Transactions on Communications*, vol. 48, no. 12, pp. 2064–2076, Dec. 2000.

94. X. Giraud and J.-C. Belfiore, "Constellations matched to the Rayleigh fading channel," *IEEE Transactions on Information Theory*, vol. 42, no. 1, pp. 106–115, Jan. 1996.

95. X. Giraud, E. Boutillon, and J.-C. Belfiore, "Algebraic tools to build modulation schemes for fading channels," *IEEE Transactions on Information Theory*, vol. 43, no. 3, pp. 938–952, May 1997.

96. D. L. Goeckel and G. Ananthaswamy, "On the design of multidimensional signal sets for OFDM systems," *IEEE Transactions on Communications*, vol. 50, no. 3, pp. 442–452, Mar. 2002.

97. A. Goldsmith, *Wireless Communications*, Cambridge University Press, New York, 2005.

98. A. Goldsmith, S. Jafar, N. Jindal, and S. Vishwanath, "Capacity limits of MIMO channels," *IEEE Journal on Selected Areas in Communications*, vol. 21, no. 5, pp. 684–702, June 2003.

99. G. H. Golub and C. F. van Loan, *Matrix Computations*, 3rd ed., Johns Hopkins University Press, Baltimore, MD, 1996.

100. D. Gore, S. Sandhu, and A. Paulraj, "Delay diversity code for frequency selective channels," *Electronics Letters*, vol. 37, no. 20, pp. 1230–1231, Sept. 2001.

101. J. Grimm, M. P. Fitz, and J. V. Krogmeier, "Further results on space-time coding for Rayleigh fading," in *Proc. of Allerton Conference*, Monticello, IL, Sept. 24, 1998, pp. 391–400.

102. M. Haardt, A. Klein, R. Koehn, S. Oestreich, M. Purat, V. Sommer, and T. Ulrich, "The TD-CDMA based UTRA TDD mode," *IEEE Journal on Selected Areas in Communications*, vol. 18, no. 8, pp. 1375–1385, Aug. 2000.

103. A. Hajimiri and T. H. Lee, *The Design of Low Noise Oscillators*, Kluwer Academic, Norwell, MA, 1996.

104. L. Hanzo, M. Münster, B. J. Choi, and T. Keller, *OFDM and MC-CDMA for Broadcasting Multi-user Communications, WLANs and Broadcasting*, IEEE Press, Piscataway, NJ/Wiley, Hoboken, NJ, 2003.

105. F. Hasegawa, J. Luo, K. R. Pattipati, P. Willett, and D. Pham, "Speed and accuracy comparison of techniques for multiuser detection in synchronous CDMA," *IEEE Transactions on Communications*, vol. 52, no. 4, pp. 540–545, Apr. 2004.

106. B. Hassibi and B. M. Hochwald, "Caley differential unitary space-time codes," *IEEE Transactions on Information Theory*, vol. 48, no. 6, pp. 1485 –1503, June 2002.

107. B. Hassibi and B. M. Hochwald, "High-rate codes that are linear in space and time," *IEEE Transactions on Information Theory*, vol. 48, pp. 1804–1824, July 2002.

108. B. Hassibi and B. M. Hochwald, "How much training is needed in multiple-antenna wireless links?" *IEEE Transactions on Information Theory*, vol. 49, no. 4, pp. 951–963, Apr. 2003.

109. B. Hassibi and H. Vikalo, "On the expected complexity of sphere decoding," in *Proc. of 35th Asilomar Conference on Signals, Systems, and Computers*, Pacific Grove, CA, Nov. 4–7, 2001, vol. 2, pp. 1051–1055.

110. B. Hassibi and H. Vikalo, "On the sphere-decoding algorithm. I. Expected complexity," *IEEE Transactions on Signal Processing*, vol. 53, no. 8, pp. 2806–2818, Aug. 2005.

111. A. Hiroike, F. Adachi, and N. Nakajima, "Combined effects of phase sweeping transmitter diversity and channel coding," *IEEE Transactions on Vehicular Technology*, vol. 41, no. 2, pp. 170 –176, May 1992.

112. P. Ho and D. Fung, "Error performance of multiple-symbol differential detection of PSK signals transmitted over correlated Rayleigh fading channels," *IEEE Transactions on Communications*, vol. 40, no. 10, pp. 1566–1569, Oct. 1992.

113. B. Hochwald, T. Marzetta, T. Richardson, W. Sweldens, and R. Urbanke, "Systematic design of unitary space-time constellations," *IEEE Transactions on Information Theory*, vol. 46, no. 6, pp. 1962–1973, Sept. 2000.

114. B. Hochwald and T. L. Marzetta, "Unitary space-time modulation for multiple-antenna communications in Rayleigh flat fading," *IEEE Transactions on Information Theory*, vol. 46, no. 2, pp. 543–564, Mar. 2000.

115. B. Hochwald, T. L. Marzetta, and C. B. Papadias, "A transmitter diversity scheme for wideband CDMA systems based on space-time spreading," *IEEE Journal on Selected Areas in Communications*, vol. 19, no. 1, pp. 48–60, Jan. 2001.

116. B. Hochwald and W. Sweldens, "Differential unitary space-time modulation," *IEEE Transactions on Communications*, vol. 48, no. 12, pp. 2041–2052, Dec. 2000.

117. B. M. Hochwald, T. L. Marzetta, and V. Tarokh, "Multi-antenna channel hardening and its implications for rate feedback and scheduling," *IEEE Transactions on Information Theory*, vol. 50, no. 9, pp. 1893–1909, Sept. 2004.

118. B. M. Hochwald and S. ten Brink, "Achieving near-capacity on a multiple-antenna channel," *IEEE Transactions on Communications*, vol. 51, no. 3, pp. 389–399, Mar. 2003.

119. A. Hottinen, O. Tirkkonen, and R. Wichman, *Multi-antenna Transceiver Techniques for 3G and Beyond*, Wiley, Hoboken, NJ, 2003.

120. B. L. Hughes, "Differential space-time modulation," *IEEE Transactions on Information Theory*, vol. 46, no. 7, pp. 2567–2578, Nov. 2000.

121. S. A. Jafar and A. Goldsmith, "On optimality of beamforming for multiple antenna systems with imperfect feedback," in *Proc. of IEEE International Symposium on Information Theory*, Washington, DC, June 2001, p. 321.

122. S. A. Jafar and A. J. Goldsmith, "Transmitter optimization and optimality of beamforming for multiple antenna systems," *IEEE Transactions on Wireless Communications*, vol. 3, no. 4, pp. 1165–1175, July 2004.

123. S. A. Jafar and S. Srinivasa, "Capacity of the isotropic fading vector channel with quantized channel direction feedback," in *Proc. of Asilomar Conference on Signals, Systems, and Computers*, Pacific Grove, CA, Nov. 2004, pp. 1178–1182.

124. S. A. Jafar, S. Vishwanath, and A. Goldsmith, "Channel capacity and beamforming for multiple transmit and receive antennas with covariance feedback," in *Proc. of International Conference on Communications*, Helsinki, Finland, June 2001, vol. 7, pp. 2266–2270.

125. H. Jafarkhani, "A quasi-orthogonal space-time block code," *IEEE Communications Letters*, vol. 49, no. 1, pp. 1–4, Jan. 2001.

126. H. Jafarkhani, *Space-Time Coding: Theory and Practice*, Cambridge University Press, New York, 2005.

127. W. C. Jakes, *Microwave Mobile Communication*, Wiley, New York, 1974.

128. J. Jalden and B. Ottersten, "On the complexity of sphere decoding in digital communications," *IEEE Transactions on Signal Processing*, vol. 53, no. 4, pp. 1474–1484, Apr. 2005.

129. J. Jalden and B. Ottersten, "An exponential lower bound on the expected complexity of sphere decoding," in *Proc. of International Conference on Acoustics, Speech, and Signal Processing*, Montreal, Quebec, Canada, May 17–21, 2004, pp. 393–396.

130. M. Jankiraman, *Space-Time Codes and MIMO Systems*, Artech House, Norwood, MA, 2004.

131. Y. Jing and B. Hassibi, "Unitary space-time modulation via Cayley transform," *IEEE Transactions on Signal Processing*, vol. 51, no. 11, pp. 2891–2904, Nov. 2003.

132. G. Jöngren, M. Skoglund, and B. Ottersten, "Combining beamforming and orthogonal space-time block coding," *IEEE Transactions on Information Theory*, vol. 48, no. 3, pp. 611–627, Mar. 2002.

133. M. Kang and M. S. Alouini, "Capacity of correlated MIMO Rician channels," *IEEE Transactions on Wireless Communications*, vol. 1, no. 5, pp. 143–155, Jan. 2006.

134. M. Kang and M. S. Alouini, "Capacity of MIMO Rician channels," *IEEE Transactions on Wireless Communications*, vol. 1, no. 5, pp. 112–122, Jan. 2006.

135. S. M. Kay, *Fundamentals of Statistical Signal Processing: Estimation Theory*, Prentice Hall, Upper Saddle River, NJ, 1993.

136. J. P. Kermoal, L. Schumacher, K. I. Pedersen, P. E. Mogensen, and F. Frederiksen, "A stochastic MIMO radio channel model with experimental validation," *IEEE Journal on Selected Areas in Communications*, vol. 20, no. 6, pp. 1211–1226, Aug. 2002.

137. M. Kisialiou and Z.-Q. Luo, "Performance analysis of quasi-maximum-likelihood detector based on semi-definite programming," in *Proc. of International Conference on Acoustics, Speech, and Signal Processing*, Philadelphia, PA, Mar. 2005, vol. 3, pp. 433–436.

138. Y. Ko and C. Tepedelenlioglu, "Orthogonal space-time block coded rate-adaptive modulation with outdated feedback," *IEEE Transactions on Wireless Communications*, vol. 5, no. 2, pp. 290–295, Feb. 2006.

139. W.-Y. Kuo and M. P. Fitz, "Design and analysis of transmitter diversity using intentional frequency offset for wireless communications," *IEEE Transactions on Vehicular Technology*, vol. 46, no. 4, pp. 871–881, Nov. 1997.

140. C. Lamy and J. Boutros, "On random rotations diversity and minimum MSE decoding of lattices," *IEEE Transactions on Information Theory*, vol. 46, no. 4, pp. 1584–1589, July 2000.

141. E. G. Larsson, G. Liu, J. Li, and G. B. Giannakis, "Joint symbol timing and channel estimation for OFDM based WLANs," *IEEE Communications Letters*, vol. 5, no. 8, pp. 325–327, Aug. 2001.

142. E. G. Larsson and P. Stoica, *Space-Time Block Coding for Wireless Communications*, Cambridge University Press, New York, 2003.

143. E. G. Larsson, P. Stoica, and J. Li, "Orthogonal space-time block codes: maximum likelihood detection for unknown channels and unstructured interferences," *IEEE Transactions on Signal Processing*, vol. 51, no. 2, pp. 362–372, Feb. 2003.

144. G. Latsoudas and N. D. Sidiropoulos, "A hybrid probabilistic data association-sphere decoding detector for multiple-input multiple-output systems," *IEEE Signal Processing Letters*, vol. 12, no. 4, pp. 309–312, Apr. 2005.

145. G. Latsoudas and N. D. Sidiropoulos, "On the performance of certain fixed-complexity multiuser detectors in FEXT-limited vector DSL systems," in *Proc. of International Conference on Acoustics, Speech, and Signal Processing*, Philadelphia, PA, Mar. 19–23, 2005.

146. V. Lau, Y. Liu, and T.-A. Chen, "On the design of MIMO block-fading channels with feedback-link capacity constraint," *IEEE Transactions on Communications*, vol. 52, no. 1, pp. 62–70, Jan. 2004.

147. A. K. Lenstra, H. W. Lenstra, and L. Lovász, "Factoring polynomials with rational coefficients," *Mathematische Annalen*, vol. 261, pp. 515–534, 1982.

148. G. Leus, S. Zhou, and G. B. Giannakis, "Orthogonal multiple access over time- and frequency-selective fading," *IEEE Transactions on Information Theory*, vol. 49, no. 8, pp. 1442–1450, Aug. 2003.

149. H. Li and J. Li, "Differential and coherent decorrelating multiuser receivers for space-time-coded CDMA systems," *IEEE Transactions on Signal Processing*, vol. 50, no. 10, pp. 2529–2537, Oct. 2002.

150. Y. Li, "Simplified channel estimation for OFDM systems with multiple transmit antennas," *IEEE Transactions on Wireless Communications*, vol. 1, no. 1, pp. 67–75, Jan. 2002.

151. Y. Li, J. C. Chung, and N. R. Sollenberger, "Transmitter diversity for OFDM systems and its impact on high-rate data wireless networks," *IEEE Journal on Selected Areas in Communications*, vol. 17, no. 7, pp. 1233–1243, July 1999.

152. X.-B. Liang, "Orthogonal designs with maximal rates," *IEEE Transactions on Information Theory*, vol. 49, no. 10, pp. 2468–2503, Oct. 2003.

153. Y.-C. Liang and F. P. S. Chin, "Downlink channel covariance matrix (DCCM) estimation and its applications in wireless DS-CDMA systems," *IEEE Journal on Selected Areas in Communications*, vol. 19, no. 2, pp. 222–232, Feb. 2001.

154. E. Lindskög and A. Paulraj, "A transmit diversity scheme for channels with intersymbol interference," in *Proc. of International Conference on Communications*, New Orleans, LA, June 2000, vol. 1, pp. 307–311.

155. H. Liu and U. Tureli, "A high-efficiency carrier estimator for OFDM communications," *IEEE Communications Letters*, vol. 2, pp. 104–106, Apr. 1998.

156. Y. Liu, M. P. Fitz, and O. Y. Takeshita, "Space-time codes for frequency selective channel: Outage probability, performance criteria, and code design," in *Proc. of 38th Annual Allerton Conference on Communication, Control, and Computing*, Monticello, IL, Oct. 2000.

157. Y. Liu, M. P. Fitz, and O. Y. Takeshita, "Space-time codes performance criteria and design for frequency selective fading channels," in *Proc. of International Conference on Communications*, June 2001, vol. 9, pp. 2800–2804.

158. Z. Liu and G. B. Giannakis, "Space-time block coded multiple access through frequency-selective fading channels," *IEEE Transactions on Communications*, vol. 49, no. 6, pp. 1033–1044, June 2001.

159. Z. Liu and G. B. Giannakis, "Block differentially encoded OFDM with maximum multipath diversity," *IEEE Transactions on Wireless Communications*, vol. 2, no. 3, pp. 420–423, May 2003.

160. Z. Liu, G. B. Giannakis, S. Barbarossa, and A. Scaglione, "Transmit-antennae space-time block coding for generalized OFDM in the presence of unknown multipath," *IEEE Journal on Selected Areas in Communications*, vol. 19, no. 7, pp. 1352–1364, July 2001.

161. Z. Liu, G. B. Giannakis, B. Muquet, and S. Zhou, "Space-time coding for broadband wireless communications," *Wireless Communications and Mobile Computing*, vol. 1, no. 1, pp. 33–53, Jan.–Mar. 2001.

162. Z. Liu, Y. Xin, and G. B. Giannakis, "Space-time-frequency coded OFDM over frequency-selective fading channels," *IEEE Transactions on Signal Processing*, vol. 50, no. 10, pp. 2465–2476, Oct. 2002.

163. Z. Liu, Y. Xin, and G. B. Giannakis, "Linear constellation precoding for OFDM with maximum multipath diversity and coding gains," *IEEE Transactions on Communications*, vol. 51, no. 3, pp. 416–427, Mar. 2003.

164. D. J. Love, Personal Webpage on Grassmannian Subspace Packing, http://dynamo.ecn.purdue.edu/~djlove/grass.html.

165. D. J. Love and R. W. Heath, Jr., "Grassmannian precoding for spatial multiplexing systems," in *Proc. of Allerton Conference on Communications, Control, and Computing*, Monticello, IL, Oct. 1–3, 2003.

166. D. J. Love and R. W. Heath, Jr., "Limited feedback precoding for spatial multiplexing systems," in *Proc. of Global Telecommunications Conference*, San Francisco, CA, Dec. 2003, vol. 4, pp. 1857–1861.

167. D. J. Love and R. W. Heath, Jr., "Limited feedback unitary precoding for orthogonal space-time block codes," *IEEE Transactions on Signal Processing*, vol. 53, no. 1, pp. 64–73, Jan. 2005.

168. D. J. Love, R. W. Heath, Jr., W. Santipach, and M. L. Honig, "What is the value of limited feedback for MIMO channels?" *IEEE Communications Magazine*, vol. 42, no. 10, pp. 54–59, Oct. 2004.

169. D. J. Love, R. W. Heath, Jr., and T. Strohmer, "Grassmannian beamforming for multiple-input multiple-output wireless systems," *IEEE Transactions on Information Theory*, vol. 49, no. 10, pp. 2735–2747, Oct. 2003.

170. B. Lu and X. Wang, "Space-time code design in OFDM systems," in *Proc. of Global Telecommunications Conference*, San Francisco, CA, 2000, vol. 2, pp. 1000–1004.

171. B. Lu, X. Wang, and Y. Li, "Iterative receivers for space-time block-coded OFDM systems in dispersive fading channels," *IEEE Transactions on Wireless Communications*, vol. 1, no. 2, pp. 213–225, Apr. 2002.

172. B. Lu, X. Wang, and K. R. Narayanan, "LDPC-based space-time coded OFDM systems over correlated fading channels: Performance analysis and receiver design," *IEEE Transactions on Communications*, vol. 50, no. 1, pp. 74–88, Jan. 2002.

173. B. Lu, G. Yue, and X. Wang, "Performance analysis and design optimization of LDPC-coded MIMO OFDM systems," *IEEE Transactions on Signal Processing*, vol. 52, no. 2, pp. 348–361, Feb. 2004.

174. J. Luo, K. R. Pattipati, P. Willett, and G. Levchuk, "Fast optimal and sub-optimal any-time algorithms for CDMA multiuser detection based on branch and bound," *IEEE Transactions on Communications*, vol. 52, no. 4, pp. 632–642, Apr. 2004.

175. J. Luo, K. R. Pattipati, P. K. Willett, and F. Hasegawa, "Near-optimal multiuser detection in synchronous CDMA using probabilistic data association," *IEEE Communications Letters*, vol. 5, no. 9, pp. 361–363, Sept. 2001.

176. Q. Ma, C. Tepedelenlioglu, and Z. Liu, "Differential space-time-frequency coded OFDM with maximum multipath diversity," *IEEE Transactions on Wireless Communications*, vol. 4, no. 5, pp. 2232–2243, Sept. 2005.

177. W.-K. Ma, T. N. Davidson, K. M. Wong, Z.-Q. Luo, and P.-C. Ching, "Quasi-maximum-likelihood multiuser detection using semi-definite relaxation with application to synchronous CDMA," *IEEE Transactions on Signal Processing*, vol. 50, no. 4, pp. 912–922, Apr. 2002.

178. X. Ma and G. B. Giannakis, "Complex field coded MIMO systems: Performance, rate, and tradeoffs," *Wireless Commmunication and Mobile Computing*, vol. 2, no. 7, pp. 693–717, Nov. 2002.

179. X. Ma and G. B. Giannakis, "Layered space-time complex field coding: Full-diversity with full-rate, and tradeoffs," in *Proc. of 2nd Sensor Array and Multichannel SP Workshop*, Rosslyn, VA, Aug. 2002, pp. 442–446.

180. X. Ma and G. B. Giannakis, "Maximum-diversity transmissions over time-selective wireless channels," in *Proc. of Wireless Communications and Networking Conference*, Orlando, FL, Mar. 2002, vol. 1, pp. 497–501.

181. X. Ma and G. B. Giannakis, "Space-time coding for doubly-selective channels," in *International Conference on Circuits and Systems*, Scottsdale, AZ, Mar. 25–29, 2002, pp. 647–650.

182. X. Ma and G. B. Giannakis, "Full-diversity full-rate complex-field space-time coding," *IEEE Transactions on Signal Processing*, vol. 51, no. 11, pp. 2917–2930, Nov. 2003.

183. X. Ma and G. B. Giannakis, "Maximum-diversity transmissions over doubly-selective wireless channels," *IEEE Transactions on Information Theory*, vol. 49, no. 7, pp. 1832–1840, July 2003.

184. X. Ma and G. B. Giannakis, "Space-time-multipath coding using digital phase sweeping or block circular delay diversity," *IEEE Transactions on Signal Processing*, vol. 53, no. 3, pp. 1121–1131, Mar. 2005.

185. X. Ma, G. B. Giannakis, and S. Barbarossa, "Non-data aided frequency-offset and channel estimation in OFDM and related block transmissions," in *Proc. of International Conference on Communications*, Helsinki, Finland, June 2001, pp. 1866–1870.

186. X. Ma, G. B. Giannakis, and B. Lu, "Block differential encoding for rapidly fading channels," *IEEE Transactions on Communications*, vol. 52, no. 3, pp. 996–1011, Mar. 2004.

187. X. Ma, G. B. Giannakis, and S. Ohno, "Optimal training for block transmissions over doubly-selective wireless fading channels," *IEEE Transactions on Signal Processing*, vol. 51, no. 5, pp. 1351–1366, May 2003.

188. X. Ma, G. Leus, and G. B. Giannakis, "Space-time-Doppler coding for correlated time-selective fading channels," *IEEE Transactions on Signal Processing*, vol. 53, no. 6, pp. 2167–2181, June 2005.

189. X. Ma, M.-K. Oh, G. B. Giannakis, and D.-J. Park, "Hopping pilots for estimation of frequency-offsets and multi-antenna channels in MIMO-OFDM," *IEEE Transactions on Communications*, vol. 53, no. 1, pp. 162–172, Jan. 2005.

190. X. Ma, C. Tepedelenlioglu, G. B. Giannakis, and S. Barbarossa, "Non-data-aided carrier offset estimations for OFDM with null subcarriers: Identifiability, algorithms, and performance," *IEEE Journal on Selected Areas in Communications*, vol. 19, no. 12, pp. 2504–2515, Dec. 2001.

191. X. Ma, L. Yang, and G. B. Giannakis, "Optimal training for MIMO frequency-selective fading channels," *IEEE Transactions on Wireless Communications*, vol. 4, no. 2, pp. 453–466, Mar. 2005.

192. J. W. Mark and W. Zhuang, *Wireless Communications and Networking*, Prentice Hall, Upper Saddle River, NJ, 2003.

193. T. L. Marzetta and B. M. Hochwald, "Capacity of a mobile multiple-antenna communication link in Rayleigh flat fading," *IEEE Transactions on Information Theory*, vol. 45, no. 1, pp. 139–157, Jan. 1999.

194. M. L. McCloud, M. Brehler, and M. K. Varanasi, "Signal design and convolutional coding for noncoherent space-time communication on the block-Rayleigh-fading channel," *IEEE Transactions on Information Theory*, vol. 48, pp. 1186–1194, May 2002.

195. M. Medard, "The effect upon channel capacity in wireless communications of perfect and imperfect knowledge of the channel," *IEEE Transactions on Information Theory*, vol. 46, no. 3, pp. 933–946, May 2000.

196. A. Medles and D. Slock, "Linear space-time coding at full rate and full diversity," in *Proc. of IEEE International Symposium on Information Theory*, Lausanne, Switzerland, June 30–July 5, 2002, p. 221.

197. A. N. Mody and G. L. Stuber, "Synchronization for MIMO OFDM systems," in *Proc. of Global Telecommunications Conference*, San Antonio, TX, Nov. 2001, pp. 509–513.

198. P. H. Moose, "A technique for orthogonal frequency division multiplexing frequency offset correction," *IEEE Transactions on Communications*, vol. 42, no. 10, pp. 2908–2914, Oct. 1994.

199. M. Morelli and U. Mengali, "Carrier-frequency estimation for transmissions over selective channels," *IEEE Transactions on Communications*, vol. 48, no. 9, pp. 1580–1589, Sept. 2000.

200. A. L. Moustakas and S. H. Simon, "Optimizing multiple-input single-output (MISO) communication systems with general Gaussian channels: Nontrivial covariance and nonzero mean," *IEEE Transactions on Information Theory*, vol. 49, no. 10, pp. 2770–2780, Oct. 2003.

201. A. L. Moustakas, S. H. Simon, and A. M. Sengupta, "MIMO capacity through correlated channels in the presence of correlated interferers and noise: A (not so) large n analysis," *IEEE Transactions on Information Theory*, vol. 49, no. 10, pp. 2545–2561, Oct. 2003.

202. K. K. Mukkavilli, A. Sabharwal, E. Erkip, and B. Aazhang, "On beamforming with finite rate feedback in multiple antenna systems," *IEEE Transactions on Information Theory*, vol. 49, no. 10, pp. 2562–2579, Oct. 2003.

203. A. Narula, M. J. Lopez, M. D. Trott, and G. W. Wornell, "Efficient use of side information in multiple-antenna data transmission over fading channels," *IEEE Journal on Selected Areas in Communications*, vol. 16, no. 8, pp. 1423–1436, Oct. 1998.

204. S. Ohno and G. B. Giannakis, "Optimal training and redundant precoding for block transmissions with application to wireless OFDM," *IEEE Transactions on Communications*, vol. 50, no. 12, pp. 2113–2123, Dec. 2002.

205. S. Ohno and G. B. Giannakis, "Capacity maximizing MMSE-optimal pilots for wireless OFDM over frequency-selective block Rayleigh-fading channels," *IEEE Transactions on Information Theory*, vol. 50, pp. 2138–2145, Sept. 2004.

206. H. Olofsson, M. Almgren, and M. Hook, "Transmitter diversity with antenna hopping for wireless communication systems," in *Proc. of Vehicular Technology Conference*, Phoenix, AZ, May 4–7, 1997, vol. 3, pp. 1743–1747.

207. H. Ozcelik, M. Herdin, W. Weichselberger, J. Wallace, and E. Bonek, "Deficiencies of 'Kronecker' MIMO radio channel model," *Electronics Letters*, vol. 39, no. 16, pp. 1209–1210, Aug. 2003.

208. A. Paulraj, R. Nabar, and D. Gore, *Introduction to Space-Time Wireless Communications*, Cambridge University Press, New York, 2003.

209. F. Petre, G. Leus, L. Deneire, M. Engels, M. Moonen, and H. De Man, "Space-time block coding for single-carrier block transmission DS-CDMA downlink," *IEEE Journal on Selected Areas in Communications*, vol. 21, no. 3, pp. 350–361, Apr. 2003.

210. D. Pham, K. R. Pattipati, P. K. Willett, and J. Luo, "An improved complex sphere decoder for V-BLAST systems," *IEEE Signal Processing Letters*, vol. 11, no. 9, pp. 748–751, Sept. 2004.

211. R. J. Piechocki, P. N. Fletcher, A. R. Nix, C. N. Canagarajah, and J. P. McGeehan, "Performance evaluation of BLAST-OFDM enhanced Hiperlan/2 using simulated and measured channel data," *Electronics Letters*, vol. 37, no. 18, pp. 1137–1139, Aug. 2001.

212. H. V. Poor, *An Introduction to Signal Detection and Estimation*, Springer-Verlag, New York, 1988.

213. H. V. Poor and S. Verdú, "Probability of error in MMSE multiuser detection," *IEEE Transactions on Information Theory*, vol. 43, no. 3, pp. 858–871, May 1997.

214. J. G. Proakis, *Digital Communications*, 4th ed., McGraw-Hill, New York, 2001.

215. M. Qin and R. S. Blum, "Properties on space-time codes for frequency selective channels," *IEEE Transactions on Signal Processing*, vol. 52, no. 3, pp. 694–702, Mar. 2004.

216. D. Rainish, "Diversity transform for fading channels," *IEEE Transactions on Communications*, vol. 44, no. 12, pp. 1653–1661, Dec. 1996.

217. G. G. Raleigh and J. M. Cioffi, "Spatio-temporal coding for wireless communication," *IEEE Transactions on Communications*, vol. 46, no. 3, pp. 357–366, Mar. 1998.

218. T. S. Rappaport, *Wireless Communications: Principles and Practice*, Prentice Hall, Upper Saddle River, NJ, 1996.

219. S. P. Reichhart, B. Youmans, and R. Dygert, "The software radio development system," *IEEE Personal Communications*, vol. 6, no. 4, pp. 20–24, Aug. 1999.

220. P. Robertson, E. Villebrun, and P. Hoeher, "A comparison of optimal and suboptimal MAP decoding algorithms operating in the log domain," in *Proc.*

of International Conference on Communications, Seattle, WA, June 1995, pp. 1009–1013.

221. S. Roman, *The Laguerre Polynomials*, Academic Press, New York, 1984.

222. S. Sandhu and A. Paulraj, "Space-time block codes: A capacity perspective," *IEEE Communications Letters*, vol. 4, no. 12, pp. 384–386, Dec. 2000.

223. H. Sari and G. Karam, "Orthogonal frequency-division multiple access and its application to CATV network," *European Transactions on Telecommunications*, vol. 9, pp. 507–516, Nov./Dec. 1998.

224. H. Sari, G. Karam, and I. Jeanclaude, "Transmission techniques for digital terrestrial TV broadcasting," *IEEE Communications Magazine*, vol. 33, no. 2, pp. 100–109, Feb. 1995.

225. A. Scaglione, G. B. Giannakis, and S. Barbarossa, "Redundant filterbank precoders and equalizers. Part I: Unification and optimal designs," *IEEE Transactions on Signal Processing*, vol. 47, pp. 1988–2006, July 1999.

226. T. M. Schmidl and D. C. Cox, "Robust frequency and timing synchronization for OFDM," *IEEE Transactions on Communications*, vol. 45, pp. 1613–1621, Dec. 1997.

227. R. Schober, W. H. Gerstacker, and J. B. Huber, "Decision-feedback differential detection of MDPSK for flat Rayleigh fading channels," *IEEE Transactions on Communications*, vol. 47, no. 7, pp. 1025–1035, July 1999.

228. R. Schober and L. H. J. Lampe, "Differential modulation diversity," *IEEE Transactions on Vehicular Technology*, vol. 51, no. 6, pp. 1431–1444, Nov. 2002.

229. R. Schober and L. H. J. Lampe, "Noncoherent receivers for differential space-time modulation," *IEEE Transactions on Communications*, vol. 50, no. 5, pp. 768–777, May 2002.

230. M. Schwartz, W. R. Benett, and S. Stein, *Communication Systems and Techniques*, IEEE Press, Piscataway, NJ, 1996.

231. N. Seshadri and J. H. Winters, "Two signaling schemes for improving the error performance of frequency-division-duplex (FDD) transmission systems using transmitter antenna diversity," in *Proc. of Vehicular Technology Conference*, Secaucus, NJ, May 18–20, 1993, pp. 508 –511.

232. C. E. Shannon, "A mathematical theory of communication," *Bell System Technical Journal*, vol. 27, pp. 623–656, Oct. 1948.

233. N. Sharma and C. B. Papadias, "Improved quasi-orthogonal codes through constellation rotation," *IEEE Transactions on Communications*, vol. 51, no. 3, pp. 332–335, Mar. 2003.

234. N. Sharma and C. B. Papadias, "Full rate full diversity linear quasi-orthogonal space-time codes for any transmit antennas," *EURASIP Journal on Applied Signal Processing*, no. 9, pp. 1246–1256, Aug. 2004.

235. D. Shiu, G. J. Foschini, M. J. Gans, and J. M. Kahn, "Fading correlation and its effect on the capacity of multi-element antenna systems," *IEEE Transactions on Communications*, vol. 48, no. 3, pp. 502–513, Mar. 2000.

236. M. K. Simon and M.-S. Alouini, *Digital Communication over Generalized Fading Channels: A Unified Approach to the Performance Analysis*, Wiley, New York, 2000.

237. S. H. Simon and A. L. Moustakas, "Optimizing MIMO antenna systems with channel covariance feedback," *IEEE Journal on Selected Areas in Communications*, vol. 21, no. 3, pp. 406–417, Apr. 2003.

238. H. Steendam and M. Moeneclaey, "Analysis and optimization of the performance of OFDM on frequency-selective time-selective fading channels," *IEEE Transactions on Communications*, vol. 47, no. 12, pp. 1811–1819, Dec. 1999.

239. B. Steingrimsson, Z.-Q. Luo, and K.-M. Wong, "Quasi-maximum-likelihood detection for multiple-antenna channels," *IEEE Transactions on Signal Processing*, vol. 51, no. 11, pp. 2710–2719, Nov. 2003.

240. D. R. Stinson, *Combinatorial Designs: Constructions and Analysis*, Springer-Verlag, New York, 2004.

241. T. Strohmer and R. W. Heath, Jr., "Grassmannian frames with applications to coding and communications," *Applied and Computational Harmonic Analysis*, vol. 14, no. 3, pp. 257–275, May 2003.

242. G. L. Stüber, *Principles of Mobile Communication*, 2nd ed., Kluwer Academic, Norwell, MA, 2001.

243. W. Su and X. G. Xia, "Signal constellations for quasi-orthogonal space time block codes with full diversity," *IEEE Transactions on Information Theory*, vol. 50, no. 10, pp. 2331–2347, Oct. 2004.

244. M. Taherzadeh, A. Mobasher, and A. K. Khandani, "Lattice-basis reduction achieves the precoding diversity in MIMO broadcast systems," in *Proc. of 39th Conference on Information Sciences and Systems*, Johns Hopkins University, Baltimore, MD, Mar. 15–18, 2005.

245. P. H. Tan and L. K. Rasmussen, "The application of semidefinite programming for detection in CDMA," *IEEE Journal on Selected Areas in Communications*, vol. 19, no. 8, pp. 1442–1449, Aug. 2001.

246. P. H. Tan and L. K. Rasmussen, "Multiuser detection in CDMA: A comparison of relaxations, exact, and heuristic search methods," *IEEE Transactions on Wireless Communications*, vol. 3, no. 5, pp. 1802–1809, Sept. 2004.

247. G. Taricco and E. Biglieri, "Exact pairwise error probability of space-time codes," *IEEE Transactions on Information Theory*, vol. 48, no. 2, pp. 510–513, Feb. 2002.

248. V. Tarokh and H. Jafarkhani, "A differential detection scheme for transmit diversity," *IEEE Journal on Selected Areas in Communications*, vol. 18, no. 7, pp. 1169–1174, July 2000.

249. V. Tarokh, H. Jafarkhani, and A. R. Calderbank, "Space-time block codes from orthogonal designs," *IEEE Transactions on Information Theory*, vol. 45, no. 5, pp. 1456–1467, July 1999.

250. V. Tarokh and M. Kim, "Existence and construction of noncoherent unitary space-time codes," *IEEE Transactions on Information Theory*, vol. 48, no. 12, pp. 3112–3117, Dec. 2002.

251. V. Tarokh, A. Naguib, N. Seshadri, and A. R. Calderbank, "Combined array processing and space-time coding," *IEEE Transactions on Information Theory*, vol. 45, no. 4, pp. 1121–1128, May 1999.

252. V. Tarokh, A. Naguib, N. Seshadri, and A. R. Calderbank, "Space-time codes for high data rate wireless communication: Performance criteria in the presence of channel estimation errors, mobility, and multiple paths," *IEEE Transactions on Communications*, vol. 47, no. 2, pp. 199–207, Feb. 1999.

253. V. Tarokh, N. Seshadri, and A. R. Calderbank, "Space-time codes for high data rate wireless communication: Performance criterion and code construction," *IEEE Transactions on Information Theory*, vol. 44, no. 2, pp. 744–765, Mar. 1998.

254. A. Tarr and M. D. Zoltowski, "Design of codes for multicode CDMA based on correlation feedback from receiver," in *Proc. of SPIE*, Orlando, FL, Apr. 2003, pp. 261–266.

255. A. Tarr and M. D. Zoltowski, "MMSE receiver with reduced cross-talk in CDMA via eigenvector codes," in *Proc. of 41st Allerton Conference*, Monticello, IL, Oct. 2003, pp. 1807–1808.

256. I. E. Telatar, "Capacity of multi-antenna Gaussian channels," Technical Memorandum, AT&T Bell Laboratories, Oct. 1995.

257. I. E. Telatar and D. N. C. Tse, "Capacity and mutual information of wideband multipath fading channels," *IEEE Transactions on Information Theory*, vol. 46, no. 4, pp. 1384–1400, July 2000.

258. S. ten Brink, J. Speidel, and R.-H. Yan, "Iterative demapping and decoding for multilevel modulation," in *Proc. of Global Telecommunications Conference*, Sydney, Australia, 1998, vol. 1, pp. 579–584.

259. O. Tirkkonen, A. Boariu, and A. Hottinen, "Minimal nonorthogonality rate 1 space-time block code for 3+ tx antennas," in *Proc. of 6th International Symposium on Spread-Spectrum Techniques and Applications*, Parsippany, NJ, Sept. 2000, pp. 429–432.

260. O. Tirkkonen and A. Hottinen, "Improved MIMO performance with non-orthogonal space-time block codes," in *Proc. of GLOBECOM Conference*, San Antonio, TX, Nov. 2001, pp. 1122–1126.

261. L. Tong, B. M. Sadler, and M. Dong, "Pilot-assisted wireless transmissions: General model, design criteria, and signal processing," *IEEE Signal Processing Magazine*, vol. 21, no. 6, pp. 12–25, Nov. 2004.

262. M. K. Tsatsanis, G. B. Giannakis, and G. Zhou, "Estimation and equalization of fading channels with random coefficients," *Signal Processing*, vol. 53, no. 2–3, pp. 211–228, 1996.

263. D. Tse and P. Viswanath, *Fundamentals of Wireless Communication*, Cambridge University Press, New York, 2005.

264. J. K. Tugnait, "Identification and deconvolution of multichannel linear non-Gaussian processes using higher-order statistics and inverse filter criteria," *IEEE Transactions on Signal Processing*, vol. 45, no. 3, pp. 658–672, Mar. 1997.

265. U. Tureli and P. J. Honan, "Modified high-efficiency carrier estimator for OFDM communications with antenna diversity," in *Proc. of Asilomar Conference on Signals, Systems, and Computers*, Pacific Grove, CA, Nov. 2001, pp. 1470–1474.

266. G. Ungerboeck, "Adaptive maximum likelihood receiver for carrier modulated data transmission systems," *IEEE Transactions on Communications*, vol. 22, no. 5, pp. 624–635, May 1974.

267. J.-J. van de Beek, M. Sandell, and P. O. Borjesson, "ML estimation of time and frequency offset in OFDM systems," *IEEE Transactions on Signal Processing*, vol. 45, no. 7, pp. 1800–1805, July 1997.

268. A.-J. van der Veen, H. Bölcskei, D. Gesbert, and C. Papadias (Eds.), *Space-Time Wireless Systems: From Array Processing to MIMO Communications*, Cambridge University Press, New York, 2006.

269. S. Verdu, *Multiuser Detection*, Cambridge Unversity Press, New York, 1998.

270. H. Vikalo, B. Hassibi, B. Hochwald, and T. Kailath, "Optimal training for frequency-selective fading channels," in *Proc. of International Conference on Acoustics, Speech, and Signal Processing*, Salt Lake City, UT, May 7–11, 2001, vol. 4, pp. 29–48.

271. H. Vikalo, B. Hassibi, and T. Kailath, "Iterative decoding for MIMO channels via modified sphere decoding," *IEEE Transactions on Wireless Communications*, vol. 3, no. 6, pp. 2299–2311, Nov. 2004.

272. S. Vishwanath, N. Jindal, and A. Goldsmith, "Duality, achievable rates and sum-rate capacity of Gaussian MIMO broadcast channels," *IEEE Transactions on Information Theory*, vol. 39, no. 10, pp. 2658–2668, Oct. 2003.

273. E. Visotsky and U. Madhow, "Space-time transmit precoding with imperfect feedback," *IEEE Transactions on Information Theory*, vol. 47, no. 6, pp. 2632–2639, Sept. 2001.

274. P. Viswanath and D. Tse, "Sum capacity of the vector Gaussian broadcast channel and uplink-downlink duality," *IEEE Transactions on Information Theory*, vol. 49, no. 8, pp. 1912–1921, Aug. 2003.

275. A. J. Viterbi, "An intuitive justification and a simplified implementation of the MAP decoder for convolutional codes," *IEEE Journal on Selected Areas in Communications*, vol. 16, no. 2, pp. 260–264, Feb. 1998.

276. E. Viterbo and J. Boutros, "A universal lattice code decoder for fading channels," *IEEE Transactions on Information Theory*, vol. 45, no. 5, pp. 1639–1642, July 1999.

277. G. M. Vitetta and D. P. Taylor, "Viterbi decoding of differentially encoded PSK signals transmitted over Rayleigh frequency-flat fading channels," *IEEE Transactions on Communications*, vol. 43, no. 2/3/4, pp. 1256–1259, Feb./Mar./April 1995.

278. P. K. Vitthaladevuni and M.-S. Alouini, "A recursive algorithm for the exact BER computation of generalized hierarchical QAM constellations," *IEEE Transactions on Information Theory*, vol. 49, no. 1, pp. 297–307, Jan. 2003.

279. J. Vogt, K. Koors, A. Finger, and G. Fettweis, "Comparison of different turbo decoder realizations for IMT-2000," in *Proc. of Global Telecommunications Conference*, Rio de Janeiro, Brazil, 1999, vol. 5, pp. 2704–2708.

280. F. W. Vook and T. A. Thomas, "Transmit diversity schemes for broadband mobile communication systems," in *Proc. of Vehicular Technology Conference*, Boston, MA, Fall 2000, vol. 6, pp. 2523–2529.

281. B. Vucetic and J. Yuan, *Space-Time Coding*, Wiley, New York, 2003.

282. G. Wang and X.-G. Xia, "On optimal multilayer cyclotomic space-time code designs," *IEEE Transactions on Information Theory*, vol. 51, no. 3, pp. 1102–1135, Mar. 2005.

283. H. Wang and P. Chang, "On verifying the first-order Markovian assumption for a Rayleigh fading channel model," *IEEE Transactions on Vehicular Technology*, vol. 45, no. 2, pp. 353–357, May 1996.

284. R. Wang and G. B. Giannakis, "Approaching MIMO channel capacity with reduced-complexity sphere decoding," *IEEE Transactions on Communications*, vol. 54, no. 4, pp. 587–590, Apr. 2006.

285. X. Wang and H. V. Poor, "Iterative (turbo) soft interference cancellation and decoding for coded CDMA," *IEEE Transactions on Communications*, vol. 46, no. 7, pp. 1046–1061, July 1999.

286. Z. Wang and G. B. Giannakis, "Wireless multicarrier communications: Where Fourier meets Shannon," *IEEE Signal Processing Magazine*, vol. 17, no. 3, pp. 29–48, May 2000.

287. Z. Wang and G. B. Giannakis, "Block precoding for MUI/ISI-resilient generalized multi-carrier CDMA with multirate capabilities," *IEEE Transactions on Communications*, vol. 49, no. 11, pp. 2016–2027, Nov. 2001.

288. Z. Wang and G. B. Giannakis, "Complex-field coding for OFDM over fading wireless channels," *IEEE Transactions on Information Theory*, vol. 49, no. 3, pp. 707–720, Mar. 2003.

289. Z. Wang and G. B. Giannakis, "A simple and general parameterization quantifying performance in fading channels," *IEEE Transactions on Communications*, vol. 51, no. 8, pp. 1389–1398, Aug. 2003.

290. Z. Wang and G. B. Giannakis, "Outage mutual information of space-time MIMO channels," *IEEE Transactions on Information Theory*, vol. 50, no. 4, pp. 657–662, Apr. 2004.

291. Z. Wang, X. Ma, and G. B. Giannakis, "OFDM or single-carrier block transmissions?" *IEEE Transactions on Communications*, vol. 52, no. 3, pp. 380–394, Mar. 2004.

292. Z. Wang, S. Zhou, and G. B. Giannakis, "Joint coding-precoding with low-complexity turbo-decoding," *IEEE Transactions on Wireless Communications*, vol. 3, no. 3, pp. 832–842, May 2004.

293. W. Weichselberger, M. Herdin, H. Ozcelik, and E. Bonek, "A stochastic MIMO channel model with joint correlation of both link ends," *IEEE Transactions on Wireless Communications*, vol. 5, no. 1, pp. 90–100, Jan. 2006.

294. H. Weingarten, Y. Steinberg, and S. Shamai, "The capacity region of the Gaussian MIMO broadcast channel," in *Proc. of IEEE International Symposium on Information Theory*, Chicago, IL, June 2004.

295. L. Welch, "Lower bounds on the maximum cross correlation of signals," *IEEE Transactions on Information Theory*, vol. 20, no. 3, pp. 397–399, May 1974.

296. S. G. Wilson, *Digital Modulation and Coding*, Prentice Hall, Upper Saddle River, NJ, 1996.

297. C. Windpassinger and R. F. H. Fischer, "Low-complexity near-maximum-likelihood detection and precoding for MIMO systems using lattice reduction," in *Proc. of Information Theory Workshop*, Paris, France, 2003, pp. 345–348.

298. A. Wittneben, "Basestation modulation diversity for digital simulcast," in *Proc. of Vehicular Technology Conference*, St. Louis, MO, 1991, pp. 848–853.

299. P. W. Wolniansky, G. J. Foschini, G. D. Golden, and R. A. Valenzuela, "V-BLAST: An architecture for realizing very high data rates over rich scattering wireless channels," in *Proc. of International Symposium on Signals, Systems, and Electronics*, Sept. 1998, pp. 295–300.

300. Y. C. Wu, S.-C. Chan, and E. Serpedin, "Symbol-timing estimation in space-time coding systems based on orthogonal training sequences," *IEEE Transactions on Wireless Communications*, vol. 4, no. 2, pp. 603–613, Mar. 2005.

301. Y. C. Wu and E. Serpedin, "Data-aided maximum likelihood symbol timing estimation in MIMO correlated fading channels," *Wireless Communications and Mobile Computing*, vol. 4, no. 7, pp. 773–791, Nov. 2004.

302. Y. C. Wu, K.-W. Yip, T.-S. Ng, and E. Serpedin, "Maximum-likelihood frame synchronization for IEEE 802.11a WLANs on frequency-selective fading channels with unknown sampling phase offset," *IEEE Transactions on Wireless Communications*, vol. 4, no. 6, pp. 2751–2763, Nov. 2005.

303. D. Wübben, R. Bohnke, V. Kuhn, and K.-D. Kammeyer, "MMSE extension of V-BLAST based on sorted QR decomposition," in *Proc. of Vehicular Technology Conference*, Orlando, FL, Oct. 2003, vol. 1, pp. 508–512.

304. D. Wübben, R. Bohnke, V. Kuhn, and K.-D. Kammeyer, "Near-maximum-likelihood detection of MIMO systems using MMSE-based lattice reduction," in *Proc. of International Conference on Communications*, Paris, France, June 2004, vol. 2, pp. 798–802.

305. P. Xia, Personal webpage on collected finite-rate-beamforming codebooks, http://spincom.ece.umn.edu/pengfei/codebooks.

306. P. Xia and G. B. Giannakis, "Design and analysis of transmit-beamforming based on limited-rate feedback," *IEEE Transactions on Signal Processing*, vol. 54, no. 5, pp. 1853–1863, May 2006.

307. P. Xia, S. Zhou, and G. B. Giannakis, "Bandwidth- and power-efficient multi-carrier multiple access," *IEEE Transactions on Communications*, vol. 51, no. 11, pp. 1828–1837, Nov. 2003.

308. P. Xia, S. Zhou, and G. B. Giannakis, "Adaptive MIMO OFDM based on partial channel state information," *IEEE Transactions on Signal Processing*, vol. 52, no. 1, pp. 202–213, Jan. 2004.

309. P. Xia, S. Zhou, and G. B. Giannakis, "Achieving the Welch bound with difference sets," *IEEE Transactions on Information Theory*, vol. 51, no. 5, pp. 1900–1907, May 2005.

310. P. Xia, S. Zhou, and G. B. Giannakis, "Multi-antenna adaptive modulation and transmit beamforming with bandwidth-constrained feedback," *IEEE Transactions on Communications*, vol. 53, no. 3, pp. 526–536, Mar. 2005.

311. Y. Xin and G. B. Giannakis, "High-rate space-time layered OFDM," *IEEE Communications Letters*, vol. 6, no. 5, pp. 187–189, May 2002.

312. Y. Xin, Z. Liu, and G. B. Giannakis, "High-rate layered space-time coding based on constellation rotation," in *Proc. of Wireless Communications and Networking Conference*, Orlando, FL, Mar. 17–21, 2002, pp. 471–476.

313. Y. Xin, Z. Wang, and G. B. Giannakis, "Space-time diversity systems based on linear constellation precoding," *IEEE Transactions on Wireless Communications*, vol. 2, no. 2, pp. 294–309, Mar. 2003.

314. Y. Xin, Z. Wang, and G. B. Giannakis, "Linear unitary precoders for maximum diversity gains with multiple transmit- and receive-antennas," in *Proc. of 34th Asilomar Conference on Signals, Systems, and Computers*, Pacific Grove, CA, Oct. 29–Nov. 1, 2000, vol. 2, pp. 1553–1557.

315. Q. Yan and R. S. Blum, "Improved space-time convolutional codes for quasi-static slow fading channels," *IEEE Transactions on Wireless Communications*, vol. 1, no. 4, pp. 563–571, Oct. 2002.

316. Y. Yao and G. B. Giannakis, "Blind carrier-frequency offset estimation of SISO, MIMO, and multi-user OFDM," *IEEE Transactions on Wireless Communications*, vol. 53, no. 1, pp. 173–183, Jan. 2005.

317. W. Yu and J. Cioffi, "Sum capacity of Gaussian vector broadcast channels," *IEEE Transactions on Information Theory*, vol. 50, no. 9, pp. 1875–1892, Sept. 2004.

318. Y. Yu and G. B. Giannakis, "Joint low-density parity-check coding and linear precoding for Rayleigh fading channels," in *Proc. of 40th Conference on Information Sciences and Systems*, Princeton University, Princeton, NJ, Mar. 2006.

319. J.-K. Zhang, G. Wang, and K.-M. Wong, "Optimal norm form integer space-time codes for two antenna MIMO systems," in *Proc. of International Conference on Acoustics, Speech, and Signal Processing*, Philadelphia, PA, Mar. 2005, pp. 1061–1064.

320. W. Zhang and X. Ma, "Performance analysis for V-BLAST systems with linear equalization," in *Proc. of 39th Conference on Information Sciences and Systems*, Johns Hopkins University, Baltimore, MD, Mar. 15–18, 2005.

321. L. Zhao, W. Mo, Y. Ma, and Z. Wang, "Diversity and multiplexing tradeoff in general fading channels," in *Proc. of 40th Conference on Information Sciences and Systems*, Princeton University, Princeton, NJ, Mar. 2006.

322. W. Zhao, X. Cai, and G. B. Giannakis, "Efficient ML decoding of linear block codes," in *Proc. of 38th Conference on Information Sciences and Systems*, Princeton University, NJ, Mar. 17–19, 2004.

323. W. Zhao and G. B. Giannakis, "Reduced complexity closest point algorithms for random lattices," *IEEE Transactions on Wireless Communications*, vol. 5, no. 1, pp. 101–111, Jan. 2006.

324. W. Zhao and G. B. Giannakis, "Sphere decoding algorithms with improved radius search," *IEEE Transactions on Communications*, vol. 53, no. 7, pp. 1104–1109, July 2005.

325. W. Zhao, G. Leus, and G. B. Giannakis, "Orthogonal design of unitary constellations for uncoded and trellis coded non-coherent space-time systems," *IEEE Transactions on Information Theory*, vol. 50, no. 6, pp. 1319–1327, June 2004.

326. L. Zheng and D. Tse, "Optimal diversity-multiplexing tradeoff in multi-antenna channels," in *Proc. of 39th Allerton Conference*, Monticello, IL, Oct. 3–5, 2001.

327. L. Zheng and D. Tse, "Diversity and multiplexing: A fundamental tradeoff in multiple-antenna channels," *IEEE Transactions on Information Theory*, vol. 49, no. 5, pp. 1073–1096, May 2003.

328. S. Zhou, Personal webpage on collected finite-rate-precoding codebooks, http://engr.uconn.edu/~shengli/codebooks.

329. S. Zhou and G. B. Giannakis, "Space-time coding with maximum diversity gains over frequency-selective fading channels," *IEEE Signal Processing Letters*, vol. 8, no. 10, pp. 269–272, Oct. 2001.

330. S. Zhou and G. B. Giannakis, "Optimal transmitter eigen-beamforming and space-time block coding based on channel mean feedback," *IEEE Transactions on Signal Processing*, vol. 50, no. 10, Oct. 2002.

331. S. Zhou and G. B. Giannakis, "Optimal transmitter eigen-beamforming and space-time block coding based on channel correlations," *IEEE Transactions on Information Theory*, vol. 49, no. 7, pp. 1673–1690, July 2003.

332. S. Zhou and G. B. Giannakis, "Single-carrier space-time block coded transmissions over frequency-selective fading channels," *IEEE Transactions on Information Theory*, vol. 49, no. 1, pp. 164–179, Jan. 2003.

333. S. Zhou and G. B. Giannakis, "Adaptive modulation for multi-antenna transmissions with channel mean feedback," *IEEE Transactions on Wireless Communications*, vol. 3, no. 5, pp. 1626–1636, Sept. 2004.

334. S. Zhou, G. B. Giannakis, and C. Le Martret, "Chip interleaved block spread code division multiple access," *IEEE Transactions on Communications*, vol. 50, no. 2, pp. 235–248, Feb. 2002.

335. S. Zhou, G. B. Giannakis, and A. Swami, "Digital multi-carrier spread spectrum versus direct-sequence spread spectrum for resistance to jamming and multipath," *IEEE Transactions on Communications*, vol. 50, no. 4, pp. 643–655, Apr. 2002.

336. S. Zhou and B. Li, "BER criterion and codebook construction for finite-rate precoded spatial multiplexing with linear receivers," *IEEE Transactions on Signal Processing*, vol. 54, no. 5, pp. 1653–1665, May 2006.

337. S. Zhou, B. Li, and P. Willett, "Recursive and trellis-based feedback reduction for MIMO-OFDM with rate-limited feedback," *IEEE Transactions on Wireless Communications*, 2006.

338. S. Zhou, X. Ma, and K. Pattipati, "A view on full-diversity modulus-preserving rate-one linear space time block codes," *Signal Processing*, vol. 86, no. 8, pp. 1968–1975, Aug. 2006.

339. S. Zhou, B. Muquet, and G. B. Giannakis, "Subspace-based (semi-)blind channel estimation for block precoded space-time OFDM," *IEEE Transactions on Signal Processing*, vol. 50, pp. 1215–1228, May 2002.

340. S. Zhou, Z. Wang, and G. B. Giannakis, "Quantifying the power-loss when transmit-beamforming relies on finite rate feedback," *IEEE Transactions on Wireless Communications*, vol. 4, no. 4, pp. 1948–1957, July 2005.

341. S. Zhou, J. Wu, Z. Wang, and M. Doroslovacki, "Quantifying the performance gain of direction feedback in a MISO system," in *Proc. of Conference on Information Sciences and Systems*, Princeton University, Princeton, NJ, Mar. 2006.

342. S. Zhou, P. Xia, G. Leus, and G. B. Giannakis, "Chip-interleaved block-spread CDMA versus DS-CDMA for cellular downlink: A comparative study," *IEEE Transactions on Wireless Communications*, vol. 3, no. 1, pp. 176–190, Jan. 2004.

343. C. Zong and J. Talbot, *Sphere Packings*, Springer-Verlag, New York, 1999.

Index

1D beamformer, 335
2D coder-beamformer, 335

a posteriori probability, 169
a priori probability, 169
accumulated metric, 61
adaptive modulation, 344, 347, 385
additive white Gaussian noise (AWGN), 26
affine model, 272
affine precoding, 163
Alamouti's code, 62
algebraic construction, 74
algebraic FDFR code, 98
analog phase sweeping, 222
antenna selection, 359
antenna switching, 71
AP-CP-only, 163
arithmetic-geometric mean inequality, 73
asymptotic union bound, 136
autoregressive (AR) model, 295

bandwidth efficiency, 419, 427
basis expansion model (BEM), 209, 212
Bessel function, 296
block circular delay diversity (BCDD), 200
block spreading, 413
bursty errors, 28

capacity, 173
capacity outage probability, 30
carrier frequency, 212
carrier frequency offset, 269, 296, 299
categories of ST codes, 48
CDMA, 2, 413
channel capacity, 29
channel coherence time, 25
channel delay spread, 25
channel diversity, 44
channel order, 152, 181
channel state information, 26, 313
Chernoff bound, 40, 246
chip-interleaved block-spread (CIBS), 413
chordal distance, 135, 371, 382
circulant matrix, 164, 181, 214, 417

circular convolution, 164, 181
closed-loop, 313, 391
coherence bandwidth, 6
coherence time, 7
coherent ST codes, 49, 51
column-orthonormal precoder, 369, 370, 381
complex orthogonal designs, 61
composite bound, 359
conference matrix, 359
conjugation, 166
constant-modulus (CM), 294, 427
constellation rotation, 69, 75
constraint length, 54
covariance feedback, 317, 328, 334, 339, 351
CP-only transmission, 162
cyclic prefix (CP), 181, 215
cyclostationary statistics, 294

D-BLAST, 88
decision-directed method, 294
decision-feedback differential detection, 232
deinterleaving, 416
delay diversity, 51, 89
delay spread, 6
delayed feedback, 317
depth-first search, 112
detection ordering (DO), 117
difference sets, 360
differential Alamouti's ST coding, 144
differential designs, 232
differential ST codes, 133
differential ST coding, 139
differential ST decoding, 140
differential unitary ST coding, 141
digital-phase sweeping, 197, 223
Dirac's delta function, 210
diversity gain, 43, 44
diversity product distance, 135
Doppler diversity, 215
Doppler effects, 296
Doppler shift, 212
Doppler spread, 7
doubly selective channels, 213
downlink, 412

EDGE, 176, 208
ergodic capacity, 30, 278
error matrix, 43
error propagation, 294
error-oriented ST code, 49
ETACS, 2
Euclidean distance, 44
Euler number, 76
extrinsic information, 127, 169

fading effect, 5
FDFR, 96
FDMA, 2
FFT, 98
finite alphabet, 279
finite impulse response, 70
finite-rate beamforming, 356
finite-rate CSI, 353
finite-rate feedback, 354
finite-rate precoding, 366
frame error rate, 160
frequency-division duplex (FDD), 318
frequency-division multiplexing (FDM), 285
frequency-selective fading channels, 151
Frobenius norm, 64, 235
Fubini-Study distance, 371
full-diversity full-rate MIMO OFDM, 201

generalized delay diversity (GDD), 157
GF-LCF FDFR, 251
GF-LCF ST codes, 242
Grassmannian line packing, 358
Grassmannian subspace packing, 371
group STF (GSTF), 189
GSM, 2
GSTF block codes, 190
GSTF trellis coding, 192

HiperLAN, 265
HiperLAN/2, 195

integer least-squares (ILS), 105
interblock interference (IBI), 181, 272
interleaver, 28, 245
interleaving, 28, 416

Jakes model, 212, 214, 223, 237, 296

Kalman filtering (KF), 297
kissing number, 43
Kronecker delta function, 280
kurtosis, 306

lattice reduction algorithm, 106
least-squares (LS), 271
Lindemann's theorem, 98
line-of-sight (LOS), 4, 316
linear constellation precoding, 190
linear minimum mean-square error (LMMSE), 271
linear ST block code, 49
list sphere decoding, 125
Lloyd's algorithm, 360, 372, 384
log-likelihood ratio (LLR), 169

maximum a posteriori (MAP), 248
mean feedback, 316, 324, 334, 335, 345, 351
MIMO OFDM, 179
minimum mean-square error (MMSE), 105
minimum polynomial, 98
ML decoding, 167
mobile broadband wireless access (MBWA), 239
modulus-preserving scheme, 79
multi-carrier system, 179
multipath propagation, 4
multipath-Doppler diversity, 216
multiple symbol detection (MSD), 232
multiple trellis-coded modulation (MTCM), 229
multiplexing-diversity trade-off, 46
multiuser interference, 409
multiuser separation, 409
mutual information, 99, 205
MWBE, 359

narrowband service, 2
noncoherent detection, 133
noncoherent ST codes, 49, 133
nonlinear ST code, 49
NTT, 2
null subspace, 302
nulling-canceling, 90, 106, 117
Nyquist's pulse shaper, 212

OFDMA, 425
one-dimensional (1D) beamforming, 334
optimal placement, 280
optimal training, 271
orthogonal frequency division multiple access (OFDMA), 425
orthogonal frequency division multiplexing (OFDM), 179
orthogonal multiple access, 409
orthogonal STBC, 169, 331
outage probability, 155, 174
outage rate, 30

pairwise coding gain, 41
pairwise diversity gain, 41
pairwise error probability, 40
partial CSI, 313
path loss, 3
PDC, 2
permutation matrix, 161, 285, 417
phase distortion, 304
phase sweeping, 222
power loss, 363
probabilistic data association algorithm, 106
product distance, 72
projection two-norm distance, 371

QR-decomposition, 90, 109
quasi-orthogonal ST block codes, 68
quasi-synchronous, 411

rate-oriented ST code, 49
Rayleigh fading, 23
residual CFO, 303
reverse cyclic shift, 161

scalar codes, 174
scattering, 3
Schnorr-Euchner (SE) SDA, 118
semidefinite programming, 106
shadowing effect, 4
Shannon capacity, 30
signal subspace, 302
signal-to-noise ratio (SNR), 26
simplex signal set, 359
single carrier, 151
single-input single-output (SISO), 277
smart-greedy codes, 193, 229
soft sphere decoder, 126
soft-input soft-output (siso), 248
soft-output Viterbi algorithm, 248
space-frequency coded OFDM, 188
space-multipath diversity, 154
space-multipath-Doppler diversity, 235
space-time coded OFDM, 188
space-time-frequency (STF), 184
space-virtual-time (SVT), 192
spatial multiplexing, 45, 366
sphere decoding algorithm (SDA), 78, 106
ST block codes, 161
ST code matrix, 24
ST code rate, 25
ST constellation, 25
ST linear complex field codes, 70
ST trellis, 53
ST trellis code, 158
statistical CSI, 315
STBC-CIBS-CDMA, 413
STF block coding, 426
STF coded OFDM, 189
STF-OFDMA, 426
subcarrier grouping, 189
sum-of-ranks criterion, 228
superimposed training, 272
symbol error rate, 362
system diversity, 44

time dispersion, 4
time reversal, 161, 417
time-division duplex (TDD), 318
time-division multiplexing (TDM), 276
time-frequency duality, 215
trace, 24
transcendental number, 98
transmission modes, 354
transmit-beamformer, 330
turbo equalization, 169
turbo principle, 248
two-dimensional coder-beamformer, 333
typical urban (TU), 176

union bound distance, 136
unitary combining, 166
unitary precoding, 73
unitary ST modulation, 134
uplink, 412

V-BLAST, 88, 91, 125
vector quantization, 372
Viterbi's algorithm, 53, 60, 105, 167, 235

Welch bound, 359
wireless communications, 2

zero-forcing (ZF), 105
zero-symmetry, 59
ZP-only transmission, 164, 221, 417

Printed in the United States
By Bookmasters